AF766398

VOLUME ONE HUNDRED AND TWENTY EIGHT

ADVANCES IN IMMUNOLOGY

Molecular Mechanisms that Orchestrate the Assembly of Antigen Receptor Loci

ASSOCIATE EDITORS

K. Frank Austen
Harvard Medical School, Boston, Massachusetts, USA

Tasuku Honjo
Kyoto University, Kyoto, Japan

Fritz Melchers
University of Basel, Basel, Switzerland

Hidde Ploegh
Massachusetts Institute of Technology, Massachusetts, USA

Kenneth M. Murphy
Washington University, St. Louis, Missouri, USA

VOLUME ONE HUNDRED AND TWENTY EIGHT

ADVANCES IN IMMUNOLOGY

Molecular Mechanisms that Orchestrate the Assembly of Antigen Receptor Loci

Edited by

CORNELIS MURRE
Department of Molecular Biology,
University of California,
San Diego, La Jolla, USA

ELSEVIER

AMSTERDAM • BOSTON • HEIDELBERG • LONDON
NEW YORK • OXFORD • PARIS • SAN DIEGO
SAN FRANCISCO • SINGAPORE • SYDNEY • TOKYO
Academic Press is an imprint of Elsevier

Academic Press is an imprint of Elsevier
225 Wyman Street, Waltham, MA 02451, USA
525 B Street, Suite 1800, San Diego, CA 92101-4495, USA
The Boulevard, Langford Lane, Kidlington, Oxford OX5 1GB, UK
125 London Wall, London, EC2Y 5AS, UK

First edition 2015

© 2015 Elsevier Inc. All rights reserved

No part of this publication may be reproduced or transmitted in any form or by any means, electronic or mechanical, including photocopying, recording, or any information storage and retrieval system, without permission in writing from the publisher. Details on how to seek permission, further information about the Publisher's permissions policies and our arrangements with organizations such as the Copyright Clearance Center and the Copyright Licensing Agency, can be found at our website: www.elsevier.com/permissions.

This book and the individual contributions contained in it are protected under copyright by the Publisher (other than as may be noted herein).

Notices
Knowledge and best practice in this field are constantly changing. As new research and experience broaden our understanding, changes in research methods, professional practices, or medical treatment may become necessary.

Practitioners and researchers must always rely on their own experience and knowledge in evaluating and using any information, methods, compounds, or experiments described herein. In using such information or methods they should be mindful of their own safety and the safety of others, including parties for whom they have a professional responsibility.

To the fullest extent of the law, neither the Publisher nor the authors, contributors, or editors, assume any liability for any injury and/or damage to persons or property as a matter of products liability, negligence or otherwise, or from any use or operation of any methods, products, instructions, or ideas contained in the material herein.

ISBN: 978-0-12-803296-1
ISSN: 0065-2776

For information on all Academic Press publications
visit our website at http://store.elsevier.com/

ELSEVIER Book Aid International
Working together to grow libraries in developing countries

www.elsevier.com • www.bookaid.org

CONTENTS

CONTRIBUTORS

Craig H. Bassing
Division of Cancer Pathobiology, Center for Childhood Cancer Research, Children's Hospital of Philadelphia, and Abramson Family Cancer Research Institute, Cell and Molecular Biology Graduate Program, Department of Pathology and Laboratory Medicine, Perelman School of Medicine at the University of Pennsylvania, Philadelphia, Pennsylvania, USA

Meinrad Busslinger
Research Institute of Molecular Pathology (IMP), Vienna Biocenter (VBC), Vienna, Austria

Zachary Carico
Department of Immunology, Duke University Medical Center, Durham, North Carolina, USA

Julie Chaumeil
Institut Curie, CNRS UMR3215, INSERM U934, Paris, France

Claudia Ribeiro de Almeida
Sir William Dunn School of Pathology, University of Oxford, Oxford, United Kingdom

Anja Ebert
Research Institute of Molecular Pathology (IMP), Vienna Biocenter (VBC), Vienna, Austria

Pierre Ferrier
Centre d'Immunologie de Marseille-Luminy, Aix-Marseille Université UM2, Inserm, U1104, CNRS UMR7280, 13288 Marseille, France

Bingtao Hao
Department of Pathology, New York University School of Medicine, New York, USA

Rudi W. Hendriks
Department of Pulmonary Medicine, Erasmus MC, Rotterdam, The Netherlands

Louisa Hill
Research Institute of Molecular Pathology (IMP), Vienna Biocenter (VBC), Vienna, Austria

Sébastien Jaeger
Centre d'Immunologie de Marseille-Luminy, Aix-Marseille Université UM2, Inserm, U1104, CNRS UMR7280, 13288 Marseille, France

Michael S. Krangel
Department of Immunology, Duke University Medical Center, Durham, North Carolina, USA

Gita Kumari
Laboratory of Molecular Biology and Immunology, National Institute on Aging, Baltimore, Maryland, USA

Kinjal Majumder
Department of Pathology and Immunology, Washington University School of Medicine, Saint Louis, Missouri, USA

Eugene M. Oltz
Department of Pathology and Immunology, Washington University School of Medicine, Saint Louis, Missouri, USA

Pernelle Outters
Centre d'Immunologie de Marseille-Luminy, Aix-Marseille Université UM2, Inserm, U1104, CNRS UMR7280, 13288 Marseille, France

Charlotte Proudhon
Department of Pathology, New York University School of Medicine, New York, USA

Ramya Raviram
Department of Pathology, New York University School of Medicine, New York, USA

David G. Schatz
Department of Immunobiology, Yale University School of Medicine, New Haven, Connecticut, and Howard Hughes Medical Institute, Chevy Chase, Maryland, USA

Ranjan Sen
Laboratory of Molecular Biology and Immunology, National Institute on Aging, Baltimore, Maryland, USA

Jane A. Skok
Department of Pathology, New York University School of Medicine, New York, USA

Ralph Stadhouders
Gene Regulation, Stem Cells and Cancer Programme, Centre for Genomic Regulation (CRG), Barcelona, Spain

Grace Teng
Department of Immunobiology, Yale University School of Medicine, New Haven, Connecticut, USA

Nancy Zaarour
Centre d'Immunologie de Marseille-Luminy, Aix-Marseille Université UM2, Inserm, U1104, CNRS UMR7208, 13288 Marseille, France

PREFACE

The adaptive immune system is composed of populations of lymphocytes that permit vertebrate organisms to defend themselves against a wide variety of invading pathogens. In the mid-twentieth century, it was Dreyer and Burnett who initially proposed a mechanism, DNA rearrangement, which underpins the mechanisms by which the adaptive immune system generates an effective response. In the decades to follow, this idea was validated by many studies revealing that receptors for antigens are organized in the genome as distinct gene segments, containing variable, diversity, and joining gene segments. During the developmental progression of lymphoid cells, the variable, diversity, and joining elements rearrange to generate a diverse antigen receptor repertoire. This volume describes the most recent advances that have provided deeper insight into the mechanisms that underpin the shuffling of DNA segments of antigen receptor loci. The volume is organized into eight chapters that describe in molecular detail how the antigen receptor gene segments are assembled and how this process is regulated. The chapters are unique from one another, but they also overlap as it relates to molecular principles that control the DNA recombination reaction for different cell types. Key to antigen receptor assembly are the RAG proteins, RAG1 and RAG2, that initiate the VDJ recombination reaction by cleaving recombination signal sequences that flank the variable, diversity, and joining gene segments. In Chapter 1, Teng and Schatz describe in molecular detail how the RAG proteins are structured, how the structure relates to its activities, how the RAG proteins find their targets across the genome, and how the RAG proteins evolved from an ancient transposase. In Chapter 2, Kumari and Sen discuss how the immunoglobulin heavy chain locus is organized and how the assembly of variable, diversity, and joining gene segments is controlled in a lineage- and developmental-specific fashion. They describe how epigenetics, changes in immunoglobulin heavy chain locus topology, as well as nuclear location contribute to the mechanism by which VDJ recombination is regulated. They point to new directions in the field that investigate how genomic encounters driven by thermal Langevin motion enable variable, diversity, and joining elements to find each other in a crowded nuclear environment. Although the initial studies that relate to mechanisms that underpin antigen receptor assembly were focused on the immunoglobulin heavy chain locus, studies involving other loci have provided additional and unique insights. In Chapters 3 and 4, Busslinger et al. and Skok et al. describe common principles that mediate antigen receptor locus contraction and nuclear localization. They describe how changes in chromosome topology permit equal playing fields for variable regions in both B and T lineage cells. In Chapters 5 and 6, Hendriks et al. and Oltz et al. describe how long-range genomic interactions associated with antigen receptor loci are regulated. They point to recently identified regulatory DNA elements that contain pairs of anchors and insulators localized in genomic regions that separate the variable from the diversity and joining gene segments in both immunoglobulin and TCR loci. Interestingly, these anchors/insulators have recently been shown to be essential in ensuring proper developmental timing and regulation of antigen receptor assembly. In Chapter 7, Carico and Krangel focus on the unique structure of the TCRa/d locus. This locus is distinct in that it undergoes DNA rearrangements of two distinct loci at distinct developmental stages. Lymphoid cells generally express a singular antigen receptor as originally proposed by the aforementioned Dreyer and Burnett, a process named allelic exclusion. Although substantial progress has been made during the past few decades, our understanding of the allelic exclusion principles remains rudimentary. In Chapter 8, Ferrier et al. review the literature and propose models to describe the allelic exclusion mechanisms in molecular and quantitative details.

Collectively, the chapters presented in this volume summarize in detail the substantial progress that has been made during the past few decades. Much more remains to be learned as to how variable, diversity, and joining segments are assembled. We suggest that knowledge obtained from these studies will not only forward our understanding of the mechanisms that underpin the enigmatic process of antigen receptor assembly but also will provide new insight into the biochemical, molecular, and physical principles that regulate gene expression across the animal kingdom.

Cornelis Murre

CHAPTER ONE

Regulation and Evolution of the RAG Recombinase

Grace Teng*, David G. Schatz*,†,1
*Department of Immunobiology, Yale University School of Medicine, New Haven, Connecticut, USA
†Howard Hughes Medical Institute, Chevy Chase, Maryland, USA
[1]Corresponding author: e-mail address: david.schatz@yale.edu

Contents

Abstract

The modular, noncontiguous architecture of the antigen receptor genes necessitates their assembly through V(D)J recombination. This program of DNA breakage and rejoining occurs during early lymphocyte development, and depends on the RAG1 and RAG2 proteins, whose collaborative endonuclease activity targets specific DNA motifs enriched in the antigen receptor loci. This essential gene shuffling reaction requires lymphocytes to traverse several developmental stages wherein DNA breakage is tolerated, while minimizing the expense to overall genome integrity. Thus, RAG activity is subject to stringent temporal and spatial regulation. The RAG proteins themselves also contribute autoregulatory properties that coordinate their DNA cleavage activity with target chromatin structure, cell cycle status, and DNA repair pathways. Even so, lapses in regulatory restriction of RAG activity are apparent in the aberrant V(D)J recombination events that underlie many lymphomas. In this review, we discuss the current understanding of the RAG endonuclease, its widespread binding in the lymphocyte genome, its noncleavage activities that restrain its enzymatic potential, and the growing evidence of its evolution from an ancient transposase.

Advances in Immunology, Volume 128
ISSN 0065-2776
http://dx.doi.org/10.1016/bs.ai.2015.07.002
© 2015 Elsevier Inc.
All rights reserved.

1. INTRODUCTION

Adaptive immunity relies on a cellular system of non-self recognition that is provided by antigen receptors on B and T lymphocytes. These receptors are expressed from a peculiar family of genes (immunoglobulin [*Ig*] and T cell receptor [*Tcr*]) that originate as nonfunctional arrays of split-gene segments. Only upon induction of the lymphoid developmental program do these genes undergo a series of targeted genetic rearrangements that create functional expression units, each conferring a distinct antigen specificity upon the resultant receptor. This diversification process, V(D)J recombination, is one means by which biology overcomes limitations on the information content encoded by the genome. It is thus that modern-day vertebrates achieve the immunological diversity that allows them to respond to virtually any antigenic epitope encountered in their lifetimes.

V, D, and J gene segments present in the *Ig* and *Tcr* loci are cut-and-pasted together through a DNA recombination mechanism reminiscent of DNA transposition (Fugmann, Lee, Shockett, Villey, & Schatz, 2000; Fig. 1). Recombination signal sequences (RSSs) adjacent to each gene

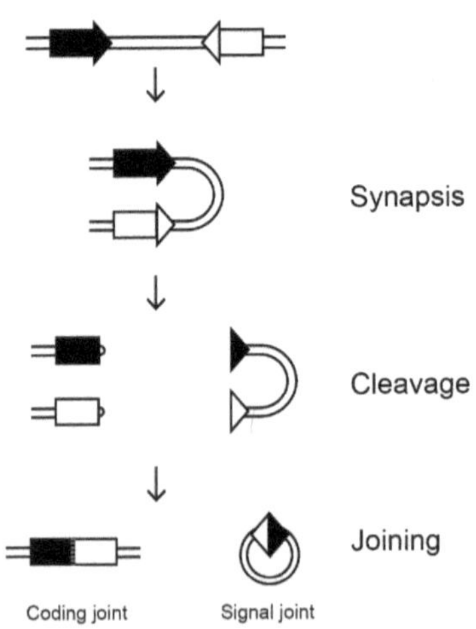

Figure 1 V(D)J recombination. Recombination signal sequences (RSSs, triangles) flank antigen receptor gene segments (rectangles). A compatible pair of 12- and 23-RSSs undergoes RAG-dependent synapsis and cleavage. This results in hairpinned coding ends and blunt signal ends. NHEJ-mediated repair joins the ends. Deletions and insertions at the coding junction (red, light gray in the print version) introduce additional diversity, while the signal ends are joined precisely. The RAG–HMGB1 recombinase complex is not depicted.

demarcate the site of cleavage, which lies immediately between the RSS and its partner gene segment. Each RSS begins with a heptamer (consensus sequence 5′-CACAGTG), followed by a spacer of 12 or 23 base pairs, and ends with a nonamer (consensus 5′-ACAAAAACC). For recombination, the RSSs partner in an asymmetrical fashion, such that high-efficiency cleavage and rejoining occurs only between a 12/23 RSS pair; this is referred to as the "12/23" rule. However, not every pair of 12- and 23-RSS is compatible; the factors that govern the "beyond 12/23" restriction are not completely understood, but include the distribution of chromatin marks, intralocus chromosome interactions, and catalytic incompatibilities between certain 12/23 RSS pairs (Banerjee & Schatz, 2014; Bassing et al., 2000; Drejer-Teel, Fugmann, & Schatz, 2007; Hughes, Tillman, Wehrly, White, & Sleckman, 2003; Olaru, Patterson, Cai, & Livak, 2004). After cleavage, the broken ends containing the V, D, or J gene segments are joined to one another through nonhomologous end-joining (NHEJ; Helmink & Sleckman, 2012), while the ends containing the RSSs are usually excised as an extrachromosomal circle (Fig. 1). In some cases, the RSS orientation necessitates inversional (as opposed to deletional) rearrangement (Lewis, 1994; Sollbach & Wu, 1995). Randomness in the choice of gene segment usage, along with imprecision of the joining process, contributes to the diversity of the final antigen receptor repertoire (Lewis, 1994).

Progression through successive stages of early lymphocyte development requires multiple rounds of V(D)J recombination, subjecting lymphoid lineages to several cycles of genotoxic stress (in the form of DNA breaks). The processes that enable lymphocytes to tolerate and sequester DNA breakage to the appropriate locations are wide ranging, and cover both pre- and postcleavage processes.

2. THE RAG RECOMBINASE

The RAG recombinase that initiates V(D)J recombination is typically characterized as a catalytic engine (RAG1), its cofactor (RAG2), and a DNA-bending protein, HMGB1 (Swanson, 2004). Recent structural studies have illustrated the architecture of the core RAG heterotetramer, which contains two units each of RAG1 and RAG2 (Grundy et al., 2009; Kim, Lapkouski, Yang, & Gellert, 2015). The paired RAG1 molecules form a Y-shaped structure with C-terminal "arms" and an intercalated N-terminal "stalk," while the two RAG2 molecules occupy the ends of the Y (Kim et al., 2015; Fig. 2).

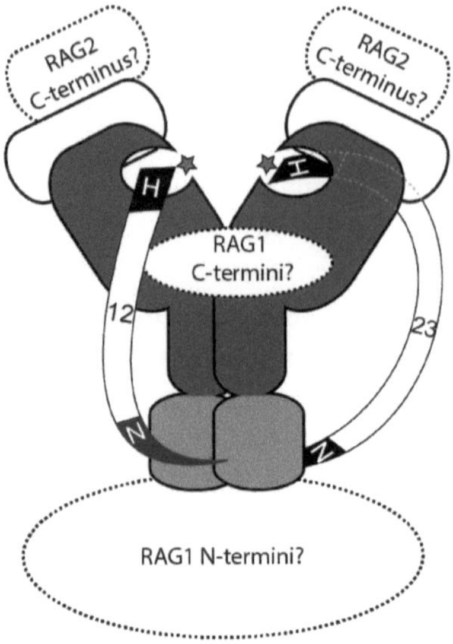

Figure 2 The core RAG heterotetramer in the synaptic complex. The RAG heterotetramer contains two RAG1 monomers (blue, with the nonamer-binding domain indicated in cyan), and two RAG2 monomers (green). When the synaptic complex forms, the heterotetramer binds one 12-RSS and one 23-RSS, which are shown schematically as ribbons. The heptamer (H) and nonamer (N) of each RSS are depicted in black. The two active sites of the complex are marked by red stars. The structural conformation and position of the noncore RAG1 and RAG2 termini are not known and are depicted as non-specific gray ovals (dashed white ovals in the print version). *Schematic derived from Kim et al. (2015).* (See the color plate.)

3. RAG1—A DNA BINDING AND CLEAVING ENZYME

The two key activities of the recombinase, DNA cleavage and sequence-specific DNA binding, are wholly contained within the RAG1 protein, specifically in the "core" domain (residues 384–1008 of 1040; Fig. 3). The active site of each RAG1 moiety is cradled within the boughs of the Y-shaped structure, where it makes contacts with RAG2 (Kim et al., 2015). The active site contains three acidic residues (D600, D708, and E962; Fugmann, Villey, Ptaszek, & Schatz, 2000; Kim, Dai, Mundy, Yang, & Oettinger, 1999; Landree, Wibbenmeyer, & Roth, 1999), which fold into a metal-coordinating motif to initiate a hydrolysis–transesterification reaction, producing a double-stranded DNA break characterized by one

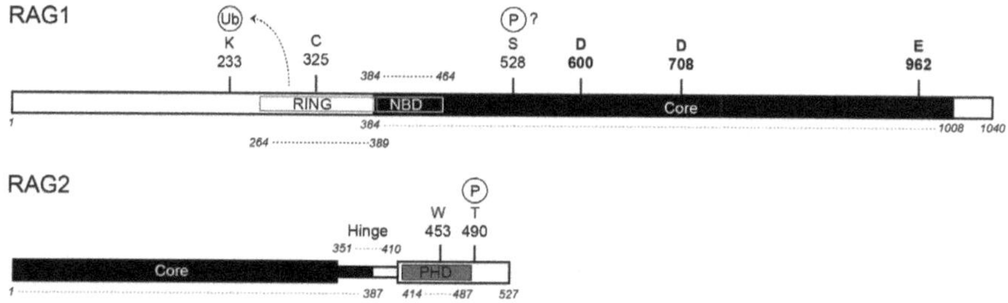

Figure 3 Schematic of RAG1 and RAG2 proteins. The RAG1 and RAG2 proteins are diagrammed. Specific domains, functional residues, and targets of posttranslational modifications (P, phosphorylation; Ub, ubiquitylation) are indicated. See text for additional details.

hairpinned and one blunt DNA end (Schatz & Swanson, 2011; Fig. 1). Phosphorylation on RAG1-S528 by AMPK has been reported to stimulate RAG1 cleavage activity in response to metabolic stress (Um et al., 2013), though it is unknown if this stimulatory modification occurs in lymphocytes.

DNA-binding activity (both sequence-specific and nonspecific) appears to be distributed among several regions of core RAG1 (Difilippantonio, McMahan, Eastman, Spanopoulou, & Schatz, 1996; Huye, Purugganan, Jiang, & Roth, 2002; Yin et al., 2009). The RAG1 N-terminal region has also been implicated in nonspecific DNA interactions (Arbuckle, Rahman, Zhao, Rodgers, & Rodgers, 2011). The best-defined RAG1–DNA interaction lies in the RAG1 nonamer-binding domain (NBD, residues 389–464), which occupies the bottom of the "stem" of the RAG tetramer as an intertwined dimer (Kim et al., 2015; Yin et al., 2009; Figs. 2 and 3). The NBD features an AT-hook motif, common to many DNA-binding proteins (Aravind & Landsman, 1998). In the context of RAG1, the AT-hook latches onto the A-rich tract of the nonamer (Yin et al., 2009), forming one of the few sequence-specific contact points. The two NBD monomers make extensive contacts with each other and with the two DNA substrates *in cis* and *in trans* (Yin et al., 2009). Substantial nonspecific interactions with DNA are also mediated by the NBD (Yin et al., 2009). Both structural and biochemical evidence suggest that additional regions along the central and C-terminal regions of RAG1 core (which create the "arms" of the Y) serve as DNA-binding surfaces for the RSS heptamer and coding flank, respectively (Arbuckle, Fauss, Simpson, Ptaszek, & Rodgers, 2001; Huye et al., 2002; Kim et al., 2015; Mo, Bailin, & Sadofsky, 2001; Peak, Arbuckle, & Rodgers, 2003). RAG2 has also been implicated in direct DNA interactions (Fugmann & Schatz,

2001), particularly near the heptamer-coding flank border (Eastman, Villey, & Schatz, 1999; Mo, Bailin, & Sadofsky, 1999; Swanson & Desiderio, 1999). RAG1 interactions with the heptamer appear to be somewhat sequence specific (Arbuckle et al., 2001). The geometry of these DNA-binding surfaces in relation their RSS substrates necessitates a substantial deformation of the DNA, with bends proposed to occur either within and/or immediately flanking the spacer sequences of the RSS (Ciubotaru et al., 2013, 2015).

Both the RAG1 active site architecture and cleavage mechanism are evocative of DDE transposases, which are thought to be evolutionary siblings of RAG1. Indeed, the structure and function of extant transposases have served as important guideposts in predicting and understanding the architecture of RAG1 (Grundy et al., 2009; Kim et al., 2015; Zhang, Shetty, Surleac, Petrescu, & Schatz, 2015).

4. RAG2—A REGULATORY COFACTOR

Genetic deficiency in either RAG1 or RAG2 results in the same early developmental block in lymphoid lineages, demonstrating the necessity of the noncatalytic RAG2 protein for proper lymphocyte development (Mombaerts et al., 1992; Shinkai et al., 1992). Though RAG2 contributes no catalytic properties, its presence enhances the ability of RAG1 to bind and cleave DNA (Akamatsu & Oettinger, 1998; Fugmann & Schatz, 2001; Swanson, 2004; Swanson & Desiderio, 1999). It is a six-bladed beta-propeller that forms a ring-like structure around a central cavity (Callebaut & Mornon, 1998; Kim et al., 2015). Within the RAG tetramer, each RAG2 molecule occupies the end of an "arm" in the Y-shaped structure (Fig. 2), dimerizing with one RAG1 molecule, and making additional contacts near the active site (Kim et al., 2015). This intimate connection to the catalytic center of the recombinase suggests that RAG2 may assist in formation or maintenance of the active site, help configure the DNA-binding portion of RAG1, or both.

RAG2 has also been experimentally whittled down into a biochemically competent "core" (residues 1–383 of 527) and a "noncore" C-terminus (Fig. 3). The RAG2 core has long been considered to be residues 1–383 (or 387) (Cuomo & Oettinger, 1994; Sadofsky, Hesse, & Gellert, 1994), but recent evidence indicates that residues 1–351 constitute the minimal region required for catalytic function of the RAG complex (Gigi et al., 2014; Kim et al., 2015) and hence should be considered to constitute the true RAG2 core. Regulatory properties, both stimulating and inhibitory, are distributed among both of these regions. Residues 394–408 have been shown to interact with core histones (West et al., 2005), but the major link between RAG2 and chromatin appears to depend on a domain located in the C-terminus. A noncanonical plant homeodomain (PHD) in this region binds to histone 3 in which lysine 4 is trimethylated (H3K4me3; Matthews et al., 2007; Ramon-Maiques et al., 2007), a modification on histone H3 that marks active transcriptional start sites (TSSs; Ruthenburg, Allis, & Wysocka, 2007). This creates a link between the RAG recombinase and accessible chromatin, and implicates RAG2 as a reader of the histone code. A conserved residue, W453, is essential for this bridging interaction (Matthews et al., 2007; Ramon-Maiques et al., 2007). Disruption of this RAG–histone interaction impairs V(D)J recombination, suggesting a role for RAG2-dependent tethering to chromatin (Liu, Subrahmanyam, Chakraborty, Sen, & Desiderio, 2007; Matthews et al., 2007; Ramon-Maiques et al., 2007). Accordingly, RAG2 localizes precisely to sites of H3K4me3 *in vivo*, not only at the antigen receptor genes, but genomewide (Ji et al., 2010; Teng et al., 2015). This localization pattern is ablated in mice expressing a RAG2 truncation mutant lacking the entire C-terminal region, which includes the PHD (Teng et al., 2015). The RAG2–H3K4me3 interaction is not merely a passive tether, however, as biochemical experiments show that H3K4me3 stimulates the catalytic activity of the RAG complex (Shimazaki, Tsai, & Lieber, 2009). Recent evidence argues that binding of the RAG2 PHD to H3K4me3 stimulates DNA binding and cleavage by RAG at least in part by relieving an autoinhibitory effect mediated by RAG2 residues 370–405 (Lu, Ward, Bettridge, Liu, & Desiderio, 2015). This suggests the possibility of a conformational change in the RAG complex upon binding of RAG2 to H3K4me3 (Lu et al., 2015).

The RAG2 C-terminus also serves as a nexus for inhibitory regulation. Abundance of RAG2 protein (and to a lesser extent, RAG1) drops considerably when lymphocytes transition to S phase (Li, Dordai, Lee, & Desiderio, 1996). Cell cycle-coupled phosphorylation on residue T490 tags the RAG2 protein for ubiquitin-dependent degradation prior to the G1–S transition. This temporal restriction precludes RAG-generated breaks from participating in replication-dependent genome instability. Interestingly, the steady-state levels of RAG1 are increased in mouse thymocytes that lack RAG2 or express C-terminally deleted RAG2 (Teng et al., 2015), suggesting that the active fraction of RAG1 in a cell may be coshepherded toward the proteasome through its interaction with RAG2. *In vivo* disruption of this ubiquitin-dependent degradation pathway unleashes V(D)J recombination from its usual cell cycle restrictions, and allows RAG-dependent breaks to persist or form *de novo* outside of G1 phase (Zhang, Reynolds, Shan, & Desiderio, 2011). This has consequences for genome integrity, illustrated by oncogenic RAG-dependent chromosomal translocations that develop in P53-deficient mice expressing a RAG2-T490A mutant or C-terminally deleted RAG2 (Deriano et al., 2011; Zhang et al., 2011).

Containment of RAG activity to nonreplicative cell cycle phases also specifies the method of postcleavage repair, as NHEJ predominates in G1 (Branzei & Foiani, 2008) and is essential for classical V(D)J recombination (Helmink & Sleckman, 2012). Truncated RAG2 mutants (FS361: a frameshift mutation at residue 361, or core RAG2: truncation at 383) allow the recombination machinery to access alternative modes of DNA repair, including alternative end-joining (A-EJ) and homologous recombination (Corneo et al., 2007; Gigi et al., 2014). Hence, RAG guides the DNA breaks it makes into the NHEJ pathway at the expense of other repair pathways (Lee, Neiditch, Salus, & Roth, 2004). Notably, if RAG is physically tethered to the I-SceI nuclease, it can direct I-SceI-mediated breaks into the NHEJ pathway (Cui & Meek, 2007). Postcleavage repair during V(D)J recombination has likely evolved to avoid pathways that rely on sequence homology (such as homologous recombination), which could limit the diversification potential of the reaction and also to avoid pathways that promote genomic instability (such as A-EJ; Zhu et al., 2002).

The repair-specifying properties of RAG2 have been mapped to an acidic region (residues 350–410) predicted to form a flexible hinge (Coussens et al., 2013; Fig. 3). The crystal structure of core RAG2 (which includes part of this hinge) confirms the disordered nature of this region (Kim et al., 2015). Charge-neutralizing mutations that diminish the predicted flexibility of the hinge also allow for aberrant repair by A-EJ (Coussens et al., 2013), similar to the phenotype observed in C-terminal RAG2 truncation mutants. It was proposed that these disruptions in RAG2 architecture destabilize the postcleavage complex (PCC), allowing for deregulated release of the broken DNA ends into alternative repair pathways (Coussens et al., 2013; Deriano et al., 2011). Interestingly, PCCs that contain RSSs that diverge from the consensus (particularly in the heptamer) are also destabilized and allow for postcleavage repair by A-EJ (Arnal, Holub, Salus, & Roth, 2010). Given that the RAG1–heptamer interaction is reported to depend on RAG2 (Swanson & Desiderio, 1999), it is appealing to think that these overlapping observations could reflect a common mechanism that enforces NHEJ-mediated repair. Observations of increased recombination between cryptic RSSs in the context of RAG2 truncation mutations may also be relevant in this regard (Gigi et al., 2014; G. Teng, unpublished data). While a stable PCC is likely important for preventing the engagement of repair pathways other than NHEJ, the ability of RAG to shunt breaks made by another nuclease into the NHEJ pathway (Cui & Meek, 2007) argues for an additional, active role in repair pathway choice. Very little is known about how this might occur. RAG has been suggested to interact with the DNA repair factors Ku70/Ku80 (Swanson, Kumar, & Raval, 2009) and MDC1 (Coster, Gold, Chen, Schatz, & Goldberg, 2012), but it has yet to be shown that these interactions serve a function during V(D)J recombination.

Recent findings also implicate the RAG2 C-terminus in autoinhibition. Though the presence of C-terminal residues 387–520 promotes the formation of RAG heterotetramers, they weaken precleavage RAG–DNA interactions and reduce hairpin formation *in vitro* (Grundy, Yang, & Gellert, 2010). The presence of H3K4me3 counteracts these inhibitory effects (Grundy et al., 2010), demonstrating a link between chromatin structure and functional release of RAG activity. RAG complexes containing RAG2 with residues 370–405 mutated to alanine (again, affecting the overall charge of the hinge region), show increased recombination activity beyond the basal level observed for wild-type RAG complexes (Lu et al., 2015). This "unleashing" of RAG activity is further stimulated upon the addition of H3K4me3, which increases the affinity of RAG for its DNA substrates as well as the catalytic turnover of the reaction, consistent with previous reports (Lu et al., 2015; Shimazaki et al., 2009). The direct interaction between H3K4me3 and RAG2-W453 is proposed to induce an allosteric change in the RAG complex mediated by the RAG2 hinge. This restructuring may allow the RAG complex to adopt a conformation that is simultaneously more catalytically active and better suited for binding to its substrates, perhaps explaining the PCC stability defects observed by Coussens et al. (2013).

Control of RAG2 subcellular localization is one additional means of regulating RAG activity, as the substrates for recombination reside in the nucleus. Association of *Ig* and *Tcr* alleles with the nuclear lamina has been shown to correlate with transcriptional and recombinational repression (Chan et al., 2013; Reddy, Zullo, Bertolino, & Singh, 2008). The suppressive nature of the nuclear periphery is partially enforced by the selective exclusion of RAG2 from this compartment (Chan et al., 2013). The subnuclear distribution of RAG2 does not fully correlate with the distribution of H3K4me3, however, indicating that RAG2–chromatin interactions cannot be solely responsible for the nonuniformity of RAG2 localization in the nucleus (Chan et al., 2013). Both RAG1 and RAG2 contain nuclear localization signals, with the latter serving an important role in the replenishment of nuclear RAG2 after each round of cell cycle-associated degradation (Arbuckle et al., 2011; Corneo, Benmerah, & Villartay, 2002; Ross, Vuica, & Desiderio, 2003). Though nuclear export of RAG2 coincides with its degradation during ubiquitin-mediated proteolysis (Mizuta, Mizuta, Araki, & Kitamura, 2002), a recent study suggests that cytoplasmic localization of RAG2 is not a passive by-product of this degradation process. Instead, genotoxic stress applied during G1 arrest expels a fraction of the nuclear RAG2 pool to the cytoplasm, where it colocalizes with the centrosome (Rodgers et al., 2015). Proteins that transduce the DNA damage signal (ATM and/or DNA-PKcs) are required for this process, leading the authors of this study to propose that RAG2 acts as a messenger between the nuclear DNA damage response and the centrosome to coordinate DNA repair and cell cycle (Rodgers et al., 2015). The details of this putative intracellular signaling pathway need further exploration, but the idea is consistent with the growing body of evidence that ATM coordinates feedback inhibition in response to RAG-induced DNA breaks, thus enforcing allelic exclusion (Hewitt et al., 2009; Steinel et al., 2013) and suppressing off-target recombination events (Teng et al., 2015).

5. RAG1—NONCLEAVAGE FUNCTIONS

Though the central enzymatic requirements are satisfied by residues in the RAG1 core, the contributions of the noncatalytic N- and C-termini of RAG1 should not be ignored. A number of studies suggest that the RAG1 N-terminus is somehow linked to ubiquitylation-dependent regulatory processes (Chao, Rothschild, & Basu, 2014). The region contains a RING motif (264–389, Fig. 3) that exhibits E3 ubiquitin ligase activity *in vitro* (Yurchenko, Xue, & Sadofsky, 2003) and *in vivo* (Jones & Gellert, 2003). The RING motif is part of an unusual RING zinc-finger domain that coordinates four zinc atoms and exists as a tight dimer (Bellon, Rodgers, Schatz, Coleman, & Steitz, 1997). Since the NBD also forms a tight dimer, this places two dimerization domains immediately adjacent to one another and suggests that for full-length RAG1, the "stalk" of the Y-shaped structure would extend further than in the crystal structure of core RAG1 (Fig. 2). A diverse range of cellular processes have been implicated as targets of RAG1-mediated ubiquitylation, ranging from nuclear import (Simkus, Makiya, & Jones, 2009), to modulation of the histone code (Deng, Liu, & Liu, 2015; Grazini et al., 2010; Jones et al., 2011), to postcleavage repair of V(D)J breaks (Grazini et al., 2010; Kassmeier et al., 2012). Interestingly, the RAG1 N-terminal region interacts with the cullin E3 ligase complex through its VprBP subunit (Kassmeier et al., 2012), and hence RAG1 has two different modalities available for mediating ubiquitylation, one intrinsic and the other extrinsic. B-lineage-specific knockout of VprBP results in a strong block in B cell development that is due in part to a defect in V(D)J recombination, perhaps because of a defect in the repair phase of the reaction (Kassmeier et al., 2012). The relevant target(s) for ubiquitylation by the cullin E3 ligase complex remain to be identified, and it has yet to be demonstrated that the RAG1–VprBP interaction is required for V(D)J recombination.

RAG1 autoubiquitylation on residue K233 has been proposed to regulate RAG1 protein turnover in a proteasome-independent manner (Jones & Gellert, 2003). This is consistent with the observation of increased levels of core RAG1 protein relative to full-length RAG1 in thymocytes (Dudley et al., 2003; McMahan, Difilippantonio, Rao, Spanopoulou, & Schatz, 1997; Teng et al., 2015). Given that RAG2 localizes to transcriptionally active chromatin genomewide (Ji et al., 2010; Teng et al., 2015), it would seem biologically prudent to constrain the amount of active RAG1 available for colocalization to these sites. Recent work from our lab suggests that cellular RAG1 and RAG2 concentrations lie below the in-solution K_D of the RAG1–RAG2 interaction, raising the intriguing possibility that only a small fraction of RAG1 molecules reside in catalytically active RAG1–RAG2 complexes (Zhang et al., 2015). Notably, the K_D was measured using separately expressed RAG1 core and RAG2 core proteins, and it is not known if the noncore regions influence the affinity of the RAG1–RAG2 interaction. Nor is it known whether coexpression (and hence cofolding) of the two proteins might alter the interaction affinity. It will now be interesting to explore whether mechanisms exist to regulate the RAG1–RAG2 interaction, for example, by posttranslational modifications. It is appealing to think that autoubiquitylation-mediated protein turnover could limit RAG levels and hence the risk of off-target RAG activity, though there is currently no direct evidence for such a mechanism.

In vivo, the substrates for V(D)J recombination are bundled within chromatin, and the RAG2 PHD provides one means by which RAG can "read" the chromatin landscape. Several reports suggest that RAG1 can also function as a "writer" of the histone code. RAG1 RING activity has been implicated in the monoubiquitylation of histone H3 (Grazini et al., 2010) and histone variant H3.3 (Jones et al., 2011). One study found that point mutations in the RAG1 RING domain (H307A and/or C325G) impair H3 monoubiquitylation activity and reduce recombination efficiency on both extrachromosomal and endogenous substrates (Grazini et al., 2010). This study concluded that the recombination defect was primarily due to altered DSB repair (based on the detection of unrepaired signal ends). More recent work suggests an alternate (but not exclusive) mechanism by which RAG1 ubiquitylation activity influences V(D)J recombination. In mice, a point mutation associated with human Omenn's syndrome (C325Y) results in a V(D)J-specific block in lymphocyte development, as does the expression of H3 mutated at the proposed ubiquitylation target, H3K79 (Deng et al., 2015). A "restrain and release" model was proposed based on two observations: (1) RAG1 preferentially interacts with unubiquitylated H3 and (2) overexpression of the noncatalytic RAG1 N-terminus can partially rescue the V(D)J defect in RAG1-C325Y mutant mice (Deng et al., 2015). According to this model, chromatin-associated RAG1 is catalytically restrained by its interaction with H3, and remains inert until released from H3 through its ubiquitylation activity. While this is an appealing hypothesis, and is consistent with the recombination defects observed in mice expressing core RAG1 (Dudley et al., 2003), several aspects of the model require further exploration. What is the mechanism by which the H3–RAG1 interaction inhibits cleavage—could it be related to the allosteric conformation change communicated by the H3K4me3–RAG2 interaction? What triggers RAG1-mediated ubiquitylation activity? What is the distribution of H3K79Ub throughout the genome? We have shown that RAG1 occupies thousands of sites outside of the antigen receptor genes in developing lymphocytes (Teng et al., 2015), and that restriction of RAG1 recruitment is not the primary mechanism of targeting V(D)J recombination activity to the antigen receptor genes. Perhaps, RAG1-mediated H3K79Ub serves as the activation code that distinguishes the chromatin structure of antigen receptor genes from the rest of the genome. The activating properties of RAG1-mediated ubiquitylation may also have implications for repertoire diversity, as noncore RAG1 regions are proposed to aid in the inclusion of Vβ gene segments that are paired with nonpreferred, low-efficiency RSSs (Horowitz & Bassing, 2014).

The RAG1 C-terminus (residues 1009–1040, Fig. 3) has been shown to collaborate in RAG2-mediated autoinhibition of RAG activity (Grundy et al., 2010). This seems biologically prudent, given the genotoxic potential of RAG-mediated cleavage. *In vitro* assays suggest that the C-termini of both RAG proteins synergize to inhibit the reaction at the hairpinning step (Grundy et al., 2010). The C-terminal region of core RAG1 (962–1008) forms a major surface surrounding the active site (Kim et al., 2015). While the structural placement of the succeeding noncore C-terminus is currently unknown, its linear proximity suggests that it too is held in close association with the RAG catalytic center, where it could affect direct or allosteric changes in the local structure. Built-in restrictions of RAG activity are an important feature of V(D)J recombination, as evidenced by the oncogenic consequences of over-exuberant RAG activity (Gostissa, Alt, & Chiarle, 2011; Larmonie et al., 2013; Papaemmanuil et al., 2014).

6. RECOMBINATION CENTERS

The temporospatial specificity of V(D)J recombination is enforced by a number of factors. RAG expression is restricted to specific bursts during lymphoid cell development, correlating exactly with the stages in which V(D)J recombination occurs (Grawunder et al., 1995; Wilson, Held, & MacDonald, 1994). In parallel, stage-specific epigenetic modifications (DNA demethylation, activating histone modifications; Cobb, Oestreich, Osipovich, & Oltz, 2006; Schatz & Ji, 2011) and dynamic intralocus interactions (Chaumeil et al., 2013; Ebert, Medvedovic, Tagoh, Schwickert, & Busslinger, 2013; Jhunjhunwala, van Zelm, Peak, & Murre, 2009; Lucas, Zhang, Dudko, & Murre, 2014; Shih & Krangel, 2013) specify the accessibility of the recombination targets to RAG activity. The antigen receptor loci also undergo repositioning events that move them between distinct subnuclear compartments that either permit or repress recombination (Chan et al., 2013; Kosak et al., 2002).

Highly focused RAG recruitment within the antigen receptor loci provides an additional layer of regulation, where the J gene segments (and the J-proximal D segments in some cases) form a preferential site for RAG accumulation (Ji et al., 2010). This "recombination center" is highly enriched in transcriptional activity and activating chromatin marks, and serves as a centralized hub for sampling and capture of partner RSSs. A number of conformational changes in these loci—reversible locus compaction, along with long-range chromosome interactions facilitated by different *cis*-acting elements allow for this capture process (Chaumeil et al., 2013; Ebert et al., 2013; Jhunjhunwala et al., 2009; Lucas et al., 2014; Shih & Krangel, 2013). Together, the epigenetic and chromosome conformation mechanisms that help to specify the recombination center enforce important properties of the recombination reaction: inclusion of distal gene segments in the receptor repertoire, allelic exclusion, and a defined recombination order (D-J rearrangement prior to V-DJ rearrangement; Brady, Steinel, & Bassing, 2010; Choi et al., 2013; Gopalakrishnan et al., 2013; Guo et al., 2011).

An interesting prediction of the recombination center model is that partner gene segments can be captured in a synaptic complex in the recombination center and then released without undergoing cleavage, allowing sampling of multiple partner RSSs before recombination actually takes place (Ji et al., 2010). This prediction is based on multiple studies from the Alt lab demonstrating that Vβ gene segment usage is strongly influenced by the strength of the RSS flanking the Vβ segment and cannot be determined simply by the frequency with which the gene segment interacts with the recombination center (Bassing et al., 2008; Ranganath et al., 2008; Wu et al., 2007). Very recently, the predicted reversibility of synaptic complex formation was validated biochemically using a single-molecule method to monitor the dynamics of RAG-mediated DNA binding and cleavage (Lovely, Brewster, Schatz, Baltimore, & Phillips, 2015). This study found that the core RAG proteins, HMGB1, and a consensus 12/23 RSS pair form a synaptic complex with a half-life of about 400 s, which goes forward to cleavage about 40% of the time. Given that endogenous RSSs typically differ from the consensus sequence at one or more positions and function less efficiently than consensus RSSs (Feeney, Goebel, & Espinoza, 2004; Hesse, Lieber, Mizuuchi, & Gellert, 1989; Yu, Taghva, & Lieber, 2002), it is tempting to think that the efficiency with which a synaptic complex containing endogenous RSSs goes forward to cleavage will usually be lower than one containing consensus RSSs.

The mechanisms that govern locus accessibility are highly lineage-, stage-, and locus-specific, and rely on transcription factors and *cis*-acting elements whose activity coincides with the developmental time frame of V(D)J recombination (Cobb et al., 2006; Hesslein & Schatz, 2001; Krangel, 2007). The epigenetic and transcriptional signals correlated with RAG recruitment, however, are much less specific. To date, there is little evidence that a singular epigenetic code distinguishes the active antigen receptor loci from the rest of the genome.

7. RAG BINDING GENOMEWIDE

Given the direct DNA-binding properties of RAG1, combined with the unique abundance of RSSs present in the antigen receptor genes, it was reasonable to assume that the site specificity of V(D)J recombination is simply enforced by preferential RAG1 recruitment to its preferred DNA substrates. RAG2, owing to its PHD and its lack of intrinsic catalytic activity, should exhibit no functional restrictions on its localization patterns. Accordingly, RAG2 binds broadly across the lymphocyte genome, occupying a major fraction (~60%) of the active TSSs, which are enriched in H3K4me3 (Ji et al., 2010; Teng et al., 2015). However, RAG1 too localizes to thousands of sites (~3500 in mouse thymocytes and pre-B cells) that are co-occupied by RAG2 and H3K4me3, encompassing a substantial portion of highly active, lineage-specific promoters and enhancers (Teng et al., 2015). This widespread localization is also observed for a catalytic mutant of RAG1 (D708A), indicating that the widespread binding of RAG1 does not reflect postcleavage accumulation (Teng et al., 2015). The RAG1-binding sites outside of antigen receptor genes showed no clear correlation with the presence of RSS-like motifs (cryptic RSSs, or cRSSs), further indicating that sequence-specific contacts and RAG-mediated cleavage activity could not explain patterns of RAG1 binding throughout the genome. Instead, markers of chromatin accessibility are the best predictors of RAG1 recruitment: H3K4me3, H3K27Ac, elevated GC content, and elevated CpG content (Teng et al., 2015).

In fact, only a handful of examples demonstrate specific RAG1 recruitment to antigen receptor genes via *trans*-factors that interact with both RAG1 and specific DNA motifs. In each case, the locus specificity of the recruitment event depends not on the RSS *per se*, but on a transcription factor binding motif that fortuitously overlaps with an RSS. The mouse *Tcrb* locus illustrates an instance where targeted RAG1 recruitment contributes to "beyond 12/23" restrictions that help maintain the correct order of recombination (Dβ-Jβ prior to Vβ-DJβ). The 3′ 23-RSSs flanking Dβ1 and Dβ2 contain conserved binding sites for c-fos, a subunit of the AP-1 transcription factor (Wang et al., 2008). In a manner independent of its transcriptional activating properties, c-fos preferentially recruits RAG1 to the 3′ Dβ RSSs, thus favoring Dβ-Jβ recombination (Wang et al., 2008). In the B-lineage, the analogous antigen receptor locus is the *Igh* gene, where direct

Vh-Dh recombination is disfavored relative to Vh-DJh recombination. In this case, however, recombination order appears to be regulated by local shifts in epigenetic marks that follow D-Jh recombination, and mark the DJ product for preferential RAG1 binding and recombination relative to germline Dh gene segments (Subrahmanyam et al., 2012). In addition, direct Vh-Dh recombination is suppressed by a sequence element known as IGCR1 that lies between the D and V gene clusters (Guo et al., 2011). IGCR1 contains two vital CTCF sites and is thought to function as an anchor point for chromatin loops that sequester the DJ portion of the Igh locus from the V portion (Guo et al., 2011; Lin, Guo, Su, Zhang, & Alt, 2015; Medvedovic et al., 2013). The human *Tcrd* locus provides another example of locus-specific RAG1 deposition, where Runx1-dependent RAG1 recruitment to Dδ2 favors a Dδ2-Dδ3 rearrangement event that precedes classical Dδ-Jδ recombination (Cieslak et al., 2014). This results in human-specific inclusion of two D gene segments in *Tcrd* rearrangements. Finally, Pax5 has been suggested to interact with RAG and facilitate RAG recruitment to Vh gene segments (Zhang et al., 2006), although it has yet to be shown that this mechanism enhances endogenous *Igh* locus assembly.

Outside of these exceptions, RAG1 binding does not appear to initiate primarily from sequence-specific recruitment events. Even within the *Tcr* and *Ig* genes, RAG1 binding extends beyond the RSS boundaries, occupying intronic regions that are enriched in H3K4me3 but devoid of RSSs (Teng et al., 2015). This is not to say that the RSSs do not play secondary roles in stabilizing or enhancing RAG1 binding postrecruitment. The *Tcra* and *Igk* genes are the top recruiters of RAG1 binding in double-positive thymocytes and pre-B cells, respectively, and defined peaks of RAG1 accumulation are clearly observed over the RSSs in recombination centers (Teng et al., 2015). The exceptional density of RSSs in the antigen receptor genes is likely one of the more important features that discriminates them from the remainder of the genome.

The genome in general, however, exhibits RSS-independent RAG1 recruitment, which likely occurs through nonspecific interactions between RAG1 and DNA that are mediated by various domains in the RAG1 core, including the NBD (Teng et al., 2015; Yin et al., 2009). In addition, indirect recruitment of RAG1 to chromatin may occur through its interaction with RAG2, though the latter cannot fully account for RAG1-binding patterns, which are partially maintained in the absence of RAG2, albeit with decreased intensity (Teng et al., 2015).

The genomic range of RAG1-binding sites, though broader than predicted, expands even further in the absence of noncore domains of either

RAG1 or RAG2. In the absence of the RAG2 C-terminal region (which includes the PHD), RAG1 diffuses toward sites with lower densities of H3K4me3, rather than homing to the most active TSSs (Teng et al., 2015). One possibility is that this results from an unmasking of additional nonspecific DNA-binding activities of RAG1, given the absence of the autoinhibitory domain that lies in the RAG2 C-terminal region (Lu et al., 2015). Another possibility is that in the absence of the RAG2 PHD domain, the RAG complex has less propensity to home to regions of high H3K4me3. Also relevant to the broader RAG1-binding pattern might be the fact that deletion of the RAG2 C-terminal region leads to upregulation of RAG1 protein levels (Teng et al., 2015). Core RAG1 also binds much more promiscuously than full-length RAG1, and intriguingly, exhibits an elevated intensity of binding at many nonantigen receptor sites but decreased intensity at the *Tcra* locus, relative to full-length RAG1(Teng et al., 2015). This might reflect changes in intrinsic DNA-binding properties of the truncated RAG1 protein, or the absence of activities encoded in the noncore termini of RAG1 (such as RING-mediated histone ubiquitylation or C-terminus-mediated autoinhibition). Also relevant might be the fact, noted above, that core RAG1 is expressed at substantially higher levels than full-length RAG1.

The RAG recombinase tolerates sequence variation not only in its substrates, particularly in the RSS spacer, but also in the heptamer and nonamer (Hesse et al., 1989; Lewis, Agard, Suh, & Czyzyk, 1997; Marculescu, Le, Simon, Jaeger, & Nadel, 2002; Ramsden, Baetz, & Wu, 1994; Zhang & Swanson, 2008). Even the RAG1–nonamer interaction mediated by the NBD relies on relatively few sequence-specific contacts (Yin et al., 2009). The RSSs that populate antigen receptor genes are diverse in their sequence content (sometimes diverging significantly from consensus heptamer and nonamer motifs), and substrate recognition by RAG must be sufficiently flexible to allow for maximal repertoire diversity. Outside of the antigen receptor genes, RAG1-binding sites show no cumulative preference for any particular sequence signature (Teng et al., 2015).

Flexible substrate recognition by RAG creates a problem for the remainder of the genome, however. The fairly minimal sequence requirements for a functional RSS (at least, by the rules that we currently understand) allow cRSSs to occur fairly frequently in vertebrate genomes (Lewis et al., 1997). The danger of fortuitous cRSSs is made apparent by their participation in RAG-dependent genetic rearrangements that contribute to lymphomagenesis (Larmonie et al., 2013). Since the majority of RAG1 binding throughout the genome is governed by nonspecific recruitment to DNA or accessible chromatin, it would seem biologically prudent to limit the availability of strong cRSSs in the vicinity of RAG1-binding sites and thus minimize the potential for off-target RAG activity. Indeed, RAG1-binding sites in mouse and human lymphocytes are preferentially depleted of cRSSs (Teng et al., 2015), and we propose that this has come about due to selective pressures imposed by genetic rearrangements occurring outside of the antigen receptor genes. Thus, the repertoire of RAG1-associated TSSs is protected from RAG activity during V(D)J recombination. We speculate that the rare cRSSs that retain activity and continue to mediate infrequent ectopic rearrangement have been maintained in the genome due to their overlap with regulatory or functional sequences in their host loci, or simply have not exerted sufficiently strong negative consequences to be selected against.

8. ABERRANT V(D)J RECOMBINATION AND GENOME STABILITY

The well-documented connections between aberrant V(D)J recombination and lymphomagenesis emphasize the genotoxic nature of RAG activity. Inappropriate genetic rearrangements can arise from the mishandling of legitimate V(D)J break intermediates and their subsequent release into inappropriate joining events (Lewis, 1994; Fig. 4). Recurrent chromosomal translocations in lymphomas often involve DNA breaks that originate during V(D)J recombination, but have been diverted away from normal repair pathways (Alt, Zhang, Meng, Guo, & Schwer, 2013; Gostissa et al., 2011). The strong regulatory elements (e.g., enhancers and locus control regions) contained within these broken ends can mediate oncogene activation when aberrantly joined to RAG-independent DSBs (Gostissa et al., 2011; Janz, 2006).

The formation of a stable PCC, consisting of RAG and four broken DNA ends, is an important step that ushers the DNA ends toward repair by NHEJ (Agrawal & Schatz, 1997; Hiom & Gellert, 1998; Jones & Gellert, 2001). As noted above, both the stability of the PCC and some other activity of RAG disfavor inappropriate repair pathways such as A-EJ or homologous recombination (Corneo et al., 2007; Cui & Meek, 2007; Lee et al., 2004). Classical NHEJ factors are essential for the resolution of RAG-dependent breaks and also guard against RAG-dependent genomic instability (Deriano & Roth, 2013; Helmink & Sleckman, 2012; Rooney,

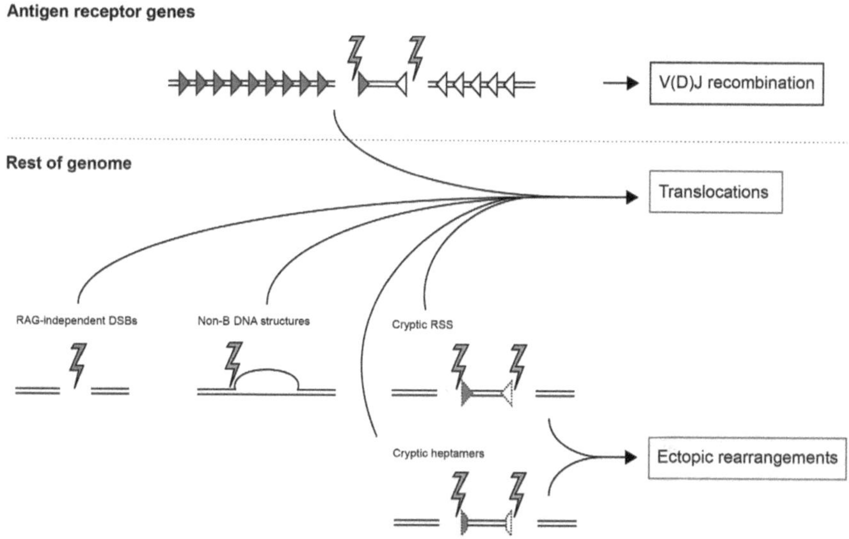

Figure 4 Outcomes of RAG activity in the genome. RAG activity in lymphocytes (indicated by blue lightning bolts) is mainly focused on the antigen receptor genes and the RSSs contained therein (solid triangles). However, breaks generated during legitimate V(D)J recombination can escape postcleavage repair and become joined to RAG-independent breaks (indicated by a yellow lightning bolt) elsewhere in the genome, leading to chromosomal translocations. The widespread binding of RAG outside of the antigen receptor genes also allows for ectopic RAG activity at non-B structures, cryptic RSSs (dashed triangles), and isolated cryptic heptamers (dashed trapezoids). Paired cryptic RSSs (and perhaps paired heptamers) can also undergo local rearrangement events that disrupt gene structures (such as deletions, depicted here, or inversions). (See the color plate.)

Chaudhuri, & Alt, 2004). The ataxia telangiectasia mutated (ATM) kinase (and DNA-PKcs, which is partially redundant with ATM), in particular, has come to the forefront as a multifunctional regulator of the postcleavage resolution phase (Gapud & Sleckman, 2011). During the cellular response to DSBs, ATM phosphorylation events transduce activating signals to downstream effector molecules (Shiloh & Ziv, 2013). In addition to fulfilling this role in the context of V(D)J recombination, ATM also contributes additional functions that maintain the fidelity of the reaction. ATM kinase activity prevents the unregulated release of coding ends from PCCs, thus preventing their participation in chromosomal translocations (Bredemeyer et al., 2006) and *trans*-rearrangements between different antigen receptor alleles (Bowen, Wangsa, Ried, Livak, & Hodes, 2013). The identity of the phosphorylation event(s) that stabilize the PCC is currently unknown.

In addition, RAG-induced breaks activate ATM-dependent signals that alter the expression of hundreds of genes, including that of RAG1 and RAG2 (Bednarski et al., 2012; Bredemeyer et al., 2008). Thus, ATM mediates a feedback inhibition pathway that senses RAG-mediated DNA damage and downregulates the V(D)J recombination machinery, thus enforcing allelic exclusion (Hewitt et al., 2009; Steinel et al., 2013). The nature of ATM-dependent responses during and after RAG-mediated cleavage is of great interest, as ATM has been implicated in pathways that extend beyond the classical DNA damage response, including cellular homeostatic mechanisms (Shiloh & Ziv, 2013).

Despite the depletion of cRSSs near sites of RAG1 occupancy, some active cRSS pairs have persisted in mouse and human genomes, and contribute to RAG-dependent genome instability in the form of interstitial deletions. Ectopic recombination events have been identified in the mouse *Notch1* and *Bcl11b* loci, and in the human *IKZF1*, *CDKN2A*, *SIL*, and *SCL* genes (Onozawa & Aplan, 2012; Sakata et al., 2004; Tsuji et al., 2009). The genetic structures derived from these rearrangements are hallmarks of lymphoid malignancies. Interestingly, these active cRSSs are not conserved between mice and humans, and generally show very low intrinsic recombination potential (Marculescu et al., 2002; Raghavan, Kirsch, & Lieber, 2001; Teng et al., 2015; Zhang & Swanson, 2008). Their apparent activity *in vivo* likely stems from selective advantages conferred by the deletion event (Ashworth et al., 2010; Jeannet et al., 2010). Isolated RSS heptamer motifs also continue to pose threats to genome stability, illustrated by the correlation between heptamers and recurrent sites of genetic variation in *ETV-RUNX1* B-ALL samples (Papaemmanuil et al., 2014). A weaker, but statistically significant, correlation between heptamers and deletion end-points has also been observed in cutaneous T cell lymphoma (Choi et al., 2015). Fittingly, heptamers (which specify sites of cleavage) are depleted from sites of RAG1 binding, whereas nonamers (which provide a DNA-binding motif) are slightly enriched, indicating that the genomewide depletion of cRSSs focuses on neutralizing the dangerous heptamer element (Teng et al., 2015). The continued existence of these illegitimate RAG targets, however, demonstrates that modern vertebrate genomes have not been completely cleansed of cRSSs, and highlight the important role of postcleavage regulatory mechanisms suppressing off-target RAG activity. For example, ATM-dependent feedback inhibition may partially suppress cryptic recombination at the mouse *Notch1* and *Bcl11b* loci (Teng et al., 2015).

Computational prediction of cRSS activity remains a challenge. Though the RSS information content (RIC) algorithm (Cowell, Davila, Kepler, & Kelsoe, 2002) has proven useful, particularly in the prediction of the activity levels of strong RSSs (Lee et al., 2003), its predictive power is less robust for weak RSS sequences. This is largely due to gaps in our knowledge of the rules that define RSS sequence identity, particularly the spacer. Since imaging, biochemical, and modeling studies show that both 12- and 23-RSSs are substantially bent in the context of the RAG complex (Ciubotaru et al., 2013, 2015; Pavlicek, Lyubchenko, & Chang, 2008; Shlyakhtenko et al., 2009), it will be important to determine the sequence constraints on the structural flexibility of the RSS. Improved versions of the RIC algorithm will also require high-throughput analysis of weak RSS activity levels, a technically nontrivial task. In addition, the *in vivo* applicability of new implementations of the RIC algorithm should take into account the chromatin context of the sequences, long-range chromatin interactions, and the presence of paired cRSSs (not just isolated cRSSs). Though RSSs do not appear to actively reposition nucleosomes to regulate V(D)J recombination *in vivo* (Kondilis-Mangum et al., 2010), their occlusion within nucleosomes has been shown to inhibit RAG cleavage (Golding, Chandler, Ballestar, Wolffe, & Schlissel, 1999; Kwon, Imbalzano, Matthews, & Oettinger, 1998). The nucleosome occupancy of cRSSs is yet another factor that could determine the accessibility of these sites for off-target RAG activity.

The danger of RAG mistargeting also extends to noncanonical substrates including non-B form DNA structures (Raghavan, Swanson, Wu, Hsieh, & Lieber, 2004), heteroduplex DNA (Nishana & Raghavan, 2012), and non-RSS sequences (Nambiar & Raghavan, 2012). Fragile sites in the *BCL2* gene, which mediate translocations to *Igh* in follicular lymphoma, provide examples of such noncanonical RAG activity *in vivo*. Particularly intriguing is the hypothesis that an entire class of translocation breakpoints (referred to as CpG type) is created by the combined action of RAG and the activation-induced deaminase (AID; Tsai et al., 2008). In this model, deamination of 5-methyl cytosine by AID leads to a T–G mismatch, which is subsequently targeted by RAG endonuclease activity (Tsai et al., 2008). A recent study that identifies a vulnerable stage of human B cell development with coexpression of RAG and AID provides substantial support for the model (Swaminathan et al., 2015). Hence, it is increasingly clear that a wide variety of potential noncanonical RAG targets exist, and it will be important to understand their genomic distribution and how they are accommodated by the DNA-binding and catalytic domains of RAG.

Temporal restriction of RAG breaks provides yet another checkpoint that maintains genome integrity. The previously discussed cell cycle restriction on RAG protein accumulation is an important component of this process, as RAG2 T490A mutant mice on a p53-deficient background exhibit increased levels of chromosomal translocations and lymphomas (Zhang et al., 2011). Similar phenotypes are observed in $P53^{-/-}$ mice expressing C-terminally truncated RAG2, though the authors of this study favor the interpretation that ATM and the C-terminus of RAG2 collaborate in maintenance of PCC stability (Deriano et al., 2011). Stage-specific restriction of RAG activity is also essential because mature B lymphocytes generate additional DSBs during *Igh* class switch recombination, thus creating yet another pool of substrates for chromosomal translocations. Indeed, NHEJ deficiency can unmask *Igh-Igl* translocations that arise in mature B cells that attempt secondary V(D)J rearrangements (Wang et al., 2009). In addition to temporal control of RAG activity, the postcleavage fate of RAG-induced breaks is also regulated. ATM prevents the propagation of V(D)J breaks beyond the appropriate lymphoid cell stages, as long-lived DSBs can form translocations and other aberrant chromosome structures that can promote lymphoma in mature lymphoid cells (Callen et al., 2007; Hu, Tepsuporn, Meyers, Gostissa, & Alt, 2014; Tepsuporn, Hu, Gostissa, & Alt, 2014).

9. FUNCTIONAL SIGNIFICANCE OF WIDESPREAD RAG BINDING

Despite the regulatory and evolutionary solutions that have arisen to restrain ectopic RAG activity, the contribution of aberrant V(D)J recombination to modern-day oncogenesis demonstrates a continued threat to organism fitness and survival. The evolutionary maintenance of widespread RAG1 binding seems confoundingly counterproductive in this light.

While RAG binding to promoters does not correlate with global changes in the expression of their associated genes (Teng et al., 2015), it is premature to rule out the possibility that small numbers of genes may respond to noncleavage RAG activities. Histone-modifying properties of RAG1 may hint at noncanonical functions in regulating chromatin identity, though the current understanding of histone ubiquitylation, particularly of histone H3, is quite limited. Additionally, the degree to which RAG interactions with histones could occlude or compete with alternate histone readers and modifiers is unknown.

A recent study of natural killer (NK) cells perhaps provides a clearer explanation for the tolerance of prevalent RAG1 binding in the genome (Karo, Schatz, & Sun, 2014). Though classical NK cell function does not derive from antigen receptors or their rearrangement during development, a subset of NK cell precursors express RAG during early ontogeny and propagate as a distinct functional class (Karo et al., 2014). The mechanisms that activate RAG expression and activity in common lymphoid progenitors are unknown. The RAG-experienced NK cell population exhibits increased fitness in the face of viral infection, stemming from enhanced DNA repair capacity (Karo et al., 2014). The fitness advantage is dependent on RAG1 cleavage function, and is also observed in the mature T lineage (Karo et al., 2014), suggesting that RAG activity has left a lasting imprint on cellular function that remains in place long after the developmental stages in which it is active. V(D)J breaks in lymphocytes have been shown to induce a diverse gene expression program that extends beyond classical DNA damage response proteins (Bednarski et al., 2012; Bredemeyer et al., 2008). It is tempting to speculate that low levels of RAG-mediated DNA breakage outside of the antigen receptor genes (perhaps as open-and-shut events) contribute to the induction of gene expression pathways that immunize cells against genome damage, particularly in cell lineages that typically do not experience *Ig* or *Tcr* gene rearrangement. Alternatively, the relevant RAG-mediated breaks might occur at antigen receptor genes, since NK cells are known to contain rearrangements at several antigen receptor loci (Borghesi et al., 2004; Fronkova et al., 2005; Pilbeam et al., 2008). The activity level associated with RAG1 binding will require direct measurement of RAG-dependent breaks genomewide and their correlation to RAG1-binding sites. Though elegant sequencing methods have been applied to detect translocation products of DSBs (Chiarle et al., 2011; Klein et al., 2011; Zhang et al., 2012), they likely capture only a fraction of the originating breakage events. A variety of approaches that directly detect broken DNA ends have been described (Crosetto et al., 2013; Dorsett et al., 2014; Leduc et al., 2011), and their application to genomewide DSB mapping in lymphocytes will likely generate more appropriate datasets.

Alternatively, perhaps the oddity of widespread RAG1 binding can be viewed as a nonfunctional side effect that can be explained by its proposed origins from an ancestral transposase. Perhaps, genomic binding preferences of RAG1 simply reflect the targeting preferences of the primordial RAG transposon. Though DNA transposases are diverse in their target-site preferences (Deininger & Roy-Engel, 2002; de Jong et al., 2014; Green, Bouchier, Fairhead, Craig, & Cormack, 2012; Jursch, Miskey, Izsvak, & Ivics, 2013; Yant et al., 2005), many DDE-family transposases prefer to integrate into open regions of chromatin. The ATAC-seq (transposase-accessible chromatin using sequencing) method actually takes advantage of this propensity, using the sites of Tn5 transposition to map open regions of the genome (Buenrostro, Giresi, Zaba, Chang, & Greenleaf, 2013). The Hermes transposase has been noted for its mechanistic and structural similarities to RAG, featuring a nick-hairpin mechanism of DNA cleavage, as well as an intertwined dimerization and DNA-binding domain similar to the RAG1 NBD (Hickman et al., 2014). A survey of Hermes transposition sites revealed some characteristics reminiscent of RAG1-binding sites: elevated GC content, and a preference for nucleosome-free regions that reside in the 5′-ends of genes (Gangadharan, Mularoni, Fain-Thornton, Wheelan, & Craig, 2010). The DNA-binding patterns and/or transposition sites of the closest known ancestors of RAG are unknown; it will be informative to compare these features to those of RAG1.

10. EVOLUTION OF RAG

An adaptive immune system borne of RAG-mediated gene rearrangement is thought to have arisen from a transpositional event (or events) that gave rise to the RAG locus itself and to a gene-splitting integration in a primordial antigen receptor locus (Cooper & Alder, 2006; Flajnik, 2014; Fugmann, 2010; Schatz, 2004; Thompson, 1995). Striking similarities between RAG and DDE transposases have been noted: common reaction chemistry for DNA cleavage, similar organization of protein domain structure, and sequence similarities between RSSs and terminal inverted repeats (TIRs) targeted by transposases (Fugmann, Lee, et al., 2000). The hypothetical ancestry of RAG was further solidified upon the demonstration of RAG-mediated transposition *in vitro* (Agrawal, Eastman, & Schatz, 1998; Hiom, Melek, & Gellert, 1998) and *in vivo* (Chatterji, Tsai, & Schatz, 2006; Curry et al., 2007; Ramsden, Weed, & Reddy, 2010; Vanura et al., 2007), though the efficiency of such reactions *in vivo* is highly disfavored relative to recombination. The reaction preference of modern RAG is partially determined by the C-terminus of RAG2, which inhibits transpositional activity (Elkin, Matthews, & Oettinger, 2003; Tsai & Schatz, 2003). The RAGs have also been shown to retarget and cleave the signal joints excised during canonical V(D)J recombination (Neiditch, Lee, Huye, Brandt, & Roth,

2002), creating a transpositional substrate that resembles the proposed DNA fragment(s) that gave rise to the split-gene structure of antigen receptor loci. Insertions of signal-end fragments can give rise to additional deletion, insertion, or translocation events (Chatterji et al., 2006; Curry et al., 2007), highlighting the importance of suppressing transpositional activity during V(D)J recombination. The structural and regulatory features that cause RAG to favor deletional/inversional recombination over transposition are not understood. An attractive idea is that this relates to how the cleaved ends (particularly the signal ends) are processed. RAG binds signal ends tightly, as would be expected of a transposase (Agrawal & Schatz, 1997), but has evolved to relinquish these ends efficiently to the NHEJ machinery, allowing for the desired outcome (signal joint formation), at least from the point of view of the host cell. For a transposase, such an outcome would represent a failure to propagate and must be avoided. Hence, one would expect that the modern-day RAG recombinase differs from its evolutionary precursor transposase in how it interfaces with the DNA repair apparatus.

Despite the evolutionary restriction of V(D)J recombination to jawed vertebrates, RAG1- and RAG2-like genes have been identified in several invertebrates that predate the origin of adaptive immunity (Fugmann, Messier, Novack, Cameron, & Rast, 2006; Huang et al., 2014; Kapitonov & Jurka, 2005; Kapitonov & Koonin, 2015; Wilson, Norton, & Fugmann, 2008); these might represent the evolutionary fossils of a once-active RAG transposon. The purple sea urchin (*Strongylocentrotus purpuratus*) genome harbors a gene cluster similar to modern RAG loci, which contain two convergently transcribed coding units that align with core RAG1 and RAG2 (Fugmann et al., 2006). The latter encodes a protein containing a PHD with preferred binding specificity for H3K4me2 (Wilson et al., 2008). The RAG-like cluster in the related green sea urchin (*Lytechinus variegatus*) genome occupies a nonsyntenic locus, suggestive of its origin from an independent transposition event (Kapitonov & Koonin, 2015). While the purple sea urchin RAG-like genes are expressed and their open reading frames are intact (Fugmann et al., 2006), their function remains undetermined, and it is not known if they possess nuclease activity. The green sea urchin RAG1-like open reading frame contains multiple inactivating mutations (Kapitonov & Koonin, 2015), consistent with signs of transposon fossilization. Neither the purple sea urchin nor the green sea urchin RAG-like cluster is flanked by discernible TIRs or target-site duplications (which arise when a transposon inserts into a new genomic location). If indeed these RAG-like clusters derive from a RAG transposon, then the TIRs and target-site duplications must have degenerated since the last transposition event. A starfish (*Patiria miniata*) genome also reveals evidence of adjacent RAG1-like and RAG2-like open reading frames, with a suggestion of flanking TIRs and target-site duplications (Kapitonov & Koonin, 2015). However, because of incomplete contig assembly, further analysis will be required to confirm the structure of the starfish element. Importantly, additional echinoderm genomes whose sequencing has yet to be completed seem to show clear evidence of RAG1/2-like genes (S. Fugmann & J. Rast, personal communication). Interestingly, all echinoderm RAG1-like and RAG2-like genes characterized to date are split into multiple exons (Fugmann et al., 2006; S. Fugmann & J. Rast, personal communication; Kapitonov & Koonin, 2015).

A RAG-related transposable element, *ProtoRAG*, has also been identified in lancelet, a chordate that also predates the origin of jawed vertebrates (Huang et al., 2014; Zhang et al., 2014). Lancelet genomes exhibit an immense degree of genomic variation (SNPs, indels, and translocations) between individuals, partially due to continued transpositional activity (Huang et al., 2014). Modern vertebrates, on the other hand, appear to have dampened genetic diversification activities in favor of maintaining genomic integrity. Antigen receptor gene rearrangement, mediated by a domesticated transposase, appears to be the one remnant of an evolutionary past characterized by extensive genetic diversification.

The Transib family of DDE-type transposons is a potential source of an ancestral RAG transposon, perhaps predating RAG1/RAG2 transposons. Transibs show sequence homology to modern core RAG1 and feature TIRs with remarkable sequence similarity to RSSs (Kapitonov & Jurka, 2005). However, Transibs usually contain symmetrical TIRs (Kapitonov & Jurka, 2005), compared to the asymmetrical 12- and 23-RSS pairs recognized by RAG, and do not contain a RAG2-like open reading frame. The Transib family is widely scattered in evolution, being found in numerous insect genomes as well as hydra, sea urchin, and even a fungus (Kapitonov & Jurka, 2005). The evolutionary relationship between Transib and RAG transposons has yet to be established.

The putative RAG transposons have only been concisely described in the literature. We currently know very little about their structural characteristics, transpositional activity, DNA-binding properties, and preferred integration sites of Transib and ProtoRAG. Future analysis of these entities may allow us to understand the evolutionary steps that connect Transib, ProtoRAG, and modern RAGs, and explain the transition from transposon to recombinase.

11. PERSPECTIVE

The RAG endonuclease is not merely a simple cut-and-paste machine, but is instead a highly *trans*- and autoregulated enzyme that coordinates its activity with cues provided by DNA and chromatin structure, cell cycle signals, and DNA repair. Beyond its critical role in generation of adaptive immunity, RAG activity appears to have more far-reaching effects on cellular fitness (which appears to extend beyond the B- and T-lineages) and genome evolution. Recent breakthroughs—crystallization of the core RAG tetramer, and discovery of transposase predecessors of RAG—have addressed two of the most long-standing questions in the field, and provide springboards for investigation of the open issues that remain unresolved. How are the noncore regions of RAG1 and RAG2 structured and positioned relative to the core complex? What is the mechanistic basis of the 12/23 rule? How does the RAG1 DNA-binding domain accommodate variable DNA sequences (for example, consensus RSSs vs. weak cRSSs), and is it restructured in response to histone modifications or noncleavage activities of RAG1? What are the relevant structural and regulatory differences that distinguish modern RAG from its ancestral transposases? Combined with our growing insights into the collaborative processes of RAG-mediated cleavage, DNA repair, and chromosome conformation dynamics, additional knowledge of the structural and regulatory features of RAG will help us understand the mechanisms that have evolved to restrict V(D)J recombination in space and time.

REFERENCES

Agrawal, A., Eastman, Q. M., & Schatz, D. G. (1998). Transposition mediated by RAG1 and RAG2 and its implications for the evolution of the immune system. *Nature*, *394*(6695), 744–751. http://dx.doi.org/10.1038/29457.

Agrawal, A., & Schatz, D. G. (1997). RAG1 and RAG2 form a stable postcleavage synaptic complex with DNA containing signal ends in V(D)J recombination. *Cell*, *89*(1), 43–53.

Akamatsu, Y., & Oettinger, M. A. (1998). Distinct roles of RAG1 and RAG2 in binding the V(D)J recombination signal sequences. *Molecular and Cellular Biology*, *18*(8), 4670–4678.

Alt, F. W., Zhang, Y., Meng, F. L., Guo, C., & Schwer, B. (2013). Mechanisms of programmed DNA lesions and genomic instability in the immune system. *Cell*, *152*(3), 417–429. http://dx.doi.org/10.1016/j.cell.2013.01.007.

Aravind, L., & Landsman, D. (1998). AT-hook motifs identified in a wide variety of DNA-binding proteins. *Nucleic Acids Research*, *26*(19), 4413–4421.

Arbuckle, J. L., Fauss, L. A., Simpson, R., Ptaszek, L. M., & Rodgers, K. K. (2001). Identification of two topologically independent domains in RAG1 and their role in macromolecular interactions relevant to V(D)J recombination. *The Journal of Biological Chemistry*, *276*(40), 37093–37101. http://dx.doi.org/10.1074/jbc.M105988200.

Arbuckle, J. L., Rahman, N. S., Zhao, S., Rodgers, W., & Rodgers, K. K. (2011). Elucidating the domain architecture and functions of non-core RAG1: The capacity of a non-core zinc-binding domain to function in nuclear import and nucleic acid binding. *BMC Biochemistry, 12*, 23. http://dx.doi.org/10.1186/1471-2091-12-23.

Arnal, S. M., Holub, A. J., Salus, S. S., & Roth, D. B. (2010). Non-consensus heptamer sequences destabilize the RAG post-cleavage complex, making ends available to alternative DNA repair pathways. *Nucleic Acids Research, 38*(9), 2944–2954. http://dx.doi.org/10.1093/nar/gkp1252.

Ashworth, T. D., Pear, W. S., Chiang, M. Y., Blacklow, S. C., Mastio, J., Xu, L., et al. (2010). Deletion-based mechanisms of Notch1 activation in T-ALL: Key roles for RAG recombinase and a conserved internal translational start site in Notch1. *Blood, 116*(25), 5455–5464. http://dx.doi.org/10.1182/blood-2010-05-286328.

Banerjee, J. K., & Schatz, D. G. (2014). Synapsis alters RAG-mediated nicking at Tcrb recombination signal sequences: Implications for the "beyond 12/23" rule. *Molecular and Cellular Biology, 34*(14), 2566–2580. http://dx.doi.org/10.1128/MCB.00411-14.

Bassing, C. H., Alt, F. W., Hughes, M. M., D'Auteuil, M., Wehrly, T. D., Woodman, B. B., et al. (2000). Recombination signal sequences restrict chromosomal V(D)J recombination beyond the 12/23 rule. *Nature, 405*(6786), 583–586. http://dx.doi.org/10.1038/35014635.

Bassing, C. H., Whitlow, S., Mostoslavsky, R., Yang-Iott, K., Ranganath, S., & Alt, F. W. (2008). Vbeta cluster sequences reduce the frequency of primary Vbeta2 and Vbeta14 rearrangements. *European Journal of Immunology, 38*(9), 2564–2572. http://dx.doi.org/10.1002/eji.200838347.

Bednarski, J. J., Nickless, A., Bhattacharya, D., Amin, R. H., Schlissel, M. S., & Sleckman, B. P. (2012). RAG-induced DNA double-strand breaks signal through Pim2 to promote pre-B cell survival and limit proliferation. *The Journal of Experimental Medicine, 209*(1), 11–17. http://dx.doi.org/10.1084/jem.20112078.

Bellon, S. F., Rodgers, K. K., Schatz, D. G., Coleman, J. E., & Steitz, T. A. (1997). Crystal structure of the RAG1 dimerization domain reveals multiple zinc-binding motifs including a novel zinc binuclear cluster. *Nature Structural Biology, 4*(7), 586–591.

Borghesi, L., Hsu, L. Y., Miller, J. P., Anderson, M., Herzenberg, L., Herzenberg, L., et al. (2004). B lineage-specific regulation of V(D)J recombinase activity is established in common lymphoid progenitors. *The Journal of Experimental Medicine, 199*(4), 491–502. http://dx.doi.org/10.1084/jem.20031800.

Bowen, S., Wangsa, D., Ried, T., Livak, F., & Hodes, R. J. (2013). Concurrent V(D)J recombination and DNA end instability increase interchromosomal trans-rearrangements in ATM-deficient thymocytes. *Nucleic Acids Research, 41*(8), 4535–4548. http://dx.doi.org/10.1093/nar/gkt154.

Brady, B. L., Steinel, N. C., & Bassing, C. H. (2010). Antigen receptor allelic exclusion: An update and reappraisal. *Journal of Immunology, 185*(7), 3801–3808. http://dx.doi.org/10.4049/jimmunol.1001158.

Branzei, D., & Foiani, M. (2008). Regulation of DNA repair throughout the cell cycle. *Nature Reviews. Molecular Cell Biology, 9*(4), 297–308. http://dx.doi.org/10.1038/nrm2351.

Bredemeyer, A. L., Helmink, B. A., Innes, C. L., Calderon, B., McGinnis, L. M., Mahowald, G. K., et al. (2008). DNA double-strand breaks activate a multi-functional genetic program in developing lymphocytes. *Nature, 456*(7223), 819–823. http://dx.doi.org/10.1038/nature07392.

Bredemeyer, A. L., Sharma, G. G., Huang, C. Y., Helmink, B. A., Walker, L. M., Khor, K. C., et al. (2006). ATM stabilizes DNA double-strand-break complexes during V(D)J recombination. *Nature, 442*(7101), 466–470. http://dx.doi.org/10.1038/nature04866.

Buenrostro, J. D., Giresi, P. G., Zaba, L. C., Chang, H. Y., & Greenleaf, W. J. (2013). Transposition of native chromatin for fast and sensitive epigenomic profiling of open chromatin, DNA-binding proteins and nucleosome position. *Nature Methods, 10*(12), 1213–1218. http://dx.doi.org/10.1038/nmeth.2688.

Callebaut, I., & Mornon, J. P. (1998). The V(D)J recombination activating protein RAG2 consists of a six-bladed propeller and a PHD fingerlike domain, as revealed by sequence analysis. *Cellular and Molecular Life Sciences, 54*(8), 880–891.

Callen, E., Jankovic, M., Difilippantonio, S., Daniel, J. A., Chen, H. T., Celeste, A., et al. (2007). ATM prevents the persistence and propagation of chromosome breaks in lymphocytes. *Cell, 130*(1), 63–75. http://dx.doi.org/10.1016/j.cell.2007.06.016.

Chan, E. A., Teng, G., Corbett, E., Choudhury, K. R., Bassing, C. H., Schatz, D. G., et al. (2013). Peripheral subnuclear positioning suppresses Tcrb recombination and segregates Tcrb alleles from RAG2. *Proceedings of the National Academy of Sciences of the United States of America, 110*(48), E4628–E4637. http://dx.doi.org/10.1073/pnas.1310846110.

Chao, J., Rothschild, G., & Basu, U. (2014). Ubiquitination events that regulate recombination of immunoglobulin Loci gene segments. *Frontiers in Immunology, 5*, 100. http://dx.doi.org/10.3389/fimmu.2014.00100.

Chatterji, M., Tsai, C. L., & Schatz, D. G. (2006). Mobilization of RAG-generated signal ends by transposition and insertion in vivo. *Molecular and Cellular Biology, 26*(4), 1558–1568. http://dx.doi.org/10.1128/MCB.26.4.1558-1568.2006.

Chaumeil, J., Micsinai, M., Ntziachristos, P., Deriano, L., Wang, J. M., Ji, Y., et al. (2013). Higher-order looping and nuclear organization of Tcra facilitate targeted rag cleavage and regulated rearrangement in recombination centers. *Cell Reports, 3*(2), 359–370. http://dx.doi.org/10.1016/j.celrep.2013.01.024.

Chiarle, R., Zhang, Y., Frock, R. L., Lewis, S. M., Molinie, B., Ho, Y. J., et al. (2011). Genome-wide translocation sequencing reveals mechanisms of chromosome breaks and rearrangements in B cells. *Cell, 147*(1), 107–119. http://dx.doi.org/10.1016/j.cell.2011.07.049.

Choi, J., Goh, G., Walradt, T., Hong, B. S., Bunick, C. G., Chen, K. B., et al. (2015). Genomic landscape of cutaneous T cell lymphoma. *Nature Genetics*. http://www.nature.com/ng/journal/vaop/ncurrent/full/ng.3356.html.

Choi, N. M., Loguercio, S., Verma-Gaur, J., Degner, S. C., Torkamani, A., Su, A. I., et al. (2013). Deep sequencing of the murine IgH repertoire reveals complex regulation of nonrandom V gene rearrangement frequencies. *Journal of Immunology, 191*(5), 2393–2402. http://dx.doi.org/10.4049/jimmunol.1301279.

Cieslak, A., Le Noir, S., Trinquand, A., Lhermitte, L., Franchini, D. M., Villarese, P., et al. (2014). RUNX1-dependent RAG1 deposition instigates human TCR-delta locus rearrangement. *The Journal of Experimental Medicine, 211*(9), 1821–1832. http://dx.doi.org/10.1084/jem.20132585.

Ciubotaru, M., Surleac, M. D., Metskas, L. A., Koo, P., Rhoades, E., Petrescu, A. J., et al. (2015). The architecture of the 12RSS in V(D)J recombination signal and synaptic complexes. *Nucleic Acids Research, 43*(2), 917–931. http://dx.doi.org/10.1093/nar/gku1348.

Ciubotaru, M., Trexler, A. J., Spiridon, L. N., Surleac, M. D., Rhoades, E., Petrescu, A. J., et al. (2013). RAG and HMGB1 create a large bend in the 23RSS in the V(D)J recombination synaptic complexes. *Nucleic Acids Research, 41*(4), 2437–2454. http://dx.doi.org/10.1093/nar/gks1294.

Cobb, R. M., Oestreich, K. J., Osipovich, O. A., & Oltz, E. M. (2006). Accessibility control of V(D)J recombination. *Advances in Immunology, 91*, 45–109. http://dx.doi.org/10.1016/S0065-2776(06)91002-5.

Cooper, M. D., & Alder, M. N. (2006). The evolution of adaptive immune systems. *Cell, 124*(4), 815–822. http://dx.doi.org/10.1016/j.cell.2006.02.001.
Corneo, B., Benmerah, A., & Villartay, J. P. (2002). A short peptide at the C terminus is responsible for the nuclear localization of RAG2. *European Journal of Immunology, 32*(7), 2068–2073. http://dx.doi.org/10.1002/1521-4141(200207)32:7<2068::AID-IMMU2068>3.0.CO;2[–]6.
Corneo, B., Wendland, R. L., Deriano, L., Cui, X., Klein, I. A., Wong, S. Y., et al. (2007). Rag mutations reveal robust alternative end joining. *Nature, 449*(7161), 483–486. http://dx.doi.org/10.1038/nature06168.
Coster, G., Gold, A., Chen, D., Schatz, D. G., & Goldberg, M. (2012). A dual interaction between the DNA damage response protein MDC1 and the RAG1 subunit of the V(D)J recombinase. *The Journal of Biological Chemistry, 287*(43), 36488–36498. http://dx.doi.org/10.1074/jbc.M112.402487.
Coussens, M. A., Wendland, R. L., Deriano, L., Lindsay, C. R., Arnal, S. M., & Roth, D. B. (2013). RAG2's acidic hinge restricts repair-pathway choice and promotes genomic stability. *Cell Reports, 4*(5), 870–878. http://dx.doi.org/10.1016/j.celrep.2013.07.041.
Cowell, L. G., Davila, M., Kepler, T. B., & Kelsoe, G. (2002). Identification and utilization of arbitrary correlations in models of recombination signal sequences. *Genome Biology, 3*(12), RESEARCH0072.
Crosetto, N., Mitra, A., Silva, M. J., Bienko, M., Dojer, N., Wang, Q., et al. (2013). Nucleotide-resolution DNA double-strand break mapping by next-generation sequencing. *Nature Methods, 10*(4), 361–365. http://dx.doi.org/10.1038/nmeth.2408.
Cui, X., & Meek, K. (2007). Linking double-stranded DNA breaks to the recombination activating gene complex directs repair to the nonhomologous end-joining pathway. *Proceedings of the National Academy of Sciences of the United States of America, 104*(43), 17046–17051. http://dx.doi.org/10.1073/pnas.0610928104.
Cuomo, C. A., & Oettinger, M. A. (1994). Analysis of regions of RAG-2 important for V(D)J recombination. *Nucleic Acids Research, 22*(10), 1810–1814.
Curry, J. D., Schulz, D., Guidos, C. J., Danska, J. S., Nutter, L., Nussenzweig, A., et al. (2007). Chromosomal reinsertion of broken RSS ends during T cell development. *The Journal of Experimental Medicine, 204*(10), 2293–2303. http://dx.doi.org/10.1084/jem.20070583.
de Jong, J., Akhtar, W., Badhai, J., Rust, A. G., Rad, R., Hilkens, J., et al. (2014). Chromatin landscapes of retroviral and transposon integration profiles. *PLoS Genetics, 10*(4), e1004250. http://dx.doi.org/10.1371/journal.pgen.1004250.
Deininger, P. L., & Roy-Engel, A. M. (2002). *Mobile DNA II* (pp. 1080–1083). (N. L. Craig, R. Craigie, M. Gellert, & A. M. Lambowitz, Eds.). Washington, D.C.: ASM Press.
Deng, Z., Liu, H., & Liu, X. (2015). RAG1-mediated ubiquitylation of histone H3 is required for chromosomal V(D)J recombination. *Cell Research, 25*(2), 181–192. http://dx.doi.org/10.1038/cr.2015.1.
Deriano, L., Chaumeil, J., Coussens, M., Multani, A., Chou, Y., Alekseyenko, A. V., et al. (2011). The RAG2 C terminus suppresses genomic instability and lymphomagenesis. *Nature, 471*(7336), 119–123. http://dx.doi.org/10.1038/nature09755.
Deriano, L., & Roth, D. B. (2013). Modernizing the nonhomologous end-joining repertoire: Alternative and classical NHEJ share the stage. *Annual Review of Genetics, 47*, 433–455. http://dx.doi.org/10.1146/annurev-genet-110711-155540.
Difilippantonio, M. J., McMahan, C. J., Eastman, Q. M., Spanopoulou, E., & Schatz, D. G. (1996). RAG1 mediates signal sequence recognition and recruitment of RAG2 in V(D)J recombination. *Cell, 87*(2), 253–262.
Dorsett, Y., Zhou, Y., Tubbs, A. T., Chen, B. R., Purman, C., Lee, B. S., et al. (2014). HCoDES reveals chromosomal DNA end structures with single-nucleotide resolution. *Molecular Cell, 56*(6), 808–818. http://dx.doi.org/10.1016/j.molcel.2014.10.024.

Drejer-Teel, A. H., Fugmann, S. D., & Schatz, D. G. (2007). The beyond 12/23 restriction is imposed at the nicking and pairing steps of DNA cleavage during V(D)J recombination. *Molecular and Cellular Biology*, *27*(18), 6288–6299. http://dx.doi.org/10.1128/MCB.00835-07.

Dudley, D. D., Sekiguchi, J., Zhu, C., Sadofsky, M. J., Whitlow, S., DeVido, J., et al. (2003). Impaired V(D)J recombination and lymphocyte development in core RAG1-expressing mice. *The Journal of Experimental Medicine*, *198*(9), 1439–1450. http://dx.doi.org/10.1084/jem.20030627.

Eastman, Q. M., Villey, I. J., & Schatz, D. G. (1999). Detection of RAG protein-V(D)J recombination signal interactions near the site of DNA cleavage by UV cross-linking. *Molecular and Cellular Biology*, *19*(5), 3788–3797.

Ebert, A., Medvedovic, J., Tagoh, H., Schwickert, T. A., & Busslinger, M. (2013). Control of antigen receptor diversity through spatial regulation of V(D)J recombination. *Cold Spring Harbor Symposia on Quantitative Biology*, *78*, 11–21. http://dx.doi.org/10.1101/sqb.2013.78.019943.

Elkin, S. K., Matthews, A. G., & Oettinger, M. A. (2003). The C-terminal portion of RAG2 protects against transposition in vitro. *The EMBO Journal*, *22*(8), 1931–1938. http://dx.doi.org/10.1093/emboj/cdg184.

Feeney, A. J., Goebel, P., & Espinoza, C. R. (2004). Many levels of control of V gene rearrangement frequency. *Immunological Reviews*, *200*, 44–56. http://dx.doi.org/10.1111/j.0105-2896.2004.00163.x.

Flajnik, M. F. (2014). Re-evaluation of the immunological Big Bang. *Current Biology*, *24*(21), R1060–R1065. http://dx.doi.org/10.1016/j.cub.2014.09.070.

Fronkova, E., Krejci, O., Kalina, T., Horvath, O., Trka, J., & Hrusak, O. (2005). Lymphoid differentiation pathways can be traced by TCR delta rearrangements. *Journal of Immunology*, *175*(4), 2495–2500.

Fugmann, S. D. (2010). The origins of the Rag genes—From transposition to V(D)J recombination. *Seminars in Immunology*, *22*(1), 10–16. http://dx.doi.org/10.1016/j.smim.2009.11.004.

Fugmann, S. D., Lee, A. I., Shockett, P. E., Villey, I. J., & Schatz, D. G. (2000). The RAG proteins and V(D)J recombination: Complexes, ends, and transposition. *Annual Review of Immunology*, *18*, 495–527. http://dx.doi.org/10.1146/annurev.immunol.18.1.495.

Fugmann, S. D., Messier, C., Novack, L. A., Cameron, R. A., & Rast, J. P. (2006). An ancient evolutionary origin of the Rag1/2 gene locus. *Proceedings of the National Academy of Sciences of the United States of America*, *103*(10), 3728–3733. http://dx.doi.org/10.1073/pnas.0509720103.

Fugmann, S. D., & Schatz, D. G. (2001). Identification of basic residues in RAG2 critical for DNA binding by the RAG1-RAG2 complex. *Molecular Cell*, *8*(4), 899–910.

Fugmann, S. D., Villey, I. J., Ptaszek, L. M., & Schatz, D. G. (2000). Identification of two catalytic residues in RAG1 that define a single active site within the RAG1/RAG2 protein complex. *Molecular Cell*, *5*(1), 97–107.

Gangadharan, S., Mularoni, L., Fain-Thornton, J., Wheelan, S. J., & Craig, N. L. (2010). DNA transposon Hermes inserts into DNA in nucleosome-free regions in vivo. *Proceedings of the National Academy of Sciences of the United States of America*, *107*(51), 21966–21972. http://dx.doi.org/10.1073/pnas.1016382107.

Gapud, E. J., & Sleckman, B. P. (2011). Unique and redundant functions of ATM and DNA-PKcs during V(D)J recombination. *Cell Cycle*, *10*(12), 1928–1935.

Gigi, V., Lewis, S., Shestova, O., Mijuskovic, M., Deriano, L., Meng, W., et al. (2014). RAG2 mutants alter DSB repair pathway choice in vivo and illuminate the nature of 'alternative NHEJ'. *Nucleic Acids Research*, *42*(10), 6352–6364. http://dx.doi.org/10.1093/nar/gku295.

Golding, A., Chandler, S., Ballestar, E., Wolffe, A. P., & Schlissel, M. S. (1999). Nucleosome structure completely inhibits in vitro cleavage by the V(D)J recombinase. *The EMBO Journal, 18*(13), 3712–3723. http://dx.doi.org/10.1093/emboj/18.13.3712.

Gopalakrishnan, S., Majumder, K., Predeus, A., Huang, Y., Koues, O. I., Verma-Gaur, J., et al. (2013). Unifying model for molecular determinants of the preselection Vbeta repertoire. *Proceedings of the National Academy of Sciences of the United States of America, 110*(34), E3206–E3215. http://dx.doi.org/10.1073/pnas.1304048110.

Gostissa, M., Alt, F. W., & Chiarle, R. (2011). Mechanisms that promote and suppress chromosomal translocations in lymphocytes. *Annual Review of Immunology, 29*, 319–350. http://dx.doi.org/10.1146/annurev-immunol-031210-101329.

Grawunder, U., Leu, T. M., Schatz, D. G., Werner, A., Rolink, A. G., Melchers, F., et al. (1995). Down-regulation of RAG1 and RAG2 gene expression in preB cells after functional immunoglobulin heavy chain rearrangement. *Immunity, 3*(5), 601–608.

Grazini, U., Zanardi, F., Citterio, E., Casola, S., Goding, C. R., & McBlane, F. (2010). The RING domain of RAG1 ubiquitylates histone H3: A novel activity in chromatin-mediated regulation of V(D)J joining. *Molecular Cell, 37*(2), 282–293. http://dx.doi.org/10.1016/j.molcel.2009.12.035.

Green, B., Bouchier, C., Fairhead, C., Craig, N. L., & Cormack, B. P. (2012). Insertion site preference of Mu, Tn5, and Tn7 transposons. *Mobile DNA, 3*(1), 3. http://dx.doi.org/10.1186/1759-8753-3-3.

Grundy, G. J., Ramon-Maiques, S., Dimitriadis, E. K., Kotova, S., Biertumpfel, C., Heymann, J. B., et al. (2009). Initial stages of V(D)J recombination: The organization of RAG1/2 and RSS DNA in the postcleavage complex. *Molecular Cell, 35*(2), 217–227. http://dx.doi.org/10.1016/j.molcel.2009.06.022.

Grundy, G. J., Yang, W., & Gellert, M. (2010). Autoinhibition of DNA cleavage mediated by RAG1 and RAG2 is overcome by an epigenetic signal in V(D)J recombination. *Proceedings of the National Academy of Sciences of the United States of America, 107*(52), 22487–22492. http://dx.doi.org/10.1073/pnas.1014958107.

Guo, C., Yoon, H. S., Franklin, A., Jain, S., Ebert, A., Cheng, H. L., et al. (2011). CTCF-binding elements mediate control of V(D)J recombination. *Nature, 477*(7365), 424–430. http://dx.doi.org/10.1038/nature10495.

Helmink, B. A., & Sleckman, B. P. (2012). The response to and repair of RAG-mediated DNA double-strand breaks. *Annual Review of Immunology, 30*, 175–202. http://dx.doi.org/10.1146/annurev-immunol-030409-101320.

Hesse, J. E., Lieber, M. R., Mizuuchi, K., & Gellert, M. (1989). V(D)J recombination: A functional definition of the joining signals. *Genes and Development, 3*(7), 1053–1061.

Hesslein, D. G., & Schatz, D. G. (2001). Factors and forces controlling V(D)J recombination. *Advances in Immunology, 78*, 169–232.

Hewitt, S. L., Yin, B., Ji, Y., Chaumeil, J., Marszalek, K., Tenthorey, J., et al. (2009). RAG-1 and ATM coordinate monoallelic recombination and nuclear positioning of immunoglobulin loci. *Nature Immunology, 10*(6), 655–664. http://dx.doi.org/10.1038/ni.1735.

Hickman, A. B., Ewis, H. E., Li, X., Knapp, J. A., Laver, T., Doss, A. L., et al. (2014). Structural basis of hAT transposon end recognition by Hermes, an octameric DNA transposase from Musca domestica. *Cell, 158*(2), 353–367. http://dx.doi.org/10.1016/j.cell.2014.05.037.

Hiom, K., & Gellert, M. (1998). Assembly of a 12/23 paired signal complex: A critical control point in V(D)J recombination. *Molecular Cell, 1*(7), 1011–1019.

Hiom, K., Melek, M., & Gellert, M. (1998). DNA transposition by the RAG1 and RAG2 proteins: A possible source of oncogenic translocations. *Cell, 94*(4), 463–470.

Horowitz, J. E., & Bassing, C. H. (2014). Noncore RAG1 regions promote Vbeta rearrangements and alphabeta T cell development by overcoming inherent inefficiency of Vbeta recombination signal sequences. *Journal of Immunology, 192*(4), 1609–1619. http://dx.doi.org/10.4049/jimmunol.1301599.

Hu, J., Tepsuporn, S., Meyers, R. M., Gostissa, M., & Alt, F. W. (2014). Developmental propagation of V(D)J recombination-associated DNA breaks and translocations in mature B cells via dicentric chromosomes. *Proceedings of the National Academy of Sciences of the United States of America, 111*(28), 10269–10274. http://dx.doi.org/10.1073/pnas.1410112111.

Huang, S., Chen, Z., Yan, X., Yu, T., Huang, G., Yan, Q., et al. (2014). Decelerated genome evolution in modern vertebrates revealed by analysis of multiple lancelet genomes. *Nature Communications, 5*, 5896. http://dx.doi.org/10.1038/ncomms6896.

Hughes, M. M., Tillman, R. E., Wehrly, T. D., White, J. M., & Sleckman, B. P. (2003). The B12/23 restriction is critically dependent on recombination signal nonamer and spacer sequences. *Journal of Immunology, 171*(12), 6604–6610.

Huye, L. E., Purugganan, M. M., Jiang, M. M., & Roth, D. B. (2002). Mutational analysis of all conserved basic amino acids in RAG-1 reveals catalytic, step arrest, and joining-deficient mutants in the V(D)J recombinase. *Molecular and Cellular Biology, 22*(10), 3460–3473.

Janz, S. (2006). Myc translocations in B cell and plasma cell neoplasms. *DNA Repair (Amst), 5*(9–10), 1213–1224. http://dx.doi.org/10.1016/j.dnarep.2006.05.017.

Jeannet, R., Mastio, J., Macias-Garcia, A., Oravecz, A., Ashworth, T., Geimer Le Lay, A. S., et al. (2010). Oncogenic activation of the Notch1 gene by deletion of its promoter in Ikaros-deficient T-ALL. *Blood, 116*(25), 5443–5454. http://dx.doi.org/10.1182/blood-2010-05-286658.

Jhunjhunwala, S., van Zelm, M. C., Peak, M. M., & Murre, C. (2009). Chromatin architecture and the generation of antigen receptor diversity. *Cell, 138*(3), 435–448. http://dx.doi.org/10.1016/j.cell.2009.07.016.

Ji, Y., Resch, W., Corbett, E., Yamane, A., Casellas, R., & Schatz, D. G. (2010). The in vivo pattern of binding of RAG1 and RAG2 to antigen receptor loci. *Cell, 141*(3), 419–431. http://dx.doi.org/10.1016/j.cell.2010.03.010.

Jones, J. M., Bhattacharyya, A., Simkus, C., Vallieres, B., Veenstra, T. D., & Zhou, M. (2011). The RAG1 V(D)J recombinase/ubiquitin ligase promotes ubiquitylation of acetylated, phosphorylated histone 3.3. *Immunology Letters, 136*(2), 156–162. http://dx.doi.org/10.1016/j.imlet.2011.01.005.

Jones, J. M., & Gellert, M. (2001). Intermediates in V(D)J recombination: A stable RAG1/2 complex sequesters cleaved RSS ends. *Proceedings of the National Academy of Sciences of the United States of America, 98*(23), 12926–12931. http://dx.doi.org/10.1073/pnas.221471198.

Jones, J. M., & Gellert, M. (2003). Autoubiquitylation of the V(D)J recombinase protein RAG1. *Proceedings of the National Academy of Sciences of the United States of America, 100*(26), 15446–15451. http://dx.doi.org/10.1073/pnas.2637012100.

Jursch, T., Miskey, C., Izsvak, Z., & Ivics, Z. (2013). Regulation of DNA transposition by CpG methylation and chromatin structure in human cells. *Mobile DNA, 4*(1), 15. http://dx.doi.org/10.1186/1759-8753-4-15.

Kapitonov, V. V., & Jurka, J. (2005). RAG1 core and V(D)J recombination signal sequences were derived from Transib transposons. *PLoS Biology, 3*(6), e181. http://dx.doi.org/10.1371/journal.pbio.0030181.

Kapitonov, V. V., & Koonin, E. V. (2015). Evolution of the RAG1-RAG2 locus: Both proteins came from the same transposon. *Biology Direct, 10*(1), 20. http://dx.doi.org/10.1186/s13062-015-0055-8.

Karo, J. M., Schatz, D. G., & Sun, J. C. (2014). The RAG recombinase dictates functional heterogeneity and cellular fitness in natural killer cells. *Cell, 159*(1), 94–107. http://dx.doi.org/10.1016/j.cell.2014.08.026.

Kassmeier, M. D., Mondal, K., Palmer, V. L., Raval, P., Kumar, S., Perry, G. A., et al. (2012). VprBP binds full-length RAG1 and is required for B-cell development and V(D)J recombination fidelity. *The EMBO Journal*, *31*(4), 945–958. http://dx.doi.org/10.1038/emboj.2011.455.
Kim, D. R., Dai, Y., Mundy, C. L., Yang, W., & Oettinger, M. A. (1999). Mutations of acidic residues in RAG1 define the active site of the V(D)J recombinase. *Genes and Development*, *13*(23), 3070–3080.
Kim, M. S., Lapkouski, M., Yang, W., & Gellert, M. (2015). Crystal structure of the V(D)J recombinase RAG1-RAG2. *Nature*, *518*(7540), 507–511. http://dx.doi.org/10.1038/nature14174.
Klein, I. A., Resch, W., Jankovic, M., Oliveira, T., Yamane, A., Nakahashi, H., et al. (2011). Translocation-capture sequencing reveals the extent and nature of chromosomal rearrangements in B lymphocytes. *Cell*, *147*(1), 95–106. http://dx.doi.org/10.1016/j.cell.2011.07.048.
Kondilis-Mangum, H. D., Cobb, R. M., Osipovich, O., Srivatsan, S., Oltz, E. M., & Krangel, M. S. (2010). Transcription-dependent mobilization of nucleosomes at accessible TCR gene segments in vivo. *Journal of Immunology*, *184*(12), 6970–6977. http://dx.doi.org/10.4049/jimmunol.0903923.
Kosak, S. T., Skok, J. A., Medina, K. L., Riblet, R., Le Beau, M. M., Fisher, A. G., et al. (2002). Subnuclear compartmentalization of immunoglobulin loci during lymphocyte development. *Science*, *296*(5565), 158–162. http://dx.doi.org/10.1126/science.1068768.
Krangel, M. S. (2007). T cell development: Better living through chromatin. *Nature Immunology*, *8*(7), 687–694. http://dx.doi.org/10.1038/ni1484.
Kwon, J., Imbalzano, A. N., Matthews, A., & Oettinger, M. A. (1998). Accessibility of nucleosomal DNA to V(D)J cleavage is modulated by RSS positioning and HMG1. *Molecular Cell*, *2*(6), 829–839.
Landree, M. A., Wibbenmeyer, J. A., & Roth, D. B. (1999). Mutational analysis of RAG1 and RAG2 identifies three catalytic amino acids in RAG1 critical for both cleavage steps of V(D)J recombination. *Genes and Development*, *13*(23), 3059–3069.
Larmonie, N. S., Dik, W. A., Meijerink, J. P., Homminga, I., van Dongen, J. J., & Langerak, A. W. (2013). Breakpoint sites disclose the role of the V(D)J recombination machinery in the formation of T-cell receptor (TCR) and non-TCR associated aberrations in T-cell acute lymphoblastic leukemia. *Haematologica*, *98*(8), 1173–1184. http://dx.doi.org/10.3324/haematol.2012.082156.
Leduc, F., Faucher, D., Bikond Nkoma, G., Gregoire, M. C., Arguin, M., Wellinger, R. J., et al. (2011). Genome-wide mapping of DNA strand breaks. *PloS One*, *6*(2), e17353. http://dx.doi.org/10.1371/journal.pone.0017353.
Lee, A. I., Fugmann, S. D., Cowell, L. G., Ptaszek, L. M., Kelsoe, G., & Schatz, D. G. (2003). A functional analysis of the spacer of V(D)J recombination signal sequences. *PLoS Biology*, *1*(1), E1. http://dx.doi.org/10.1371/journal.pbio.0000001.
Lee, G. S., Neiditch, M. B., Salus, S. S., & Roth, D. B. (2004). RAG proteins shepherd double-strand breaks to a specific pathway, suppressing error-prone repair, but RAG nicking initiates homologous recombination. *Cell*, *117*(2), 171–184.
Lewis, S. M. (1994). The mechanism of V(D)J joining: Lessons from molecular, immunological, and comparative analyses. *Advances in Immunology*, *56*, 27–150.
Lewis, S. M., Agard, E., Suh, S., & Czyzyk, L. (1997). Cryptic signals and the fidelity of V(D)J joining. *Molecular and Cellular Biology*, *17*(6), 3125–3136.
Li, Z., Dordai, D. I., Lee, J., & Desiderio, S. (1996). A conserved degradation signal regulates RAG-2 accumulation during cell division and links V(D)J recombination to the cell cycle. *Immunity*, *5*(6), 575–589.
Lin, S. G., Guo, C., Su, A., Zhang, Y., & Alt, F. W. (2015). CTCF-binding elements 1 and 2 in the Igh intergenic control region cooperatively regulate V(D)J recombination.

Proceedings of the National Academy of Sciences of the United States of America, *112*(6), 1815–1820. http://dx.doi.org/10.1073/pnas.1424936112.

Liu, Y., Subrahmanyam, R., Chakraborty, T., Sen, R., & Desiderio, S. (2007). A plant homeodomain in RAG-2 that binds hypermethylated lysine 4 of histone H3 is necessary for efficient antigen-receptor-gene rearrangement. *Immunity*, *27*(4), 561–571. http://dx.doi.org/10.1016/j.immuni.2007.09.005.

Lovely, G. A., Brewster, R. C., Schatz, D. G., Baltimore, D., & Phillips, R. (2015). Single-molecule analysis of RAG-mediated V(D)J DNA cleavage. *Proceedings of the National Academy of Sciences of the United States of America*, *112*(14), E1715–E1723. http://dx.doi.org/10.1073/pnas.1503477112.

Lu, C., Ward, A., Bettridge, J., Liu, Y., & Desiderio, S. (2015). An autoregulatory mechanism imposes allosteric control on the V(D)J recombinase by histone H3 methylation. *Cell Reports*, *10*(1), 29–38. http://dx.doi.org/10.1016/j.celrep.2014.12.001.

Lucas, J. S., Zhang, Y., Dudko, O. K., & Murre, C. (2014). 3D trajectories adopted by coding and regulatory DNA elements: First-passage times for genomic interactions. *Cell*, *158*(2), 339–352. http://dx.doi.org/10.1016/j.cell.2014.05.036.

Marculescu, R., Le, T., Simon, P., Jaeger, U., & Nadel, B. (2002). V(D)J-mediated translocations in lymphoid neoplasms: A functional assessment of genomic instability by cryptic sites. *The Journal of Experimental Medicine*, *195*(1), 85–98.

Matthews, A. G., Kuo, A. J., Ramon-Maiques, S., Han, S., Champagne, K. S., Ivanov, D., et al. (2007). RAG2 PHD finger couples histone H3 lysine 4 trimethylation with V(D)J recombination. *Nature*, *450*(7172), 1106–1110. http://dx.doi.org/10.1038/nature06431.

McMahan, C. J., Difilippantonio, M. J., Rao, N., Spanopoulou, E., & Schatz, D. G. (1997). A basic motif in the N-terminal region of RAG1 enhances V(D)J recombination activity. *Molecular and Cellular Biology*, *17*(8), 4544–4552.

Medvedovic, J., Ebert, A., Tagoh, H., Tamir, I. M., Schwickert, T. A., Novatchkova, M., et al. (2013). Flexible long-range loops in the VH gene region of the Igh locus facilitate the generation of a diverse antibody repertoire. *Immunity*, *39*(2), 229–244. http://dx.doi.org/10.1016/j.immuni.2013.08.011.

Mizuta, R., Mizuta, M., Araki, S., & Kitamura, D. (2002). RAG2 is down-regulated by cytoplasmic sequestration and ubiquitin-dependent degradation. *The Journal of Biological Chemistry*, *277*(44), 41423–41427. http://dx.doi.org/10.1074/jbc.M206605200.

Mo, X., Bailin, T., & Sadofsky, M. J. (1999). RAG1 and RAG2 cooperate in specific binding to the recombination signal sequence in vitro. *The Journal of Biological Chemistry*, *274*(11), 7025–7031.

Mo, X., Bailin, T., & Sadofsky, M. J. (2001). A C-terminal region of RAG1 contacts the coding DNA during V(D)J recombination. *Molecular and Cellular Biology*, *21*(6), 2038–2047. http://dx.doi.org/10.1128/MCB.21.6.2038-2047.2001.

Mombaerts, P., Iacomini, J., Johnson, R. S., Herrup, K., Tonegawa, S., & Papaioannou, V. E. (1992). RAG-1-deficient mice have no mature B and T lymphocytes. *Cell*, *68*(5), 869–877.

Nambiar, M., & Raghavan, S. C. (2012). Mechanism of fragility at BCL2 gene minor breakpoint cluster region during t(14;18) chromosomal translocation. *The Journal of Biological Chemistry*, *287*(12), 8688–8701. http://dx.doi.org/10.1074/jbc.M111.307363.

Neiditch, M. B., Lee, G. S., Huye, L. E., Brandt, V. L., & Roth, D. B. (2002). The V(D)J recombinase efficiently cleaves and transposes signal joints. *Molecular Cell*, *9*(4), 871–878.

Nishana, M., & Raghavan, S. C. (2012). A non-B DNA can replace heptamer of V(D)J recombination when present along with a nonamer: Implications in chromosomal translocations and cancer. *The Biochemical Journal*, *448*(1), 115–125. http://dx.doi.org/10.1042/BJ20121031.

Olaru, A., Patterson, D. N., Cai, H., & Livak, F. (2004). Recombination signal sequence variations and the mechanism of patterned T-cell receptor-beta locus rearrangement. *Molecular Immunology*, *40*(16), 1189–1201.

Onozawa, M., & Aplan, P. D. (2012). Illegitimate V(D)J recombination involving non-antigen receptor loci in lymphoid malignancy. *Genes, Chromosomes & Cancer*, *51*(6), 525–535. http://dx.doi.org/10.1002/gcc.21942.

Papaemmanuil, E., Rapado, I., Li, Y., Potter, N. E., Wedge, D. C., Tubio, J., et al. (2014). RAG-mediated recombination is the predominant driver of oncogenic rearrangement in ETV6-RUNX1 acute lymphoblastic leukemia. *Nature Genetics*, *46*(2), 116–125. http://dx.doi.org/10.1038/ng.2874.

Pavlicek, J. W., Lyubchenko, Y. L., & Chang, Y. (2008). Quantitative analyses of RAG-RSS interactions and conformations revealed by atomic force microscopy. *Biochemistry*, *47*(43), 11204–11211. http://dx.doi.org/10.1021/bi801426x.

Peak, M. M., Arbuckle, J. L., & Rodgers, K. K. (2003). The central domain of core RAG1 preferentially recognizes single-stranded recombination signal sequence heptamer. *The Journal of Biological Chemistry*, *278*(20), 18235–18240. http://dx.doi.org/10.1074/jbc.M302041200.

Pilbeam, K., Basse, P., Brossay, L., Vujanovic, N., Gerstein, R., Vallejo, A. N., et al. (2008). The ontogeny and fate of NK cells marked by permanent DNA rearrangements. *Journal of Immunology*, *180*(3), 1432–1441.

Raghavan, S. C., Kirsch, I. R., & Lieber, M. R. (2001). Analysis of the V(D)J recombination efficiency at lymphoid chromosomal translocation breakpoints. *The Journal of Biological Chemistry*, *276*(31), 29126–29133. http://dx.doi.org/10.1074/jbc.M103797200.

Raghavan, S. C., Swanson, P. C., Wu, X., Hsieh, C. L., & Lieber, M. R. (2004). A non-B-DNA structure at the Bcl-2 major breakpoint region is cleaved by the RAG complex. *Nature*, *428*(6978), 88–93. http://dx.doi.org/10.1038/nature02355.

Ramon-Maiques, S., Kuo, A. J., Carney, D., Matthews, A. G., Oettinger, M. A., Gozani, O., et al. (2007). The plant homeodomain finger of RAG2 recognizes histone H3 methylated at both lysine-4 and arginine-2. *Proceedings of the National Academy of Sciences of the United States of America*, *104*(48), 18993–18998. http://dx.doi.org/10.1073/pnas.0709170104.

Ramsden, D. A., Baetz, K., & Wu, G. E. (1994). Conservation of sequence in recombination signal sequence spacers. *Nucleic Acids Research*, *22*(10), 1785–1796.

Ramsden, D. A., Weed, B. D., & Reddy, Y. V. (2010). V(D)J recombination: Born to be wild. *Seminars in Cancer Biology*, *20*(4), 254–260. http://dx.doi.org/10.1016/j.semcancer.2010.06.002.

Ranganath, S., Carpenter, A. C., Gleason, M., Shaw, A. C., Bassing, C. H., & Alt, F. W. (2008). Productive coupling of accessible Vbeta14 segments and DJbeta complexes determines the frequency of Vbeta14 rearrangement. *Journal of Immunology*, *180*(4), 2339–2346.

Reddy, K. L., Zullo, J. M., Bertolino, E., & Singh, H. (2008). Transcriptional repression mediated by repositioning of genes to the nuclear lamina. *Nature*, *452*(7184), 243–247. http://dx.doi.org/10.1038/nature06727.

Rodgers, W., Byrum, J. N., Sapkota, H., Rahman, N. S., Cail, R. C., Zhao, S., et al. (2015). Spatio-temporal regulation of RAG2 following genotoxic stress. *DNA Repair (Amsterdam)*, *27*, 19–27. http://dx.doi.org/10.1016/j.dnarep.2014.12.008.

Rooney, S., Chaudhuri, J., & Alt, F. W. (2004). The role of the non-homologous end-joining pathway in lymphocyte development. *Immunological Reviews*, *200*, 115–131. http://dx.doi.org/10.1111/j.0105-2896.2004.00165.x.

Ross, A. E., Vuica, M., & Desiderio, S. (2003). Overlapping signals for protein degradation and nuclear localization define a role for intrinsic RAG-2 nuclear uptake in dividing cells. *Molecular and Cellular Biology*, *23*(15), 5308–5319.

Ruthenburg, A. J., Allis, C. D., & Wysocka, J. (2007). Methylation of lysine 4 on histone H3: Intricacy of writing and reading a single epigenetic mark. *Molecular Cell, 25*(1), 15–30. http://dx.doi.org/10.1016/j.molcel.2006.12.014.

Sadofsky, M. J., Hesse, J. E., & Gellert, M. (1994). Definition of a core region of RAG-2 that is functional in V(D)J recombination. *Nucleic Acids Research, 22*(10), 1805–1809.

Sakata, J., Inoue, J., Ohi, H., Kosugi-Okano, H., Mishima, Y., Hatakeyama, K., et al. (2004). Involvement of V(D)J recombinase in the generation of intragenic deletions in the Rit1/Bcl11b tumor suppressor gene in gamma-ray-induced thymic lymphomas and in normal thymus of the mouse. *Carcinogenesis, 25*(6), 1069–1075. http://dx.doi.org/10.1093/carcin/bgh094.

Schatz, D. G. (2004). Antigen receptor genes and the evolution of a recombinase. *Seminars in Immunology, 16*(4), 245–256. http://dx.doi.org/10.1016/j.smim.2004.08.004.

Schatz, D. G., & Ji, Y. (2011). Recombination centres and the orchestration of V(D)J recombination. *Nature Reviews. Immunology, 11*(4), 251–263. http://dx.doi.org/10.1038/nri2941.

Schatz, D. G., & Swanson, P. C. (2011). V(D)J recombination: Mechanisms of initiation. *Annual Review of Genetics, 45*, 167–202. http://dx.doi.org/10.1146/annurev-genet-110410-132552.

Shih, H. Y., & Krangel, M. S. (2013). Chromatin architecture, CCCTC-binding factor, and V(D)J recombination: Managing long-distance relationships at antigen receptor loci. *Journal of Immunology, 190*(10), 4915–4921. http://dx.doi.org/10.4049/jimmunol.1300218.

Shiloh, Y., & Ziv, Y. (2013). The ATM protein kinase: Regulating the cellular response to genotoxic stress, and more. *Nature Reviews. Molecular Cell Biology, 14*(4), 197–210.

Shimazaki, N., Tsai, A. G., & Lieber, M. R. (2009). H3K4me3 stimulates the V(D)J RAG complex for both nicking and hairpinning in trans in addition to tethering in cis: Implications for translocations. *Molecular Cell, 34*(5), 535–544. http://dx.doi.org/10.1016/j.molcel.2009.05.011.

Shinkai, Y., Rathbun, G., Lam, K. P., Oltz, E. M., Stewart, V., Mendelsohn, M., et al. (1992). RAG-2-deficient mice lack mature lymphocytes owing to inability to initiate V(D)J rearrangement. *Cell, 68*(5), 855–867.

Shlyakhtenko, L. S., Gilmore, J., Kriatchko, A. N., Kumar, S., Swanson, P. C., & Lyubchenko, Y. L. (2009). Molecular mechanism underlying RAG1/RAG2 synaptic complex formation. *The Journal of Biological Chemistry, 284*(31), 20956–20965. http://dx.doi.org/10.1074/jbc.M109.028977.

Simkus, C., Makiya, M., & Jones, J. M. (2009). Karyopherin alpha 1 is a putative substrate of the RAG1 ubiquitin ligase. *Molecular Immunology, 46*(7), 1319–1325. http://dx.doi.org/10.1016/j.molimm.2008.11.009.

Sollbach, A. E., & Wu, G. E. (1995). Inversions produced during V(D)J rearrangement at IgH, the immunoglobulin heavy-chain locus. *Molecular and Cellular Biology, 15*(2), 671–681.

Steinel, N. C., Lee, B. S., Tubbs, A. T., Bednarski, J. J., Schulte, E., Yang-Iott, K. S., et al. (2013). The ataxia telangiectasia mutated kinase controls Igkappa allelic exclusion by inhibiting secondary Vkappa-to-Jkappa rearrangements. *The Journal of Experimental Medicine, 210*(2), 233–239. http://dx.doi.org/10.1084/jem.20121605.

Subrahmanyam, R., Du, H., Ivanova, I., Chakraborty, T., Ji, Y., Zhang, Y., et al. (2012). Localized epigenetic changes induced by DH recombination restricts recombinase to DJH junctions. *Nature Immunology, 13*(12), 1205–1212. http://dx.doi.org/10.1038/ni.2447.

Swaminathan, S., Klemm, L., Park, E., Papaemmanuil, E., Ford, A., Kweon, S. M., et al. (2015). Mechanisms of clonal evolution in childhood acute lymphoblastic leukemia. *Nature Immunology, 16*, 766–774.

Swanson, P. C. (2004). The bounty of RAGs: Recombination signal complexes and reaction outcomes. *Immunological Reviews*, *200*, 90–114. http://dx.doi.org/10.1111/j.0105-2896.2004.00159.x.

Swanson, P. C., & Desiderio, S. (1999). RAG-2 promotes heptamer occupancy by RAG-1 in the assembly of a V(D)J initiation complex. *Molecular and Cellular Biology*, *19*(5), 3674–3683.

Swanson, P. C., Kumar, S., & Raval, P. (2009). Early steps of V(D)J rearrangement: Insights from biochemical studies of RAG-RSS complexes. *Advances in Experimental Medicine and Biology*, *650*, 1–15.

Teng, G., Maman, Y., Resch, W., Kim, M., Yamane, A., Qian, J., et al. (2015). RAG represents a widespread threat to the lymphocyte genome. *Cell*. (in press).

Tepsuporn, S., Hu, J., Gostissa, M., & Alt, F. W. (2014). Mechanisms that can promote peripheral B-cell lymphoma in ATM-deficient mice. *Cancer Immun. Res*, *2*(9), 857–866. http://dx.doi.org/10.1158/2326-6066.CIR-14-0090.

Thompson, C. B. (1995). New insights into V(D)J recombination and its role in the evolution of the immune system. *Immunity*, *3*(5), 531–539.

Tsai, A. G., Lu, H., Raghavan, S. C., Muschen, M., Hsieh, C. L., & Lieber, M. R. (2008). Human chromosomal translocations at CpG sites and a theoretical basis for their lineage and stage specificity. *Cell*, *135*(6), 1130–1142. http://dx.doi.org/10.1016/j.cell.2008.10.035.

Tsai, C. L., & Schatz, D. G. (2003). Regulation of RAG1/RAG2-mediated transposition by GTP and the C-terminal region of RAG2. *The EMBO Journal*, *22*(8), 1922–1930. http://dx.doi.org/10.1093/emboj/cdg185.

Tsuji, H., Ishii-Ohba, H., Noda, Y., Kubo, E., Furuse, T., & Tatsumi, K. (2009). Rag-dependent and Rag-independent mechanisms of Notch1 rearrangement in thymic lymphomas of Atm(−/−) and scid mice. *Mutation Research*, *660*(1–2), 22–32. http://dx.doi.org/10.1016/j.mrfmmm.2008.10.002.

Um, J. H., Brown, A. L., Singh, S. K., Chen, Y., Gucek, M., Lee, B. S., et al. (2013). Metabolic sensor AMPK directly phosphorylates RAG1 protein and regulates V(D)J recombination. *Proceedings of the National Academy of Sciences of the United States of America*, *110*(24), 9873–9878. http://dx.doi.org/10.1073/pnas.1307928110.

Vanura, K., Montpellier, B., Le, T., Spicuglia, S., Navarro, J. M., Cabaud, O., et al. (2007). In vivo reinsertion of excised episomes by the V(D)J recombinase: A potential threat to genomic stability. *PLoS Biology*, *5*(3), e43. http://dx.doi.org/10.1371/journal.pbio.0050043.

Wang, J. H., Gostissa, M., Yan, C. T., Goff, P., Hickernell, T., Hansen, E., et al. (2009). Mechanisms promoting translocations in editing and switching peripheral B cells. *Nature*, *460*(7252), 231–236. http://dx.doi.org/10.1038/nature08159.

Wang, X., Xiao, G., Zhang, Y., Wen, X., Gao, X., Okada, S., et al. (2008). Regulation of Tcrb recombination ordering by c-Fos-dependent RAG deposition. *Nature Immunology*, *9*(7), 794–801. http://dx.doi.org/10.1038/ni.1614.

West, K. L., Singha, N. C., De Ioannes, P., Lacomis, L., Erdjument-Bromage, H., Tempst, P., et al. (2005). A direct interaction between the RAG2 C terminus and the core histones is required for efficient V(D)J recombination. *Immunity*, *23*(2), 203–212. http://dx.doi.org/10.1016/j.immuni.2005.07.004.

Wilson, A., Held, W., & MacDonald, H. R. (1994). Two waves of recombinase gene expression in developing thymocytes. *The Journal of Experimental Medicine*, *179*(4), 1355–1360.

Wilson, D. R., Norton, D. D., & Fugmann, S. D. (2008). The PHD domain of the sea urchin RAG2 homolog, SpRAG2L, recognizes dimethylated lysine 4 in histone H3 tails. *Developmental and Comparative Immunology*, *32*(10), 1221–1230. http://dx.doi.org/10.1016/j.dci.2008.03.012.

Wu, C., Ranganath, S., Gleason, M., Woodman, B. B., Borjeson, T. M., Alt, F. W., et al. (2007). Restriction of endogenous T cell antigen receptor beta rearrangements to Vbeta14 through selective recombination signal sequence modifications. *Proceedings of the National Academy of Sciences of the United States of America*, *104*(10), 4002–4007. http://dx.doi.org/10.1073/pnas.0700081104.

Yant, S. R., Wu, X., Huang, Y., Garrison, B., Burgess, S. M., & Kay, M. A. (2005). High-resolution genome-wide mapping of transposon integration in mammals. *Molecular and Cellular Biology*, *25*(6), 2085–2094. http://dx.doi.org/10.1128/MCB.25.6.2085-2094.2005.

Yin, F. F., Bailey, S., Innis, C. A., Ciubotaru, M., Kamtekar, S., Steitz, T. A., et al. (2009). Structure of the RAG1 nonamer binding domain with DNA reveals a dimer that mediates DNA synapsis. *Nature Structural & Molecular Biology*, *16*(5), 499–508. http://dx.doi.org/10.1038/nsmb.1593.

Yu, K., Taghva, A., & Lieber, M. R. (2002). The cleavage efficiency of the human immunoglobulin heavy chain VH elements by the RAG complex: Implications for the immune repertoire. *The Journal of Biological Chemistry*, *277*(7), 5040–5046. http://dx.doi.org/10.1074/jbc.M109772200.

Yurchenko, V., Xue, Z., & Sadofsky, M. (2003). The RAG1 N-terminal domain is an E3 ubiquitin ligase. *Genes and Development*, *17*(5), 581–585. http://dx.doi.org/10.1101/gad.1058103.

Zhang, Z., Espinoza, C. R., Yu, Z., Stephan, R., He, T., Williams, G. S., et al. (2006). Transcription factor Pax5 (BSAP) transactivates the RAG-mediated V(H)-to-DJ(H) rearrangement of immunoglobulin genes. *Nature Immunology*, 7(6), 616–624. http://dx.doi.org/10.1038/ni1339.

Zhang, Y., McCord, R. P., Ho, Y. J., Lajoie, B. R., Hildebrand, D. G., Simon, A. C., et al. (2012). Spatial organization of the mouse genome and its role in recurrent chromosomal translocations. *Cell*, *148*(5), 908–921. http://dx.doi.org/10.1016/j.cell.2012.02.002.

Zhang, L., Reynolds, T. L., Shan, X., & Desiderio, S. (2011). Coupling of V(D)J recombination to the cell cycle suppresses genomic instability and lymphoid tumorigenesis. *Immunity*, *34*(2), 163–174. http://dx.doi.org/10.1016/j.immuni.2011.02.003.

Zhang, Y. H., Shetty, K., Surleac, M. D., Petrescu, A. J., & Schatz, D. G. (2015). Mapping and quantitation of the interaction between the recombination activating gene proteins RAG1 and RAG2. *The Journal of Biological Chemistry*, *290*(19), 11802–11817. http://dx.doi.org/10.1074/jbc.M115.638627.

Zhang, M., & Swanson, P. C. (2008). V(D)J recombinase binding and cleavage of cryptic recombination signal sequences identified from lymphoid malignancies. *The Journal of Biological Chemistry*, *283*(11), 6717–6727. http://dx.doi.org/10.1074/jbc.M710301200.

Zhang, Y., Xu, K., Deng, A., Fu, X., Xu, A., & Liu, X. (2014). An amphioxus RAG1-like DNA fragment encodes a functional central domain of vertebrate core RAG1. *Proceedings of the National Academy of Sciences of the United States of America*, *111*(1), 397–402. http://dx.doi.org/10.1073/pnas.1318843111.

Zhu, C., Mills, K. D., Ferguson, D. O., Lee, C., Manis, J., Fleming, J., et al. (2002). Unrepaired DNA breaks in p53-deficient cells lead to oncogenic gene amplification subsequent to translocations. *Cell*, *109*(7), 811–821.

CHAPTER TWO

Chromatin Interactions in the Control of Immunoglobulin Heavy Chain Gene Assembly

Gita Kumari, Ranjan Sen[1]
Laboratory of Molecular Biology and Immunology, National Institute on Aging, Baltimore, Maryland, USA
[1]Corresponding author: e-mail address: senranja@grc.nia.nih.gov

Contents

Advances in Immunology, Volume 128
ISSN 0065-2776
http://dx.doi.org/10.1016/bs.ai.2015.08.001
© 2015 Elsevier Inc.
All rights reserved.

Abstract

Expression of antibody heavy chain occurs via precisely timed developmental activation of the immunoglobulin heavy chain (*IgH*) gene locus during B cell development. *IgH* locus activation permits coordinated gene rearrangements that assemble variable (V_H), diversity (D_H), and joining (J_H) gene segments into functional genes. Chromosomal conformation changes and epigenetic mechanisms play critical roles in ensuring rearrangement fidelity while minimizing hazardous consequences of broken DNA ends generated during recombination. In this review, we summarize the current status of regulatory mechanisms that underpin effective *IgH* gene assembly. For this, the germline locus is divided into two parts: a 2.4 Mb 5′ part that contains all V_H gene segments and a 300 kb 3′ domain that contains D_H and J_H gene segments, as well as exons that encode *IgH* isotypes. Structural features of each part are discussed individually, followed by consideration of how the two parts come together to complete *IgH* recombination. Throughout we emphasize current insights, propose plausible mechanisms, and highlight key questions for future studies.

Developmentally regulated gene expression results in activation or repression of a small subset of genes from the entire genome. Specificity is achieved by modulating structure and organization of the genome without affecting its genetic information content, that is, by epigenetic means. Genes that encode antigen receptors of B and T lymphocytes have long been used to study mechanisms of tissue- and developmental stage-specific gene regulation. In accordance with the epigenome concept, recent advances in our understanding of their tissue-specific regulation have been in the area of epigenetic regulation. This review focuses on mechanisms that activate the immunoglobulin heavy chain (*IgH*) gene locus. We start with a brief general description of epigenetic phenomena and then elaborate on aspects that impinge on the *IgH* locus.

1. GENERAL FEATURES OF CHROMOSOME ORGANIZATION

1.1 Covalent Modification of Histones

The fundamental structural unit of mammalian chromosomes is the nucleosome, which comprises approximately 150 base pairs (bp) of DNA on the surface of a histone octamer. All four histones constituting the core nucleosome (H2A, H2B, H3, and H4) can be covalently modified (Bannister & Kouzarides, 2011; Tropberger & Schneider, 2013), leading to many possible distinctions between nucleosomes that carry the same DNA sequence. To date, 15 kinds of chemical modifications of histones have been identified

(Huang, Sabari, Garcia, Allis, & Zhao, 2014). Since a single gene is associated with multiple nucleosomes, the diversity of histone modifications can, in principle, generate tremendous heterogeneity even in this basic form of gene organization. The heterogeneity, in turn, can impact whether a gene is transcriptionally active, can be induced to be transcribed or is transcriptionally inert. The information content possible in the combination of histone modifications has been referred to as the histone code (Chi, Allis, & Wang, 2010; Jenuwein & Allis, 2001).

While there is a range of possibilities, many studies have directed attention to a subset of histone modifications that are especially pertinent to gene expression. Acetylation of multiple residues in the N-terminal tails of histones H3 and H4 were the earliest marks to be positively correlated with gene activity (Allfrey, Faulkner, & Mirsky, 1964; Pogo, Allfrey, & Mirsky, 1966). Later, genome-wide studies demonstrated that histone acetylation was more often a mark of genes that were "poised" for activation, rather than a mark of transcriptional activity *per se* (Grunstein, 1997; Mikkelsen et al., 2007; Struhl, 1998). The relationship of histone acetylation to gene activation was strongly corroborated by the demonstration that transcriptional coactivators, such as p300 and CBP, turned out to be histone acetyltransferases (HATs) (Eckner et al., 1994; Kwok et al., 1994). Conversely, mono-, di-, or trimethylation of lysine 9 or 27 on histone H3 (H3K9me(1–3) or H3K27me(1–3)) is best correlated with gene inactivity (Arthur et al., 2014; Pauler et al., 2009; Stewart, Li, & Wong, 2005; Young et al., 2011). Histone methylation is carried out by histone methyl transferases (HMTs) that contain SET domains and work as multiprotein complexes. These complexes vary greatly in the specificity of their target lysine residues as well as in the number of methyl groups they transfer (Black, Van Rechem, & Whetstine, 2012; Martin & Zhang, 2005). Both repressive histone methylation marks occur on residues that are also associated with gene activity in different ways. Acetylation of histone H3 on lysine 9 (H3K9ac) correlates closely with genes that can be expressed, and H3K27ac has been shown to mark active transcriptional enhancers in the genome (Creyghton et al., 2010; Heintzman et al., 2009). Clearly, activation and repression via H3K9 and H3K27 modifications are mutually exclusive.

Mechanisms by which H3K9me(1–3) and H3K27me(1–3) suppress transcription are distinct. H3K9me(1–3) is usually associated with constitutive heterochromatin found at centromeres and repeated elements in the genome (Maison et al., 2002; Martens et al., 2005; Peters et al., 2001). This modification serves as a docking site for bromodomain-containing heterochromatin proteins (HP1α, β, and γ) that lead to compaction of the associated chromosomal region (Azzaz et al., 2014; Verschure et al., 2005). H3K27me(1–3) is associated with facultative heterochromatin that can switch between inactive and active (euchromatic) states (Cao et al., 2002; Trojer & Reinberg, 2007). This mark is deposited by the polycomb repressive complex-2 (PRC2) and leads to gene repression via binding PRC1 (Beisel & Paro, 2011; Lavigne, Francis, King, & Kingston, 2004; Trojer & Reinberg, 2007). Interconversion between transcriptionally active and repressed states requires that activation-specific acetylation and repressive methylation of histones must be dynamic. This is facilitated by several histone deacetylases (De Ruijter, Van Gennip, Caron, Kemp, & Van Kuilenburg, 2003; Haberland, Montgomery, & Olson, 2009; Ng & Bird, 2000) and histone demethylases (Pedersen & Helin, 2010; Shi & Whetstine, 2007; Verrier, Vandromme, & Trouche, 2011), that remove acetyl and methyl group, respectively, from histones.

Marks most closely correlated with active transcription are trimethylation of lysine 4 and 36 on H3 (H3K4me3 and H3K36me3) (Guenther, Levine, Boyer, Jaenisch, & Young, 2007; Wagner & Carpenter, 2012). These marks are also introduced by SET domain-containing multiprotein complexes. H3K4me3 is usually located at the 5′ ends of genes and is deposited by HMT complexes that associate with the initiating form of RNA polymerase II (Pol II) (Ardehali et al., 2011; Bernstein et al., 2005; Guenther et al., 2007). H3K36me3 is found primarily in the gene body toward the 3′ ends of genes and is deposited by HMT complexes that move with the elongating form of RNA Pol II (Barski et al., 2007; Kwak & Lis, 2013; Pokholok et al., 2005). It is useful to keep in mind that these generalizations were reached on the basis of genome-wide studies. Therefore, they represent statistically significant correlations that need not apply universally to all genes.

Two other histone modifications are directly pertinent to gene expression. These are monomethylation of lysine 4 on histone H3 (H3K4me1) and H3K27ac. These marks have been associated with transcriptional enhancers (Creyghton et al., 2010; Heintzman et al., 2009; Zentner, Tesar, & Scacheri, 2011). Because regions that contain these do not overlap completely in genome-wide studies, the working model is that H3K4me1 marks both active and incipient enhancers, whereas H3K27ac only marks enhancers that are active in a particular cell type. In addition, genome-wide binding of p300/CBP to sites not associated with gene promoters also correlates with locations of enhancers (Rada-Iglesias et al., 2011; Visel et al., 2009).

Whether enhancers that are associated with specific combinations of these marks are functionally different is unclear at present.

1.2 Chromatin Remodeling

One of the earliest defined chromatin structural changes associated with tissue-specific gene expression was the generation of chromosomal domains that were sensitive to digestion by endonucleases such as DNase I or micrococcal nuclease (Deal & Henikoff, 2010; Garel & Axel, 1976; Weintraub & Groudine, 1976). DNase I-sensitive chromatin domains often contain more restricted regions that are hypersensitive to DNase digestion; these are referred to as DNase I hypersensitive sites (DHSs). Essentially, all transcriptional enhancers coincide with DHS in cell types where the enhancer is active. However, not all DHS mark transcriptional enhancers. Generation of DNase I-sensitive chromatin domains as well as induction of DHS indicate that transcriptionally active genes undergo remodeling of their chromatin state. This process can be viewed somewhat abstractly as a "loosening" of repressed chromatin structure that permits transcription factors and RNA Pol II to access genes to initiate transcription. For antigen receptor genes, the notion of locus accessibility is also pertinent from the perspective of V(D)J recombination. Proteins encoded by recombinase activation genes 1 and 2 (RAG1 and RAG2) rearrange all seven known antigen receptor loci in a tightly regulated manner, such that B cell receptor genes undergo recombination only in B lineage cells, whereas T cell receptor genes undergo recombination only in T lineage. To explain this specificity, Alt and colleagues proposed that antigen receptor loci were made selectively accessible to RAG proteins only in cells of the appropriate lineage and developmental stage (Sleckman, Gorman, & Alt, 1996; Yancopoulos & Alt, 1985). The molecular basis of RAG accessibility remains an important question in molecular immunology.

Chromatin alterations that increase or decrease locus accessibility to nuclear enzymes are mediated by chromatin remodeling complexes (Lorch, Maier-Davis, & Kornberg, 2010; Lusser & Kadonaga, 2003; Narlikar, Sundaramoorthy, & Owen-Hughes, 2013). These are multi-subunit protein complexes that utilize ATP hydrolysis to alter chromatin state. Biochemical studies show that chromatin remodeling complexes work in different ways. One category, exemplified by the SWItch/Sucrose non-fermentable (SWI/SNF) complex, moves nucleosomes around to expose or hide specific DNA sequences. A second category, exemplified by the imitation SWItch (ISWI) complex, works by ejecting or incorporating nucleosomes onto specific sequences. A third category, exemplified by the chromatin assembly factor (CAF) complex, replenishes histones after DNA replication. DHSs presumably reflect acute alterations of localized chromatin structure by such enzymes. However, direct experimental evidence for the role of chromatin remodeling complexes in inducing DHSs is only just beginning to be explored (Bossen et al., 2015). The close interplay between histone modification states and chromatin remodeling is evident from the characterization of multisubunit enzymes, such as the nucleosome remodeling factor (NURF) complex, that contain both histone modifying and ATP-dependent remodeling activities (Petty & Pillus, 2013; Vignali, Hassan, Neely, & Workman, 2000). There is also accumulating evidence that histone modification states modulate the efficiency of chromatin remodeling complexes (Lalonde, Cheng, & Cote, 2014).

1.3 Noncoding RNA Control of Chromatin Structure

Recent studies show that a large fraction of the mammalian genome is transcribed into noncoding RNAs that do not encode proteins. ncRNAs share many features with protein coding mRNAs (Birney et al., 2007; Carninci et al., 2005; Katayama et al., 2005). They range in size between 200 bp and more than 100 kb, and undergo splicing and polyadenylation to generate mature forms (Guttman et al., 2009; Kapranov et al., 2007). A small number of tissue-specific ncRNAs have been identified and there are several that are inducibly expressed in specific cell types (Nakaya et al., 2007). ncRNAs are important for gene regulation especially in the context of chromatin structure. A small number of studies conclude that ncRNAs play an important regulatory role in regulation of chromatin structure during development and differentiation by interacting or recruiting chromatin-modifying complexes to loci (Chu, Qu, Zhong, Artandi, & Chang, 2011; Dinger et al., 2008; Guttman et al., 2009). One of the best-characterized examples of ncRNA-mediated chromatin remodeling is during dosage compensation where antisense ncRNA Xist, transcribed from the X chromosome, recruits PRC2 to repress the entire X chromosome (Lee, Davidow, & Warshawsky, 1999; Zhao, Sun, Erwin, Song, & Lee, 2008). Xist RNA is believed to recruit PRC2 to DNA as PRC2 does not possess a DNA-binding domain. Another example of ncRNA-based epigenetic regulation is in imprinting control of Kcnq, where the paternally expressed Kcnq1ot1 ncRNA silences Kcnq on the paternal chromosome by controlling chromatin modification at that locus (Pandey et al., 2008).

ncRNAs can be involved in locus activation as well. Kevin et al. have shown that HOTTIP, a ncRNA transcribed from the *Hoxa* locus, recruits H3K4 trimethylation complex via binding to WDR5. HOTTIP knockdown in fibroblasts results in decreased HOXA expression (Wang et al., 2011).

1.4 3D Chromatin Structure

Development of chromosome conformation capture (3C) techniques has revolutionized our view of the 3D organization of the genome (Dekker, 2003; Dekker, Rippe, Dekker, & Kleckner, 2002). This technique is based on ligating genomic sequences that are in spatial proximity, followed by identification of specific interactions by polymerase chain reaction. Several variations of the original 3C method have been developed (Gibcus & Dekker, 2013; Gondor, Rougier, & Ohlsson, 2008). The most general form of chromosome capture, Hi-C, interrogates all possible proximity effects in the genome. A fundamental concept that has emerged from Hi-C studies is that the genome is organized into topologically associated domains (TADs) that range in size of a few megabases (Dixon et al., 2012, 2015; Lieberman-Aiden et al., 2009; Nagano et al., 2013; Rao et al., 2014; Fig. 1). By definition, sequences within a TAD interact more frequently with one another compared to sequences that lie outside the TAD. Patterns of proximity-induced cross-linking within a TAD are not random and reveal additional levels of organization. The basic TAD structure appears to be invariant between cell types; however, fine structure within TADs varies considerably. Genome-wide studies of factor binding show that some factors such as CCCTC binding factor (CTCF), TFIIIC, and components of cohesin and condensin complexes are enriched at TAD boundaries (Dixon et al., 2012; Van Bortle et al., 2014). However, no simple generalization for how TAD boundaries are created has emerged yet. Changes in TAD organization have been examined closely in embryonic stem cells and their more differentiated derivatives (Phillips-Cremins et al., 2013), as well as in the transition of pre-pro-B cells to pro-B cells (Lin et al., 2012). In general, increased interactions within TADs correlate with increased gene transcription and induction of new transcriptional regulatory elements (Dixon et al., 2015). A simple working model that has emerged is that structural changes that accompany tissue-specific gene regulation occur within more invariant chromatin folding structures. In this conceptual framework, interactions between tissue-specific transcriptional enhancers and the promoters they activate may provide a substantial amount of sub-TAD fine structure.

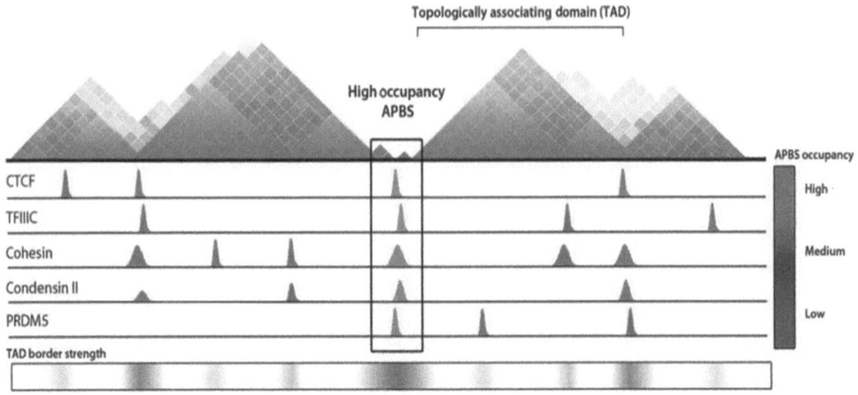

Figure 1 Organization of the genome into topologically associating domains (TADs). TADs, shown as triangles, are defined as basic structural building blocks of the genome. TAD organization is largely invariant between cell types. Sequences within TADs interact more with each other (darker pink) than between TADs. Regions with higher interaction within TADs are referred to as sub-TADs and define intra-TAD fine structure that varies between cell types. TAD borders (where two triangles meet) are bound by various architectural proteins as schematized. TAD border strength varies from strong to weak/intermediate depending on the degree of inter-TAD interactions. Changes in the number and colocalization of architectural proteins have been used to define TAD border strength between cell types. When CTCF, cohesin, condensin, PRDM5, and TFIIIC are present together at a TAD border, it is considered to be a strong border (indicated as high occupancy and shown in red) with limited inter-TAD interactions. Reduced numbers of architectural proteins makes a TAD border weak/intermediate (shown in green) and correlates with increased inter-TAD interactions. *This figure is adapted from Van Bortle et al. (2014).* (See the color plate.)

The CCCTC-binding factor, CTCF, has gained prominence as a mediator of chromatin loops (Merkenschlager & Odom, 2013; Ong & Corces, 2014). However, the role of this transcription factor remains complex. First, CTCF binds to a subset of sites genome-wide in a tissue-unrestricted manner, whereas binding to other sites is highly tissue specific. The bases for this distinction, as well as possible functional differences between the two kinds of sites, are not clear. Second, a significant number of CTCF-bound sites in the genome also bind components of the cohesin complex (Holwerda & de Laat, 2013; Parelho et al., 2008; Sofueva et al., 2013). This complex, which was identified as a complex that holds sister chromatids together after DNA replication, consists of six components. Its cobinding with CTCF is usually assessed by chromatin immunoprecipitation experiments using anti-Rad21 or anti-Smc1 antibodies. Cohesin components are recruited to specific genomic locations via the sequence-specific DNA-binding activity of

CTCF. Accordingly, both CTCF and cohesin are lost from chromatin in cells that lack CTCF. The converse is not true, with CTCF remaining chromatin-bound in cells that lack cohesin components. What makes some CTCF-binding sites competent to recruit cohesin and others not, remains to be determined. Because of cohesin's previously established role in mediating interaction between chromatin fibers, CTCF/cohesin complexes provide a plausible mechanism for CTCF-dependent chromatin looping. CTCF-bound sites involved in looping are often associated with epigenetic marks associated with transcriptional enhancers such as H3K4me1, H3K27ac, and H3K4me3 (Dixon et al., 2012; Holwerda & de Laat, 2013; Shen et al., 2012), leading to the proposal that a major function of CTCF is to mediate interactions between promoters and enhancers. However, it should be noted that these conclusions are derived largely from comparison of embryonic stem (ES) cells with more differentiated cells derived from ES cells.

A subset of CTCF-dependent looping sites also serves as boundary elements to demarcate epigenetically distinct chromatin domains. This has been most clearly demonstrated in the mammalian *Hoxa* (1–13) gene cluster (Kim, Cecchini, & Kim, 2011; Narendra et al., 2015). In ES cells, this approximately 120 kb gene cluster is uniformly marked with repressive H3K27me3 and not associated with active transcription. In motor neurons differentiated from ES cells, only *Hoxa* (1–6) genes become transcriptionally active and marked with H3K4me3, and the H3K27me3 domain recedes to encompass only *Hoxa* (7–13). A sharp distinction between the H3K4me3- and H3K27me3-marked domains coincides with a CTCF-binding site that has been shown to be involved in loop formation. Deletion of this site by clustered regularly interspaced short palindromic repeats (CRISPR) technology led to no difference in chromatin structure in ES cells. However, upon differentiation to motor neurons, the pattern of gene expression extended into *Hoxa* 7 and *9* genes, and the chromatin boundary between H3K4me3- and H3K27me3-marked domains shifted to an adjacent CTCF site. In contrast, point mutation of CTCF sites within the intergenic control region 1 (IGCR1) of the *IgH* locus did not affect epigenetic domains while disrupting chromatin looping (see below for more extensive discussion). It is not clear what features of CTCF-bound sites in the genome determine whether they will demarcate epigenetic boundaries or only serve as sites of looping interactions.

Further extending the diversity of possible functional roles for CTCF is the observation that TAD boundaries in mammalian cells are enriched for CTCF and cohesin binding. However, only 15% of CTCF-bound sites lie at TAD boundaries indicating that the presence of CTCF is not sufficient to create a boundary (Dixon et al., 2012; Rao et al., 2014; Fig. 1). Whether CTCF or cohesin binding is necessary to generate TADs has been explored in two recent publications. Seitan et al. deleted the cohesin component RAD21 in $CD4^{+}CD8^{+}$ (DP) thymocytes by breeding $RAD21^{lox/lox}$ to a CD4 cre transgene (Seitan et al., 2013). Because the majority of DP cells are in G1 phase of the cell cycle, the choice of cell type for the analysis circumvented interpretation problems caused by the effects of cohesin depletion during cell division. Hi-C analysis in RAD21-sufficient and RAD21-deficient DP thymocytes showed that TAD organization was not altered in RAD21-deficient cells. However, long-range genomic interactions within TADs, as well as gene expression patterns were changed. Similar results were obtained by Zuin et al. in human embryonic kidney epithelial (HEK) 293 cells (Zuin et al., 2014). Since 293 cells divide continuously, the authors used a clever protein degradation scheme to rapidly deplete RAD21 or reduce expression of CTCF by means of siRNA in these cells. Hi-C analyses indicated that TAD organization was largely intact in CTCF- or RAD21-depleted cells. Interestingly, the effects of deleting CTCF or cohesin were not identical, indicating that each factor had unique functions that were independent of the other. Disruption of CTCF or cohesin reduced many intradomain interactions though the average size of affected interactions differed between the two factors. Loci that were affected by RAD21 depletion fell in the 100–200 kb range, whereas loci that were affected by CTCF were on average less than 100 kb. Additionally, increased incidence of inter-TAD interactions were noted in CTCF-depleted cells.

Appropriate activation of gene expression requires coordination between genetics and epigenetics. Genetics refers to regulatory sequences associated with genes. It is increasingly clear that changes in such sequences, whether by design in model organisms or in naturally occurring mutations in humans, lead to changes in gene expression. Indeed, occurrences of many human-disease-associated single nucleotide polymorphisms outside protein coding regions have been interpreted to mean that these changes modulate gene expression (Farh et al., 2015). Close correlation of such sites with DHSs has strongly corroborated this notion (Onengut-Gumuscu et al., 2015; Stergachis et al., 2013). However, appropriate read-out of the genetic information requires epigenetic mechanisms to specify when and where a gene must be activated. How the multiple levels of epigenetics are coordinated by *cis*-regulatory sequences remains a challenge. Because both aspects have been extensively studied at antigen receptor loci (Feeney, 2011; Jung & Alt, 2004;

Sen & Oltz, 2006), we are optimistic that continued analyses of these genes will provide fundamental insights into basic mechanisms of gene regulation. In this review, we summarize our current understanding of the immunoglobulin heavy chain (*IgH*) gene locus.

2. ORGANIZATION OF THE *IgH* LOCUS

The germline *IgH* locus in mouse spans approximately 2.8 Mb close to the telomere on the long arm of chromosome 12. The distance to the telomere from the 5′ end of *IgH* is approximately 4 Mb which contains six annotated genes. Bulk of the *IgH* locus comprises variable region (V_H) gene segments (Fig. 2A) that are divided into 16 gene families based on sequence homology. The largest gene families, J558 (52 functional genes) and 3609 (8 functional genes), are located at the 5′ end of the locus and overlap considerably (Choi & Feeney, 2014; Johnston, Wood, Bolland, & Corcoran, 2006). The V_H7183 gene family with 10 functional genes lies at the 3′ end of the V_H region. Together, the J558, 3609, and 7183 V_H genes make up 70 out of the 110 functional genes in C57BL/6 mice. All other gene families constitute 40 members and are located between J558/3609 and 7183 families. The relationship between gene location and rearrangements is discussed below.

The mouse genome contains 8–12 diversity (D_H) gene segments, depending on the strain (Ye, 2004). The 5′-most D_H gene segment, DFL16.1, is located approximately 100 kb 3′ of V_H7183.2.3 (the 3′-most functional V_H gene segment). In C57BL/6 mice, DFL16.1 is followed by a series of DSP gene segments; 129 and Balb/c mice contain an additional DFL16.2 gene segment. DFL16.1 and DSP gene segments are located in the middle of an approximately 4 kb repeat sequence (Bolland et al., 2007; Chakraborty et al., 2007). The DST4 gene segment lies immediately after the repeat organization of DFL16.1/DSP gene segments (Ye, 2004). Another DST4-related gene segment, DST4.2, has been identified in the V_H7183.23.1–DFL16.1 intergenic region. It is located close to the V_H region and participates infrequently in D_H recombination. The 3′-most D_H gene segment, DQ52, is located 28 kb 3′ of DST4 and less than 1 kb 5′ of the first joining gene segment J_H1. Organization of D_H gene segments in humans has interesting similarities and dissimilarities. The human D_H gene cluster also contains repeated units, however in this case, the repeat length is 9–10 kb rather than the 4 kb repeat in mouse, and each repeat contains four D_H gene segments rather than one D_H gene segment per repeat in

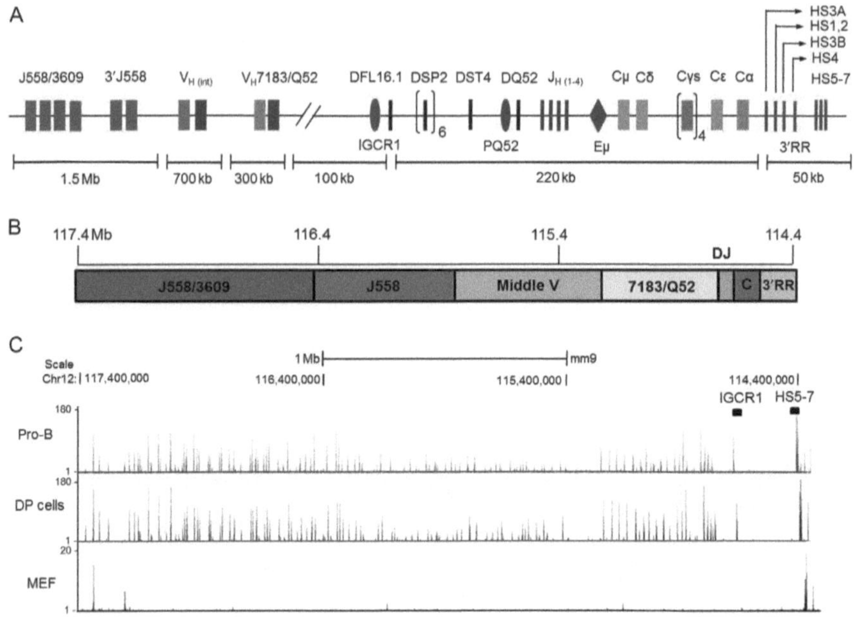

Figure 2 Gene organization of the mouse *IgH* locus. (A) Schematic organization of *IgH* locus. V_H gene segments are divided into proximal (V_H7183 and V_HQ52), middle/intermediate, proximal J558, and distal (V_H3609 and V_HJ558) gene families, of approximate lengths as noted. The 5′-most D_H gene segment DFL16.1, middle DSP and DST4, and 3′-most DQ52 are indicated, followed by J_H gene segments. Exons that encode *IgH* constant region include Cμ, Cδ, Cγ3, Cγ1, Cγ2a, Cγ2b, Cε, and Cα. Regulatory sequences shown in red are the IGCR1 close to DFL16.1, a promoter associated with DQ52 (PQ52), the intronic enhancer (Eμ), and the 3′ regulatory region (comprising a cluster of DHS-labeled HS1-4). (B) Scale representation of the murine *IgH* locus based on mm9. V_H gene families are indicated. Pink box labeled DJ contains all D_H and J_H gene segments, and green box labeled C contains exons of all *IgH* isotypes. (C) Genome browser tracks of CTCF Chip-seq in pro-B cells, $CD4^+CD8^+$ (DP) thymocytes, and mouse embryonic fibroblasts (MEF). Chip-seq track shows that CTCF binding to *IgH* locus is lymphoid-specific. MEF Chip-seq track is derived from Encode (Yue et al., 2014), pro-B and DP thymocytes tracks are derived from Shih et al. (2012). *PNAS*, 109(50), 3494–3502. (See the color plate.)

mouse (Ichihara, Matsuoka, & Kurosawa, 1988). The 3′-most D_H gene segment in humans is also located after a gap of several kb from the 5′ D_H genes and within a kb of the first J_H gene segment.

The mouse genome contains four joining gene segments, J_H1-4, spread over 2.8 kb followed by exons that encode constant regions of IgM and IgD antibody isotypes. The remaining 250 kb of the locus encodes constant regions of other antibody isotypes that are generated during immune responses via class switch recombination (CSR) (Li, Woo, Iglesias-Ussel, Ronai, & Scharff, 2004; Stavnezer, Guikema, & Schrader, 2008). In mouse, these include: IgG3, IgG1, IgG2A, IgG2B, IgE, and IgA. Functional *IgH* genes are assembled during B cell development in the bone marrow by the two independently regulated recombination events (Corcoran & Feeney, 2012; Jung & Alt, 2004). The first joins a D_H and J_H gene segment to form a DJ_H junction, and the second rearranges a functional V_H gene segment to the DJ_H junction to form a VDJ_H junction that can encode IgM protein. D_H gene segments are flanked by recombination signal sequences (RSS) that contain 12 bp spacers (12-RSS). In principle, therefore, D_H gene segments can recombine with 23-RSS-associated J_H gene segments by either inversion or deletion, yet the majority of D_H recombination occurs by deletion though inversional recombination has been detected in functional genes (Meek, Hasemann, & Capra, 1989). The first recombination event produces a DJ_H-associated 12-RSS that serves as the substrate for V_H recombination via a V_H-associated 23-RSS. All functional V_H genes are oriented in a manner such that V_H recombination proceeds only by deletion of intervening DNA.

3. *IgH* LOCUS ACTIVATION

3.1 Subnuclear Compartmentalization

Fluorescence *in situ* hybridization (FISH) experiments show that the *IgH* locus is located close to the nuclear periphery in non-B lineage cells (Kosak et al., 2002). This subcompartment is associated with gene repression (Finlan et al., 2008; Guelen et al., 2008). The DamID assay, that was developed to identify sequences throughout the genome that are located in close proximity to the nuclear envelope (Kind & van Steensel, 2010), shows that the entire *IgH* locus comprises a lamin-associated domain in NIH3T3 fibroblasts (Zullo et al., 2012). Tethering to the nuclear lamina occurs via sequences located within the 5′ end of the V_H region that contains V_HJ558 gene segments (Zullo et al., 2012). Evidence for this first came from FISH studies using bacterial artificial chromosome (BAC) probes derived from different parts of the locus. In these studies, V_HJ558 region probes were more frequently located close to the nuclear periphery compared to probes from the 3′ end of the locus (Yang, Riblet, & Schildkraut, 2005). Zullo and colleagues mapped sequences responsible for peripheral tethering in 3T3 cells and found that this activity resided in distal V_H segments. These sequences were enriched for a GAGA motif that bound the transcription factor c-Krox. This protein has been previously shown to interact with repressive chromatin-modifying machinery such as the NURF complex (Melnick et al., 2002). One of the first steps of lineage-specific locus activation relocates *IgH* away from the nuclear periphery to occupy a more central nuclear position. Most of *IgH* gene assembly and expression occurs at this location. Consistent with the proposed role of c-Krox in tethering the locus to the nuclear lamina, binding of this factor to the *IgH* locus is reduced in pro-B cells coincident with radial repositioning.

This raises the question as to what lineage-specific event triggers loss of c-Krox (and associated repressive machinery) from the V_H locus. Of several transcription factors that are essential for early B cell differentiation (Bain et al., 1997; Busslinger & Urbanek, 1995; Liu, Schmidt-Supprian, et al., 2007; Treiber et al., 2010), EBF1 and E2A may be involved in this process because *IgH* alleles remain peripheral in EBF1- or E2A-deficient pro-B cells (Hewitt, Chaumeil, & Skok, 2010; Lin et al., 2012; Sayegh, Jhunjhunwala, Riblet, & Murre, 2005). The *IgH* locus is also peripherally located in IL-7 receptor-deficient pro-B cells (Kosak et al., 2002), as are *IgH* alleles that lack the intronic enhancer Eμ located between J_H gene segments and Cμ exons (Guo, Gerasimova, et al., 2011). Though the connection between these factors is unclear, several hypotheses can be considered. First, loss of EBF1 and IL-7R blocks B cell development at a very early stage. One possibility is that EBF1 and IL-7R-deficient pro-B cells may not have differentiated far enough for relocalization to occur. That is, effects of EBF1 and IL-7R deficiency may impinge indirectly on *IgH* location via developmental stage of residual pro-B cells in these strains. Alternatively, IL-7-dependent activation of V_HJ558 genes may help disrupt interactions with the repressive nuclear lamina (Chowdhury & Sen, 2001, 2003; Corcoran, Riddell, Krooshoop, & Venkitaraman, 1998). It should be noted that IL-7 responsiveness of distal V_H genes has been questioned, (Malin et al., 2010), and additional studies are required to firmly establish connections between IL-7/IL-7R signaling and V_H J558 gene activation. The similar phenotype of E2A-deficient pro-B cells and Eμ-deficient alleles could be mechanistically connected since E2A proteins bind to and are likely to be essential for Eμ activity. We have proposed that Eμ activation disassociates *IgH* alleles from the nuclear periphery, perhaps by relocating *IgH* alleles to transcription factories located more centrally in the nucleus. Repressive factors such as c-Krox may then be lost from the locus in the active environment of a transcription factory. Alternatively, Eμ activity may displace c-Krox and associated repressive complexes by direct interactions with the V_H region. In this regard, there is evidence that Eμ interacts with proximal and distal parts of the V_HJ558 locus (Guo, Gerasimova, et al., 2011; Verma-Gaur et al., 2012). Loss of c-Krox from the *IgH* locus would permit it to relocate away from the repressive nuclear peripheral microenvironment.

3.2 Chromatin Organization of the Germline *IgH* Locus

The *IgH* locus can be divided into two parts to better understand its structural organization and regulatory mechanisms that act upon it. A 5′-part that contains all the variable region gene segments and encompasses most of the locus (approximately 2.5 Mb) and a 3′-part of approximately 280 kb that includes all D_H and J_H gene segments, and exons that encode the various *IgH* isotypes. We first discuss features of each domain individually and then consider interactions between these domains as they impact *IgH* expression.

4. 3′ *IgH* DOMAIN

4.1 DHSs Flank the 3′ *IgH* Domain

The 3′ *IgH* domain extends from a cluster of B lineage-specific DHS located 4–6 kb 5′ of DFL16.1 to a cluster of DHSs located approximately 40 kb 3′ of the last coding exons of the IgA isotype. Two of the three DHSs located in 5′ of DFL16.1 contains CTCF-binding sites and each has been shown to confer insulator activity in reporter-based transfection assays (Featherstone, Wood, Bowen, & Corcoran, 2010; Fig. 2C). Deletion of a 4 kb region that encompasses these DHS or mutation of both CTCF-binding sites, leads to aberrant V_H recombination (Guo, Yoon, et al., 2011). The most striking feature of the resultant dysregulation is recombination of V_H7183.2.3, the 3′-most V_H gene segment, to unrearranged DQ52 gene segments, thereby breaking the normal order of VDJ recombination in the *IgH* locus. Based on the critical role of this region, and associated CTCF binding, in ordering VDJ recombination, this region has been named IGCR1 (Guo, Yoon, et al., 2011). How the IGCR1 enforces recombination order is further discussed below.

CTCF-bound sequences have been shown to have insulator or boundary functions (Bushey, Dorman, & Corces, 2008; Gaszner & Felsenfeld, 2006). Very briefly, insulator activity refers to the ability of a sequence to prevent promoter/enhancer communication when the two lie on different sides of the insulator. Boundary function is mediated by a CTCF-bound sequence when it prevents spreading of a chromatin domain from one side of the boundary element to the other side. It is unclear at present whether recombination regulation by CTCF-bound IGCR1 reflects its insulator property as assessed by reporter assays or a boundary function. Analyses of a limited number of histone modifications by ChIP-Seq has not revealed extension of chromatin domains from either side of the IGCR1 into flanking regions as might be expected if the IGCR1 served a boundary function. This may be in part because both sides of the intact IGCR1 are usually marked with repressive histone modifications in pro-B cells. Since IGCR1 does not separate active and inactive chromatin domains, where spreading of one into the other can be visualized, we do not consider it to be a classical boundary element. Instead, recent studies provide evidence in favor of an insulation function for IGCR1. The intronic enhancer Eμ has been shown to interact with the region 5′ of DFL16.1 on wild-type (WT) *IgH* alleles to generate a 70 kb chromatin loop (Guo, Gerasimova, et al., 2011; Verma-Gaur et al., 2012). Analyses of pro-B cell lines with mutated IGCR1 show that Eμ does not loop to the 5′ DFL16.1 region in the absence of CTCF binding. Instead, Eμ interacts with a region close to V_H7183.2.3 located 110 kb 5′ of DFL16.1 on IGCR1-mutated alleles to generate an alternate 180 kb loop (unpublished observation). Eμ interactions with this distal sequence result in localized induction of active chromatin modifications and transcriptional activity. These observations are consistent with a model in which IGCR1 works as an insulator to "sequester" Eμ function and thereby prevent it from activating parts of the locus located further 5′ of DFL16.1.

4.2 3′ End of the *IgH* Locus

At the 3′ end of the *IgH* locus, a cluster of DHSs defines a complex regulatory region (Birshtein, 2014; Khamlichi, Pinaud, Decourt, Chauveau, & Cogne, 2000), that is made up of two parts. A 28 kb 5′ portion that encompasses an inverted repeat sequence is located 13 kb from the Cα exons. This region was first identified in myeloma cell lines where it is marked by four DHSs labeled HS3a, 1, 2, 3b, and 4 (Dariavach, Williams, Campbell, Petersson, & Neuberger, 1991; Michaelson, Giannini, & Birshtein, 1995). Sequences encompassing HS4 site confer transcriptional enhancer activity in transfection assays (HS4 is also referred to as Eα). Transcription factors Pax5, NF-κB, and YY1 have been implicated in the function of Eα. DHS1-4 comprises an inducible regulatory element that is referred to as the 3′ regulatory region (3′RR). The main function of the 3′RR is in activated mature B cells to control CSR as demonstrated by BAC transgenesis and germline deletion of combinations of DHSs (Dunnick, Shi, Zerbato, Fontaine, & Collins, 2011; Pinaud et al., 2001; Rouaud et al., 2014). The BAC transgenic assays utilized a functionally rearranged *IgH* locus that extended till the 3′RR to study CSR. These studies showed that the 3′RR was essential to induce isotype-specific germline transcription and CSR in *in vitro*-activated B cells. Deletion of HS3a/b, 4 of the 3′RR in the mouse germline showed that this region was not required for B cell development and VDJ recombination (Bebin et al., 2010; Morvan, Pinaud, Decourt, Cuvillier, & Cogne, 2003). However, HS3a/b, 4 were essential to induce Cγ germline transcription and CSR to this isotype (Dunnick, Shi, Graves, & Collins, 2005). In keeping with function of the 3′RR as an inducible regulatory element in mature B cells, HS1-4 are not active in pro-B cells.

At the very 3′ end of the *IgH* locus, a 10 kb region located approximately 5 kb from HS4 of the 3′RR contains three DHSs, HS5-7. These DHSs are not tissue-specific, based on the observation that they are present in pro-B cells and mature B cells as well as ES cells (Yue et al., 2014). Each DHS has multiple CTCF-binding sites and, like the DHS themselves, CTCF binding is not restricted to the B lineage (Fig. 2C). Deletion of HS5-7 has little effect on B cell development or CSR (Volpi et al., 2012). One of the relatively minor effects noted during development was that proximal V_H7183 gene family was utilized at twofold greater frequency. CTCF-binding characteristics and lack of tissue specificity of HS5-7 strongly suggest that it is a boundary element that demarcates the 3′ end of the *IgH* locus. However, Volpi et al. found that expression of the neighboring *Tmem21* gene was only modestly increased in pro-B cells that lacked HS5-7, suggesting that HS5-7 did not serve as a boundary element. It remains possible that boundary functions could be more evident at other developmental stages or tissue types. Histone modification status on either sides of HS5-7 in normal or deleted alleles has not been systemically examined to identify other aspects of boundary function.

4.3 Eμ and Other DHS in the Germline *IgH* Locus

In addition to clusters of DHSs at either end of the 3′ *IgH* domain, DHSs mark the intronic enhancer Eμ and a promoter 5′ of DQ52 (PQ52) in pro-B cells (Fig. 2A). These sites are strictly B lineage-specific. Studies with Eμ-deleted *IgH* alleles show that the PQ52 DHS is Eμ dependent, likely corresponding to Eμ-dependent transcriptional activity of the promoter (Chakraborty et al., 2009). Deletion of PQ52 has little effect on *IgH* recombination or expression during B cell development (Afshar, Pierce, Bolland, Corcoran, & Oltz, 2006; Nitschke, Kestler, Tallone, Pelkonen, & Pelkonen, 2001). In contrast, Eμ plays an important role in both processes (Afshar et al., 2006; Perlot, Alt, Bassing, Suh, & Pinaud, 2005). D_H recombination is reduced five- to eightfold on Eμ-deleted *IgH* alleles and V_H recombination is reduced even further. While it has been argued that the virtual absence of V_H recombination on Eμ-deleted alleles may simply reflect reduced availability of DJ_H substrates, our working hypothesis is that Eμ activity participates directly in V_H recombination by activating DJ_H junctions and altering *IgH* locus conformation (see below).

Sense- and antisense-directed transcripts initiate within Eμ (Fig. 3A). Sense-directed Iμ transcripts (toward Cμ) initiate at multiple sites 3′ of an octamer element located within the 220 bp core enhancer (Lennon & Perry, 1985; Su & Kadesch, 1990). The start site for antisense transcription

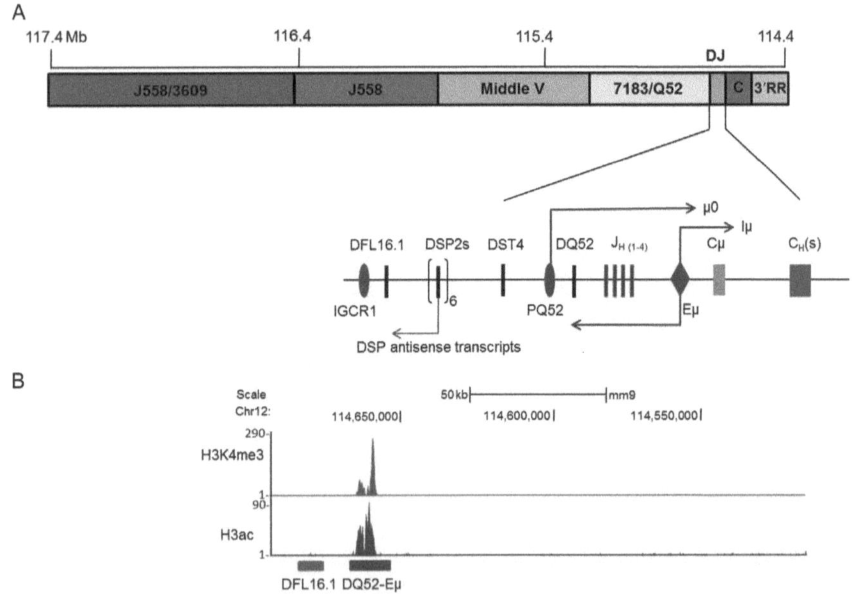

Figure 3 Transcriptional and epigenetic features of the 3′ *IgH* domain. (A) Scale representation of the murine *IgH* locus. The 3′ part of the *IgH* locus is expanded below to identify noncoding transcripts in the 3′ *IgH* domain. μ0 transcripts initiate at the DQ52 promoter, PQ52. Sense-directed Iμ transcripts and antisense transcription initiate within the Eμ enhancer. Low-level antisense transcripts are present at DSP repeats. (B) Genome browser tracks of H3K4me3 and H3ac Chip-seq in 3′ part of the *IgH* locus (derived from Choi et al., 2013). The track shows that high levels of activating histone modifications are concentrated in the region between DQ52 and Eμ. (See the color plate.)

has not been precisely mapped, but it is also likely to initiate within the Eμ core (Bolland et al., 2007). Eμ also activates transcription from PQ52 in the sense orientation; this transcript is referred to as μ0 (Schlissel, Corcoran, & Baltimore, 1991). Iμ and μ0 transcripts are spliced and polyadenylated, and constitute the major sterile transcripts in the unrearranged *IgH* locus in pro-B cells. In addition, the genomic region containing DFL16.1 and DSP gene segments has low levels of antisense-oriented transcripts (that extend toward the V_H domain). Bolland et al. have proposed that D_H antisense transcripts are derived from antisense transcription initiated at Eμ (Bolland et al., 2007). Since the size of Eμ-initiated antisense transcripts has not been determined, it is not clear whether they extend 70 kb to DFL16.1/DSP2.9 located at the 5′ end of the D_H region. We have suggested that bidirectional promoters associated with DSP gene segments may be the source of D_H region associated antisense transcripts (Chakraborty et al., 2007). Interestingly, the level of antisense D_H transcripts increases several-fold upon deletion of the DHS cluster 5′ of DFL16.1 (Guo, Yoon, et al., 2011). Even higher levels of antisense D_H transcripts were noted when the 100 kb intervening sequence between V_H and DFL16.1 was deleted (Giallourakis et al., 2010).

It is interesting to note that because of sense-directed μ0 transcription from PQ52 and Eμ-initiated antisense transcription, the region between DQ52 and Eμ that contains the four J_H gene segments is transcribed on both DNA strands (Fig. 3A). Simultaneous passage of RNA Pol II on both strands may impart a special structure to this region that could, in part, explain the high sensitivity of this entire 4 kb region to DNase I digestion (Maes et al., 2006) This region also contains highest levels of transcription-associated H3K4me3 mark and H3K9 acetylation making it quite distinct from all other parts of the *IgH* locus (Fig. 3B). Whether bidirectional transcription through this region is of functional importance remains to be determined.

A relative strong DHS was recently identified in the intergenic region between exons that encode γ2b and ϵ isotopes (Revilla-i-Domingo et al., 2012). This site coincides closely with Pax5 and E2A binding; however, no function has been attributed to this region yet.

4.4 Histone Modifications

The entire 3′ *IgH* domain is uniformly marked with repressive H3K9me2 and H3K27me3 modifications in non-B lineage and ES cells (Yue et al., 2014). In the transition to pro-B cells, this pattern is interrupted by a patchwork of activation-associated histone modifications that result in a markedly heterogeneous epigenetic state. Histone modifications such as H3K9 acetylation and H3K4 dimethylation (H3Kme2), that correlate best with the availability of a locus for transcription rather than transcription *per se*, are located just 5′ of DFL16.1, the region spanning DQ52 through the Cμ exons, and the region close to the 3′RR (Chakraborty et al., 2007; Morshead, Ciccone, Taverna, Allis, & Oettinger, 2003; Fig. 3B). These regions map closely with location of DHSs described in the preceding section. The levels of modifications vary greatly between these regions. Specifically, the DQ52-Eμ region, that contains four J_H gene segments, has especially high levels of H3ac, whereas the DFL16.1 region is relatively low for both H3ac and H3K4me2. Other than these three regions, all other parts of the 3′ *IgH* domain remain marked with repressive histone modifications. Each of the regions that contain acetylated histones H3 or H4 is also marked by transcription-associated H3K4me3. Highest levels of this modification also occur in the DQ52-Eμ region, possibly as a result of bidirectional transcription through it (Fig. 3B). Much lower levels of H3K4me3 are found at DFL16.1 and the 3′RR, presumably reflecting low transcriptional activity of these parts of the locus. H3K36me3 is usually found at the 3′ ends of transcribed genes and has been proposed to get there in association with the elongating form of RNA Pol II (Kolasinska-Zwierz et al., 2009; Wagner & Carpenter, 2012). Accordingly, this mark is located mainly at 3′ of Eμ toward the Cμ region. All forms of B lineage-specific activating and transcription-associated histone modifications reduce substantially by the Cδ exons located approximately 11 kb 3′ of the Cμ exons.

The observed epigenetic heterogeneity of the 3′ *IgH* domain leads to two fundamental questions. First, how is the heterogeneity established and, second, is it functionally important? Close concordance between DHS and pockets of activating histone modifications suggests that the punctate pattern is established by localized chromatin activation by factors that lead to formation of the DHSs. In this model, DHS-binding transcription factors recruit chromatin-modifying enzymes, such as HATs, to initiate the process. The only extended domain of histone acetylation covers approximately 10 kb from DQ52 to an imprecisely defined region between Cμ and Cδ. This domain of activation is lost on Eμ-deleted alleles, indicating that it is established by Eμ-binding transcription factors (Chakraborty et al., 2009). Among Eμ-binding proteins, E2A has been shown to interact with CBP/p300, a well-established HAT. The extent to which histone acetylation spreads from the point of initiation is dependent upon several factors. These include the state of neighboring chromatin (which could determine the "ease" with which the initiating event is propagated), the presence of boundary elements (which act as a barrier to the propagation of a defined chromatin state), and competition with other chromatin-modifying activities located in flanking regions (which could, for example, deacetylate newly acetylated histones to prevent further propagation of the active domain). The relatively abrupt 5′ end of the DQ52–Cμ domain suggests the presence of a boundary element at this end. The most likely candidate is the DQ52 promoter or a closely associated sequence. The gradual loss of histone acetylation at the 3′ end of the DQ52–Cμ domain suggests that the influence of Eμ in inducing these changes is counteracted by opposing enzymatic activities associated with maintenance of a repressive (H3K9me2-marked) chromatin state further downstream. Activating histone modifications at DFL16.1 are also substantially reduced on Eμ-deleted alleles. The most likely explanation is that these modifications are induced by Eμ-associated transcription factors that are brought into proximity of DFL16.1 via chromatin looping (see below).

We have previously proposed that the heterochromatic state of the DSP region may be actively maintained by repeat-induced gene silencing (Chakraborty et al., 2007). This idea was based on the observation that each DSP gene segments in the mouse is part of an approximately 4 kb repeated sequence, and the region is marked by antisense transcription. Both features are reminiscent of subtelomeric repeats in fission yeast which have been shown to be actively heterochromatinized by an RNAi-directed process (Grewal & Jia, 2007; Hall, Noma, & Grewal, 2003). In addition to the RNAi machinery, maintenance of heterochromatin at these repeats requires ongoing histone deacetylation, H3K9 methylation, and protection of H3K9me2 by chromodomain proteins of the HP1 family. In extending the analogy of DSP heterochromatinization to repeat-induced gene silencing in *S. pombe*, we found that the inhibition of type I histone deacetylases in pro-B cell lines with trichostatin A resulted in spreading of H3 acetylation from the DFL16.1 patch into DSP repeat region (Chakraborty et al., 2007). More definitive evidence of a role for the RNAi machinery in this process was provided by Koralov et al. in their analysis of *Dicer*-deficient mice (Koralov et al., 2008). They found increased transcription of DSP repeats in *Dicer*-deficient pro-B cells, suggestive of reduced heterochromatinization of this region. Concomitantly, DSP recombination was significantly increased at the cost of DFL16.1 utilization in these cells, consistent with increased accessibility of RAG1/2 recombinase to DSP-associated RSSs. These observations are very similar to reduced heterochromatin and increased transcription of subtelomeric repeats in RNAi-deficient strains of *S. Pombe* (Hall et al., 2003). Although histone modifications were not assayed in *Dicer*-deficient pro-B cells, the simplest interpretation of these observations is that DSP heterochromatinization is reduced in *Dicer*-deficient pro-B cells. Taken together with the model for establishment of the active DQ52–Cμ domain and DFL16.1 patch, we hypothesize that continuous generation of heterochromatin (in the DSP region) and passive maintenance of heterochromatin (between Cδ and the 3′RR) determine the epigenetic fine structure of the 3′ *IgH* domain.

4.5 DNA Methylation

Epigenetic heterogeneity of the 3′ *IgH* domain is also reflected in its DNA methylation status (Selimyan et al., 2013). On germline alleles on pro-B cells only two pockets of tissue-specific CpG demethylation are found that correspond to Eμ and the DQ52-associated promoter. It is interesting to note that the intervening region between these two regulatory elements, which contains J_H gene segments and highest levels of H3ac and H3K4me3, maintains their CpGs in a hypermethylated state in pro-B cells. Thus, there is no correlation between levels of activating histone modifications and activating CpG demethylation in the 3′ *IgH* domain. CpGs that remain on Eμ-deleted alleles are hypermethylated, indicating that Eμ is a DNA demethylating element. However, B lineage-specific CpG demethylation of DQ52 does not require Eμ despite loss of the DQ52 DHS on Eμ-deficient alleles, indicating that DQ52 promoter has autonomous DNA demethylating activity. Partial DNA demethylation also occurs at the 5′ DFL16.1 region and in the DHS5-7 region of the 3′RR on WT *IgH* alleles. These observations demonstrate that the sites of tissue-specific DNA hypomethylation in the 3′ *IgH* domain correspond closely with the identified DHSs in the locus. The mechanism of highly localized demethylation at DQ52 and Eμ is not clear. One possibility is that the demethylated state is generated by passive loss of CpG methylation due to the presence of enhancer binding proteins that block access of maintenance methyl transferases to the region. Alternatively, Tet enzymes may be recruited by enhancer binding proteins to carry out active local demethylation (Kohli & Zhang, 2013; Pastor, Aravind, & Rao, 2013; Serandour et al., 2012; Wu & Zhang, 2011). Distinction between these models awaits development of an *in vitro* system for inducible demethylation of *IgH* regulatory sequences.

4.6 Establishing the Epigenetic State

Analyses of Eμ-deficient *IgH* alleles suggest an epigenetic hierarchy in establishing the fully active state in pro-B cells (Chakraborty et al., 2009). The key observations leading to this model are (1) Eμ-deficient alleles have substantially reduced levels of H3ac and transcription-associated H3K4me3 and H3K36me3 marks in the DQ52–Cμ region and lack these marks at 5′ DFL16.1, (2) pro-B-specific gain of H3K4me2 is only partially reduced in the DQ52–Cμ domain on Eμ-deleted alleles, and (3) loss of repressive H3K9me2 is comparable to that of WT *IgH* alleles. Additionally, pro-B-specific recruitment of CTCF to sites across the locus is also unchanged on Eμ-deleted alleles (Guo, Gerasimova, et al., 2011). We infer that Eμ-deleted alleles are in a partially active state, having achieved a subset of lineage-specific epigenetic changes.

Our working model is that an Eμ-independent, but B lineage-specific, regulatory mechanism leads to loss of repressive histone modifications from the locus and gain of a subset of activating histone modifications. We hypothesize that this step of locus activation permits ubiquitously expressed factors, such as CTCF, to gain tissue-specific access to developmentally regulated loci. This step may also increase accessibility of tissue-specific *cis*-regulatory sequences to binding proteins such as YY1 and TFE3 that are ubiquitously expressed, yet gain access to Eμ only in B lineage cells. Other Eμ-binding proteins, like E2A and PU.1, are more tissue-restricted but also expressed in cells where *IgH* is inactive. Binding of these proteins to Eμ may be regulated by the first, Eμ-independent, locus activation step as well.

Because Eμ-deleted alleles are located close to the nuclear periphery (Guo, Gerasimova, et al., 2011), the simplest scenario is that this first step of locus activation does not require radial repositioning. However, at present, it is impossible to rule out a model where activation occurs on Eμ-deficient alleles that have moved away from the periphery, which reestablish peripheral localization in the course of FISH experiments used to measure nuclear positioning. Regardless of the exact sequence, it is quite clear that peripherally located *IgH* alleles can have several hallmarks of lineage-specific activation. We propose that the first step of locus activation converts a "silent" *IgH* locus that contains H3K9me2 to one that is permissive for transcription.

Binding of transcription factors to Eμ converts such a transcriptionally competent locus to a transcriptionally active locus. Eμ-enhancer binding proteins and associated HATs, such as CBP/p300, lead to an H3/H4 acetylated locus that can recruit RNA polymerase and be transcribed. Transcriptionally active polymerases and associated chromatin-modifying activities lead to the decoration of the locus with H3K4me3 and H3K36me3 and the fully active epigenetic state characteristic of pro-B cells. It is possible that radial repositioning represents the migration of transcriptionally silent alleles located at the nuclear periphery to more centrally located transcription factories.

4.7 3D Organization of the 3′ *IgH* Domain

The apparently sporadic peaks of active histone modifications interspersed with regions that maintain distinctively repressive epigenetic signatures within a locus that is ready to initiate V(D)J recombination begins to make more "sense" when considered in the context of the spatial configuration of the 3′ *IgH* domain. Chromosome conformation capture on chip (ChIP-loop) studies that queried Eμ-interacting regions show that Eμ interacts with sequences 5′ of DFL16.1 and with HS5-7 region of the 3′RR in pro-B cells (Guo, Gerasimova, et al., 2011). Taken together with the fact that Eμ activates the DQ52 promoter, the 3′ *IgH* domain can be visualized as a three-loop structure (Fig. 4). The smallest of these loops extends from Eμ to DQ52 (~5 kb). This loop coincides closely with the region that contains highest

Figure 4 Model illustrating Eμ-dependent interactions in 3′ part of the *IgH* locus. Regulatory elements are shown as colored diamonds, Eμ (red), PQ52 (yellow), 3′RR (green), and IGCR1 (blue). Eμ–PQ52 interaction forms the smallest loop (shown in green), which contains DQ52 and four J_H gene segments. This region contains highest levels of activation-specific histone modifications and forms the RAG1/2 rich RC. Eμ–IGCR1 interaction forms the intermediate-sized loop (70 kb; shown in red), which contains all D_H gene segments. DFL16.1 is located near to the base of the loop close to IGCR1. Eμ–3′RR interactions generate the largest loop (200 kb; shown in red). This loop contains all exons that encode *IgH* isotypes. (See the color plate.)

levels of H3K4me3 and H3/H4 acetylation, and bidirectional transcription through its entire length.

The next biggest loop (70 kb) brings together Eμ and IGCR1 that is located less than 10 kb 5′ of DFL16.1 (Verma-Gaur et al., 2012). Eμ–IGCR1 interactions are lost in CTCF-depleted pro-B cells (unpublished observation) or on alleles in which two CTCF-binding sites within IGCR1 have been mutated (unpublished observation). At the other end of this loop, YY1 bound to Eμ has been proposed to mediate interaction with IGCR1, since the Eμ–IGCR1 loop is disrupted in YY1-deficient pro-B cells (unpublished observation). However, it remains to be directly demonstrated by mutational analysis that the YY1-binding site within Eμ is essential for this interaction. Most of this 70 kb loop, which contains all the D_H gene segments (other than DQ52), is marked with repressive histone modifications and contains low levels of antisense-oriented transcripts. Based on this spatial configuration, the Eμ-dependent peak of activating histone modifications near DFL16.1 may be considered to be the result of its spatial proximity to Eμ. Interaction of Eμ with sequences close to the 5′-most (DFL16.1) and 3′-most (DQ52) D_H gene segments makes these D_H gene segments the only D_H gene to be marked with activating histone modifications. This epigenetic feature may, in part, explain dominant utilization of DFL16.1 and DQ52 gene segments in VDJ recombination compared to the several DSP gene segments that lie in between.

Interaction of Eμ with the HS5-7 generates the largest of the three loops (~200 kb). Formation of this loop requires both CTCF and YY1 (Medvedovic et al., 2013), and most of it is also marked with repressive histone modifications in pro-B cells. Inference of a three-loop configuration on the basis of 3C/4C studies is not intended to imply that this is a stable structure or that it occurs in all pro-B cells. Rather, the bases of these loops (Eμ, IGCR1, PQ52, and HS5-7) should be viewed as genomic regions that have the potential to interact because they are bound by architectural transcription factors. It is likely that in a population of cells all single- and double-loop configurations will coexist with the unique three-loop configuration. Conversely, we imagine that each of these configurations exists in dynamic equilibrium within a single cell, resulting in a state where loop formation and disruption is in constant flux. It is quite remarkable that all DHSs identified within the 3′ *IgH* domain, all sites of active histone modifications, and all sites of DNA demethylation are spatially clustered in the three-loop configuration. Eμ is essential to hold this configuration together since all interactions, including IGCR1–HS5-7 interaction, are lost in Eμ-deficient alleles.

It is interesting to consider the possibility that sequences 5′ of DFL16.1 and HS5-7 may be only partially demethylated because these sites come into contact with Eμ in only a subset of cells.

The two smaller loops serve distinct functions. The smallest loop, which is structurally and epigenetically unique, has been shown to bind the highest levels of recombination activating gene products (RAG) 1 and 2 in the *IgH* locus (Schatz & Ji, 2011). Schatz and colleagues refer to this, and similar high-density regions of RAG1/2 at other antigen receptor loci, as recombination centers (RC). Formation of the RC may be facilitated by high levels of H3K4me3 in this region that can recruit RAG2 via interaction with its plant homeodomain (Liu, Subrahmanyam, Chakraborty, Sen, & Desiderio, 2007; Matthews et al., 2007). It is likely that a major function of this smallest loop is to generate a structure that permits formation of the *IgH* RC. The Eμ–IGCR1 loop encapsulates all D_H gene segments. Its close juxtaposition to the RC may be a major mechanism by which the order of recombination at the *IgH* locus is established. In this view, *IgH* recombination initiates at J_H-RSSs located within the RC. Thereafter, synapsis with a complementary RSS in accordance with the 12/23 rule would induce double-strand DNA breaks. In the three-loop configuration, RSSs associated with DFL16.1 and DQ52 would be closest to the RC, and therefore be selected most frequently for recombination, leading to their prominent utilization in VDJ junctions.

Two pathways can be imagined to recruit DSP gene segments for recombination. One possibility is that RAG1/2 located close to DFL16.1 "spills over" into proximal DSP gene segments, to engage DSP-RSSs for recombination. Since the probability of encountering DSP-RSSs would reduce with increasing distance from DFL16.1, this would lead to greater use of DSP gene segments located closer to DFL16.1 than of those located in the middle of the DSP cluster. Lack of D_H gene segments close to DQ52 would preclude RAG1/2 spill-over at the DQ52 end of the 70 kb loop from having a significant effect on D_H recombination. Alternatively, DSP rearrangements may require transient formation of alternate loops. For example, the Eμ–IGCR1 interaction may be occasionally replaced by interaction between Eμ and a DSP-associated promoter. Such an interaction would transiently place the interacting DSP gene segment close to the RC, thereby permitting its recombination.

The importance of clustering of all D_H gene segments within the Eμ–IGCR1 loop is revealed dramatically with IGCR1-mutated alleles. On these alleles, the 3′ most V_H gene segment, V_H7183.2.3, recombines with unrearranged DQ52 in a breakdown of the normal order of recombination. While addressing the mechanistic basis for the dysregulation, we found that Eμ loops to a site close to V_H7183.2.3 when IGCR1 is mutated (unpublished observation). That is, disruption of the Eμ–IGCR1 loop results in the formation of an alternate Eμ–V_H7183.2.3 loop. Now, V_H7183.2.3 occupies a position analogous to that occupied by DFL16.1 on WT alleles, placing it close to the RC. Since there is considerable RAG1/2 density near DQ52, recombination initiated at the DQ52-RSS can synapse with a spatially proximal V_H7183.2.3-RSS leading to the observed recombination product. In this model, rearrangement of a V_H gene segment to a germline D_H gene segment is the consequence of altered 3D chromatin configuration of the 3′ *IgH* domain.

The function of the large Eμ–HS5-7 loop in *IgH* gene assembly is less clear. Deletion of HS5-7 by Birshtein and colleagues has a surprisingly modest phenotype (Volpi et al., 2012). During B cell development, utilization of proximal V_H7183 genes was elevated twofold and DQ52 to J_H3 rearrangements were increased in ΔHS5-7 mice compared to WT controls. In terms of locus structure, V_H region locus compaction was somewhat reduced and DFL16.1 to 3′RR looping was reduced twofold. Though the Eμ–HS5-7 loop was not directly examined, it is likely that it would be adversely affected on ΔHS5-7 alleles. What do these observations say about the function of the largest loop? One possibility is that Eμ–HS5-7 does not serve an essential function in *IgH* locus rearrangements. Alternatively, as discussed by Volpi et al., there are several CTCF-binding sites located 3′ of the deleted region that might compensate for loss of HS5-7 (Fig. 2C). Perhaps Eμ loops to the next available CTCF-binding sites in the absence of HS5-7, thereby maintaining a relatively large (though not identical) loop in the 3′ *IgH* domain on ΔHS5-7 alleles. What function could such a loop serve in pro-B cells? One possibility is that it may contribute to the dynamic structure of the 3′ *IgH* domain by "sequestering" Eμ activity away from its other looping interactions. Indeed, the slight increase in V_H7183 gene recombination noted by Volpi et al. is reminiscent of the much greater increase of V_H7183 gene recombination on IGCR1-mutated alleles. On those alleles, as discussed above, Eμ loops to a site within the 3′ V_H7183 genes, causing the observed increase in rearrangements involving these genes. Perhaps Eμ interaction with 3′ V_H7183 genes is increased on ΔHS5-7 alleles because of reduced looping to the 3′ end of the *IgH* locus. The effect is much less than what has been observed on IGCR1 mutated alleles in large part because IGCR1–Eμ interactions on ΔHS5-7 constrain the structure of the 5′ end of the 3′ *IgH* domain. Additional studies are needed to test the sequestration model and distinguish between these possibilities.

4.8 Regulation of D_H Recombination

Once the RC and the Eμ–IGCR1 loop are established, the *IgH* locus is poised to undergo the first step of antigen receptor gene assembly, the rearrangement of a D_H gene segment to a J_H gene segment to form a DJ_H junction. D_H recombination has been proposed to occur simultaneously on both alleles (Chen & Alt, 1993; Yancopoulos & Alt, 1986). This is based largely on the observation that in WT mice, both alleles are found to have DJ_H rearrangements in pre-B cells (and mature B cells), whereas only one of the two alleles may have a V_H recombination (Dudley et al., 2003; Ehlich, Martin, Muller, & Rajewsky, 1994; Talukder, Dudley, Alt, Takahama, & Akamatsu, 2004). However, a careful kinetic study to truly determine whether both alleles rearrange D_H gene segments simultaneously has not been reported. An observation that is suggestive of some asynchrony in D_H recombination is that mature B cells that emerge in mice containing mutated RAG1 or RAG2 genes often contain one completely germline *IgH* allele together with the functionally rearranged second allele (Dudley et al., 2003; Liang et al., 2002; Talukder et al., 2004). The observed genotype indicates that D_H recombination occurred on one allele at a time, at least when accentuated by reduced efficiency of VDJ recombination inherent to the mutated RAG protein.

Two other features of D_H recombination must been explained in molecular terms. First, the 5′-most and 3′-most D_H gene segments, DFL16.1 and DQ52, are used more frequently in D_H recombination than the numerically larger numbers of DSP (and the lone DST4) gene segments that lie between them (Atkinson et al., 1994; Bangs, Sanz, & Teale, 1991; Chang, Paige, & Wu, 1992; Tsukada, Sugiyama, Oka, & Kishimoto, 1990). This tendency was first observed in immunoglobulin sequences derived from mature B cells and has recently been solidly substantiated by deep sequencing of the VDJ repertoire of primary pro-B cells (Choi et al., 2013). A possible caveat in both datasets with regard to D_H utilization is that these trends were derived from the frequency of D_H gene segments found in VDJ recombined alleles. It is possible, therefore, that the observed trends may be skewed by selection events during the second recombination step or during B cell maturation. Primary D_H recombination frequencies were carefully quantitated in one study in which a temperature shift was used to activate D_H recombination in an Abelson virus-transformed pro-B cell line (Tsukada et al., 1990). The authors found that DFL16.1 and DQ52 were utilized most frequently; thereafter, frequency of DSP gene usage correlated closely with the distance of the various DSP family members from DFL16.1. That is, DSP gene segments located closer to DFL16.1 rearranged more frequently than those buried in the middle of the cluster. While a quantitative analysis of D_H usage in DJ_H junctions from primary pro-B cells is required, it is unlikely that the general trends described above will change qualitatively. As described above, the looped structure of the 3′ *IgH* domain provides a plausible mechanism for the observed frequency of D_H gene recombination.

A second characteristic of D_H recombination (in mouse) is that it occurs largely by deletion rather than inversion of the intervening DNA between the D_H and J_H gene segments (Gauss & Lieber, 1992; Pan, Lieber, & Teale, 1997; Sollbach & Wu, 1995). Because D_H gene segments are flanked by 23-bp RSSs, recombination can occur by either mechanism in accordance with the 12/23 rule. Yet, deletion has been found to be the preferred pathway for D_H recombination (though infrequent inversional events have been clearly documented (Pan et al., 1997)). Lieber and colleagues showed that part of the preference for deletion could be explained by the nature of the RSSs themselves (Gauss & Lieber, 1992). That is, RSSs on the 3′ side of D_H gene segments, whose use in J_H recombination led to deletion, were preferred recombination substrates in transient transfection assays. However, the selectivity observed in these studies still left a large part of the deletional bias unexplained. The growing realization that conformational states of germline and partially rearranged (see below) *IgH* loci define recombinational choice in major ways, leads us to speculate that structure of DJ_H rearranged alleles generated by inversion rather than deletion may be deleterious for secondary V_H recombination. For example, alleles that have undergone DJ_H recombination by deletion retain two major Eμ-interacting sequences—the incoming D_H-associated promoter and the IGCR1 5′ of DFL16.1. This configuration is reminiscent of the germline structure where Eμ interacts with the DQ52 promoter and IGCR1. Alleles that undergo DJ_H recombination by inversion would retain the DQ52 promoter and one or more unrearranged J_H gene segments between the IGCR1 and the promoter associated with the rearranged DJ_H. This would result in a third Eμ-interacting sequence (the DQ52 promoter) between the DJ_H rearrangement and IGCR1. Such an organization may hamper Eμ activation of V_H recombination. It will be interesting to engineer (or isolate) *IgH* alleles that have undergone DJ_H recombination by inversion to determine the epigenetic state and chromosomal architecture of such alleles.

4.9 Chromatin Structural Changes that Accompany DJ_H Recombination

DSP and DQ52 rearrangements have one or more D_H gene segments in germline configuration 5′ of the DJ_H junction. The recombined D_H gene segment gains several characteristics of active chromatin and becomes epigenetically distinct from germline D_H gene segments on the same allele (Subrahmanyam et al., 2012). This is most apparent for DSP2 gene segments that are in heterochromatic state in pro-B cells prior to initiation of *IgH* rearrangements. Rearranged DSP2 gene segments lose repressive histone modifications and gain activation-associated H3/H4 acetylation, H3K4me2, and H3K4me3 modifications. Presence of H3K4me3 correlates with increased levels of bidirectional transcription from the promoter associated with the rearranged gene segment. Second, DJ_H junctions become hypersensitive to DNase I compared to the same gene segment in its germline location. Presumably, this feature also reflects promoter activation of the rearranged D_H gene segment. Third, DJ_H junctions are selectively CpG hypomethylated (Selimyan et al., 2013). Recall that DQ52 is the only D_H gene segment that is CpG demethylated in the unrearranged configuration; all other D_H and J_H gene segments are CpG hypermethylated in pro-B cells. This means that generation of DJ_H junction converts hypermethylated D_H and J_H gene segments into hypomethylated recombined products. It remains unclear whether this happens prior to, during, or after joining of D_H and J_H gene segments. In the absence of data directly addressing this question, our working hypothesis is that demethylation occurs after formation of a DJ_H junction. Remarkably, all of these epigenetic changes are highly localized to the DJ_H junction and do not extend to the first unrearranged D_H gene segment that is located less than 5 kb 5′ to the DJ_H junction. Thus, unrearranged D_H gene segments that lie in the region between DFL16.1 and the DJ_H junction remain in repressive chromatin state. The resulting chromatin configuration is very similar to the state of the germline *IgH* locus prior to initiation of rearrangements, with the unrearranged DQ52 gene segment "replaced" by a DJ_H junction. As described in more detail in the next section, we hypothesize that selective activation of DJ_H junctions primes them for V_H recombination.

Chromatin structural changes that impact DJ_H junctions are dependent on Eμ (Subrahmanyam et al., 2012). We have proposed that D_H

recombination brings the associated promoter under the influence of Eμ leading to its epigenetic activation and transcription. As noted above, in the case of the germline locus, Eμ activity with regard to these functions appears to be relatively local, which also explains why DJ_H junction activation does not spread to germline D_H segments on recombined alleles. The role of Eμ in directing changes at DJ_H junctions and thereby affecting V_H recombination has received further support from recent studies in which the imprinting control region (ICR) from the *IgF2/H19* locus was inserted into the *IgH* locus between J_H and Eμ (Puget et al., 2015). This CTCF-binding *cis*-regulatory sequence is known to have enhancer-blocking (insulator) activity at its normal location (Ferguson-Smith, 2011; Singh, Lee, & Szabo, 2012). The ICR-modified *IgH* was unimpaired for D_H recombination; however, V_H recombination was significantly reduced. Analysis of DJ_H junction transcription and DNA methylation showed that both were impaired on ICR-containing alleles, leading to the proposal that the ICR blocked Eμ-dependent activation of DJ_H junctions for subsequent V_H recombination.

5. V_H REGION (5′ *IgH* DOMAIN)

5.1 Organization of V_H Region

Starting approximately 100 kb 5′ of DFL16.1, the largest part of the *IgH* locus encodes V_H gene segments (Fig. 5). Johnston et al.'s map of the V_H region in C57BL/6 mice is, at present, the gold standard for organization of V_H gene segments (Johnston et al., 2006). In this strain, there are 195 V_H gene segments of which 110 are considered to be functional based on the presence of recognizable RSSs and no translational termination codons in the coding region. The remaining pseudogenes either have defective RSSs (therefore cannot rearrange) or have premature termination codons (therefore cannot encode functional heavy chain protein). V_H gene segments have been organized into 16 gene families based on DNA sequence similarity. Members of the V_H7183 and V_HQ52 gene families are interspersed within approximately 300 kb and lie closest to the 3′ *IgH* domain (Fig. 5). These gene families are referred to as proximal V_H genes. The largest family of V_H genes comprises the V_HJ558 family. Functional- and pseudo-V_HJ588 gene segments account for 89 gene segments that span a little over 1.5 Mb at the 5′ end of the V_H locus. 16 members of the 3609 gene family are interspersed among V_HJ558 genes in the 5′ 1 Mb. V_HJ558 genes are referred to as distal V_H genes. The intervening 0.6 Mb between

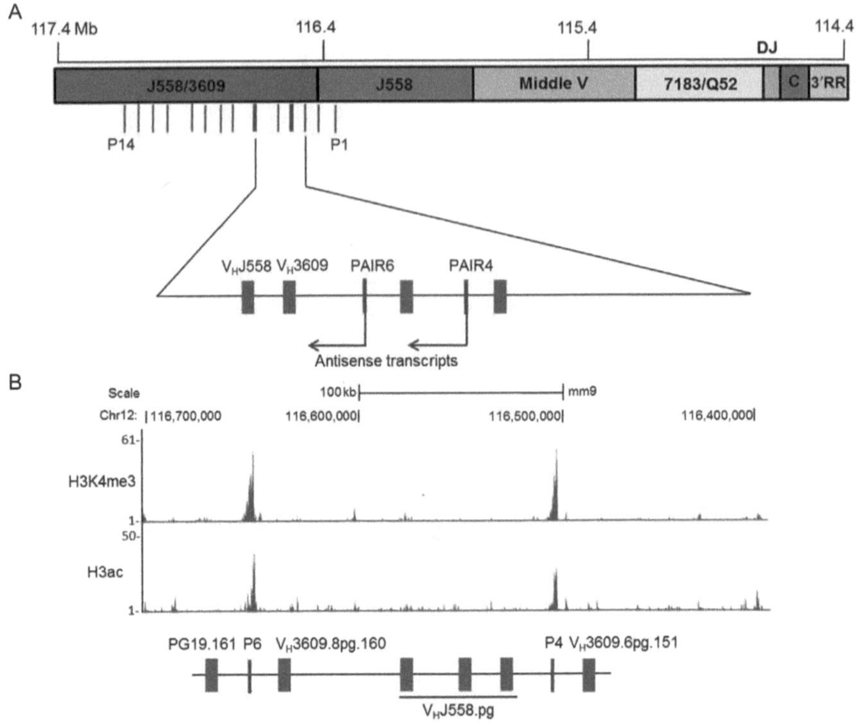

Figure 5 Transcriptional and epigenetic features of the V_H region. (A) Scale representation of the murine *IgH* locus. The distal V_H region contains 14 Pax5-activated intergenic repeat elements (PAIR elements, P1–P14). PAIR elements are shown as red lines within distal V_H gene segments. The region between PAIR3 and PAIR6 is expanded to show that PAIR4 and PAIR6 are sites of antisense transcription and contain activation-specific histone modifications. (B) Genome browser track of H3K4me3 and H3ac Chip-seq in distal part of the locus (Chip-seq track is derived from Choi et al. (2013)). Schematics below the tracks show relative disposition of P4/P6 and the most closely located V_H gene segments (approximately to scale). All six indicated gene segments are pseudogenes. (See the color plate.)

V_HJ558/3609 and V_H7183/Q52 gene families encodes most of the 12 intermediate V_H gene families, each of which has two to four family members (inclusive of pseudogenes). Most functional V_H gene segments have been shown to rearrange, though frequency of rearrangements varies considerably (see below).

30 kb upstream of the 5′-most V_H gene segments, Brodeur and colleagues identified a cluster of B lineage-specific DHSs (Pawlitzky et al., 2006). These sites may serve as the 5′ boundary of the *IgH* locus. However, deletion of these sites had relatively minor effects on *IgH* gene assembly and

early B cell development (Perlot et al., 2010), reminiscent of the phenotype of ΔHS5-7 mice described above. The function of these putative regulatory elements remains to be defined.

The distal V_H region also contains a conserved sequence motif of approximately 470 bp, known as Pax5-associated intergenic repeat (PAIR) (Ebert et al., 2011). There are 14 PAIR elements spread over approximately 1 Mb (Fig. 5A). As implied by its name, these regions were identified as Pax5-binding regions. However, extended homology beyond the Pax5-binding site led to the discovery that a subset of PAIR elements also had consensus binding sequences for the transcription factors E2A and CTCF. Analysis of PAIR 4 and 7 by transfection assays showed that these elements had promoter activity oriented in the antisense direction (that is, transcripts extended toward the telomere). These observations led to the proposal that PAIR elements were the major source of antisense transcription that had been previously discovered in the V_HJ558 region (Bolland et al., 2004). More recent RNA-Seq analysis of primary pro-B cells confirm that PAIR4 and PAIR6 have the highest levels of antisense transcripts in the entire V_H region (Verma-Gaur et al., 2012). As the predominant sites of Pax5 interaction in the V_H region, PAIR elements have also been proposed to underpin Pax5-dependent locus contraction (see below).

5.2 Regulation of V_H Recombination

Four levels of regulation impact the second step of *IgH* gene assembly. First, V_H recombination always follows D_H recombination on WT alleles. Second, V_H recombination occurs precisely to the RSS corresponding to the DJ_H junction, despite availability of multiple unrearranged D_H-associated RSSs 5′ of the DJ_H junction. Third, V_H recombination occurs preferentially on DJ_H rearranged alleles. Fourth, V_H recombination is subject to feedback inhibition after generation of a functional VDJ recombinant allele. Based on epigenetic studies of DJ_H rearranged alleles, we proposed the following working model for the developmental timing and precision of V_H recombination (Subrahmanyam et al., 2012).

DJ_H junctions are selectively marked by high levels of activation-associated histone modification (such as H3/H4ac and H3K4me3), transcription from the D_H-associated promoter, CpG demethylation, and increased sensitivity to DNase I. All these characteristics are highly localized and do not spread to unrearranged D_Hs located further 5′. Additionally, RAG1/2 protein density is maximal at the DJ_H junction, and also does not extend to unrearranged D_H gene segments that lie 5′ to the DJ_H junction. Since the RAG1/2 RC is located over the J_H gene segments on germline *IgH* alleles and does not extend to unrearranged D_H gene segments, this is the first chance for a D_H-associated RSS to initiate the recombination reaction. In other words, D_H recombination brings a D_H-RSS into the RC and positions it to initiate recombination. As a consequence, V_H to DJ_H recombination always follows D_H recombination. Since the RAG1/2 density does not extend to 5′ unrearranged D_H gene segments, only the DJ_H-associated RSS can participate in this step. As a consequence, V_H gene segments from as far as 2 Mb recombine only to DJ_H junctions.

We imagine that RAG1/2 recognition of the DJ_H-RSS leads to nicking, followed by search for a 12/23 compatible RSS which, in this case, would be one associated with a V_H gene segment. Once such an RSS is found, synapsis and double-strand DNA breaks are introduced, followed by resolution of the double-stand break by non-homologous end joining (NHEJ) to complete VDJ recombination. The only exception to D_H-RSS availability prior recombination may be the RSS associated with DQ52. This RSS lies very close to, if not within, the J_H-recombination center and could potentially initiate recombination on germline alleles. Because of the three-loop structure of WT alleles, however, the only compatible RSSs available would be those of J_H gene segments, leading to DQ52–J_H joining. That is, reaction products of J_H RSS-initiated recombination or DQ52-RSS-initiated recombination of WT *IgH* alleles are indistinguishable.

The products can be very different if a V_H-RSS is available to the reaction initiated at DQ52. Such a mechanism may explain two situations where V_H gene segments have been found to rearrange to DQ52. In the first example, a V_H gene segment was inserted by homologous recombination just 5′ of DFL16.1 in the *IgH* locus (Bates, Cado, Nolla, & Schlissel, 2007). This gene segment recombined with unrearranged DQ52 in pro-B cells as well as in $CD4^+CD8^+$ thymocytes. Current conformational views of the germline *IgH* locus indicate that the inserted V_H gene segment is located next to IGCR1, and therefore, close to the base of the Eμ–IGCR1 loop. This spatial position places it close to the DQ52–J_H-associated RC within the D_H loop, thereby making it available for synapsis with DQ52-nicked RSS for recombination. This model also explains why the inserted V_H gene recombines predominantly to DQ52. As described above, IGCR1-mutated *IgH* alleles also undergo V_H to DQ52 rearrangements. In this case, a specific V_H gene segments V_H7183.2.3 is brought closer to the DQ52–J_H-associated RC by alternative looping, leading to its use during the first step of *IgH* gene assembly.

The unique epigenetic status of DJ_H recombinant alleles provide a plausible hypothesis for several regulatory aspects of V_H recombination as described above. The model does not explain, however, the mechanism by which RAG proteins located near the DJ_H-associated RSS find a complementary V_H-associated RSS to synapse with to complete the cleavage reaction. The question of how V_H gene segments interact with DJ_H has been explored via several independent lines of investigation that we synthesize into a working model.

5.3 V_H Gene Choice

An essential aspect of this inquiry has been to determine whether some parts of the V_H locus are more prone to recombine. V_H recombination frequency was recently reevaluated in a large scale and unbiased manner by deep-sequencing VDJ junctions from pro-B cell mRNA (Choi et al., 2013). The results show that a small subset of V_H gene segments (<20) are represented at particularly high frequency in the pro-B cell repertoire. These gene segments are found within proximal V_H7183, distal V_HJ558, as well as intervening V_H gene families, resulting in considerable variation of gene utilization even within gene families. Interestingly, the location of even the highest recombining gene segments did not correlate with sites of V_H gene transcription. To understand the basis for differential V_H gene utilization, Choi et al. correlated recombination frequency with an array of epigenetic marks and CTCF binding using a newly developed computational algorithm. The analysis led to the definition of four domains in the V_H region. Domains 1 and 2 correspond closely with proximal and intervening regions, constituting V_H7183 and non-V_HJ558/3609 genes, respectively. The larger V_HJ558/3609 domain gets further subdivided into a 3′ portion (of approximately 0.5 Mb) that contains mostly V_HJ558 genes and a 5′ portion (of approximately 1 Mb) that contains remaining V_HJ558 genes and interspersed V_H3609 genes.

The epigenetic characteristics of each domain are quite distinct. This is especially true for domains 1 and 4. Domain 1 has lowest levels of activation-associated histone modifications (such as H3ac, H3K4me2, and H3K4me3) and is, instead, marked by "repressive" H3K27me3. In contrast, domain 4 has highest levels of activating histone modifications (among V_H gene segments) and no detectable H3K27me3. Additionally, V_H gene segments in domain 1 have closely positioned CTCF-binding sites (Lucas, Bossen, & Murre, 2011). That both domains 1 and 4 have highly rearranging genes, yet are disparately epigenetically marked indicates that neither histone modifications nor transcriptional activity *per se* are dominant determinants of V_H recombination choice.

This is in marked contrast to features of the 3′ *IgH* domain, where presence of activating epigenetic marks and sterile transcription coincide closely with RAG1/2 binding and D_H recombination. However, there are qualitative and quantitative differences in the transcriptional and epigenetic status of 3′ *IgH* domain and the V_H region. First, levels of H3ac and H3K4me3 are much higher in the DQ52–J_H4 region compared to V_H region. It is possible that the levels in the V_H region may not reach a threshold required to recruit RAG1/2 proteins. Second, transcriptional profiles in the 3′ *IgH* region are quite different from sterile V_H transcription in terms of level as well as orientation. As discussed above, the DQ52–J_H region contains bidirectional transcripts through exactly the same genomic region that has high H3ac/H3K4me3. These characteristics may impart special characteristics to DQ52–J_H that may be essential for RAG1/2 recruitment. In contrast, sterile transcription in the V_H region at P4 and P6 occurs at very low levels and is oriented only in the antisense direction. Third, the presence of clustered RSSs corresponding to DQ52 and J_H1-4 has been proposed to be critical for RAG1/2 recruitment to form the RC (Schatz & Ji, 2011). The only sites of H3K4me3 and sterile transcription in the V_H locus (P4 and P6) are not near a cluster of RSSs, thereby precluding RAG1/2 recruitment to this region (Fig. 5B).

While these differences between the 3′ *IgH* and V_H regions may account for the distinct epigenetic characteristics of the two domains, what can we infer about regulation of recombination? An insight of the RC concept was that antigen receptor genes apparently needed only one RC to be competent to rearrange. That is, RAG1/2 needed to bind (stably) to only one of a pair of RSSs for effective recombination. However, the possibility that RAG1/2 also bound to different V_H gene sequences in individual cells within a population could not be ruled out. Beyond simply recruiting RAG2, interaction of RAG2 with H3K4me3 has been shown to allosterically activate recombinase activity (Lu, Ward, Bettridge, Liu, & Desiderio, 2015; Shimazaki, Tsai, & Lieber, 2009). This might suggest that H3K4me3 would be required at both RSSs to initiate recombination. An intriguing possibility from the relationship between epigenetic status and V_H utilization is that H3K4me3 may not be needed close to the V_H-RSS in order to promote efficient recombination. This could be because RAG2 binding to H3K4me3 associated with one RSS is sufficient to activate the complex for subsequent synapsis with a complementary RSS to initiate DNA cleavage. However, we note the caveat that transient, or cell-specific differences in H3K4me3, at V_H gene segments cannot be ruled out from the current data. Thus, what constitutes RAG1/2 accessibility at the V_H locus remains an open question.

5.4 Spatial Organization of V_H Gene Segments

Regardless of the mechanistic relationship between epigenetic marks, transcriptional status and V_H recombination, one step that is essential for the reaction to proceed is for V_H gene segments to come into physical proximity of the DJ_H part of the locus. This was first visualized by Kosak et al. (2002), using FISH and BAC probes specific to the 5′ and 3′ ends of the *IgH* locus. They showed that FISH signals from these widely separated BACs colocalized in nuclei of pro-B cells but not in non-B cells, suggesting that the two ends of the locus were in spatial proximity. This phenomenon is referred to as locus compaction or contraction. Its relevance for antigen receptor gene assembly is inferred from the strong correlation between absence of compaction and lack of distal V_H recombination in several genetically mutant mouse strains. These include strains deficient for the transcription factors Pax5 (Fuxa et al., 2004), E2A (Sayegh et al., 2005), Ikaros, and YY1 (Liu, Schmidt-Supprian, et al., 2007; Reynaud et al., 2008). Because E2A and Ikaros deficiencies have broad effects on B cell development, effects specific to the *IgH* locus are difficult to deconvolute. However, in Pax5- and YY1-deficient mice, B cell development till the pro-B cell stage is not severely affected, yet distal V_H recombination and locus compaction are significantly attenuated. These observations implicate these two transcription factors in mediating locus compaction. In contrast to the close correlation between locus decompaction and lack of distal V_H gene recombination in Pax5- and YY1-deficient pro-B cells, a recent study showed that distal V_H recombination was severely reduced, despite normal locus compaction on *IgH* alleles that carry point mutations in CTCF-binding sites within the IGCR1 (Medvedovic et al., 2013). This is the first "inconsistency" between locus compaction (as assayed by microscopy) and V_H gene utilization and suggests that locus compaction is not sufficient to ensure effective utilization of distal V_H gene segments.

6. INTERACTION BETWEEN 5′ AND 3′ *IgH* DOMAINS

A high-resolution FISH-based structure for the *IgH* locus was first undertaken by Jhunjhunwala et al. (2008). Using spatial distance measurements derived from FISH and computational modeling, they proposed that the *IgH* locus matched most closely to the multiloop subcompartment (MLS) model of chromatin folding in E2A-deficient pre-pro-B cells. Each subcompartment was imagined as a rosette, having a base from which multiple chromatin loops emanated. The entire locus comprised several such rosettes, which provides a level of compaction. Similar studies in E2A-sufficient pro-B cells revealed that the rosettes were spatially "compressed" so that distal and proximal V_H genes were at comparable distance from the 3′ end of the locus. The altered configuration between pre-pro and pro-B cells was seen as making the large repertoire of germline V_H genes equally available for recombination.

The first indication of how the locus transitioned between these two states came from studies that revealed that Eμ contacts sequences located at the 3′ end of V_HJ558 genes and 5′ end of V_H7183 genes (Guo, Gerasimova, et al., 2011). Such interactions were visualized as drawing-in the distal V_H gene segments toward the DJ_H part of the locus. Eμ has also been shown to loop to P4/P6 located in the 5′ part of the V_HJ558 genes (Verma-Gaur et al., 2012), suggesting that Eμ contacts several sites in the V_H region. These interactions are lost on Eμ-deleted *IgH* alleles implicating a direct role of Eμ in mediating long-range contacts. Since E2A is likely to be an essential component of Eμ activity, the MLS model for the *IgH* locus in E2A-deficient pre-pro-B cells may be the one in which Eμ-dependent locus compaction has not occurred. However, the state of the locus in E2A-deficient pre-pro-B cells may also reflect Eμ-independent features of locus compactions that are developmental stage-specific. The simplest model of how Eμ mediates this form of locus compaction is via chromosome looping. The transcription factor YY1 has been proposed to be the key Eμ-binding protein for this function (Guo, Gerasimova, et al., 2011). Accordingly, Eμ interaction with sites in the V_HJ558 genes as well as with the 5′7183 site is disrupted in YY1-deficient pro-B cells (unpublished observation). Site-specific mutation of the YY1-binding site in Eμ is underway to further confirm or refute this model.

An intriguing alternative hypothesis for Eμ-dependent looping, especially to P4/P6, has been put forward by Verma-Gaur et al. (2012). Noting that both "ends" of this loop have transcriptional activity, they proposed that interaction between these elements could reflect their independent recruitment to the same transcription factory (Cook, 2010; Schoenfelder, Clay, & Fraser, 2010), in order to activate transcription. Increased interaction between these sequences, as scored by chromosome conformation capture assays (3C), would be because they were located in the same factory, rather than because of direct interaction via looping. This model was supported by the observation that transcriptional inhibition resulted in reduced interaction frequency between Eμ and P4/P6, presumably because one or both elements dropped out of the shared transcription factory. Such a mechanism would not apply to the observed looping between Eμ and 3′558 or 5′7183 because these sites do not have significant transcriptional activity. The relationships between eRNAs produced at Eμ and looping of this element to different sites in the V_H region remains to be explored.

Eμ contacts with specific sites in the V_H locus provide, at best, the means to recruit a few V_H gene segments that lie nearby to the 3′ *IgH* domain. These interactions do not readily explain how the majority of V_H genes are used for recombination. Moreover, V_H genes that recombine at highest levels are not located close to the identified sites of Eμ contact. These considerations indicate that there must be other levels of structural organization in the V_H region. Based on 4C assays carried out with genomic DNA that was immunoprecipitated with anti-CTCF antibodies, Guo et al. found that the V_H locus contained multiple looping interactions mediated by CTCF (Guo, Gerasimova, et al., 2011). Examples of such loops are found in both the proximal and distal V_H genes. These interactions are B lineage-specific but Eμ-independent, suggesting a structural hierarchy in establishing *IgH* locus conformation. These CTCF-dependent domains may form the rosettes postulated by Murre and colleagues based on FISH studies (Lucas et al., 2011).

Our working model is that the V_H region folds into CTCF-dependent domains whose size ranges in the hundreds of kilobases (the prototypical loops identified by anti-CTCF ChIP-loop are 250–400 kb). Together with structural features of the 3′ *IgH* domain, these loops provide the first level of chromatin compaction to the *IgH* locus. Since the epigenetic and transcriptional state of the V_H region is quite distinct from the 3′ *IgH* domain, there is little in common between these structural units other than size. Moreover, while the 3′ *IgH* domain has limited structural flexibility, CTCF domains in the V_H region may be much more dynamic. Though the three-loop configuration at the 3′ *IgH* domain is dynamic to the extent that it can break apart and re-form, the major sites of interaction remain the same. Whereas, because there are many CTCF-bound sites in the V_H region, when a CTCF loop breaks apart each CTCF site may loop to a different CTCF-bound site to form a different loop. In other words, CTCF loops in the V_H region may not be discrete entities, but represent an array of structures that share the common feature of having CTCF bound at the bases of the loops. We hypothesize that Eμ-dependent steps of locus compaction work on this prefolded structure via specific interaction sites discussed above. Because of the prefolded state each Eμ-dependent interaction brings more than just a single V_H gene segment associated with each Eμ-interacting site (such as 5′ 7183, 3′ 558, and P4/P6) to the 3′ end of the locus. In this way, many more V_H gene segments gain access to the DJ_H part for each Eμ-dependent interaction, ultimately resulting in relatively uniform utilization of all V_H genes for recombination.

This Eμ-centric view of *IgH* locus compaction has been challenged by Medvedovic et al. (2013). Using extensive 4C-seq and FISH assays to probe chromosome structure these authors found no role for Eμ in *IgH* locus compaction. Since some decompaction was evident on IGCR1-mutated alleles (that lack CTCF binding to IGCR1) and the largest effect was that of Pax5 deficiency, Medvedovic et al. proposed that locus compaction is driven by protein–protein interactions between Pax5 and CTCF. Lack of specific looping sites with the V_H region was interpreted to indicate a "continuum of flexible loops" in this part of the locus that permits all V_H genes to recombine comparably. The basis for differences in effects of Eμ need further study.

The latest advance in understanding how V_H genes interact with the 3′ *IgH* domain comes from live cell imaging and computational modeling of *IgH* locus movements (Lucas, Zhang, Dudko, & Murre, 2014). For this, *IgH* alleles were engineered to contain many copies of tetracycline operator (TetO) sequences either in the J_H–Cμ intron or within the V_HJ558 genes. Recruitment of green fluorescent protein (GFP) fused to the tetracycline repressor DNA-binding domain to these TetO repeats permitted visualization of *IgH* alleles in primary pro-B cells to determine kinetic parameters. Incorporation of these parameters into dynamic modeling showed that the V_H region comes close to the DJ_H part of the locus in order of minutes. Lucas et al. proposed that these fly-by encounters create circumstances where V_H gene segments can be captured for synapsis by recombinase machinery located at DJ_H junctions.

7. OVERVIEW AND FUTURE DIRECTIONS

In summary, multiple levels of epigenetic events activate the *IgH* locus in pro-B cells. The first and most mysterious is Eμ-independent, but B lineage-specific, activation that likely occurs while the locus is still at the nuclear periphery. This step is accompanied by removal of repressive

H3K9me2/3 from selected parts, deposition of H3K4me2, and accessibility of ubiquitously expressed transcription factors such as CTCF to the locus. We hypothesize that this step also permits protein binding to tissue-specific regulatory sequences such as Eμ, the DQ52 promoter, IGCR1, and the PAIR elements. Transcriptional activation leads to additional histone modifications and 3D conformational changes that fully establish the pre-rearrangement configuration of the locus. Whether or not transcriptional activity contributes to locus conformation is an intriguing and important question that requires further study. Special features of the 3′ *IgH* domain result in generation of a RAG1/2-rich RC, and the locus is poised to initiate VDJ recombination. With V_H genes sequestered away from the three-loop configured 3′ *IgH* domain, RSSs associated with J_Hs, DFL16.1, and DQ52 are the most readily available to RAG proteins, providing a plausible mechanism for D_H recombination occurring first and the preponderant use of DFL16.1 and DQ52 in VDJ recombined alleles. Formation of DJ_H junctions relocates 5′ D_H-RSSs to the RC thereby permitting RAG1/2 to complete *IgH* gene assembly by recombining V_H genes to DJ_H junctions. We propose that secondary availability of 5′ D_H-RSSs for recombination may explain why V_H rearrangements always follow D_H rearrangements. Since Eμ/IGCR1 interactions prevent precocious V_H rearrangements, and IGCR1 remains intact on all DJ_H recombined alleles, we imagine that IGCR1's Eμ sequestration function must be attenuated to get effective V_H recombination. How this happens is unclear. Finally, the mechanism by which RAG1/2 associated with DJ_H-RSSs synapse with V_H RSSs located 0.1–2 Mb away remains to be understood. While the 3D structure of the locus likely serves as an essential scaffold, chemical bond length proximity of paired RSSs ultimately relies on molecular motion. Such studies are just beginning and likely to be a major direction in the future.

ACKNOWLEDGMENT

This work was supported entirely by the Intramural Research Program of the NIH, National Institute on Aging, Baltimore, MD.

REFERENCES

Afshar, R., Pierce, S., Bolland, D. J., Corcoran, A., & Oltz, E. M. (2006). Regulation of IgH gene assembly: Role of the intronic enhancer and 5′ DQ52 region in targeting DHJH recombination. *Journal of Immunology*, *176*(4), 2439–2447.

Allfrey, V. G., Faulkner, R., & Mirsky, A. E. (1964). Acetylation+methylation of histones+their possible role in regulation of RNA synthesis. *Proceedings of the National Academy of Sciences of the United States of America*, *51*(5), 786–794.

Ardehali, M. B., Mei, A., Zobeck, K. L., Caron, M., Lis, J. T., & Kusch, T. (2011). Drosophila Set1 is the major histone H3 lysine 4 trimethyltransferase with role in transcription. *The EMBO Journal*, *30*(14), 2817–2828.

Arthur, R. K., Ma, L. J., Slattery, M., Spokony, R. F., Ostapenko, A., Negre, N., et al. (2014). Evolution of H3K27me3-marked chromatin is linked to gene expression evolution and to patterns of gene duplication and diversification. *Genome Research*, *24*(7), 1115–1124.

Atkinson, M. J., Chang, Y. H., Celler, J. W., Huang, C., Paige, C. J., & Wu, G. E. (1994). Overusage of mouse DH gene segment, DFL16.1, is strain-dependent and determined by cis-acting elements. *Developmental Immunology*, *3*(4), 283–295.

Azzaz, A. M., Vitalini, M. W., Thomas, A. S., Price, J. P., Blacketer, M. J., Cryderman, D. E., et al. (2014). Human heterochromatin protein 1 alpha promotes nucleosome associations that drive chromatin condensation. *Journal of Biological Chemistry*, *289*(10), 6850–6861.

Bain, G., Maandag, E. C. R., Riele, H., Feeney, A. J., Sheehy, A., Schlissel, M., et al. (1997). Both E12 and E47 allow commitment to the B cell lineage. *Immunity*, *6*(2), 145–154.

Bangs, L. A., Sanz, I. E., & Teale, J. M. (1991). Comparison of DH, JH, and junctional diversity in the fetal, adult, and aged B-cell repertoires. *Journal of Immunology*, *146*(6), 1996–2004.

Bannister, A. J., & Kouzarides, T. (2011). Regulation of chromatin by histone modifications. *Cell Research*, *21*(3), 381–395.

Barski, A., Cuddapah, S., Cui, K. R., Roh, T. Y., Schones, D. E., Wang, Z. B., et al. (2007). High-resolution profiling of histone methylations in the human genome. *Cell*, *129*(4), 823–837.

Bates, J. G., Cado, D., Nolla, H., & Schlissel, M. S. (2007). Chromosomal position of a V-H gene segment determines its activation and inactivation as a substrate for V(D)J recombination. *Journal of Experimental Medicine*, *204*(13), 3247–3256.

Bebin, A. G., Carrion, C., Marquet, M., Cogne, N., Lecardeur, S., Cogne, M., et al. (2010). In vivo redundant function of the 3′ IgH regulatory element Hs3b in the mouse. *Journal of Immunology*, *184*(7), 3710–3717.

Beisel, C., & Paro, R. (2011). Silencing chromatin: Comparing modes and mechanisms. *Nature Reviews. Genetics*, *12*(2), 123–135.

Bernstein, B. E., Kamal, M., Lindblad-Toh, K., Bekiranov, S., Bailey, D. K., Huebert, D. J., et al. (2005). Genomic maps and comparative analysis of histone modifications in human and mouse. *Cell*, *120*(2), 169–181.

Birney, E., Stamatoyannopoulos, J. A., Dutta, A., Guigo, R., Gingeras, T. R., Margulies, E. H., et al. (2007). Identification and analysis of functional elements in 1% of the human genome by the encode pilot project. *Nature*, *447*(7146), 799–816.

Birshtein, B. K. (2014). Epigenetic regulation of individual modules of the immunoglobulin heavy chain locus 3′ regulatory region. *Frontiers in Immunology*, *5*, 1–9.

Black, J. C., Van Rechem, C., & Whetstine, J. R. (2012). Histone lysine methylation dynamics: Establishment, regulation, and biological impact. *Molecular Cell*, *48*(4), 491–507.

Bolland, D. J., Wood, A. L., Afshar, R., Featherstone, K., Oltz, E. M., & Corcoran, A. E. (2007). Antisense intergenic transcription precedes Igh D-to-J recombination and is controlled by the intronic enhancer Emu. *Molecular and Cellular Biology*, *27*(15), 5523–5533.

Bolland, D. J., Wood, A. L., Johnston, C. M., Bunting, S. F., Morgan, G., Chakalova, L., et al. (2004). Antisense intergenic transcription in V(D)J recombination. *Nature Immunology*, *5*(6), 630–637.

Bossen, C., Murre, C. S., Chang, A. N., Mansson, R., Rodewald, H. R., & Murre, C. (2015). The chromatin remodeler Brg1 activates enhancer repertoires to establish B cell identity and modulate cell growth. *Nature Immunology*, *16*(7), 775–784.

Bushey, A. M., Dorman, E. R., & Corces, V. G. (2008). Chromatin insulators: Regulatory mechanisms and epigenetic inheritance. *Molecular Cell, 32*(1), 1–9.

Busslinger, M., & Urbanek, P. (1995). The role of BSAP (Pax-5) in B-cell development. *Current Opinion in Genetics & Development, 5*(5), 595–601.

Cao, R., Wang, L. J., Wang, H. B., Xia, L., Erdjument-Bromage, H., Tempst, P., et al. (2002). Role of histone H3 lysine 27 methylation in polycomb-group silencing. *Science, 298*(5595), 1039–1043.

Carninci, P., Kasukawa, T., Katayama, S., Gough, J., Frith, M. C., Maeda, N., et al. (2005). The transcriptional landscape of the mammalian genome. *Science, 309*(5740), 1559–1563.

Chakraborty, T., Chowdhury, D., Keyes, A., Jani, A., Subrahmanyam, R., Ivanova, I., et al. (2007). Repeat organization and epigenetic regulation of the DH-Cmu domain of the immunoglobulin heavy-chain gene locus. *Molecular Cell, 27*(5), 842–850.

Chakraborty, T., Perlot, T., Subrahmanyam, R., Jani, A., Goff, P. H., Zhang, Y., et al. (2009). A 220-nucleotide deletion of the intronic enhancer reveals an epigenetic hierarchy in immunoglobulin heavy chain locus activation. *Journal of Experimental Medicine, 206*(5), 1019–1027.

Chang, Y., Paige, C. J., & Wu, G. E. (1992). Enumeration and characterization of DJH structures in mouse fetal liver. *The EMBO Journal, 11*(5), 1891–1899.

Chen, J. Z., & Alt, F. W. (1993). Gene rearrangement and B-cell development. *Current Opinion in Immunology, 5*(2), 194–200.

Chi, P., Allis, C. D., & Wang, G. G. (2010). Covalent histone modifications—Miswritten, misinterpreted and mis-erased in human cancers. *Nature Reviews. Cancer, 10*(7), 457–469.

Choi, N. M., & Feeney, A. J. (2014). CTCF and ncRNA regulate the three-dimensional structure of antigen receptor loci to facilitate V(D)J recombination. *Frontiers in Immunology, 5*, 1–8.

Choi, N. M., Loguercio, S., Verma-Gaur, J., Degner, S. C., Torkamani, A., Su, A. I., et al. (2013). Deep sequencing of the murine IgH repertoire reveals complex regulation of nonrandom V gene rearrangement frequencies. *Journal of Immunology, 191*(5), 2393–2402.

Chowdhury, D., & Sen, R. (2001). Stepwise activation of the immunoglobulin mu heavy chain gene locus. *The EMBO Journal, 20*(22), 6394–6403.

Chowdhury, D., & Sen, R. (2003). Transient IL-7/IL-7R signaling provides a mechanism for feedback inhibition of immunoglobulin heavy chain gene rearrangements. *Immunity, 18*(2), 229–241.

Chu, C., Qu, K., Zhong, F. L., Artandi, S. E., & Chang, H. Y. (2011). Genomic maps of long noncoding RNA occupancy reveal principles of RNA-chromatin interactions. *Molecular Cell, 44*(4), 667–678.

Cook, P. R. (2010). A model for all genomes: The role of transcription factories. *Journal of Molecular Biology, 395*(1), 1–10.

Corcoran, A. E., & Feeney, A. J. (2012). Lymphocyte development. *Current Opinion in Immunology, 24*(2), 129–131.

Corcoran, A. E., Riddell, A., Krooshoop, D., & Venkitaraman, A. R. (1998). Impaired immunoglobulin gene rearrangement in mice lacking the Il-7 receptor. *Nature, 391*(6670), 904–907.

Creyghton, M. P., Cheng, A. W., Welstead, G. G., Kooistra, T., Carey, B. W., Steine, E. J., et al. (2010). Histone H3K27ac separates active from poised enhancers and predicts developmental state. *Proceedings of the National Academy of Sciences of the United States of America, 107*(50), 21931–21936.

Dariavach, P., Williams, G. T., Campbell, K., Petersson, S., & Neuberger, M. S. (1991). The mouse IgH 3′-enhancer. *European Journal of Immunology, 21*(6), 1499–1504.

De Ruijter, A. J. M., Van Gennip, A. H., Caron, H. N., Kemp, S., & Van Kuilenburg, A. B. P. (2003). Histone deacetylases (HDACs): Characterization of the classical HDAC family. *Biochemical Journal, 370*, 737–749.

Deal, R. B., & Henikoff, S. (2010). Catching a glimpse of nucleosome dynamics. *Cell Cycle, 9*(17), 3389–3390.

Dekker, J. (2003). A closer look at long-range chromosomal interactions. *Trends in Biochemical Sciences, 28*(6), 277–280.

Dekker, J., Rippe, K., Dekker, M., & Kleckner, N. (2002). Capturing chromosome conformation. *Science, 295*(5558), 1306–1311.

Dinger, M. E., Amaral, P. P., Mercer, T. R., Pang, K. C., Bruce, S. J., Gardiner, B. B., et al. (2008). Long noncoding RNAs in mouse embryonic stem cell pluripotency and differentiation. *Genome Research, 18*(9), 1433–1445.

Dixon, J. R., Jung, I., Selvaraj, S., Shen, Y., Antosiewicz-Bourget, J. E., Lee, A. Y., et al. (2015). Chromatin architecture reorganization during stem cell differentiation. *Nature, 518*(7539), 331–336.

Dixon, J. R., Selvaraj, S., Yue, F., Kim, A., Li, Y., Shen, Y., et al. (2012). Topological domains in mammalian genomes identified by analysis of chromatin interactions. *Nature, 485*(7398), 376–380.

Dudley, D. D., Sekiguchi, J., Zhu, C. M., Sadofsky, M. J., Whitlow, S., DeVido, J., et al. (2003). Impaired V(D)J recombination and lymphocyte development in core Rag1-expressing mice. *Journal of Experimental Medicine, 198*(9), 1439–1450.

Dunnick, W. A., Shi, J., Graves, K. A., & Collins, J. T. (2005). The 3′ end of the heavy chain constant region locus enhances germline transcription and switch recombination of the four gamma genes. *Journal of Experimental Medicine, 201*(9), 1459–1466.

Dunnick, W. A., Shi, J., Zerbato, J. M., Fontaine, C. A., & Collins, J. T. (2011). Enhancement of antibody class-switch recombination by the cumulative activity of four separate elements. *Journal of Immunology, 187*(9), 4733–4743.

Ebert, A., McManus, S., Tagoh, H., Medvedovic, J., Salvagiotto, G., Novatchkova, M., et al. (2011). The distal V-H gene cluster of the Igh locus contains distinct regulatory elements with Pax5 transcription factor-dependent activity in pro-B cells. *Immunity, 34*(2), 175–187.

Eckner, R., Ewen, M. E., Newsome, D., Gerdes, M., Decaprio, J. A., Lawrence, J. B., et al. (1994). Molecular-cloning and functional anal of the adenovirus E1A-associated 300-KD protein (P300) reveals a protein with properties of a transcriptional adapter. *Genes & Development, 8*(8), 869–884.

Ehlich, A., Martin, V., Muller, W., & Rajewsky, K. (1994). Analysis of the B-cell progenitor compartment at the level of single cells. *Current Biology, 4*(7), 573–583.

Farh, K. K. H., Marson, A., Zhu, J., Kleinewietfeld, M., Housley, W. J., Beik, S., et al. (2015). Genetic and epigenetic fine mapping of causal autoimmune disease variants. *Nature, 518*(7539), 337–343.

Featherstone, K., Wood, A. L., Bowen, A. J., & Corcoran, A. E. (2010). The mouse immunoglobulin heavy chain V-D intergenic sequence contains insulators that may regulate ordered V(D)J recombination. *Journal of Biological Chemistry, 285*(13), 9327–9338.

Feeney, A. J. (2011). Epigenetic regulation of antigen receptor gene rearrangement. *Current Opinion in Immunology, 23*(2), 171–177.

Ferguson-Smith, A. C. (2011). Genomic imprinting: The emergence of an epigenetic paradigm. *Nature Reviews. Genetics, 12*(8), 565–575.

Finlan, L. E., Sproul, D., Thomson, I., Boyle, S., Kerr, E., Perry, P., et al. (2008). Recruitment to the nuclear periphery can alter expression of genes in human cells. *PLoS Genetics, 4*(3), 1–13.

Fuxa, M., Skok, J., Souabni, A., Salvagiotto, G., Roldan, E., & Busslinger, M. (2004). Pax5 induces V to DJ rearrangements and locus contraction of the immunoglobulin heavy-chain gene. *Genes & Development, 18*(4), 411–422.

Garel, A., & Axel, R. (1976). Selective digestion of transcriptionally active ovalbumin genes from oviduct nuclei. *Proceedings of the National Academy of Sciences of the United States of America, 73*(11), 3966–3970.

Gaszner, M., & Felsenfeld, G. (2006). Insulators: Exploiting transcriptional and epigenetic mechanisms. *Nature Reviews. Genetics*, 7(9), 703–713.

Gauss, G. H., & Lieber, M. R. (1992). The basis for the mechanistic bias for deletional over inversional V(D)J recombination. *Genes & Development, 6*(8), 1553–1561.

Giallourakis, C. C., Franklin, A., Guo, C. G., Cheng, H. L., Yoon, H. S., Gallagher, M., et al. (2010). Elements between the IgH variable (V) and diversity (D) clusters influence antisense transcription and lineage-specific V(D)J recombination. *Proceedings of the National Academy of Sciences of the United States of America, 107*(51), 22207–22212.

Gibcus, J. H., & Dekker, J. (2013). The hierarchy of the 3D genome. *Molecular Cell, 49*(5), 773–782.

Gondor, A., Rougier, C., & Ohlsson, R. (2008). High-resolution circular chromosome conformation capture assay. *Nature Protocols, 3*(2), 303–313.

Grewal, S. I. S., & Jia, S. T. (2007). Heterochromatin revisited. *Nature Reviews. Genetics, 8*(1), 35–46.

Grunstein, M. (1997). Histone acetylation in chromatin structure and transcription. *Nature, 389*(6649), 349–352.

Guelen, L., Pagie, L., Brasset, E., Meuleman, W., Faza, M. B., Talhout, W., et al. (2008). Domain organization of human chromosomes revealed by mapping of nuclear lamina interactions. *Nature, 453*(7197), 948–951.

Guenther, M. G., Levine, S. S., Boyer, L. A., Jaenisch, R., & Young, R. A. (2007). A chromatin landmark and transcription initiation at most promoters in human cells. *Cell, 130*(1), 77–88.

Guo, C. Y., Gerasimova, T., Hao, H. P., Ivanova, I., Chakraborty, T., Selimyan, R., et al. (2011). Two forms of loops generate the chromatin conformation of the immunoglobulin heavy-chain gene locus. *Cell, 147*(2), 332–343.

Guo, C. G., Yoon, H. S., Franklin, A., Jain, S., Ebert, A., Cheng, H. L., et al. (2011). CTCF-binding elements mediate control of V(D)J recombination. *Nature, 477*(7365), 424–430.

Guttman, M., Amit, I., Garber, M., French, C., Lin, M. F., Feldser, D., et al. (2009). Chromatin signature reveals over a thousand highly conserved large non-coding RNAs in mammals. *Nature, 458*(7235), 223–227.

Haberland, M., Montgomery, R. L., & Olson, E. N. (2009). The many roles of histone deacetylases in development and physiology: Implications for disease and therapy. *Nature Reviews. Genetics, 10*(1), 32–42.

Hall, I. M., Noma, K., & Grewal, S. I. S. (2003). RNA interference machinery regulates chromosome dynamics during mitosis and meiosis in fission yeast. *Proceedings of the National Academy of Sciences of the United States of America, 100*(1), 193–198.

Heintzman, N. D., Hon, G. C., Hawkins, R. D., Kheradpour, P., Stark, A., Harp, L. F., et al. (2009). Histone modifications at human enhancers reflect global cell-type-specific gene expression. *Nature, 459*(7243), 108–112.

Hewitt, S. L., Chaumeil, J., & Skok, J. A. (2010). Chromosome dynamics and the regulation of V(D)J recombination. *Immunological Reviews, 237*, 43–54.

Holwerda, S. J. B., & de Laat, W. (2013). CTCF: The protein, the binding partners, the binding sites and their chromatin loops. *Philosophical Transactions of the Royal Society of London. Series B, Biological Sciences, 368*(1620), 1–8.

Huang, H., Sabari, B. R., Garcia, B. A., Allis, C. D., & Zhao, Y. M. (2014). Snapshot: Histone modifications. *Cell, 159*(2), 458.

Ichihara, Y., Matsuoka, H., & Kurosawa, Y. (1988). Organization of human-immunoglobulin heavy-chain diversity gene loci. *The EMBO Journal*, 7(13), 4141–4150.

Jenuwein, T., & Allis, C. D. (2001). Translating the histone code. *Science, 293*(5532), 1074–1080.

Jhunjhunwala, S., van Zelm, M. C., Peak, M. M., Cutchin, S., Riblet, R., van Dongen, J. J., et al. (2008). The 3D structure of the immunoglobulin heavy-chain locus: Implications for long-range genomic interactions. *Cell*, *133*(2), 265–279.

Johnston, C. M., Wood, A. L., Bolland, D. J., & Corcoran, A. E. (2006). Complete sequence assembly and characterization of the C57bl/6 mouse Ig heavy chain V region. *Journal of Immunology*, *176*(7), 4221–4234.

Jung, D., & Alt, F. W. (2004). Unraveling V(D)J recombination: Insights into gene regulation. *Cell*, *116*(2), 299–311.

Kapranov, P., Cheng, J., Dike, S., Nix, D. A., Duttagupta, R., Willingham, A. T., et al. (2007). RNA maps reveal new RNA classes and a possible function for pervasive transcription. *Science*, *316*(5830), 1484–1488.

Katayama, S., Tomaru, Y., Kasukawa, T., Waki, K., Nakanishi, M., Nakamura, M., et al. (2005). Antisense transcription in the mammalian transcriptome. *Science*, *309*(5740), 1564–1566.

Khamlichi, A. A., Pinaud, E., Decourt, C., Chauveau, C., & Cogne, M. (2000). The 3′ IgH regulatory region: A complex structure in a search for a function. *Advances in Immunology*, *75*(75), 317–345.

Kim, Y. J., Cecchini, K. R., & Kim, T. H. (2011). Conserved, developmentally regulated mechanism couples chromosomal looping and heterochromatin barrier activity at the homeobox gene A locus. *Proceedings of the National Academy of Sciences of the United States of America*, *108*(47), 19096(Vol. 108, Pg. 7391, 2011).

Kind, J., & van Steensel, B. (2010). Genome-nuclear lamina interactions and gene regulation. *Current Opinion in Cell Biology*, *22*(3), 320–325.

Kohli, R. M., & Zhang, Y. (2013). Tet enzymes, Tdg and the dynamics of DNA demethylation. *Nature*, *502*(7472), 472–479.

Kolasinska-Zwierz, P., Down, T., Latorre, I., Liu, T., Liu, X. S., & Ahringer, J. (2009). Differential chromatin marking of introns and expressed exons by H3K36me3. *Nature Genetics*, *41*(3), 376–381.

Koralov, S. B., Muljo, S. A., Galler, G. R., Krek, A., Chakraborty, T., Kanellopoulou, C., et al. (2008). Dicer ablation affects antibody diversity and cell survival in the B lymphocyte lineage. *Cell*, *132*(5), 860–874.

Kosak, S. T., Skok, J. A., Medina, K. L., Riblet, R., Le Beau, M. M., Fisher, A. G., et al. (2002). Subnuclear compartmentalization of immunoglobulin loci during lymphocyte development. *Science*, *296*(5565), 158–162.

Kwak, H., & Lis, J. T. (2013). Control of transcriptional elongation. *Annual Review of Genetics*, *47*, 483–508.

Kwok, R. P. S., Lundblad, J. R., Chrivia, J. C., Richards, J. P., Bachinger, H. P., Brennan, R. G., et al. (1994). Nuclear-protein CBP is a coactivator for the transcription factor CREB. *Nature*, *370*(6486), 223–226.

Lalonde, M. E., Cheng, X., & Cote, J. (2014). Histone target selection within chromatin: An exemplary case of teamwork. *Genes & Development*, *28*(10), 1029–1041.

Lavigne, M., Francis, N. J., King, I. F. G., & Kingston, R. E. (2004). Propagation of silencing: Recruitment and repression of naive chromatin—In trans by polycomb repressed chromatin. *Molecular Cell*, *13*(3), 415–425.

Lee, J. T., Davidow, L. S., & Warshawsky, D. (1999). Tsix, a gene antisense to Xist at the X-inactivation centre. *Nature Genetics*, *21*(4), 400–404.

Lennon, G. G., & Perry, R. P. (1985). C mu-containing transcripts initiate heterogeneously within the IgH enhancer region and contain a novel 5′-nontranslatable exon. *Nature*, *318*(6045), 475–478.

Li, Z. Q., Woo, C. J., Iglesias-Ussel, M. D., Ronai, D., & Scharff, M. D. (2004). The generation of antibody diversity through somatic hypermutation and class switch recombination. *Genes & Development*, *18*(1), 1–11.

Liang, H. E., Hsu, L. Y., Cado, D., Cowell, L. G., Kelsoe, G., & Schlissel, M. S. (2002). The "dispensable" portion of Rag2 is necessary for efficient V to DJ rearrangement during B and T cell development. *Immunity*, *17*(5), 639–651.
Lieberman-Aiden, E., van Berkum, N. L., Williams, L., Imakaev, M., Ragoczy, T., Telling, A., et al. (2009). Comprehensive mapping of long-range interactions reveals folding principles of the human genome. *Science*, *326*(5950), 289–293.
Lin, Y. C., Benner, C., Mansson, R., Heinz, S., Miyazaki, K., Miyazaki, M., et al. (2012). Global changes in the nuclear positioning of genes and intra- and interdomain genomic interactions that orchestrate B cell fate. *Nature Immunology*, *13*(12), 1196–1204.
Liu, H. F., Schmidt-Supprian, M., Shi, Y. J., Hobeika, E., Barteneva, N., Jumaa, H., et al. (2007). Yin Yang 1 is a critical regulator of B-cell development. *Genes & Development*, *21*(10), 1179–1189.
Liu, Y., Subrahmanyam, R., Chakraborty, T., Sen, R., & Desiderio, S. (2007). A plant homeodomain in Rag-2 that binds hypermethylated lysine 4 of histone H3 is necessary for efficient antigen-receptor-gene rearrangement. *Immunity*, *27*(4), 561–571.
Lorch, Y., Maier-Davis, B., & Kornberg, R. D. (2010). Mechanism of chromatin remodeling. *Proceedings of the National Academy of Sciences of the United States of America*, *107*(8), 3458–3462.
Lu, C., Ward, A., Bettridge, J., Liu, Y., & Desiderio, S. (2015). An autoregulatory mechanism imposes allosteric control on the V(D)J recombinase by histone H3 methylation. *Cell Reports*, *10*(1), 29–38.
Lucas, J. S., Bossen, C., & Murre, C. (2011). Transcription and recombination factories: Common features? *Current Opinion in Cell Biology*, *23*(3), 318–324.
Lucas, J. S., Zhang, Y., Dudko, O. K., & Murre, C. (2014). 3D trajectories adopted by coding and regulatory DNA elements: First-passage times for genomic interactions. *Cell*, *158*(2), 339–352.
Lusser, A., & Kadonaga, J. T. (2003). Chromatin remodeling by ATP-dependent molecular machines. *Bioessays*, *25*(12), 1192–1200.
Maes, J., Chappaz, S., Cavelier, P., Neill, L., Turner, B., Rougeon, F., et al. (2006). Activation of V(D)J recombination at the IgH chain JH locus occurs within a 6-kilobase chromatin domain and is associated with nucleosomal remodeling. *Journal of Immunology*, *176*(9), 5409–5417.
Maison, C., Bailly, D., Peters, A., Quivy, J. P., Roche, D., Taddei, A., et al. (2002). Higher-order structure in pericentric heterochromatin involves a distinct pattern of histone modification and an RNA component. *Nature Genetics*, *30*(3), 329–334.
Malin, S., McManus, S., Cobaleda, C., Novatchkova, M., Delogu, A., Bouillet, P., et al. (2010). Role of Stat5 in controlling cell survival and immunoglobulin gene recombination during pro-B cell development. *Nature Immunology*, *11*(2), 171–179.
Martens, J. H. A., O'Sullivan, R. J., Braunschweig, U., Opravil, S., Radolf, M., Steinlein, P., et al. (2005). The profile of repeat-associated histone lysine methylation states in the mouse epigenome. *The EMBO Journal*, *24*(4), 800–812.
Martin, C., & Zhang, Y. (2005). The diverse functions of histone lysine methylation. *Nature Reviews. Molecular Cell Biology*, *6*(11), 838–849.
Matthews, A. G. W., Kuo, A. J., Ramon-Maiques, S., Han, S. M., Champagne, K. S., Ivanov, D., et al. (2007). Rag2 Phd finger couples histone H3 lysine 4 trimethylation with V(D)J recombination. *Nature*, *450*(7172), 1106–1110.
Medvedovic, J., Ebert, A., Tagoh, H., Tamir, I. M., Schwickert, T. A., Novatchkova, M., et al. (2013). Flexible long-range loops in the VH gene region of the Igh locus facilitate the generation of a diverse antibody repertoire. *Immunity*, *39*(2), 229–244.
Meek, K. D., Hasemann, C. A., & Capra, J. D. (1989). Novel rearrangements at the immunoglobulin-D locus—Inversions and fusions add to Igh somatic diversity. *Journal of Experimental Medicine*, *170*(1), 39–57.

Melnick, A., Carlile, G., Ahmad, K. F., Kiang, C. L., Corcoran, C., Bardwell, V., et al. (2002). Critical residues within the Btb domain of plzf and Bcl-6 modulate interaction with corepressors. *Molecular and Cellular Biology*, *22*(6), 1804–1818.

Merkenschlager, M., & Odom, D. T. (2013). CTCF and cohesin: Linking gene regulatory elements with their targets. *Cell*, *152*(6), 1285–1297.

Michaelson, J. S., Giannini, S. L., & Birshtein, B. K. (1995). Identification of 3′alpha-Hs4, a novel Ig heavy-chain enhancer element regulated at multiple stages of B-cell differentiation. *Nucleic Acids Research*, *23*(6), 975–981.

Mikkelsen, T. S., Ku, M. C., Jaffe, D. B., Issac, B., Lieberman, E., Giannoukos, G., et al. (2007). Genome-wide maps of chromatin state in pluripotent and lineage-committed cells. *Nature*, *448*(7153), 553–560.

Morshead, K. B., Ciccone, D. N., Taverna, S. D., Allis, C. D., & Oettinger, M. A. (2003). Antigen receptor loci poised for V(D)J rearrangement are broadly associated with Brg1 and flanked by peaks of histone H3 dimethylated at lysine 4. *Proceedings of the National Academy of Sciences of the United States of America*, *100*(20), 11577–11582.

Morvan, C. L., Pinaud, E., Decourt, C., Cuvillier, A., & Cogne, M. (2003). The immunoglobulin heavy-chain locus Hs3b and Hs4 3′ enhancers are dispensable for VDJ assembly and somatic hypermutation. *Blood*, *102*(4), 1421–1427.

Nagano, T., Lubling, Y., Stevens, T. J., Schoenfelder, S., Yaffe, E., Dean, W., et al. (2013). Single-cell Hi-C reveals cell-to-cell variability in chromosome structure. *Nature*, *502*(7469), 59–64.

Nakaya, H. I., Amaral, P. P., Louro, R., Lopes, A., Fachel, A. A., Moreira, Y. B., et al. (2007). Genome mapping and expression analyses of human intronic noncoding RNAs reveal tissue-specific patterns and enrichment in genes related to regulation of transcription. *Genome Biology*, *8*(3), 1–25.

Narendra, V., Rocha, P. P., An, D. S., Raviram, R., Skok, J. A., Mazzoni, E. O., et al. (2015). CTCF establishes discrete functional chromatin domains at the Hox clusters during differentiation. *Science*, *347*(6225), 1017–1021.

Narlikar, G. J., Sundaramoorthy, R., & Owen-Hughes, T. (2013). Mechanisms and functions of ATP-dependent chromatin-remodeling enzymes. *Cell*, *154*(3), 490–503.

Ng, H. H., & Bird, A. (2000). Histone deacetylases: Silencers for hire. *Trends in Biochemical Sciences*, *25*(3), 121–126.

Nitschke, L., Kestler, J., Tallone, T., Pelkonen, S., & Pelkonen, J. (2001). Deletion of the DQ52 element within the Ig heavy chain locus leads to a selective reduction in VDJ recombination and altered D gene usage. *Journal of Immunology*, *166*(4), 2540–2552.

Onengut-Gumuscu, S., Chen, W. M., Burren, O., Cooper, N. J., Quinlan, A. R., Mychaleckyj, J. C., et al. (2015). Fine mapping of type 1 diabetes susceptibility loci and evidence for colocalization of causal variants with lymphoid gene enhancers. *Nature Genetics*, *47*(4), 381–386.

Ong, C. T., & Corces, V. G. (2014). CTCF: An architectural protein bridging genome topology and function. *Nature Reviews. Genetics*, *15*(4), 234–246.

Pan, P. Y., Lieber, M. R., & Teale, J. M. (1997). The role of recombination signal sequences in the preferential joining by deletion in DH-JH recombination and in the ordered rearrangement of the IgH locus. *International Immunology*, *9*(4), 515–522.

Pandey, R. R., Mondal, T., Mohammad, F., Enroth, S., Redrup, L., Komorowski, J., et al. (2008). Kcnq1ot1 antisense noncoding RNA mediates lineage-specific transcriptional silencing through chromatin-level regulation. *Molecular Cell*, *32*(2), 232–246.

Parelho, V., Hadjur, S., Spivakov, M., Leleu, M., Sauer, S., Gregson, H. C., et al. (2008). Cohesins functionally associate with CTCF on mammalian chromosome arms. *Cell*, *132*(3), 422–433.

Pastor, W. A., Aravind, L., & Rao, A. (2013). Tetonic shift: Biological roles of Tet proteins in DNA demethylation and transcription. *Nature Reviews. Molecular Cell Biology*, *14*(6), 341–356.

Pauler, F. M., Sloane, M. A., Huang, R., Regha, K., Koerner, M. V., Tamir, I., et al. (2009). H3K27me3 forms blocs over silent genes and intergenic regions and specifies a histone banding pattern on a mouse autosomal chromosome. *Genome Research*, *19*(2), 221–233.

Pawlitzky, I., Angeles, C. V., Siegel, A. M., Stanton, M. L., Riblet, R., & Brodeur, P. H. (2006). Identification of a candidate regulatory element within the 5′ flanking region of the mouse Igh locus defined by pro-B cell-specific hypersensitivity associated with binding of Pu.1, Pax5, and E2a. *Journal of Immunology*, *176*(11), 6839–6851.

Pedersen, M. T., & Helin, K. (2010). Histone demethylases in development and disease. *Trends in Cell Biology*, *20*(11), 662–671.

Perlot, T., Alt, F. W., Bassing, C. H., Suh, H., & Pinaud, E. (2005). Elucidation of IgH intronic enhancer functions via germline deletion. *Proceedings of the National Academy of Sciences of the United States of America*, *102*(40), 14362–14367.

Perlot, T., Pawlitzky, I., Manis, J. P., Zarrin, A. A., Brodeur, P. H., & Alt, F. W. (2010). Analysis of mice lacking DNAseI hypersensitive sites at the 5′ end of the IgH locus. *PLoS One*, *5*(11), 1–10.

Peters, A., O'Carroll, D., Scherthan, H., Mechtler, K., Sauer, S., Schofer, C., et al. (2001). Loss of the Suv39h histone methyltransferases impairs mammalian heterochromatin and genome stability. *Cell*, *107*(3), 323–337.

Petty, E., & Pillus, L. (2013). Balancing chromatin remodeling and histone modifications in transcription. *Trends in Genetics*, *29*(11), 621–629.

Phillips-Cremins, J. E., Sauria, M. E. G., Sanyal, A., Gerasimova, T. I., Lajoie, B. R., Bell, J. S. K., et al. (2013). Architectural protein subclasses shape 3D organization of genomes during lineage commitment. *Cell*, *153*(6), 1281–1295.

Pinaud, E., Khamlichi, A. A., Le Morvan, C., Drouet, M., Nalesso, V., Le Bert, M., et al. (2001). Localization of the 3′ IgH locus elements that affect long-distance regulation of class switch recombination. *Immunity*, *15*(2), 187–199.

Pogo, B. G. T., Allfrey, V. G., & Mirsky, A. E. (1966). RNA synthesis and histone acetylation during course of gene activation in lymphocytes. *Proceedings of the National Academy of Sciences of the United States of America*, *55*(4), 805–812.

Pokholok, D. K., Harbison, C. T., Levine, S., Cole, M., Hannett, N. M., Lee, T. I., et al. (2005). Genome-wide map of nucleosome acetylation and methylation in yeast. *Cell*, *122*(4), 517–527.

Puget, N., Hirasawa, R., Hu, N. S., Laviolette-Malirat, N., Feil, R., & Khamlichi, A. A. (2015). Insertion of an imprinted insulator into the IgH locus reveals developmentally regulated, transcription-dependent control of V(D)J recombination. *Molecular and Cellular Biology*, *35*(3), 529–543.

Rada-Iglesias, A., Bajpai, R., Swigut, T., Brugmann, S. A., Flynn, R. A., & Wysocka, J. (2011). A unique chromatin signature uncovers early developmental enhancers in humans. *Nature*, *470*(7333), 279–283.

Rao, S. S. P., Huntley, M. H., Durand, N. C., Stamenova, E. K., Bochkov, I. D., Robinson, J. T., et al. (2014). A 3D map of the human genome at kilobase resolution reveals principles of chromatin looping. *Cell*, *159*(7), 1665–1680.

Revilla-i-Domingo, R., Bilic, I., Vilagos, B., Tagoh, H., Ebert, A., Tamir, I. M., et al. (2012). The B-cell identity factor Pax5 regulates distinct transcriptional programmes in early and late B lymphopoiesis. *The EMBO Journal*, *31*(14), 3130–3146.

Reynaud, D., Demarco, I. A., Reddy, K. L., Schjerven, H., Bertolino, E., Chen, Z. S., et al. (2008). Regulation of B cell fate commitment and immunoglobulin heavy-chain gene rearrangements by Ikaros. *Nature Immunology*, *9*(8), 927–936.

Rouaud, P., Saintamand, A., Saad, F., Carrion, C., Lecardeur, S., Cogne, M., et al. (2014). Elucidation of the enigmatic IgD class-switch recombination via germline deletion of the Igh 3′ regulatory region. *Journal of Experimental Medicine*, *211*(5), 975–985.

Sayegh, C., Jhunjhunwala, S., Riblet, R., & Murre, C. (2005). Visualization of looping involving the immunoglobulin heavy-chain locus in developing B cells. *Genes & Development*, *19*(3), 322–327.

Schatz, D. G., & Ji, Y. H. (2011). Recombination centres and the orchestration of V(D)J recombination. *Nature Reviews. Immunology*, *11*(4), 251–263.

Schlissel, M. S., Corcoran, L. M., & Baltimore, D. (1991). Virus-transformed pre-B cells show ordered activation but not inactivation of immunoglobulin gene rearrangement and transcription. *Journal of Experimental Medicine*, *173*(3), 711–720.

Schoenfelder, S., Clay, I., & Fraser, P. (2010). The transcriptional interactome: Gene expression in 3D. *Current Opinion in Genetics & Development*, *20*(2), 127–133.

Seitan, V. C., Faure, A. J., Zhan, Y., McCord, R. P., Lajoie, B. R., Ing-Simmons, E., et al. (2013). Cohesin-based chromatin interactions enable regulated gene expression within preexisting architectural compartments. *Genome Research*, *23*(12), 2066–2077.

Selimyan, R., Gerstein, R. M., Ivanova, I., Precht, P., Subrahmanyam, R., Perlot, T., et al. (2013). Localized DNA demethylation at recombination intermediates during immunoglobulin heavy chain gene assembly. *PLoS Biology*, *11*(1), 1–14.

Sen, R., & Oltz, E. (2006). Genetic and epigenetic regulation of Igh gene assembly. *Current Opinion in Immunology*, *18*(3), 237–242.

Serandour, A. A., Avner, S., Oger, F., Bizot, M., Percevault, F., Lucchetti-Miganeh, C., et al. (2012). Dynamic hydroxymethylation of deoxyribonucleic acid marks differentiation-associated enhancers. *Nucleic Acids Research*, *40*(17), 8255–8265.

Shen, Y., Yue, F., McCleary, D. F., Ye, Z., Edsall, L., Kuan, S., et al. (2012). A map of the cis-regulatory sequences in the mouse genome. *Nature*, *488*(7409), 116–120.

Shi, Y., & Whetstine, J. R. (2007). Dynamic regulation of histone lysine methylation by demethylases. *Molecular Cell*, *25*(1), 1–14.

Shih, H. Y., Verma-Gaur, J., Torkamani, A., Feeney, A. J., Galjart, N., & Krangel, M. S. (2012). TCRa gene recombination is supported by a TCRa enhancer- and CTCF-dependent chromatin hub. *Proceedings of the National Academy of Sciences of the United States of America*, *109*(50), 3492–3502.

Shimazaki, N., Tsai, A. G., & Lieber, M. R. (2009). H3k4me3 stimulates the V(D)J Rag complex for both nicking and hairpinning in trans in addition to tethering in cis: Implications for translocations. *Molecular Cell*, *34*(5), 535–544.

Singh, P., Lee, D.-H., & Szabo, P. E. (2012). More than insulator: Multiple roles of CTCF at the H19-Igf2 imprinted domain. *Frontiers in Genetics*, *3*, 1–9.

Sleckman, B. P., Gorman, J. R., & Alt, F. W. (1996). Accessibility control of antigen-receptor variable-region gene assembly: Role of cis-acting elements. *Annual Review of Immunology*, *14*, 459–481.

Sofueva, S., Yaffe, E., Chan, W. C., Georgopoulou, D., Rudan, M. V., Mira-Bontenbal, H., et al. (2013). Cohesin-mediated interactions organize chromosomal domain architecture. *The EMBO Journal*, *32*(24), 3119–3129.

Sollbach, A. E., & Wu, G. E. (1995). Inversions produced during V(D)J rearrangement at IgH, the immunoglobulin heavy-chain locus. *Molecular and Cellular Biology*, *15*(2), 671–681.

Stavnezer, J., Guikema, J. E. J., & Schrader, C. E. (2008). Mechanism and regulation of class switch recombination. *Annual Review of Immunology*, *26*, 261–292. Palo Alto: Annual Reviews.

Stergachis, A. B., Haugen, E., Shafer, A., Fu, W. Q., Vernot, B., Reynolds, A., et al. (2013). Exonic transcription factor binding directs codon choice and affects protein evolution. *Science*, *342*(6164), 1367–1372.

Stewart, M. D., Li, J., & Wong, J. (2005). Relationship between histone H3 lysine 9 methylation, transcription repression, and heterochromatin protein 1 recruitment. *Molecular and Cellular Biology, 25*(7), 2525–2538.
Struhl, K. (1998). Histone acetylation and transcriptional regulatory mechanisms. *Genes & Development, 12*(5), 599–606.
Su, L. K., & Kadesch, T. (1990). The immunoglobulin heavy-chain enhancer functions as the promoter for I mu sterile transcription. *Molecular and Cellular Biology, 10*(6), 2619–2624.
Subrahmanyam, R., Du, H., Ivanova, I., Chakraborty, T., Ji, Y., Zhang, Y., et al. (2012). Localized epigenetic changes induced by DH recombination restricts recombinase to DJH junctions. *Nature Immunology, 13*(12), 1205–1212.
Talukder, S. R., Dudley, D. D., Alt, F. W., Takahama, Y., & Akamatsu, Y. (2004). Increased frequency of aberrant V(D)J recombination products in core Rag-expressing mice. *Nucleic Acids Research, 32*(15), 4539–4549.
Treiber, T., Mandel, E. M., Pott, S., Gyory, I., Firner, S., Liu, E. T., et al. (2010). Early B cell factor 1 regulates B cell gene networks by activation, repression, and transcription-independent poising of chromatin. *Immunity, 32*(5), 714–725.
Trojer, P., & Reinberg, D. (2007). Facultative heterochromatin: Is there a distinctive molecular signature? *Molecular Cell, 28*(1), 1–13.
Tropberger, P., & Schneider, R. (2013). Scratching the (lateral) surface of chromatin regulation by histone modifications. *Nature Structural & Molecular Biology, 20*(6), 657–661.
Tsukada, S., Sugiyama, H., Oka, Y., & Kishimoto, S. (1990). Estimation of D-segment usage in initial-D to JH joinings in a murine immature B-cell line—Preferential usage of DFL16.1, the most 5′ D segment and DQ52, the most JH-proximal D segment. *Journal of Immunology, 144*(10), 4053–4059.
Van Bortle, K., Nichols, M. H., Li, L., Ong, C. T., Takenaka, N., Qin, Z. S., et al. (2014). Insulator function and topological domain border strength scale with architectural protein occupancy. *Genome Biology, 15*(6), 1–18.
Verma-Gaur, J., Torkamani, A., Schaffer, L., Head, S. R., Schork, N. J., & Feeney, A. J. (2012). Noncoding transcription within the Igh distal V-H region at pair elements affects the 3D structure of the Igh locus in pro-B cells. *Proceedings of the National Academy of Sciences of the United States of America, 109*(42), 17004–17009.
Verrier, L., Vandromme, M., & Trouche, D. (2011). Histone demethylases in chromatin cross-talks. *Biology of the Cell, 103*(8), 381–401.
Verschure, P. J., van der Kraan, I., de Leeuw, W., van der Vlag, J., Carpenter, A. E., Belmont, A. S., et al. (2005). In vivo Hp1 targeting causes large-scale chromatin condensation and enhanced histone lysine methylation. *Molecular and Cellular Biology, 25*(11), 4552–4564.
Vignali, M., Hassan, A. H., Neely, K. E., & Workman, J. L. (2000). ATP-dependent chromatin-remodeling complexes. *Molecular and Cellular Biology, 20*(6), 1899–1910.
Visel, A., Blow, M. J., Li, Z. R., Zhang, T., Akiyama, J. A., Holt, A., et al. (2009). Chip-seq accurately predicts tissue-specific activity of enhancers. *Nature, 457*(7231), 854–858.
Volpi, S. A., Verma-Gaur, J., Hassan, R., Ju, Z. L., Roa, S., Chatterjee, S., et al. (2012). Germline deletion of Igh 3′ regulatory region elements Hs 5, 6, 7 (HS5-7) affects B cell-specific regulation, rearrangement, and insulation of the Igh locus. *Journal of Immunology, 188*(6), 2556–2566.
Wagner, E. J., & Carpenter, P. B. (2012). Understanding the language of Lys36 methylation at histone H3. *Nature Reviews. Molecular Cell Biology, 13*(2), 115–126.
Wang, K. C., Yang, Y. W., Liu, B., Sanyal, A., Corces-Zimmerman, R., Chen, Y., et al. (2011). A long noncoding RNA maintains active chromatin to coordinate homeotic gene expression. *Nature, 472*(7341), 120–124.
Weintraub, H., & Groudine, M. (1976). Chromosomal subunits in active genes have an altered conformation. *Science, 193*(4256), 848–856.

Wu, H., & Zhang, Y. (2011). Mechanisms and functions of tet protein-mediated 5-methylcytosine oxidation. *Genes & Development*, *25*(23), 2436–2452.

Yancopoulos, G. D., & Alt, F. W. (1985). Developmentally controlled and tissue-specific expression of unrearranged VH gene segments. *Cell*, *40*(2), 271–281.

Yancopoulos, G. D., & Alt, F. W. (1986). Regulation of the assembly and expression of variable-region genes. *Annual Review of Immunology*, *4*, 339–368.

Yang, Q. X., Riblet, R., & Schildkraut, C. L. (2005). Sites that direct nuclear compartmentalization are near the 5′ end of the mouse immunoglobulin heavy-chain locus. *Molecular and Cellular Biology*, *25*(14), 6021–6030.

Ye, J. (2004). The immunoglobulin IGHD gene locus in C57bl/6 mice. *Immunogenetics*, *56*(6), 399–404.

Young, M. D., Willson, T. A., Wakefield, M. J., Trounson, E., Hilton, D. J., Blewitt, M. E., et al. (2011). Chip-seq analysis reveals distinct H3K27me3 profiles that correlate with transcriptional activity. *Nucleic Acids Research*, *39*(17), 7415–7427.

Yue, F., Cheng, Y., Breschi, A., Vierstra, J., Wu, W. S., Ryba, T., et al. (2014). A comparative encyclopedia of DNA elements in the mouse genome. *Nature*, *515*(7527), 355–364.

Zentner, G. E., Tesar, P. J., & Scacheri, P. C. (2011). Epigenetic signatures distinguish multiple classes of enhancers with distinct cellular functions. *Genome Research*, *21*(8), 1273–1283.

Zhao, J., Sun, B. K., Erwin, J. A., Song, J. J., & Lee, J. T. (2008). Polycomb proteins targeted by a short repeat RNA to the mouse X chromosome. *Science*, *322*(5902), 750–756.

Zuin, J., Dixon, J. R., van der Reijden, M., Ye, Z., Kolovos, P., Brouwer, R. W. W., et al. (2014). Cohesin and CTCF differentially affect chromatin architecture and gene expression in human cells. *Proceedings of the National Academy of Sciences of the United States of America*, *111*(3), 996–1001.

Zullo, J. M., Demarco, I. A., Pique-Regi, R., Gaffney, D. J., Epstein, C. B., Spooner, C. J., et al. (2012). DNA sequence-dependent compartmentalization and silencing of chromatin at the nuclear lamina. *Cell*, *149*(7), 1474–1487.

CHAPTER THREE

Spatial Regulation of V–(D)J Recombination at Antigen Receptor Loci

Anja Ebert, Louisa Hill, Meinrad Busslinger[1]
Research Institute of Molecular Pathology (IMP), Vienna Biocenter (VBC), Vienna, Austria
[1]Corresponding author: e-mail address: busslinger@imp.ac.at

Contents

Advances in Immunology, Volume 128
ISSN 0065-2776
http://dx.doi.org/10.1016/bs.ai.2015.07.006
© 2015 Elsevier Inc.
All rights reserved.

Abstract

Lymphocytes express a diverse repertoire of antigen receptors, which are able to recognize a large variety of foreign pathogens. Functional antigen receptor genes are assembled by V(D)J recombination in immature B cells (*Igh* and *Igk*) and T cells (*Tcrb* and *Tcra/d*). V(D)J recombination takes place in the 3′ proximal domain containing the D, J, and C gene segments, whereas 31 (*Tcrb*) to 200 (*Igh*) V genes are spread over a large region of 0.67 (*Tcrb*) to 3 (*Igk*) megabase pairs. The spatial regulation of V(D)J recombination has been best studied for the *Igh* locus, which undergoes reversible contraction by long-range looping in pro-B cells. This large-scale contraction brings distantly located V_H genes into close proximity of the DJ_H-rearranged gene segment, which facilitates V_H–DJ_H recombination. The B-cell-specific Pax5, ubiquitous YY1, and architectural CTCF/cohesin proteins regulate *Igh* locus contraction in pro-B cells by binding to multiple sites in the V_H gene cluster. These regulators also control the pro-B-cell-specific activity of the distally located PAIR elements, which may be involved in the regulation of V_H–DJ_H recombination by promoting locus contraction. Moreover, the large V_H gene cluster of the *Igh* locus undergoes flexible long-range looping, which guarantees similar participation of all V_H genes in V_H–DJ_H recombination to generate a diverse antibody repertoire. Importantly, long-range looping is a more general regulatory principle, as other antigen receptor loci also undergo reversible contraction at the developmental stage, where they engage in V–(D)J recombination.

1. DEVELOPMENTAL REGULATION OF V(D)J RECOMBINATION

1.1 Antigen Receptor Diversity Through V(D)J Recombination

The acquired immune system has a shear unlimited capacity to respond to foreign pathogens. This enormous adaptive potential is achieved as every newly generated lymphocyte recognizes a unique antigen, which results in an overall diversity of lymphocytes that is great enough to counteract any possible antigen. To this end, B and T cells express lineage-specific antigen receptors that mediate antibody-dependent humoral or cellular immunity, respectively. The B cell receptor (BCR) consists of an immunoglobulin heavy chain (Igh) and light chain (Igκ or Igλ). T cells of the αβ lineage, which comprise the majority of T lymphocytes in mouse and man, express the T cell receptor (TCR) β polypeptide in association with the TCRα chain, while the functionally distinct γδ T cells contain TCRγ paired with TCRδ on their cell surface. These antigen receptor proteins are encoded by large gene loci containing discontinuous variable (V), diversity (D), and joining (J) gene segments, which are assembled into one functional gene by V(D)J recombination during lymphocyte development (Bossen, Mansson, & Murre, 2012; Perlot & Alt, 2008; Fig. 1). The multiplicity of D, J, and especially V gene segments, combined with the randomness of their recombination, is responsible for the virtually unlimited diversity of the immune repertoire (Bassing, Swat, & Alt, 2002).

The mechanism of V(D)J recombination at the DNA level is rather simple. All V, D, and J gene segments are flanked by recombination signal sequences (RSSs), which consist of relatively conserved heptamer and nonamer elements separated by a spacer of either 12 or 23 base pairs (bp). The lymphoid-specific recombinase proteins RAG1 and RAG2, assisted by high-mobility group proteins, assemble 12- and 23-bp RSSs into a synaptic complex and then generate double-strand DNA breaks between the RSSs and coding segments. These DNA breaks are subsequently processed and religated by ubiquitous repair factors of the nonhomologous end joining machinery to form coding and signal joints (Bassing et al., 2002).

1.2 Developmental Regulation of V(D)J Recombination by DNA Accessibility Control of the RAG Recombinase

The simplicity of the V(D)J recombination process at the DNA template level poses logistic problems for the assembly of the different antigen receptors, as the RAG proteins are expressed in all lymphoid progenitors (Igarashi, Gregory, Yokota, Sakaguchi, & Kincade, 2002) and immature B and T cells (Jankovic, Casellas, Yannoutsos, Wardemann, & Nussenzweig, 2004). Hence, stringent regulation is required to restrict the access of RAG proteins to only specific subsets of all the recombination substrates (Stanhope-Baker, Hudson, Shaffer, Constantinescu, & Schlissel, 1996; Yancopoulos & Alt, 1985). To achieve this specificity, V(D)J recombination is tightly controlled in a lineage- and stage-specific manner. Within the B-lymphoid lineage, the *Igh* locus is rearranged at the pro-B cell stage prior to recombination of *Igk* and *Igl* genes in pre-B cells of the bone marrow, whereas the *Tcrb* and *Tcra* genes are rearranged in $CD4^{-}CD8^{-}$ double-negative (DN) pro-T cells and $CD4^{+}CD8^{+}$ double-positive (DP) T cells of the thymus, respectively. Moreover, V(D)J recombination of the *Igh* gene occurs in a defined temporal order with D_H–J_H rearrangements preceding V_H–DJ_H recombination. Rearrangements of the *Tcrb* locus also proceed in the same order (D_β–J_β before V_β–DJ_β recombination) during pro-T cell development. Therefore, control mechanisms must exist to shield V genes from RAG-mediated cleavage during D–J recombination and to facilitate rearrangement of only one out of a hundred V genes during V–DJ recombination. Consequently,

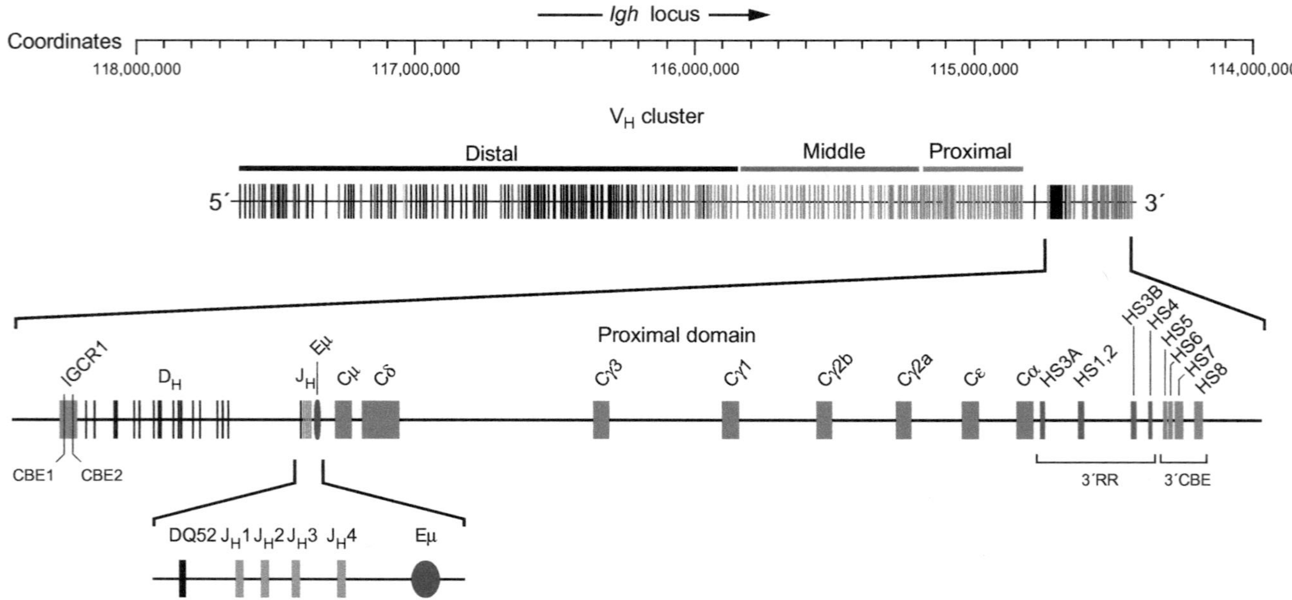

Figure 1 Structure of the *Igh* locus. Schematic diagram of the *Igh* locus. The distinct V_H gene families (different colors) in the distal, middle, and proximal V_H gene regions (Johnston, Wood, Bolland, & Corcoran, 2006) are shown together with the D_H (black), J_H (orange), C_H (blue) gene elements, the E_μ enhancer (red), and 3′ regulatory region (3′RR; red), and the IGCR1 and 3′CBE insulators (gray) in the 3′ proximal *Igh* domain. The mm9 genomic coordinates of mouse chromosome 12 are indicated. HS, DNase I hypersensitive sites; IGCR1, intergenic control region 1; 3′RR, 3′ regulatory region; CBE, CTCF-binding element. (See the color plate.)

the process of antigen receptor generation entirely depends on accurate regulation of the accessibility of RSSs for the RAG1/2 endonuclease.

The developmental and locus-specific constraints on V(D)J recombination are largely imposed at the epigenetic level (Jhunjhunwala, van Zelm, Peak, & Murre, 2009). In nonlymphoid cells, the *Ig* and *Tcr* genes are present in inaccessible chromatin, as exogenously expressed RAG proteins readily cleave transfected episomal recombination substrates but not endogenous antigen receptor genes in kidney cells (Romanow et al., 2000). Moreover, recombinant RAG proteins added to isolated lymphocyte nuclei can only cleave the *Ig* or *Tcr* gene that is actively undergoing V(D)J recombination at the developmental stage used for nuclei preparation (Stanhope-Baker et al., 1996). Hence, the lineage specificity and temporal ordering of gene rearrangements are caused by the sequential opening of local chromatin regions that render specific RSSs accessible to the V(D)J recombinase. This sequential chromatin accessibility at *Ig* and *Tcr* gene segments is controlled by multiple processes including subnuclear relocation (Kosak et al., 2002), DNA demethylation (Mostoslavsky et al., 1998), chromatin remodeling (Kwon, Morshead, Guyon, Kingston, & Oettinger, 2000), histone modification (Bossen et al., 2012), intergenic antisense RNA expression (Bolland et al., 2004; Ebert et al., 2011), and germline transcription (Yancopoulos & Alt, 1985).

1.3 Distinct Regulation of the V Gene Cluster and Proximal Domain of Antigen Receptor Loci

Antigen receptor loci consist of a large V gene cluster ranging from 31 (*Tcrb*) to 200 (*Igh*) V genes and a 3′ proximal domain containing the D, J, and constant (C) gene segments together with a potent enhancer (Fig. 1). These two regions are subject to distinct control mechanisms, as chromatin changes, germline transcription, and D–J recombination in the 3′ proximal domain strictly depend on the local enhancer, whereas chromatin accessibility and germline transcription at the V genes are regulated independently of this enhancer (Afshar, Pierce, Bolland, Corcoran, & Oltz, 2006; Chakraborty et al., 2009; Hawwari & Krangel, 2005; Mathieu et al., 2003; Perlot, Alt, Bassing, Suh, & Pinaud, 2005). Another distinctive feature is the size of the two domains. The D and J segments of the *Igh* and *Tcrb* loci are located within a relatively short region of 50 kilobase pairs (kb), which readily facilitates D–J recombination. In contrast, the V genes of all major antigen receptor loci are spread over a large region of 0.67 (*Tcrb*), 1.5 (*Tcra*), 2.4 (*Igh*), or 3 (*Igk*) megabase pairs (Mb) and are separated by a large spacer region of

100–260 kb from the proximal domain, indicating that the V–(D)J recombination step must be spatially restricted (Fig. 1). Participation of all V genes in V–(D)J recombination is therefore only possible, if the entire locus undergoes contraction by looping, which brings the V genes into close spatial proximity of the (D)J segments. In this review, we discuss the epigenetic and transcriptional mechanisms of this spatial regulation with a special focus on the *Igh* locus, which has been most extensively studied in this regard.

2. REVERSIBLE CONTRACTION AND ALLELIC EXCLUSION OF THE *Igh* LOCUS

2.1 Contraction of the *Igh* Locus Facilitates V_H–DJ_H Recombination in Pro-B Cells

Locus contraction was discovered when the nuclear conformation of the *Igh* locus was analyzed by the three-dimensional DNA fluorescence *in situ* hybridization (3D DNA-FISH) method with probes specific for both ends of the locus. These studies revealed that the *Igh* locus is present in an extended configuration in nuclei of non-B cells and lymphoid progenitors, as evidenced by a relatively large separation of the two FISH signals originating from the proximal and distal end of the *Igh* locus (Fuxa et al., 2004; Kosak et al., 2002; Fig. 2). In contrast, the two FISH signals were close together or coinciding in the nuclei of committed pro-B cells, indicating that both *Igh* alleles are contracted at the developmental stage, when V_H–DJ_H rearrangements take place (Fuxa et al., 2004; Kosak et al., 2002; Fig. 2). These data led to the concept that long-range contraction of the *Igh* locus juxtaposes distally located V_H genes next to the rearranged 3′ proximal DJ_H domain, which facilitates V_H–DJ_H rearrangements due to the close proximity of both elements (Fuxa et al., 2004; Kosak et al., 2002; Fig. 2). Subsequently, it was shown that contraction of the *Igh* locus is caused by looping of the V_H gene cluster in such a way that the different V_H genes should be able to recombine with similar frequency, which is essential for the generation of a highly diverse immunoglobulin repertoire (Jhunjhunwala et al., 2008; Roldán et al., 2005; Sayegh, Jhunjhunwala, Riblet, & Murre, 2005).

2.2 Decontraction and Allelic Exclusion of the *Igh* Locus in Response to Pre-BCR Signaling

Successful V_H–DJ_H recombination of the *Igh* gene leads to expression of the Igμ protein as part of the pre-B cell receptor (pre-BCR) complex, which

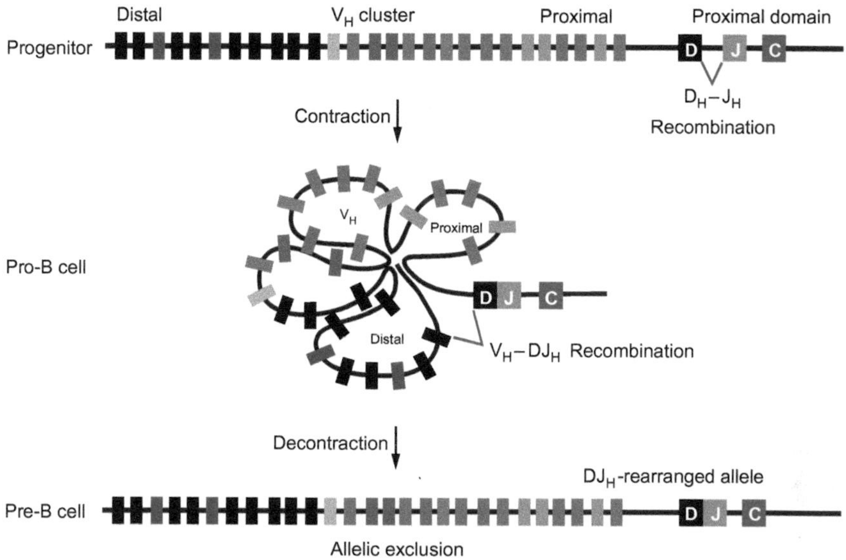

Figure 2 Reversible contraction of the *Igh* locus. The *Igh* locus is shown in an extended configuration in uncommitted progenitors, which allows D_H–J_H recombination to take place in the 3′ proximal domain. In committed pro-B cells, all V_H genes participate in V_H–DJ_H rearrangements due to contraction of the *Igh* locus by looping. The V_H genes of the incompletely rearranged *Igh* allele are no longer available for V_H–DJ_H recombination due to decontraction of the *Igh* locus in pre-B cells, which contributes to the maintenance of *Igh* allelic exclusion. (See the color plate.)

acts as an important checkpoint promoting proliferative cell expansion, maintenance of *Igh* allelic exclusion, and differentiation to small pre-B cells undergoing Ig light-chain gene recombination (Melchers, 2005). Allelic exclusion refers to the feedback inhibition of V_H–DJ_H recombination at the second, DJ_H-rearranged allele in small pre-B cells and thus ensures that each B lymphocyte expresses only one monospecific BCR. Importantly, pre-BCR signaling leads to decontraction of the nonfunctional *Igh* allele in pre-B cells (Fig. 2), which physically separates the distal V_H genes from the 3′ proximal *Igh* domain, thus preventing further recombination of the second, DJ_H-rearranged *Igh* allele in pre-B cells (Roldán et al., 2005). Decontraction may therefore be an important mechanism contributing to the maintenance of *Igh* allelic exclusion in small pre-B cells undergoing Ig light-chain gene rearrangements.

Both alleles of the *Igh* locus undergo D_H–J_H recombination in lymphoid progenitors (Alt et al., 1984) and subsequent contraction in pro-B cells (Fuxa et al., 2004). Hence, the two *Igh* alleles could in principle engage in V_H–DJ_H

recombination at the same time during pro-B cell development, which would result in the expression of two different functional Igμ proteins in a sizeable fraction of pro-B cells. However, biallelic Igμ expression is extremely rare (Barreto & Cumano, 2000), indicating that V_H–DJ_H recombination occurs at the two *Igh* alleles in a sequential and thus monoallelic manner, which is regulated by a process referred to as initiation of allelic exclusion (Mostoslavsky, Alt, & Rajewsky, 2004). It has been suggested that the initiation of allelic exclusion may be determined by differential epigenetic marking of the two *Igh* alleles during embryogenesis, which could select one allele to be the first to undergo V_H–DJ_H recombination in pro-B cells (Mostoslavsky et al., 2001). This hypothesis was recently tested by reconstitution of sublethally irradiated mice with a single hematopoietic stem cell, which demonstrated that each of the two *Igh* alleles had a similar probability to be the first to engage in V_H–DJ_H recombination (Alves-Pereira et al., 2014). Hence, this elegant experiment does not support differential epigenetic marking of the two *Igh* alleles in hematopoietic stem cells.

An alternative mechanism to initiate allelic exclusion may involve transient pairing of the two *Igh* alleles (Hewitt et al., 2009). This RAG-dependent process introduces a DNA break followed by V_H–DJ_H recombination at only one of the two paired *Igh* alleles, whereas the DNA break-induced activation of the repair–checkpoint kinase ATM leads to transient repositioning and thus sequestration of the second *Igh* allele at pericentromeric heterochromatin (Hewitt et al., 2009). As the choice of one of the paired *Igh* alleles to undergo RAG-dependent DNA cleavage is likely to be random, the pairing mechanism could potentially explain the stochastic initiation of allelic exclusion at the *Igh* locus (Alves-Pereira et al., 2014).

2.3 Subnuclear Relocation of the *Igh* Locus During B Cell Development

The nuclear periphery and pericentromeric heterochromatin are two repressive compartments that are important for propagating the inactive state of genes (Deniaud & Bickmore, 2009). Depending on their activity state, genes are repositioned between these repressive compartments and central nuclear positions that facilitate gene transcription. As shown by 3D DNA-FISH analysis, the *Igh* and *Igk* loci are located in their default state at the nuclear periphery in all non-B cells including uncommitted lymphoid progenitors (Kosak et al., 2002; Fig. 3). In this state, the *Igh* locus is anchored at the nuclear periphery via the distal V_H genes and is oriented with the 3′

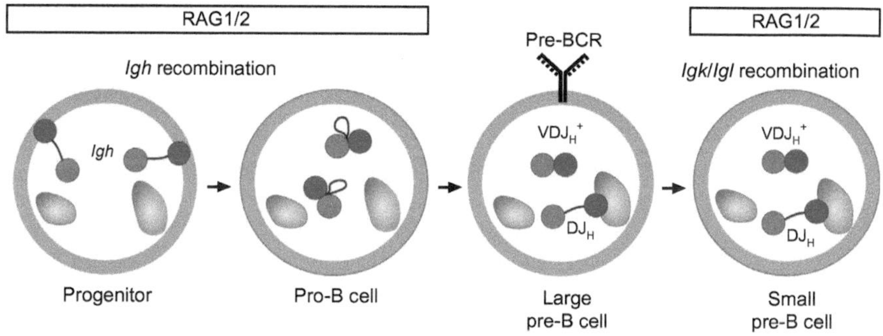

Figure 3 Subnuclear location of the *Igh* locus during early B lymphopoiesis. The subnuclear location of the *Igh* locus is shown at different stages of early B cell development. The distal V_H region (red) and proximal J_H–C_H domain (green) of the *Igh* locus are indicated together with their location relative to the repressive compartments at the nuclear periphery (gray) and pericentromeric heterochromatin (blue). The contraction and decontraction of the *Igh* alleles are schematically shown together with the expression of the RAG1/2 proteins. (See the color plate.)

proximal *Igh* domain toward the center of the nucleus, which facilitates D_H–J_H rearrangements in lymphoid progenitors (Fuxa et al., 2004; Fig. 3). At the onset of B cell development, the two *Igh* alleles together with both *Igk* loci are relocated from the nuclear periphery to more central positions within the nucleus (Kosak et al., 2002; Fig. 3). This subnuclear repositioning coincides with chromatin opening and germline transcription of the V_H genes in preparation for V_H–DJ_H recombination.

While both alleles of the *Igh* and *Igk* loci are repositioned to central nuclear positions in pro-B cells (Fuxa et al., 2004; Kosak et al., 2002), the two alleles behave differently at the next developmental stage in pre-B cells. Following successful V_H–DJ_H rearrangement of one of the two *Igh* alleles, pre-BCR signaling leads to repositioning of the second, incompletely rearranged *Igh* allele to repressive pericentromeric heterochromatin, whereas the functionally rearranged *Igh* allele remains in a central nuclear position consistent with its continuous expression in pre-B cells (Roldán et al., 2005; Fig. 3). Notably, one of the two *Igk* alleles is also recruited to pericentromeric heterochromatin in pre-B cells prior to *Igk* rearrangement, which favors V_κ–J_κ recombination on the second *Igk* allele that is located in euchromatic regions of the nucleus (Hewitt et al., 2008). Interestingly, pre-BCR signaling does not only lead to the relocation of the nonfunctional *Igh* allele and one *Igk* allele to pericentromeric heterochromatin (Roldán et al., 2005), but also results in interchromosomal pairing of these pericentromerically located *Igh* and *Igk* alleles (Hewitt et al., 2008). Moreover, the decontraction of the incompletely rearranged *Igh* allele may depend on the *Igh–Igk* pairing in a manner that requires a functional 3′ *Igk* enhancer (Hewitt et al., 2008). Future studies, involving live imaging, will be needed to carefully examine the role of pairing in the allelic exclusion process.

3. TRANSCRIPTION FACTORS AND REGULATORY ELEMENTS INVOLVED IN *Igh* LOCUS CONTRACTION

3.1 Transcription Factors Controlling *Igh* Locus Contraction

To date, only few *trans*-acting factors have been implicated in the control of locus contraction. 3D DNA-FISH analyses identified the B cell commitment factor Pax5 as the first transcriptional regulator to be involved in the pro-B-cell-specific contraction of the *Igh* locus (Fuxa et al., 2004). Pax5 was previously shown to control V_H–DJ_H rearrangements, as members of the large V_HJ558 gene family in the distal *Igh* region (black lines in Fig. 1) undergo recombination only with a ~100-fold lower efficiency in Pax5-deficient progenitors compared to wild-type pro-B cells, despite normal D_H–J_H recombination in both cell types (Nutt, Urbánek, Rolink, & Busslinger, 1997). Subsequent analyses revealed that only the first most proximal of the 200 V_H genes efficiently undergo V_H–DJ_H recombination in the absence of locus contraction, although all V_H genes are accessible at the chromatin level and give rise to germline V_H gene transcripts in Pax5-deficient progenitors (Hesslein et al., 2003; Roldán et al., 2005). A similar V_H–DJ_H recombination phenotype and lack of *Igh* locus contraction were observed in YY1-deficient pro-B cells, indicating that the ubiquitous transcription factor YY1, like Pax5, contributes to long-range looping at the *Igh* locus (Liu et al., 2007). Another potential regulator is the CCCTC-binding factor (CTCF), which has been implicated, through its association with the cohesin complex, in long-range chromatin looping at several complex loci (Merkenschlager & Odom, 2013). Indeed, shRNA knockdown experiments have suggested a role for the CTCF/cohesin complex in the regulation of looping at the *Igh* locus in pro-B cells (Degner et al., 2011). Consistent with this finding, CTCF and cohesin bind together to multiple sites along the V_H gene cluster of the *Igh* locus in lymphoid progenitors and committed pro-B cells irrespective of locus contraction (Degner et al., 2011; Ebert et al., 2011; Ebert, Medvedovic, Tagoh, Schwickert, & Busslinger,

2013; Medvedovic et al., 2013). Pax5 also interacts with numerous sites in the *Igh* locus (Ebert et al., 2011, 2013) as well as with CTCF (Medvedovic et al., 2013), suggesting that this protein–protein interaction may contribute to long-range looping in committed pro-B cells. In contrast, YY1 binds less frequently to the *Igh* locus, where most of its binding sites are located in the distal V_H gene region (Medvedovic et al., 2013). Interestingly however, YY1 also has the potential to interact with CTCF (Donohoe, Zhang, Xu, Shi, & Lee, 2007).

3.2 Putative Regulatory Elements Involved in *Igh* Locus Contraction

The V_H gene region likely contains *cis*-regulatory elements that mediate *Igh* locus contraction. For a long time however, the V_H gene promoters were the only known regulatory elements in the entire V_H gene cluster. By performing a systematic search for intergenic regions with active chromatin, we identified 14 Pax5-activated intergenic repeat (PAIR) elements in the distal V_H gene region of the *Igh* locus (Ebert et al., 2011; Fig. 4). Eleven of these PAIR elements are located upstream of a V_H gene belonging to the V_H3609 family and can be grouped into four subfamilies that evolved together with their associated V_H3609 genes (Ebert et al., 2011). The PAIR elements are bound by Pax5, E2A, YY1, CTCF, and cohesin, carry Pax5-dependent active chromatin, and give rise to Pax5-regulated noncoding antisense RNA exclusively in pro-B cells (Ebert et al., 2011; Medvedovic et al., 2013; Fig. 4). These antisense transcripts also depend on YY1 (Medvedovic et al., 2013; Verma-Gaur et al., 2012), but not on CTCF, as reduced CTCF expression upon shRNA-mediated knockdown leads to increased antisense transcription (Degner et al., 2011). Importantly, Pax5 no longer binds to PAIR elements in pre-B cells consistent with the loss of antisense RNA expression in these cells. The pro-B-cell-specific and Pax5-dependent activity therefore implicates the PAIR elements in the regulation of distal V_H–DJ_H recombination possibly by promoting *Igh* locus contraction (Ebert et al., 2011). As PAIR elements 4, 6, and 11 exhibit the highest transcriptional activity within the V_H gene cluster, they have been postulated to induce long-range looping by relocating the PAIR-containing V_H gene sequences to transcription factories where they interact with the 3′ proximal *Igh* domain containing the highly active E_μ enhancer (Verma-Gaur et al., 2012). As a consequence, these three PAIR elements may constitute interaction hot spots in the distal V_H gene region. However, 4C-sequencing experiments (see below) failed to confirm this hypothesis, as all V_H gene

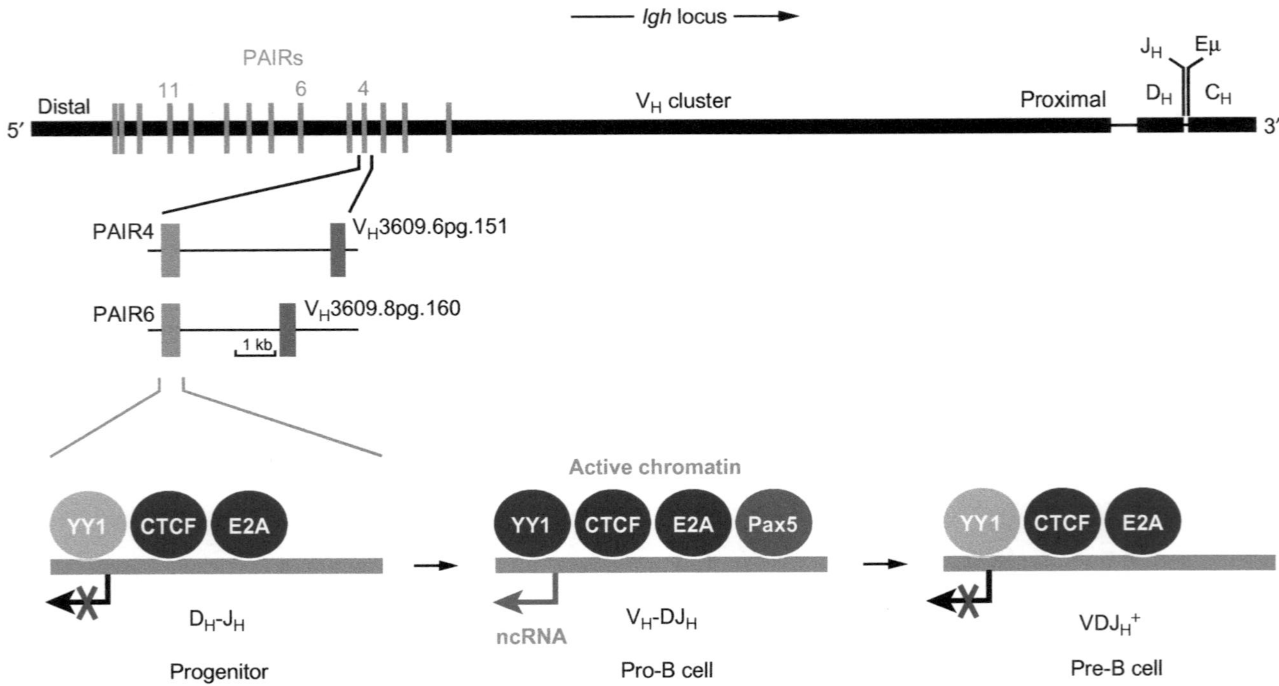

Figure 4 Pro-B-cell-specific activity of the PAIR elements in the distal *Igh* region. The location of the 14 PAIR elements is indicated together with their chromatin state, transcription factor occupancy, and transcriptional activity during early B cell development in uncommitted progenitors, committed pro-B cells, and small pre-B cells (Ebert et al., 2011). PAIR4, 6, and 11 are the three elements with the highest transcriptional activity (Verma-Gaur et al., 2012). The association of PAIR elements with a downstream located member of the V_H3609 gene family is exemplified for PAIR4 and 6. CTCF was shown to interact with cohesin at the PAIR elements (Ebert et al., 2011). The binding of YY1 was so far analyzed only in pro-B cells by ChIP-seq. Binding of YY1 in the progenitors and pre-B cells (light blue) is currently hypothetical and only inferred due to the ubiquitous expression of this transcription factor. ncRNA, noncoding RNA. (See the color plate.)

sequences contact the 3′ proximal *Igh* domain with similar probabilities (Medvedovic et al., 2013; Fig. 5).

The E_{μ} enhancer in the 3′ *Igh* domain has been implicated in the control of *Igh* locus contraction and thus V_H–DJ_H recombination (Guo, Gerasimova, et al., 2011). This conclusion was mainly based on 3D DNA-FISH analyses demonstrating that the *Igh* locus is present in an extended conformation in $E_{\mu}^{-/-}$ *Rag2*$^{-/-}$ pro-B cells in contrast to its contracted state in *Rag2*$^{-/-}$ pro-B cells (Guo, Gerasimova, et al., 2011). However, by performing 3D DNA-FISH and 4C-sequencing experiments examining pro-B cells of the same genotypes, a role for the E_{μ} enhancer in the regulation of *Igh* locus contraction could not be confirmed (Medvedovic et al., 2013). Previous analyses revealed an essential role of the E_{μ} enhancer in controlling the chromatin accessibility and transcriptional activation of the 3′ proximal *Igh* domain, as active histone marks (H3K4me3 and H3K9ac), germline transcripts (μ^0 and I_{μ}), and D_H–J_H recombination are severely reduced in $E_{\mu}^{-/-}$ pro-B cells (Afshar et al., 2006; Chakraborty et al., 2009; Perlot et al., 2005). Hence, the observed block of V_H–DJ_H recombination is likely a secondary consequence of the strongly reduced number of DJ_H-rearranged *Igh* alleles in $E_{\mu}^{-/-}$ pro-B cells (Afshar et al., 2006; Perlot et al., 2005). Notably, the recombination of proximal and distal V_H genes is equally affected by deletion of the E_{μ} enhancer (Perlot et al., 2005) in contrast to the preferential loss of distal V_H–DJ_H recombination in the absence of locus contraction (Fuxa et al., 2004). Moreover, germline V_H gene transcription is unaffected in $E_{\mu}^{-/-}$ pro-B cells, further indicating that the activity of the E_{μ} enhancer is restricted to the 3′ proximal *Igh* domain (Afshar et al., 2006; Perlot et al., 2005).

CTCF has been implicated not only in chromosome looping but also in chromatin boundary formation and transcriptional insulation (Phillips & Corces, 2009). A large array of regularly spaced CTCF-binding sites characterizes the V gene clusters of all antigen receptor loci, whereas the 3′ proximal domains contain CTCF-binding sites only at their 5′ and 3′ boundaries (Degner et al., 2011; Ebert et al., 2011, 2013; Ribeiro de Almeida et al., 2011; Seitan et al., 2011; Shih et al., 2012). The CTCF-binding sites at the 5′ end of the proximal domain act as insulators. This is best exemplified by the intergenic control region 1 (IGCR1), which contains two CTCF-binding elements (CBEs) located 2 kb upstream of the D_HFL16.1 gene segment in the intervening V_H–D_H region of the *Igh* locus (Guo, Yoon, et al., 2011; Fig. 1). Specific mutation of both or individual CBEs demonstrated that the two CTCF-binding sites cooperatively function as an insulator

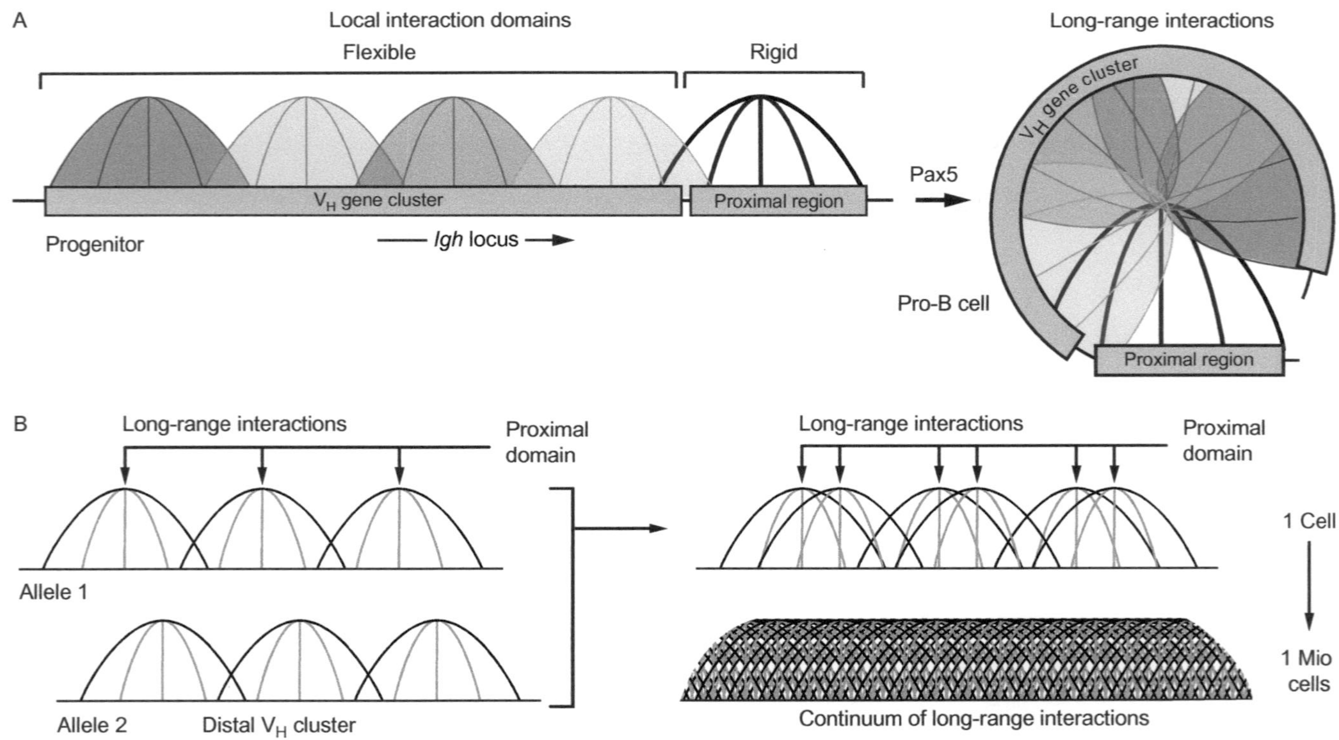

Figure 5 Flexible long-range loops across the V_H gene cluster. (A) Cartoon depicting the *Igh* locus in its "extended" conformation with overlapping local interaction domains and in its contracted state with long-range interactions connecting the local interaction domains. (B) Simplified diagram indicating how flexible local interactions in the V_H gene region may result in a continuum of chromatin loops across the V_H gene cluster in a population of 1 million pro-B cells, which is normally used for 4C-sequencing. (See the color plate.)

element to regulate ordered and lineage-specific V(D)J recombination (Guo, Yoon, et al., 2011; Lin, Guo, Su, Zhang, & Alt, 2015). In uncommitted lymphoid progenitors, the IGCR1/CBE insulator restricts the potent E_{μ} enhancer activity as well as genomic interactions to the 3′ proximal domain, thereby generating a local chromatin structure that promotes D_H–J_H recombination on both *Igh* alleles in lymphoid progenitors (Guo, Yoon, et al., 2011). During normal development, the IGCR1/CBE insulator is only neutralized in committed pro-B cells, possibly by Pax5-dependent contraction of the *Igh* locus. In IGCR1/CBE mutant mice, the activity of the Eμ enhancer is, however, no longer restricted to the 3′ proximal domain and can therefore induce premature V_H–DJ_H recombination involving the most proximal V_H genes (Guo, Yoon, et al., 2011). As a consequence of the precocious proximal V_H gene activity in lymphoid progenitors, fewer DJ_H-rearranged *Igh* alleles are available for distal V_H–DJ_H recombination in committed pro-B cells undergoing *Igh* locus contraction, which leads to a strong skewing of the *Igh* repertoire toward proximal V_H gene rearrangements. Hence, the CTCF-regulated IGCR1 activity and Pax5-mediated locus contraction fulfill opposing roles by preventing or promoting the participation of V_H genes in V(D)J recombination in lymphoid progenitors or committed pro-B cells, respectively.

4. LONG-RANGE INTERACTIONS CONTROLLING V_H–DJ_H RECOMBINATION IN PRO-B CELLS

4.1 Overlapping Topological Domains at the *Igh* Locus in Uncommitted Progenitors

The spatial conformation of the *Igh* locus has largely been studied in single cells at low resolution by DNA-FISH analysis. High-resolution analysis is, however, required to learn more about the pattern of chromatin loops that cause *Igh* locus contraction. Comprehensive mapping of chromatin loops at high resolution is usually performed by different variants of the chromosome conformation capture technique (Dekker, Rippe, Dekker, & Kleckner, 2002), such as Hi-C and 4C-sequencing (Lieberman-Aiden et al., 2009; van de Werken et al., 2012). However, these high-resolution methods do currently not allow the analysis of single cells, as they can only be performed with a relative large cell population (Lieberman-Aiden et al., 2009; van de Werken et al., 2012). Our recent 4C-seq analyses revealed that the *Igh* locus in its "extended" configuration consists of distinct local interaction domains ranging from 0.5 Mb (3′ proximal region) to 1.3 Mb (5′ PAIR region) in thymocytes and uncommitted *Pax5*$^{-/-}$ progenitors (Medvedovic et al., 2013; Fig. 5A). These local interaction domains may consist of rosettes of chromatin loops, which have been predicted to be the organizational units of the *Igh* locus in B cell progenitors based on comparison of experimental and simulated spatial distance measurements (Jhunjhunwala et al., 2008). Global Hi-C analyses of embryonic stem, pre-pro-B, and pro-B cells revealed that the entire mouse genome is organized into local chromatin interaction regions referred to as topological domains, which range from 0.5 to 3 Mb in size (Dixon et al., 2012; Lin et al., 2012; Nora et al., 2012). The local interaction domains of the *Igh* locus also fit this size range, exhibit a similar 4C-seq pattern in lymphoid progenitors and thymocytes, and may thus correspond to topological domains, which are largely invariant between different cell types (Dixon et al., 2012; Lin et al., 2012). Hence, the local interaction domains may constitute the default folding state of the "extended" *Igh* locus in non-B cells.

Previous 3C analyses of the 3′ proximal *Igh* domain identified loops between the IGCR1, E_{μ}, 3′ regulatory region (3′RR), and 3′CBE sequences (Degner et al., 2011; Guo, Gerasimova, et al., 2011; Guo, Yoon, et al., 2011; see Fig. 1). In addition to the known chromatin loops, 4C-seq analysis detected novel interactions of these regulatory elements with the Cγ1-Cγ2b intervening region as well as with the first proximal V_H genes in thymocytes, lymphoid progenitors, and committed pro-B cells (Medvedovic et al., 2013). The incorporation of these V_H genes into the local 3D architecture of the 3′ proximal *Igh* domain explains why the first proximal V_H genes are able to undergo V_H–DJ_H recombination in the absence of locus contraction in Pax5-deficient progenitors (Hesslein et al., 2003; Roldán et al., 2005). The same V_H genes also interact with the 3′ proximal *Igh* domain in thymocytes (Medvedovic et al., 2013), although they do not have the ability to undergo V_H–DJ_H recombination in developing thymocytes, unless the two CBEs of the IGCR1 insulator are inactivated (Guo, Yoon, et al., 2011). Hence, we suggest that a functional IGCR1 element prevents V_H–DJ_H recombination even if V_H genes participate in looping with the 3′ proximal *Igh* domain in thymocytes.

4.2 Flexible Long-Range Loops Across the V_H Gene Region in Pro-B Cells

In committed pro-B cells, all local interaction domains engage in long-range looping across the *Igh* locus, which depends on Pax5, YY1, and CTCF as shown by 4C-seq analysis (Medvedovic et al., 2013; Fig. 5A). The 5′ and

3′ boundaries near the first 5′ V_HJ558 gene and at the 3′CBE region (Fig. 1) restrict these long-range interactions to the *Igh* locus, which constitutes one large chromosome-folding unit (Medvedovic et al., 2013). Notably, locus contraction did not affect the interactions within the local domains, but rather provided an additional layer of long-range looping between these domains (Fig. 5A). As shown by 4C-sequencing, the 3′ proximal region, which contains the recombination center of the *Igh* locus (Ji et al., 2010), assumes a rigid chromatin structure, where most long-range loops from the V_H gene region contact the IGCR1 and 3′CBE regions containing the only CTCF/cohesin-binding sites in the 3′ proximal region (Medvedovic et al., 2013). Unexpectedly however, these long-range genomic interactions appear to initiate at a multitude of sites in the V_H gene region, which results in a continuum of flexible long-range interactions across the entire V_H gene cluster in a population of pro-B cells (Medvedovic et al., 2013; Fig. 5B). Given the critical role of the CTCF/cohesin complex in chromatin looping (Merkenschlager & Odom, 2013), it is conceivable that the regularly spaced CTCF/cohesin-binding sites anchor the long-range loops in the V_H gene region. Importantly, the pattern of chromatin loops, which define the local interaction domains within the distal V_H gene region in Pax5-deficient progenitors (prior to locus contraction), is similar to the corresponding long-range interaction pattern observed for the same V_H gene regions in committed pro-B cells (undergoing locus contraction) (Medvedovic et al., 2013). This evidence strongly suggests that the observed flexible long-range looping may be a characteristic feature of the local interaction domains in the V_H gene region (Fig. 5B). We therefore hypothesize that the flexible long-range interactions provide, at the pro-B cell population level, a similar probability for each V_H gene to be juxtaposed next to the DJ_H-rearranged gene segment in the 3′ proximal *Igh* domain. By promoting equal participation of all V_H genes in V_H–DJ_H recombination, flexible long-range loops likely contribute to the generation of a diverse antibody repertoire, which is essential for an effective immune response to foreign pathogens. It is, however, important to note that individual V_H genes rearrange at different frequencies in pro-B cells (Choi et al., 2013). Hence, in addition to the essential requirement for locus contraction, the recombination rate of a given V_H gene is also determined by other parameters, including the functionality of the V_H RSS elements (Choi et al., 2013).

4C-sequencing and 3D DNA-FISH analyses furthermore revealed that individual mutation of the IGCR1 element, E_μ, or 3′RR enhancer in the 3′

proximal *Igh* domain does not affect locus contraction (Medvedovic et al., 2013). These data further support the notion that the V_H gene region and 3′ proximal domain of the *Igh* locus fulfill distinct roles in V(D)J recombination. The primary function of the flexible long-range loops in the V_H gene cluster may be to present the V_H gene repertoire to the 3′ proximal domain, where V_H–DJ_H recombination takes place.

The spatial movement and proximity of DNA elements within the *Igh* locus were recently investigated by live imaging of pro-B cells containing insertions of Tet-operator binding sites in the distal V_H gene region or upstream of the Cμ exon in the 3′ proximal *Igh* domain (Lucas, Zhang, Dudko, & Murre, 2014). The 3D trajectories of these Tet-operator sequences in pro-B cell nuclei were visualized by binding of a Tet-repressor-GFP protein, which revealed that these elements display fractional Langevin motion due to the viscoelastic hindrance from the surrounding network of V_H and D_HJ_H segments, neighboring DNA regions, and other factors associated with the chromatin fiber (Lucas et al., 2014). Mathematical modeling indicated that the D_HJ_H elements encounter a V_H gene within minutes in this viscoelastic environment, which is strongly determined by the spatial confinement imposed by contraction of the *Igh* locus in pro-B cells. According to this model, the viscoelastic nature of the nuclear environment allows the distinct *Igh* elements to bounce back and forth in a spring-like fashion until specific contacts are established (Lucas et al., 2014).

4.3 Control of Sequential Rearrangements at the Recombination Center

The accessibility hypothesis, first reported 30 years ago, postulates that the initial step in the control of V(D)J recombination must be chromatin changes, which render the different gene segments of the antigen receptor loci accessible to the V(D)J recombination machinery (Yancopoulos & Alt, 1985). In the meantime, it is well established that the 3′ proximal domain of antigen receptor loci is activated at the chromatin level by enhancer-mediated recruitment of chromatin-remodeling and histone-modifying enzymes (Chakraborty et al., 2009; Mathieu, Hempel, Spicuglia, Verthuy, & Ferrier, 2000; McMurry & Krangel, 2000). As a consequence, the intronic $E_μ$ enhancer and adjacent J_H segments are characterized by an abundance of the three active histone marks H3K4me2, H3K4me3, and H3K9ac in pro-B cells (Chakraborty et al., 2009; Malin et al., 2010; Fig. 6), resulting in strong expression of the germline transcripts $μ^0$ and $I_μ$ (Verma-Gaur et al., 2012). However, it seems less clear what defines the accessible state of the V_H genes prior to V_H–DJ_H

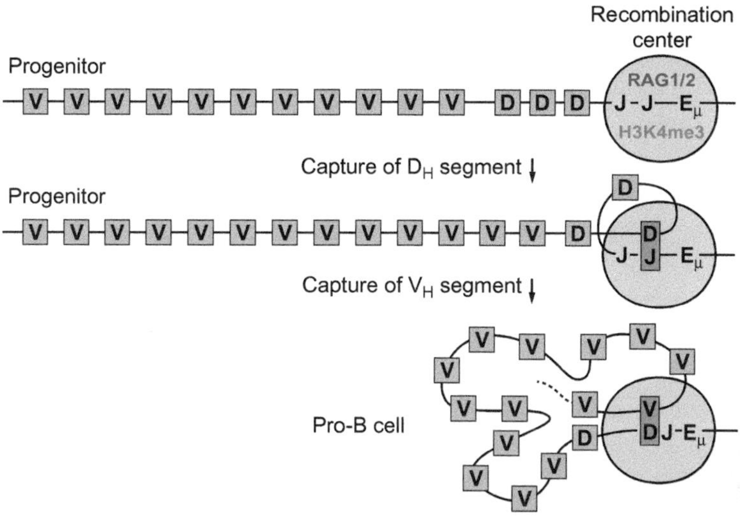

Figure 6 Active chromatin and the control of sequential rearrangements at the *Igh* locus. According to the recombination center model (Ji et al., 2010), the E_{μ} enhancer activates in lymphoid progenitors the proximal J_H gene region of the *Igh* locus as a recombination center embedded in active chromatin. The RAG1/2 complex (red; gray in the print version) is recruited into the recombination center through binding of the RAG2 PHD finger to the active histone modification H3K4me3 (green; light gray in the print version). The tethered RAG1/2 complex captures one of the D_H gene segments leading to D_H–J_H recombination in progenitors followed by the subsequent capture of one of the many V_H gene segments upon *Igh* locus contraction in pro-B cells.

recombination. For instance, the V_H genes of the *Igh* locus are characterized by nucleosomal reorganization and increased DNase I hypersensitivity in *Rag2*$^{-/-}$ pro-B cells (Maes et al., 2001), which may explain the observed low level of germline V_H gene transcription (Verma-Gaur et al., 2012; Yancopoulos & Alt, 1985). Initial chromatin immunoprecipitation (ChIP) experiments implicated histone acetylation as an important mechanism for activating chromatin across the V gene region (Chowdhury & Sen, 2001; Tripathi, Jackson, & Krangel, 2002). However, systematic chromatin profiling of the *Igh* locus by ChIP-chip and ChIP-seq analyses subsequently revealed that the active histone marks H3K4me2, H4K4me3, and H3K9ac are either absent or only present at low levels at most V_H genes except for members of the distal V_H3609 gene family that carry detectable levels of H3K4me2 and H3K9ac (Choi et al., 2013; Malin et al., 2010). Hence, most V_H genes do not exhibit an active histone modification signature (H3K4me2$^+$ H4K4me3$^+$ H3K9ac$^+$) characteristic of expressed genes, and yet their chromatin must be accessible enough to facilitate V_H–DJ_H recombination in pro-B cells.

A possible explanation for this asymmetric distribution of active chromatin at the *Igh* locus and other antigen receptor loci may be provided by the fact that the RAG2 protein contains at its C-terminus a plant homeodomain finger that specifically binds the active histone mark H3K4me3 (Matthews et al., 2007) at active promoters throughout the genome (Ji et al., 2010). Importantly, RAG2 also binds across islands of H3K4me3 that are associated with J gene segments of the different antigen receptor loci (*Igh*, *Igk*, *Tcrb*, and *Tcra/d*) (Ji et al., 2010; Fig. 6). In contrast, the RAG1 protein interacts in a more restrictive manner only with the RSS elements of the accessible J gene segments and thus further contributes to the focal and stable recruitment of the RAG1/2 complex to the active chromatin region in the 3′ proximal region of antigen receptor loci, which is referred to as the recombination center (Ji et al., 2010; Fig. 6). As exemplified for the *Igh* locus, the RAG1/2 complex, once tethered to the J_H segments at the recombination center, can capture a partner RSS element of a D_H segment to undergo synapse formation (Fig. 6). Upon completion of D_H–J_H recombination in lymphoid progenitors, the rearranged D_H element is also incorporated into the active chromatin domain and is bound by the RAG1/2 complex that can now capture an RSS element of one of the many V_H gene segments to undergo V_H–DJ_H rearrangement in pro-B cells. The H3K4me3-mediated focal targeting of the RAG1/2 endonuclease to the recombination center is a possible reason why active chromatin is asymmetrically enriched at the proximal J_H region and not spread throughout the D_H and V_H gene clusters in pro-B cells. Importantly, the localized activity of the RAG1/2 complex at the proximal chromatin hub is strictly dependent on locus contraction to present the V genes to the recombination center for subsequent generation of a functional V(D)J-rearranged antigen receptor gene (Fig. 6).

5. LOCUS CONTRACTION AS A GENERAL PRINCIPLE CONTROLLING RECOMBINATION AT ANTIGEN RECEPTOR LOCI

5.1 *Igk* Locus Contraction

Due to their large size, other antigen receptor loci (*Igk*, *Tcrb*, and *Tcra/d*) also exhibit reversible contraction at the developmental stage, where they undergo V–(D)J recombination (Roldán et al., 2005; Skok et al., 2007). For instance, the mouse *Igk* locus with its 140 V_κ genes spread over a 3-Mb region is contracted in small pre-B cells and early immature B cells undergoing primary and secondary V_κ–J_κ rearrangements, respectively, and undergoes locus decontraction already in late immature B cells of the bone marrow and mature B cells of the spleen (Roldán et al., 2005). Although the *Igk* locus was initially reported to be in an extended configuration in pro-B cells (Roldán et al., 2005), it was later shown to be contracted in *ex vivo* sorted pro-B cells (Hewitt, Chaumeil, & Skok, 2010). The latter result is consistent with the observation that a small fraction (~15%) of pro-B cells undergo V_κ–J_κ rearrangements (Novobrantseva et al., 1999). The intervening sequences between the V_κ and J_κ segments contain six DNase I hypersensitive sites (HS), which define the Cer element (HS1–2) (Xiang, Park, & Garrard, 2013) and the more proximally located Sis element (HS3–6) (Liu et al., 2006). Both of these elements contain CTCF-binding sites and control the recombination frequency of proximal V_κ genes (Xiang et al., 2013; Xiang, Zhou, Hewitt, Skok, & Garrard, 2011). The Sis element functions as a classical insulator, as its deletion strongly increases the recombination of proximal V_κ genes at the expense of distal V_κ rearrangements (Xiang et al., 2011). Inactivation of the Cer element resulted in an even stronger increase of proximal V_κ gene recombination and a corresponding decrease of distal V_κ rearrangements (Xiang et al., 2013). Loss of the Cer element furthermore gave rise to partial *Igk* decontraction in pre-B cells, leading to the hypothesis that this element may control long-range looping at the *Igk* locus (Xiang et al., 2013). Surprisingly however, deletion of the Cer element also led to the appearance of oligoclonal V_κ–J_κ rearrangements in DP thymocytes (Xiang et al., 2013) similar to the presence of proximal V_H–DJ_H rearrangements in T cells of IGCR1/CBE mutant mice (Guo, Yoon, et al., 2011). As the V_κ–J_κ rearrangements detected in thymocytes likely originate in lymphoid progenitors seeding the thymus, it is conceivable that the Cer element also functions as a strong insulator to prevent premature V_κ–J_κ rearrangements in early lymphopoiesis in analogy to the role of the IGCR1/CBE element at the *Igh* locus. Similar to loss of the CTCF-binding Cer and Sis elements, conditional inactivation of *Ctcf* also skewed the repertoire of V_κ–J_κ rearrangements by strongly increasing the interactions of the iE_κ and $3'E_\kappa$ enhancers with proximal V_κ genes (Ribeiro de Almeida et al., 2011). These data suggest that the Cer and Sis elements function as insulators to negatively regulate proximal V_κ gene recombination by blocking the long-range activity of the E_κ enhancers.

5.2 *Tcrb* Locus Contraction

The mouse *Tcrb* locus with a length of 670 kb contains 29 of its 31 V_β genes in a central cluster that is separated from the two $DJC_\beta1$ and $DJC_\beta2$ regions in the *Tcrb* 3′ proximal domain by a 250-kb spacer sequence comprising 10 trypsinogen genes (Jackson & Krangel, 2006). Contraction of the *Tcrb* locus takes place only in DN pro-T cells in the thymus by looping out the trypsinogen spacer region, which likely facilitates V_β–DJ_β recombination (Skok et al., 2007). In analogy to the *Igh* locus, decontraction at the next developmental stage likely contributes to allelic exclusion of the *Tcrb* locus in DP thymocytes undergoing V_α–J_α rearrangements at the *Tcra/d* locus (Skok et al., 2007). Unexpectedly, the distal V_β genes (V_β1–15) differ from the more proximal V_β genes (V_β16–30) in the central cluster with regard to their looping to the recombination center located in the $DJC_\beta1$ and $DJC_\beta2$ regions (Majumder et al., 2015). Only the distal V_β genes (V_β1–15) depend for these long-range interactions on the function of two newly identified elements that are located upstream of the $D_\beta1$ promoter in the trypsinogen spacer region. The 5′ located CBE (termed 5′PC) appears to tether the distal V_β genes to the recombination center possibly through interactions with CTCF/cohesin-binding sites in the distal V_β gene region. The more proximally located element functions as a chromatin boundary to prevent inactivation of the upstream tether element by inhibiting the spreading of hyperactive chromatin from the recombination center (Majumder et al., 2015).

5.3 *Tcra/d* Locus Contraction

The *Tcra/d* locus with a size of 1.6 Mb has a unique and complex structure, as the D_δ, J_δ, and C_δ segments of *Tcrd* are embedded within the sequences of *Tcra* (Krangel, Carabana, Abbarategui, Schlimgen, & Hawwari, 2004). Moreover, *Tcrd* undergoes V(D)J recombination together with the *Tcrg* locus in DN pro-T cells, where functional rearrangement of both loci initiates γδ T cell development. In contrast, *Tcra* rearranges under the control of the E_α enhancer in DP thymocytes of the αβ T cell lineage (Krangel et al., 2004). Approximately 100 V genes are located in a 1.5-Mb region upstream of the DJC_δ and JC_α clusters. A smaller subset of these V genes take part in V_δ–DJ_δ rearrangements in DN pro-T cells that undergo contraction along the entire *Tcra/d* locus (Shih & Krangel, 2010). In contrast, all V genes participate in V_α–J_α recombination in DP thymocytes, where the *Tcra/d* locus is also contracted by looping, which is followed by decontraction at the subsequent CD4 single-positive (SP) and CD8 SP thymocyte stages (Skok et al., 2007). Interestingly however, only the 3′ half of the *Tcra/d* locus is efficiently contracted in DP thymocytes as shown by 3D DNA-FISH analysis with different probes (Shih & Krangel, 2010; Skok et al., 2007). At face value, this contraction pattern seems to support the model of coordinated bidirectional V_α–J_α recombination, which predicts that the proximal (3′) V_α genes undergo primary and secondary rearrangements preferentially with 5′ J_α segments followed by further rearrangements of middle and distal (5′) V_α genes involving central and 3′ J_α elements (Krangel et al., 2004; Pasqual et al., 2002). As each V_α–J_α recombination leads to the deletion of the intervening sequences, more distally located V_α genes are brought under the regulatory influence of the E_α enhancer and can thus participate in secondary V_α–J_α rearrangements. The restrictions imposed by the coordinated bidirectional recombination model should result in a limited repertoire of *Tcra* rearrangements, as only a subset of V_α and J_α segments are allowed to recombine with each other (Pasqual et al., 2002). RNA sequencing of naïve $CD8^+$ T cells revealed, however, a highly diverse *Tcra* repertoire, which comprised nearly all possible V_α–J_α combinations and can thus efficiently contribute to protective immunity (Genolet, Stevenson, Farinelli, Osteras, & Luescher, 2012a, 2012b). How the preferential contraction observed in the 3′ half of the *Tcra/d* locus contributes to the generation of this highly diverse *Tcra* repertoire in DP thymocytes remains enigmatic and will require further investigation.

A unique feature of the *Tcra* locus is the presence of a CTCF/cohesin-binding site at the E_α enhancer (Seitan et al., 2011; Shih et al., 2012), which is contrasted by the absence of CTCF/cohesin-binding sites at enhancers of all other antigen receptor loci (Degner et al., 2011; Ebert et al., 2011; Ribeiro de Almeida et al., 2011). Conditional inactivation of *Ctcf* or the cohesin subunit gene *Rad21* in DP thymocytes revealed that a major role of the CTCF-cohesin complex is to mediate long-range interactions between the E_α enhancer and upstream promoter elements (Seitan et al., 2011; Shih et al., 2012). These interactions shape the chromatin architecture in the 3′ proximal domain, promote *Tcra* transcription, induce H3K4 trimethylation, and recruit the recombination machinery to induce V_α–J_α recombination (Seitan et al., 2011; Shih et al., 2012). Unexpectedly however, contraction of the *Tcra/d* locus still takes place in CTCF-deficient DP thymocytes, suggesting that the CTCF–cohesin complex may not play a major role in controlling *Tcra/d* contraction (Shih et al., 2012).

In summary, all major antigen receptor loci (*Igh*, *Igk*, *Tcrb*, and *Tcra/d*) undergo reversible contraction at the developmental stage, where they undergo V–(D)J recombination (Roldán et al., 2005; Skok et al., 2007). Hence, reversible locus contraction by looping is a general regulatory principle that promotes the spatial communication between V genes and the 3′ proximal domain, allowing the participation of a wide spectrum of V genes in the V–(D)J recombination reaction.

6. PERSPECTIVE

Despite the essential role of locus contraction for the generation of a diverse antigen receptor repertoire, little is yet known about the molecular mechanisms controlling locus contraction. Although Pax5, YY1, and the CTCF/cohesin complex have been characterized as crucial regulators of *Igh* locus contraction, it is at present unclear how they contribute to this process. For instance, the binding pattern of the CTCF/cohesin complex is similar across the *Igh* locus in the presence as well as the absence of locus contraction, raising the question what signals or factors allow this complex to participate in long-range interactions only in pro-B cells. Although the B-cell-specific regulator Pax5 is a prime candidate for this activity, the mechanism of Pax5 action is also largely unknown except for the finding that CTCF and Pax5 may bind each other *in vivo*. Further insight will require the identification of essential Pax5-interacting proteins that mediate the role of Pax5 in locus contraction. Moreover, the role of the PAIR elements in V_H–DJ_H recombination needs to be clarified by deletion of these sequences in the endogenous *Igh* locus. Finally, we note that even less is known about *trans*-acting factors or *cis*-regulatory elements that regulate locus contraction at other antigen receptor loci. Elucidating the molecular mechanism of locus contraction will remain an exciting and challenging task for future studies.

ACKNOWLEDGMENTS

This work was supported by Boehringer Ingelheim, the Vienna Science and Technology Fund (WWTF), and an ERC Advanced Grant (291740-LymphoControl) from the European Community's Seventh Framework Program.

REFERENCES

Afshar, R., Pierce, S., Bolland, D. J., Corcoran, A., & Oltz, E. M. (2006). Regulation of IgH gene assembly: Role of the intronic enhancer and 5′D_{Q52} region in targeting D_HJ_H recombination. *The Journal of Immunology*, *176*(4), 2439–2447.

Alt, F. W., Yancopoulos, G. D., Blackwell, T. K., Wood, C., Thomas, E., Boss, M., et al. (1984). Ordered rearrangement of immunoglobulin heavy chain variable region segments. *The EMBO Journal*, *3*(6), 1209–1219.

Alves-Pereira, C. F., de Freitas, R., Lopes, T., Gardner, R., Marta, F., Vieira, P., et al. (2014). Independent recruitment of *Igh* alleles in V(D)J recombination. *Nature Communications, 5*, 5623–5637.

Barreto, V., & Cumano, A. (2000). Frequency and characterization of phenotypic Ig heavy chain allelically included IgM-expressing B cells in mice. *The Journal of Immunology, 164*(2), 893–899.

Bassing, C. H., Swat, W., & Alt, F. W. (2002). The mechanism and regulation of chromosomal V(D)J recombination. *Cell, 109*(Suppl), S45–S55.

Bolland, D. J., Wood, A. L., Johnston, C. M., Bunting, S. F., Morgan, G., Chakalova, L., et al. (2004). Antisense intergenic transcription in V(D)J recombination. *Nature Immunology, 5*(6), 630–637.

Bossen, C., Mansson, R., & Murre, C. (2012). Chromatin topology and the regulation of antigen receptor assembly. *Annual Review of Immunology, 30*, 337–356.

Chakraborty, T., Perlot, T., Subrahmanyam, R., Jani, A., Goff, P. H., Zhang, Y., et al. (2009). A 220-nucleotide deletion of the intronic enhancer reveals an epigenetic hierarchy in immunoglobulin heavy chain locus activation. *The Journal of Experimental Medicine, 206*(5), 1019–1027.

Choi, N. M., Loguercio, S., Verma-Gaur, J., Degner, S. C., Torkamani, A., Su, A. I., et al. (2013). Deep sequencing of the murine *Igh* repertoire reveals complex regulation of nonrandom V gene rearrangement frequencies. *The Journal of Immunology, 191*(5), 2393–2402.

Chowdhury, D., & Sen, R. (2001). Stepwise activation of the immunoglobulin μ heavy chain gene locus. *The EMBO Journal, 20*(22), 6394–6403.

Degner, S. C., Verma-Gaur, J., Wong, T. P., Bossen, C., Iverson, G. M., Torkamani, A., et al. (2011). CCCTC-binding factor (CTCF) and cohesin influence the genomic architecture of the *Igh* locus and antisense transcription in pro-B cells. *Proceedings of the National Academy of Sciences of the United States of America, 108*(23), 9566–9571.

Dekker, J., Rippe, K., Dekker, M., & Kleckner, N. (2002). Capturing chromosome conformation. *Science, 295*(5558), 1306–1311.

Deniaud, E., & Bickmore, W. A. (2009). Transcription and the nuclear periphery: Edge of darkness? *Current Opinion in Genetics & Development, 19*(2), 187–191.

Dixon, J. R., Selvaraj, S., Yue, F., Kim, A., Li, Y., Shen, Y., et al. (2012). Topological domains in mammalian genomes identified by analysis of chromatin interactions. *Nature, 485*(7398), 376–380.

Donohoe, M. E., Zhang, L. F., Xu, N., Shi, Y., & Lee, J. T. (2007). Identification of a Ctcf cofactor, Yy1, for the X chromosome binary switch. *Molecular Cell, 25*(1), 43–56.

Ebert, A., McManus, S., Tagoh, H., Medvedovic, J., Salvagiotto, G., Novatchkova, M., et al. (2011). The distal V_H gene cluster of the *Igh* locus contains distinct regulatory elements with Pax5 transcription factor-dependent activity in pro-B cells. *Immunity, 34*(2), 175–187.

Ebert, A., Medvedovic, J., Tagoh, H., Schwickert, T. A., & Busslinger, M. (2013). Control of antigen receptor diversity through spatial regulation of V(D)J recombination. *Cold Spring Harbor Symposia on Quantitative Biology, 78*, 11–21.

Fuxa, M., Skok, J., Souabni, A., Salvagiotto, G., Roldán, E., & Busslinger, M. (2004). Pax5 induces *V*-to-*DJ* rearrangements and locus contraction of the *immunoglobulin heavy-chain* gene. *Genes and Development, 18*, 411–422.

Genolet, R., Stevenson, B. J., Farinelli, L., Osteras, M., & Luescher, I. F. (2012a). Highly diverse TCRα chain repertoire of pre-immune $CD8^+$ T cells reveals new insights in gene recombination. *The EMBO Journal, 31*(7), 1666–1678.

Genolet, R., Stevenson, B. J., Farinelli, L., Osteras, M., & Luescher, I. F. (2012b). Highly diverse TCRα chain repertoire of pre-immune $CD8^+$ T cells reveals new insights in gene recombination. *The EMBO Journal, 31*(21), 4247–4248.

Guo, C., Gerasimova, T., Hao, H., Ivanova, I., Chakraborty, T., Selimyan, R., et al. (2011). Two forms of loops generate the chromatin conformation of the immunoglobulin heavy-chain gene locus. *Cell, 147*(2), 332–343.

Guo, C., Yoon, H. S., Franklin, A., Jain, S., Ebert, A., Cheng, H. L., et al. (2011). CTCF-binding elements mediate control of V(D)J recombination. *Nature, 477*(7365), 424–430.

Hawwari, A., & Krangel, M. S. (2005). Regulation of TCR δ and α repertoires by local and long-distance control of variable gene segment chromatin structure. *The Journal of Experimental Medicine, 202*(4), 467–472.

Hesslein, D. G. T., Pflugh, D. L., Chowdhury, D., Bothwell, A. L. M., Sen, R., & Schatz, D. G. (2003). *Pax5* is required for recombination of transcribed, acetylated, 5′ IgH V gene segments. *Genes and Development, 17*(1), 37–42.

Hewitt, S. L., Chaumeil, J., & Skok, J. A. (2010). Chromosome dynamics and the regulation of V(D)J recombination. *Immunological Reviews, 237*(1), 43–54.

Hewitt, S. L., Farmer, D., Marszalek, K., Cadera, E., Liang, H. E., Xu, Y., et al. (2008). Association between the *Igk* and *Igh* immunoglobulin loci mediated by the 3′ *Igk* enhancer induces 'decontraction' of the *Igh* locus in pre-B cells. *Nature Immunology, 9*(4), 396–404.

Hewitt, S. L., Yin, B., Ji, Y., Chaumeil, J., Marszalek, K., Tenthorey, J., et al. (2009). RAG-1 and ATM coordinate monoallelic recombination and nuclear positioning of immunoglobulin loci. *Nature Immunology, 10*(6), 655–664.

Igarashi, H., Gregory, S. C., Yokota, T., Sakaguchi, N., & Kincade, P. W. (2002). Transcription from the RAG1 locus marks the earliest lymphocyte progenitors in bone marrow. *Immunity, 17*(2), 117–130.

Jackson, A. M., & Krangel, M. S. (2006). Turning T-cell receptor β recombination on and off: More questions than answers. *Immunological Reviews, 209*, 129–141.

Jankovic, M., Casellas, R., Yannoutsos, N., Wardemann, H., & Nussenzweig, M. C. (2004). RAGs and regulation of autoantibodies. *Annual Review of Immunology, 22*, 485–501.

Jhunjhunwala, S., van Zelm, M. C., Peak, M. M., Cutchin, S., Riblet, R., van Dongen, J. J. M., et al. (2008). The 3D structure of the immunoglobulin heavy-chain locus: Implications for long-range genomic interactions. *Cell, 133*(2), 265–279.

Jhunjhunwala, S., van Zelm, M. C., Peak, M. M., & Murre, C. (2009). Chromatin architecture and the generation of antigen receptor diversity. *Cell, 138*(3), 435–448.

Ji, Y., Resch, W., Corbett, E., Yamane, A., Casellas, R., & Schatz, D. G. (2010). The in vivo pattern of binding of RAG1 and RAG2 to antigen receptor loci. *Cell, 141*(3), 419–431.

Johnston, C. M., Wood, A. L., Bolland, D. J., & Corcoran, A. E. (2006). Complete sequence assembly and characterization of the C57BL/6 mouse Ig heavy chain V region. *The Journal of Immunology, 176*(7), 4221–4234.

Kosak, S. T., Skok, J. A., Medina, K. L., Riblet, R., Le Beau, M. M., Fisher, A. G., et al. (2002). Subnuclear compartmentalization of immunoglobulin loci during lymphocyte development. *Science, 296*(5565), 158–162.

Krangel, M. S., Carabana, J., Abbarategui, I., Schlimgen, R., & Hawwari, A. (2004). Enforcing order within a complex locus: Current perspectives on the control of V(D)J recombination at the murine T-cell receptor α/δ locus. *Immunological Reviews, 200*, 224–232.

Kwon, J., Morshead, K. B., Guyon, J. R., Kingston, R. E., & Oettinger, M. A. (2000). Histone acetylation and hSWI/SNF remodeling act in concert to stimulate V(D)J cleavage of nucleosomal DNA. *Molecular Cell, 6*(5), 1037–1048.

Lieberman-Aiden, E., van Berkum, N. L., Williams, L., Imakaev, M., Ragoczy, T., Telling, A., et al. (2009). Comprehensive mapping of long-range interactions reveals folding principles of the human genome. *Science, 326*(5950), 289–293.

Lin, Y. C., Benner, C., Mansson, R., Heinz, S., Miyazaki, K., Miyazaki, M., et al. (2012). Global changes in the nuclear positioning of genes and intra- and interdomain genomic interactions that orchestrate B cell fate. *Nature Immunology, 13*(12), 1196–1204.

Lin, S. G., Guo, C., Su, A., Zhang, Y., & Alt, F. W. (2015). CTCF-binding elements 1 and 2 in the *Igh* intergenic control region cooperatively regulate V(D)J recombination. *Proceedings of the National Academy of Sciences of the United States of America, 112*(6), 1815–1820.

Liu, H., Schmidt-Supprian, M., Shi, Y., Hobeika, E., Barteneva, N., Jumaa, H., et al. (2007). Yin Yang 1 is a critical regulator of B-cell development. *Genes and Development, 21*(10), 1179–1189.

Liu, Z., Widlak, P., Zou, Y., Xiao, F., Oh, M., Li, S., et al. (2006). A recombination silencer that specifies heterochromatin positioning and Ikaros association in the immunoglobulin κ locus. *Immunity, 24*(4), 405–415.

Lucas, J. S., Zhang, Y., Dudko, O. K., & Murre, C. (2014). 3D trajectories adopted by coding and regulatory DNA elements: First-passage times for genomic interactions. *Cell, 158*(2), 339–352.

Maes, J., O'Neill, L. P., Cavelier, P., Turner, B. M., Rougeon, F., & Goodhardt, M. (2001). Chromatin remodeling at the Ig loci prior to V(D)J recombination. *The Journal of Immunology, 167*(2), 866–874.

Majumder, K., Koues, O. I., Chan, E. A., Kyle, K. E., Horowitz, J. E., Yang-Iott, K., et al. (2015). Lineage-specific compaction of *Tcrb* requires a chromatin barrier to protect the function of a long-range tethering element. *The Journal of Experimental Medicine, 212*(1), 107–120.

Malin, S., McManus, S., Cobaleda, C., Novatchkova, M., Delogu, A., Bouillet, P., et al. (2010). Role of STAT5 in controlling cell survival and immunoglobulin gene recombination during pro-B cell development. *Nature Immunology, 11*(2), 171–179.

Mathieu, N., Hempel, W. M., Spicuglia, S., Verthuy, C., & Ferrier, P. (2000). Chromatin remodeling by the T cell receptor (TCR)-β gene enhancer during early T cell development: Implications for the control of TCR-β locus recombination. *The Journal of Experimental Medicine, 192*(5), 625–636.

Mathieu, N., Spicuglia, S., Gorbatch, S., Cabaud, O., Fernex, C., Verthuy, C., et al. (2003). Assessing the role of the T cell receptor β gene enhancer in regulating coding joint formation during V(D)J recombination. *The Journal of Biological Chemistry, 278*(20), 18101–18109.

Matthews, A. G., Kuo, A. J., Ramón-Maiques, S., Han, S., Champagne, K. S., Ivanov, D., et al. (2007). RAG2 PHD finger couples histone H3 lysine 4 trimethylation with V(D)J recombination. *Nature, 450*(7172), 1106–1110.

McMurry, M. T., & Krangel, M. S. (2000). A role for histone acetylation in the developmental regulation of VDJ recombination. *Science, 287*(5452), 495–498.

Medvedovic, J., Ebert, A., Tagoh, H., Tamir, I. M., Schwickert, T. A., Novatchkova, M., et al. (2013). Flexible long-range loops in the V_H gene region of the *Igh* locus facilitate the generation of a diverse antibody repertoire. *Immunity, 39*(2), 229–244.

Melchers, F. (2005). The pre-B-cell receptor: Selector of fitting immunoglobulin heavy chains for the B-cell repertoire. *Nature Reviews. Immunology, 5*(7), 578–584.

Merkenschlager, M., & Odom, D. T. (2013). CTCF and cohesin: Linking gene regulatory elements with their targets. *Cell, 152*(6), 1285–1297.

Mostoslavsky, R., Alt, F. W., & Rajewsky, K. (2004). The lingering enigma of the allelic exclusion mechanism. *Cell, 118*(5), 539–544.

Mostoslavsky, R., Singh, N., Kirillov, A., Pelanda, R., Cedar, H., Chess, A., et al. (1998). κ chain monoallelic demethylation and the establishment of allelic exclusion. *Genes and Development, 12*(12), 1801–1811.

Mostoslavsky, R., Singh, N., Tenzen, T., Goldmit, M., Gabay, C., Elizur, S., et al. (2001). Asynchronous replication and allelic exclusion in the immune system. *Nature, 414*(6860), 221–225.

Nora, E. P., Lajoie, B. R., Schulz, E. G., Giorgetti, L., Okamoto, I., Servant, N., et al. (2012). Spatial partitioning of the regulatory landscape of the X-inactivation centre. *Nature, 485*(7398), 381–385.

Novobrantseva, T. I., Martin, V. M., Pelanda, R., Muller, W., Rajewsky, K., & Ehlich, A. (1999). Rearrangement and expression of immunoglobulin light chain genes can precede heavy chain expression during normal B cell development in mice. *The Journal of Experimental Medicine, 189*(1), 75–87.

Nutt, S. L., Urbánek, P., Rolink, A., & Busslinger, M. (1997). Essential functions of Pax5 (BSAP) in pro-B cell development: Difference between fetal and adult B lymphopoiesis and reduced *V*-to-*DJ* recombination at the *IgH* locus. *Genes and Development, 11*, 476–491.

Pasqual, N., Gallagher, M., Aude-Garcia, C., Loiodice, M., Thuderoz, F., Demongeot, J., et al. (2002). Quantitative and qualitative changes in V-J α rearrangements during mouse thymocytes differentiation: Implication for a limited T cell receptor α chain repertoire. *The Journal of Experimental Medicine, 196*(9), 1163–1173.

Perlot, T., & Alt, F. W. (2008). Cis-regulatory elements and epigenetic changes control genomic rearrangements of the *IgH* locus. *Advances in Immunology, 99*, 1–32.

Perlot, T., Alt, F. W., Bassing, C. H., Suh, H., & Pinaud, E. (2005). Elucidation of IgH intronic enhancer functions via germ-line deletion. *Proceedings of the National Academy of Sciences of the United States of America, 102*(40), 14362–14367.

Phillips, J. E., & Corces, V. G. (2009). CTCF: Master weaver of the genome. *Cell, 137*(7), 1194–1211.

Ribeiro de Almeida, C., Stadhouders, R., de Bruijn, M. J., Bergen, I. M., Thongjuea, S., Lenhard, B., et al. (2011). The DNA-binding protein CTCF limits proximal Vκ recombination and restricts κ enhancer interactions to the immunoglobulin κ light chain locus. *Immunity, 35*(4), 501–513.

Roldán, E., Fuxa, M., Chong, W., Martinez, D., Novatchkova, M., Busslinger, M., et al. (2005). Locus 'decontraction' and centromeric recruitment contribute to allelic exclusion of the immunoglobulin heavy-chain gene. *Nature Immunology, 6*(1), 31–41.

Romanow, W. J., Langerak, A. W., Goebel, P., Wolvers-Tettero, I. L. M., van Dongen, J. J. M., Feeney, A. J., et al. (2000). E2A and EBF act in synergy with the V(D)J recombinase to generate a diverse immunoglobulin repertoire in nonlymphoid cells. *Molecular Cell, 5*, 343–353.

Sayegh, C., Jhunjhunwala, S., Riblet, R., & Murre, C. (2005). Visualization of looping involving the immunoglobulin heavy-chain locus in developing B cells. *Genes and Development, 19*(3), 322–327.

Seitan, V. C., Hao, B., Tachibana-Konwalski, K., Lavagnolli, T., Mira-Bontenbal, H., Brown, K. E., et al. (2011). A role for cohesin in T-cell-receptor rearrangement and thymocyte differentiation. *Nature, 476*(7361), 467–471.

Shih, H.-Y., & Krangel, M. S. (2010). Distinct contracted conformations of the *Tcra*/*Tcrd* locus during *Tcra* and *Tcrd* recombination. *The Journal of Experimental Medicine, 207*(9), 1835–1841.

Shih, H.-Y., Verma-Gaur, J., Torkamani, A., Feeney, A. J., Galjart, N., & Krangel, M. S. (2012). *Tcra* gene recombination is supported by a *Tcra* enhancer- and CTCF-dependent chromatin hub. *Proceedings of the National Academy of Sciences of the United States of America, 109*(50), E3493–E3502.

Skok, J. A., Gisler, R., Novatchkova, M., Farmer, D., de Laat, W., & Busslinger, M. (2007). Reversible contraction by looping of the *Tcrb* and *Tcra* loci in rearranging thymocytes. *Nature Immunology, 8*(4), 378–387.

Stanhope-Baker, P., Hudson, K. M., Shaffer, A. L., Constantinescu, A., & Schlissel, M. S. (1996). Cell type-specific chromatin structure determines the targeting of V(D)J recombinase activity in vitro. *Cell, 85*(6), 887–897.

Tripathi, R., Jackson, A., & Krangel, M. S. (2002). A change in the structure of Vβ chromatin associated with TCR β allelic exclusion. *The Journal of Immunology, 168*(5), 2316–2324.
van de Werken, H. J., de Vree, P. J., Splinter, E., Holwerda, S. J., Klous, P., de Wit, E., et al. (2012). 4C technology: Protocols and data analysis. *Methods in Enzymology, 513*, 89–112.
Verma-Gaur, J., Torkamani, A., Schaffer, L., Head, S. R., Schork, N. J., & Feeney, A. J. (2012). Noncoding transcription within the *Igh* distal V_H region at PAIR elements affects the 3D structure of the *Igh* locus in pro-B cells. *Proceedings of the National Academy of Sciences of the United States of America, 109*(42), 17004–17009.
Xiang, Y., Park, S.-K., & Garrard, W. T. (2013). Vκ gene repertoire and locus contraction are specified by critical DNase I hypersensitive sites within the Vκ-Jκ intervening region. *The Journal of Immunology, 190*(4), 1819–1826.
Xiang, Y., Zhou, X., Hewitt, S. L., Skok, J. A., & Garrard, W. T. (2011). A multifunctional element in the mouse *Igκ* locus that specifies repertoire and Ig loci subnuclear location. *The Journal of Immunology, 186*(9), 5356–5366.
Yancopoulos, G. D., & Alt, F. W. (1985). Developmentally controlled and tissue-specific expression of unrearranged V_H gene segments. *Cell, 40*, 271–281.

CHAPTER FOUR

Long-Range Regulation of V(D)J Recombination

Charlotte Proudhon*[,1], Bingtao Hao*[,1], Ramya Raviram*, Julie Chaumeil[†], Jane A. Skok*[,2]

*Department of Pathology, New York University School of Medicine, New York, USA
[†]Institut Curie, CNRS UMR3215, INSERM U934, Paris, France
[2]Corresponding author: e-mail address: jane.skok@nyumc.org

Contents

[1] These authors contributed equally.

Advances in Immunology, Volume 128
ISSN 0065-2776
http://dx.doi.org/10.1016/bs.ai.2015.07.003

© 2015 Elsevier Inc.
All rights reserved.

Abstract

Given their essential role in adaptive immunity, antigen receptor loci have been the focus of analysis for many years and are among a handful of the most well-studied genes in the genome. Their investigation led initially to a detailed knowledge of linear structure and characterization of regulatory elements that confer control of their rearrangement and expression. However, advances in DNA FISH and imaging combined with new molecular approaches that interrogate chromosome conformation have led to a growing appreciation that linear structure is only one aspect of gene regulation and in more recent years, the focus has switched to analyzing the impact of locus conformation and nuclear organization on control of recombination. Despite decades of work and intense effort from numerous labs, we are still left with an incomplete picture of how the assembly of antigen receptor loci is regulated. This chapter summarizes our advances to date and points to areas that need further investigation.

1. OVERVIEW OF V(D)J RECOMBINATION

In total, there are seven antigen receptor loci, four T cell receptor (*Tcr*) loci (*Tcrg*, *Tcrd*, *Tcrb*, and *Tcra*), and three B cell-specific immunoglobulin genes (*Igh*, *Igk*, and *Igl*). B and T cells make use of this modest investment in DNA to generate an almost infinite assortment of different specificity receptors that can be used to combat a wide variety of invading pathogens. Somatic rearrangement of variable (V), diversity (D), and joining (J) gene segments arrayed along each locus generates this receptor diversity enabling specific recognition of foreign antigen, which is a fundamental feature of the adaptive immune response (Helmink & Sleckman, 2012; Tonegawa, 1983).

1.1 RAG Binding

Recombination is mediated by the lymphoid-specific recombinase, consisting of RAG1 and RAG2 (the protein products of the *recombination activating genes 1* and *2*). The RAG1 protein, which harbors the endolytic activity, functions in conjunction with RAG2, a cofactor that is essential for recombinase activity (Mombaerts et al., 1992; Shinkai et al., 1992; Spanopoulou et al., 1994). The RAG1 protein cleaves specifically at highly conserved recombination signal sequences (RSSs) made up of heptamer and nonamer motifs separated by nonconserved spacers of either 12 or 23 bps (Kim, Dai, Mundy, Yang, & Oettinger, 1999; Landree, Wibbenmeyer, & Roth, 1999). The RAG1/2 complex preferentially binds two RSS sites of different spacer lengths, brings them together, and cuts at the borders of these elements generating double strand breaks (DSBs). RSSs, which flank the individual V, D, and J gene segments, are distributed throughout each antigen receptor locus and synapse formation and cleavage can occur between regions that are many kilobases apart. The four broken ends (two coding ends and two signal ends) are held together in a RAG postcleavage complex that directs repair through the nonhomologous end-joining pathway, which is important for the maintenance of genome stability (Deriano et al., 2011; Helmink & Sleckman, 2012; Lee, Neiditch, Salus, & Roth, 2004; Schatz & Swanson, 2011). Recent ground-breaking analyses of the crystal structures of these two proteins indicate that the RAG1–RAG2 heterotetramer is Y-shaped, with a RAG1–RAG2 heterodimer constituting each arm (Kim, Lapkouski, Yang, & Gellert, 2015). The structure explains numerous mutations known to be associated with immunodeficiencies.

According to ChIP-seq analysis, the binding profile of RAG1 and RAG2 overlaps with that of H3K4me3 (Ji et al., 2010). Promiscuous genome-wide binding to this active chromatin mark is mediated via a plant homeodomain in RAG2 (Liu, Subrahmanyam, Chakraborty, Sen, & Desiderio, 2007; Matthews et al., 2007). However, each RAG protein can bind in the absence of the other, and when RAG1 is bound without RAG2, it binds in an RSS-specific manner and is not found at H3K4me3-enriched promoters (Ji et al., 2010). This finding suggests that binding of the proteins can occur individually at differential locations or together as a preformed RAG1/2 complex that directs both proteins to RSSs as well as H3K4me3-enriched regions.

The question of how and what controls RAG targeting at the locus and allele-specific level on the individual antigen receptor loci continues to be an area under investigation. Moreover, there is the puzzle about how, in normal circumstances, other genes in the genome with the appropriate or cryptic recognition sequences are protected from being cleaved. Since cryptic RSSs are found every 1–2 kb in the genome, promiscuous RAG1 binding could contribute to off-target cleavage occurring within nonantigen receptor loci. Indeed, RAG targeting has been linked to genetic defects in *IKZF1*, *Notch1*, *SIL-SCL*, *Bcl11b*, *PTEN*, *ETV6*, *BTG1*, *TBL1XR1*, and *CDKN2A–CDKN2B* that are associated with numerous B- and T-acute lymphoblastic leukemias (Mendes et al., 2014; Mullighan et al., 2008; Onozawa & Aplan, 2012; Papaemmanuil et al., 2014).

1.2 Lineage- and Stage-Specific Rearrangement

Given the risks entailed by repeated cutting and pasting, V(D)J recombination is tightly regulated with respect to target gene accessibility, RAG expression, and the activities of the DNA damage signaling and repair pathways. As the recombinase machinery (the RAG proteins) and the DNA targets (RSSs) are the same for each antigen receptor locus in both lineages, lymphocytes restrict recombination by controlling the accessibility of the individual loci (Fig. 1). First, rearrangement is restricted by lineage: *Ig* gene segments complete rearrangement only in B cells, and *Tcr* gene segments rearrange only in T cells. Second, rearrangement is ordered by stage within a given lineage: the *Ig heavy chain* (*Igh*) is rearranged at the pre-pro and pro-B cell stage of development prior to *Ig light chain* (*kappa* or *lambda*) rearrangement in pre-B cells. Furthermore, D_H-to-J_H recombination at the *Igh* locus must take place in pre-pro-B cells before V_H-to-DJ_H rearrangement can begin in pro-B cells.

In T cells, the situation is more complex as productive rearrangement of the different *Tcr* loci gives rise to two distinct lineages: *Tcrg/Tcrd* and *Tcrb/Tcra* recombination leads to γδ and αβ T cells, respectively (Ciofani & Zuniga-Pflucker, 2010; Krangel, 2009). Nonetheless, recombination of the different loci overlaps such that *Tcrg*, *Tcrd*, and *Tcrb* are all rearranged at the early $CD4^-CD8^-$ double-negative DN2/3 stage of development, while *Tcra* recombination occurs later in double-positive (DP) cells after successful *Tcrb* rearrangement (Livak, Tourigny, Schatz, & Petrie, 1999). In addition, promiscuous D_H-to-J_H rearrangement of the *Igh* locus occurs at low level in T lineage cells (predominantly the DN cell stage) (Chaumeil, Micsinai, Ntziachristos, Roth, et al., 2013; Kurosawa et al., 1981). Multilocus rearrangement in the same developmental compartment increases the risks associated with recombination and the probability of aberrant repair (Chaumeil, Micsinai, Ntziachristos, Roth, et al., 2013).

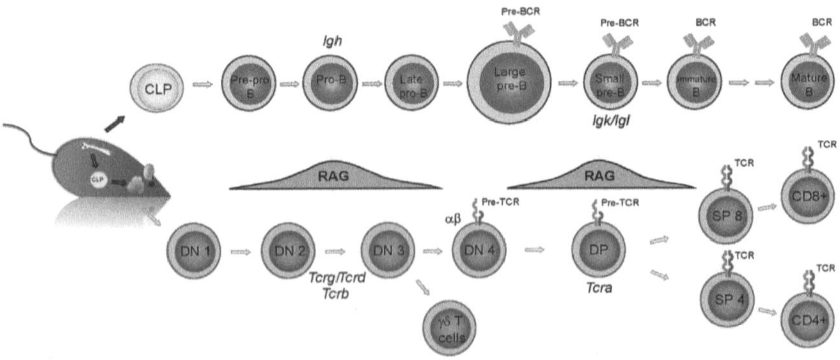

Figure 1 Scheme showing the different stages of B and T cell development where rearrangement of the *Ig* or *Tcr* loci takes place.

Regulation of recombination is further complicated because *Tcrd* and *Tcra* which are rearranged in DN and DP cell stages, respectively, share the same chromosomal location with *Tcrd* embedded between the Vα and Jα gene segments of *Tcra*.

2. LINEAR STRUCTURE OF THE ANTIGEN RECEPTOR LOCI

Antigen receptor loci consist of large arrays of V gene segments (ranging from 34 in *Tcrb* to 183 segments in *Igh* that are dispersed over 0.67 and 2.4 Mb, respectively). A much smaller proximal domain containing D, J, and C gene segments that encompass potent enhancers occupies genomic regions in the kb range (4 kb in *Igk*, 25 kb in *Tcrb*, 70 kb in *Tcra*, and 26 kb in *Igh*). Although all the loci are comprised of the same basic units (V, D, and J gene segments) that are flanked by RSSs and a constant region, each antigen receptor locus has a unique structure that impacts their regulation (Figs. 2–5).

2.1 *Igh*

The murine *Igh* locus spans 2.75 Mb (nearly a quarter of the yeast genome) and is located at the distal end of chromosome 12 in mouse. It contains a total of 113 functional V_H segments that are dispersed over 2.4 Mb. *Igh* holds 10–15 functional D_H segments (depending on the mouse strain), 4 J_H gene segments, and 8 different constant regions that are all preceded by switch regions with the exception of Cδ. These are used as substrates for class switch recombination (CSR) which generates different *Ig* isotypes that streamline antibody effector function after encounter with an antigen (IgE, IgG, IgA, etc.) (Fig. 2A).

2.2 *Igk*

The *Igk* light chain locus is located on the mouse chromosome 6. It spans 3.17 Mb and contains 92 functional Vκ segments, 4 functional Jκs, and 1 Cκ region. In contrast to *Igh*, *Igk* does not contain any D gene segments (Fig. 3A). Another feature of the *Igk* locus is that half of the Vκs are in reverse orientation and are rearranged by nondestructive inversion, which leads to retention of the segments located between the joining Vκ and Jκ segments. This conserves Vκ segments for (i) secondary rearrangements that can occur with the remaining downstream Jκs in the event of nonproductive rearrangement and (ii) receptor editing which functions to eliminate

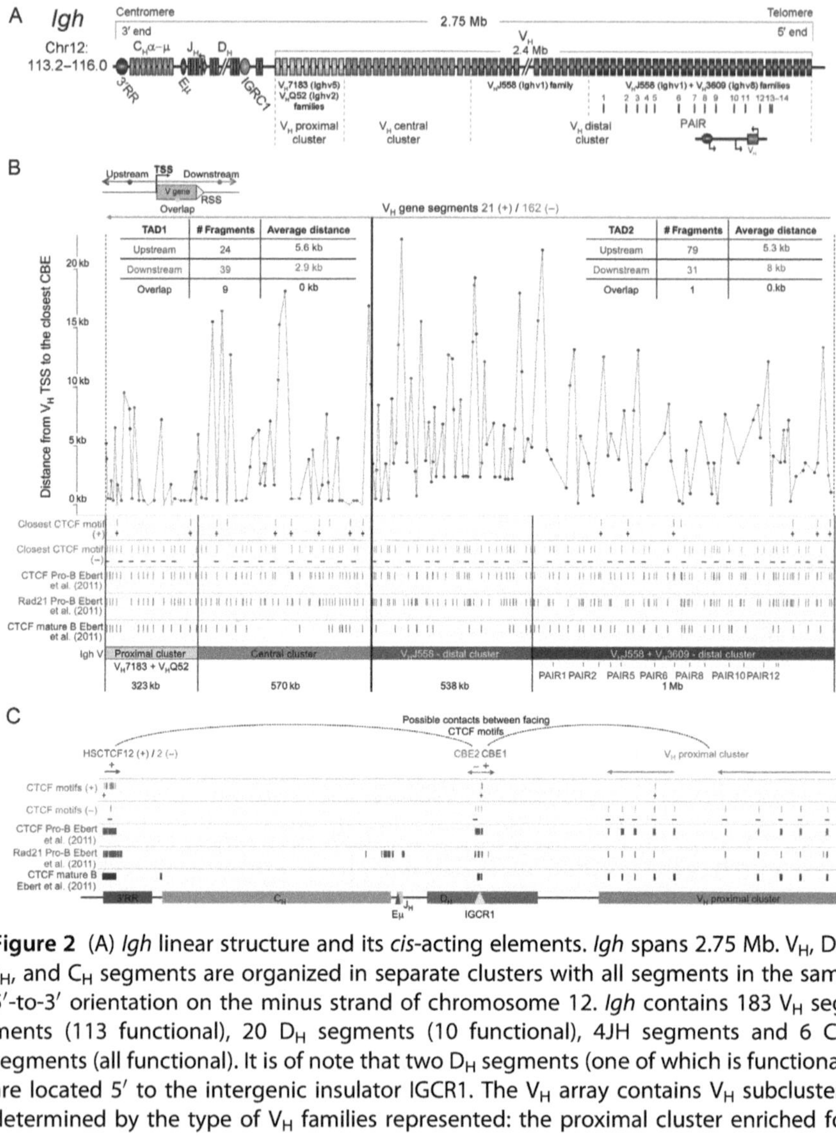

Figure 2 (A) *Igh* linear structure and its *cis*-acting elements. *Igh* spans 2.75 Mb. V_H, D_H, J_H, and C_H segments are organized in separate clusters with all segments in the same 5′-to-3′ orientation on the minus strand of chromosome 12. *Igh* contains 183 V_H segments (113 functional), 20 D_H segments (10 functional), 4JH segments and 6 CH segments (all functional). It is of note that two D_H segments (one of which is functional) are located 5′ to the intergenic insulator IGCR1. The V_H array contains V_H subclusters determined by the type of V_H families represented: the proximal cluster enriched for V_H7183 (IghV5) and V_HQ52 (IghV2) segments; the central cluster which does not include specific family types and the distal cluster enriched for V_HJ558 (IghV1) and V_H3609 (IghV8) segments. (B) Distribution of CTCF-binding elements (CBEs) within the *Igh* V_H gene region. Distance from the TSS of each V_H gene to the closest CBE is reported and marked as upstream, downstream, or overlapping depending on the CBE location relative to the TSS. Closest CBEs have been selected among motifs falling within CTCF peaks called from pro-B cell ChIP-seq data (Ebert et al., 2011). CTCF motifs were called using FIMO (part of the MEME suite) with *p*-value <10e−4. This analysis identifies 125

self-reactive receptors or enable IGK to associate with IGH (Feddersen, Martin, & Van Ness, 1990; Halverson, Torres, & Pelanda, 2004; Pelanda et al., 1997; Prak & Weigert, 1995; Tiegs, Russell, & Nemazee, 1993). Ongoing rearrangement and receptor editing are possible because of the lack of D gene segments and recombination on each allele can continue until all the Jκ gene segments are used up. Based on the delayed activation of *Igl*, it is estimated that three rounds of rearrangement are possible for each *Igk* allele (Arakawa, Shimizu, & Takeda, 1996), which corresponds to the number of functional Jκ gene segments. While no specific order is determined for Vκ rearrangement (Nadel, Tang, Escuro, Lugo, & Feeney, 1998), primary rearrangement generally involves the most 5′Jκ segment, Jκ1 (Yamagami, ten Boekel, Andersson, Rolink, & Melchers, 1999).

2.3 *Tcrb*

The T cell receptor beta locus, *Tcrb*, is encoded by 700 kb of DNA on mouse chromosome 6. The vast majority of the locus (~624 kb) is comprised of 22 functional Vβ gene segments. Except for Vβ31, which is localized downstream of the proximal domain in an inverted orientation, all the Vβ genes are located upstream of a duplicated cluster of "1 Dβ, 7 Jβ, and 1 Cβ" of which 11 of the 14 Jβs are functional. In addition to this atypical proximal

CBEs (54 in TAD1 and 71 in TAD2). The vast majority of V_H segments (162 of the 183) are associated with a CBE on the minus strand (+), pointing toward the 3′end of the *Igh* locus. The average distance between a V_H segment and its closest CBE is around 5 kb and overall there is no relationship to upstream or downstream localization of the motif. However, the Murre lab described two subdomains constituting the V_H array (annotated TAD1 and TAD2 here) (Jhunjhunwala et al., 2008) which show specificity in localization of the closest CBE associated with different distances. TAD1 contains more downstream sites which are closer than upstream ones. In contrast, TAD2 contains more upstream sites which are closer than the downstream ones. (C) Zoomed in region of the 3′end of *Igh* highlighting the orientation of CBEs. CTCF-binding motifs have been selected by intersection with CTCF peaks called from published pro-B cell ChIP-seq data (Ebert et al., 2011). CBE1 and CBE2 are pointing away from one another (Lin, Guo, Su, Zhang, & Alt, 2015) potentially enabling loop formation between convergent motifs on CBE1 and the 5′ V_H cluster and CBE2 and the 3′RR. Segment annotations with coordinates, strand orientation, and functional status as well as coordinates for regulatory elements are provided in Table 1. V_H segments (green), D_H segments (red), J_H segments (orange), C_H constant region (blue), enhancers (purple), and insulators (aqua). (See the color plate.)

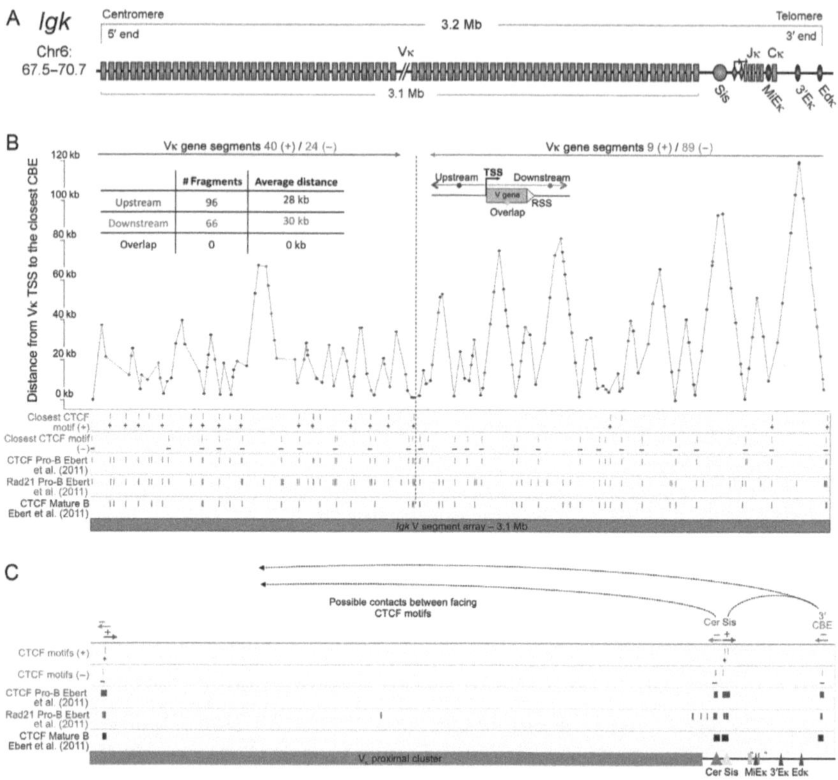

Figure 3 (A) *Igk* linear structure and its *cis*-acting regulatory elements. *Igk* spans 3.17 Mb on the murine chromosome 6. It contains 162 Vκ segments (92 functional), 5 Jκs (4 functional, Jκ3 is a pseudogene with a mutated RSS not recognized by RAG), and a single Cκ region. Half of the Vκs are positioned in reverse orientation and are rearranged by non-destructive inversion. All the other segments of the locus follow a (+) strand orientation. (B) CTCF-binding element (CBE) distribution within the Vκ region. Closest CTCF-binding motifs from the TSS of each Vκ gene segment were called using CTCF peaks from pro-B cell ChIP-seq data (Ebert et al., 2011). The Vκ cluster harbors 59 closest CBEs which display alternative orientation, but there is an enrichment of Vκ segments associated to (+) motifs at the distal part and (−) motifs at the proximal part of the Vκ cluster (40(+)/24(−) versus 9(+)/89(−), respectively). It should be noted that several segments can be associated with the same CTCF motif. The fact that Vκ rearrangement can occur via inversion could explain this nonspecific orientation that is in contrast to what is seen for the *Igh* locus. There is no apparent correlation between the CTCF motif orientation and the Vκ segment orientation. However, there are more segments associated to a CTCF motif with the same orientation (62%) as opposed to an inverse orientation (38%). The average distance between a Vκ segment and its closest CTCF motif is much larger than for *Igh*—30 versus 5 kb—and there is no relationship to upstream or downstream localization of the motif. (C) Zoomed in region of the proximal domain of *Igk* to highlight the orientation of CBEs. The CTCF motif at the 3′ boundary is directed toward the *Igk* locus, which could facilitate intralocus contacts. *Cer* and *Sis* display a similar organization to

duplication, two clusters of trypsinogen genes, that are inactive in lymphocytes, separate the bulk of the Vβ array from the first Dβ segment on the 3′ side (separation of 250 kb) as well as from the first Vβ segment, Vβ1 located at the 5′end of the locus (Fig. 4A).

2.4 *Tcra*

As mentioned above, the most striking feature of the *Tcra* locus is that *Tcrd* is embedded within it and the two loci share a subset of V genes (Fig. 5A). The whole locus spans 1.6 Mb in the 129 mouse strain and 2.0 Mb in C57BL/6. These differences stem from repeat regions within the V gene cluster of which there are two in strain 129 and three in C57BL/6. The *Tcrd* locus (which is located between the Vα and Jα gene segments of *Tcra*) harbors two Dδ, two Jδ genes, and one Cδ gene segment. There are five Vδ-specific genes located in the 3′ unique Vα cluster and 1 Vδ gene in reverse orientation that is located downstream of Cδ (Vδ5) adjacent to the Jα array that is comprised of 60 gene segments.

Tcrd rearrangement occurs in DN cells prior to *Tcra* rearrangement in DP cells. This order is important because the first round of *Tcra* rearrangement deletes the *Tcrd* gene. Unlike the other loci that contain D gene segments (*Igh* and *Tcrb*), *Tcrd* is not subjected to ordered rearrangement. Thus, Vδ-to-Dδ and Dδ-to-Jδ rearrangement occur at the same time, which enables Dδ gene segments to recombine together to form DDδ gene rearrangements (Monroe et al., 1999). *Tcrd* makes use of only a subset of V gene segments including several *Tcrd*-specific V genes (TRDV1, TRDV2-1, TRDV2-2, TRDV4, and TRDV5) and some V gene segments that are shared with *Tcra* (TRAV21/DV12, TRAV13-4/DV7, TRAV6-7/DV9, TRAV4-4/DV10, TRAV14D-3/DV8, TRAV16D/DV11, and four members of the TRAV15/DV6 family) (Hawwari & Krangel, 2005). TRDV4 rearranges specifically in fetal thymocytes and is repressed in adult thymocytes by constitutively high levels of a suppressive modification, H3K9me2 (Hao & Krangel, 2011). The local accessibility of Vδ gene

IGCR1, which encompasses CBE1 and 2 in *Igh* that have opposite orientations pointing away from each other toward the 5′end and the 3′end, respectively. *Sis* and the 3′ boundary anchor display constitutive CTCF binding throughout B cell development. They contain motifs with a head-to-head orientation that could promote proximal domain segregation, which is important for limiting proximal Vκ recombination and restricting *Igk* enhancer interactions to the *Igk* locus outside of its rearrangement stages (Ribeiro de Almeida, Stadhouders, Thongjuea, Soler, & Hendriks, 2012; Ribeiro de Almeida et al., 2011; Xiang, Zhou, Hewitt, Skok, & Garrard, 2011). (See the color plate.)

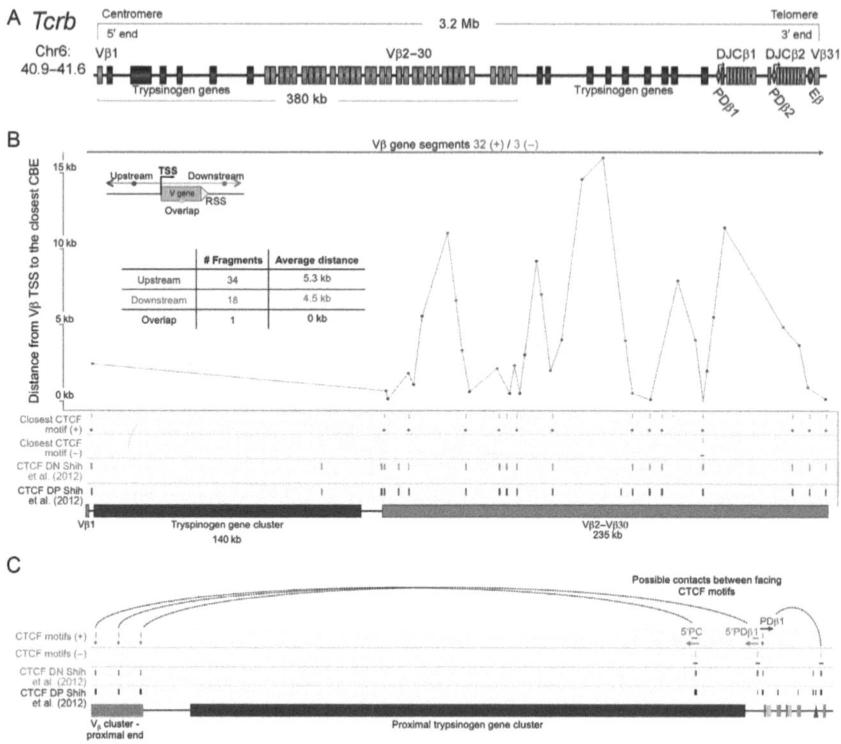

Figure 4 (A) *Tcrb* linear structure and its *cis*-acting regulatory elements. *Tcrb* encompasses 700 kb on the murine chromosome 6. It contains 35 Vβ gene segments (22 functional) spread out over 624 kb, with the exception of Vβ31 that is localized at the 3′end of the locus in an inverted orientation. All the other segment of the locus follow a (+) strand orientation. The proximal domain is duplicated with a total of 2 Dβ, 14 Jβ (11 functional), and 2 Cβ gene segments. Two clusters of trypsinogen genes separate the bulk of Vβ genes from the first Dβ segment on the 3′ side (separation of 250 kb) as well as from the Vβ1 segment located at the 5′end. (B) CTCF-binding element (CBE) distribution along *Tcrb* Vβ gene segments (excluding Vβ31). Closest CTCF-binding motifs from the TSS of each Vβ gene segment were called using CTCF peaks from DN cell ChIP-seq data (Shih et al., 2012). The Vβ cluster harbors 21 closest CBEs. 33 of the 35 Vβ segments are associated to CTCF motifs in the same orientation ((+) orientation for Vβ1 to Vβ30; (−) orientation for Vβ31). Motifs associated with Vβ1 to Vβ30 point toward the 3′end of the *Tcrb* locus and could establish contact with the facing motif in the 5′PDβ1 region. The average distance between a Vβ segment and its closest CTCF motif is around 5 kb with no relationship to upstream or downstream localization of the motif. (C) Zoomed in region of the proximal domain of *Tcrb* to highlight the orientation of CBEs. The CTCF motif associated to PDβ1 (+) is facing the 3′end motif (−) located between Eβ and Vβ31. In contrast, the 5′PDβ1 and 5′PC (5′Prss2-CTCF) motifs (−) face motifs located at Vβ segments (+). This is one more example where the region between the V array and the proximal domain harbors a composite element containing CTCF motifs pointing away from each other toward the 5′end and the 3′end of the locus. (See the color plate.)

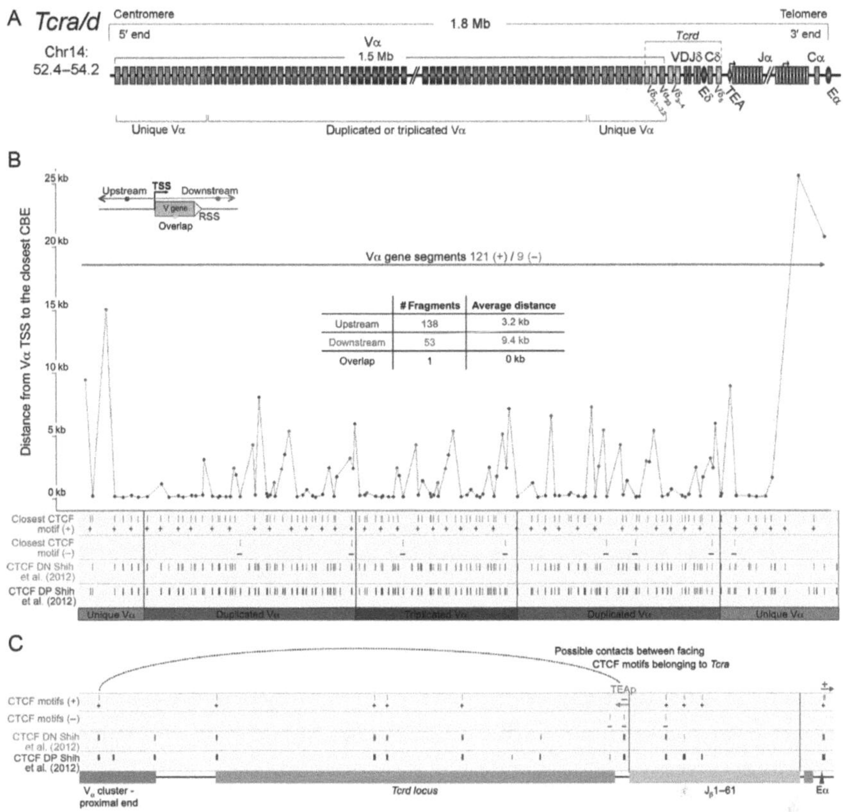

Figure 5 (A) *Tcra* linear structure and its *cis*-acting regulatory elements. *Tcra* spans 1.65 Mb on the murine chromosome 14. It contains 130 Vα segments (108 functional) spread out over 1.55 Mb and located upstream of 60 Jα genes (38 functional) and a single Cα gene. In the C57BL/6 background, the Vα array is composed of triplicated clusters located in the center with 8 and 10 unique segments on each side, respectively. *Tcra* shares V segments with the *Tcrd* locus that is embedded within its locus. These 10 V segments, annotated Trav*-dv*, rearrange either to Jα or to Dδ. (B) CTCF-binding element (CBE) distribution along *Tcra* V segments array. Closest CTCF-binding motifs from the TSS of each Vα gene segment were called using CTCF peaks from DP cell ChIP-seq data (Shih et al., 2012). The Vα cluster harbors 124 closest CBEs. 121 of the 130 Vα segments are associated to CTCF motifs in the same orientation ((+) orientation) pointing toward the 3′end of the *Tcra* locus. The average distance between a Vα segment and its closest CTCF motif is around 3 kb for motifs located upstream and 9 kb for motifs located downstream. (C) Zoomed in region of the proximal domain of *Tcra* to highlight the orientation of CBEs. The CTCF motif associated to TEAp (−) is facing the (+) motifs of the V segments toward the 5′end of the locus. (See the color plate.)

segments determines which genes undergo rearrangement in DN cells and a recent study showed that replacing the promoter of a *Tcra*-specific Vα gene, TRAV12 with the TRAV15/DV6 promoter increases the usage of the TRAV12 in *Tcrd* recombination (Hao & Krangel, 2011; Naik, Hawwari, & Krangel, 2015). *Tcra* recombination occurs after *Tcrd* recombination in DP cells and it has recently been shown that IL-7 signaling contributes to the control of stage specificity by preventing premature *Tcra* rearrangement in DN4 cells (Boudil et al., 2015).

3. NUCLEAR ORGANIZATION AND ITS IMPACT ON ACCESSIBILITY

With the exception of the stages in development where they recombine, antigen receptor loci are by default inaccessible to the RAG proteins. Opening up of the loci for rearrangement occurs at multiple levels including DNA demethylation, activation of chromatin, initiation of sense and antisense germline transcription, nucleosome repositioning, relocation of the loci from inaccessible repressive nuclear locations (peripheral lamin-associated domains or pericentromeric heterochromatin (PCH)) to accessible euchromatin, and locus contraction (which brings distal V gene segments into close physical contact with the proximal DJC domain enabling recombination between widely separated gene segments) as reviewed in Chaumeil and Skok (2013), Hewitt, Chaumeil, and Skok (2010), and Johnson, Chaumeil, and Skok (2010). All of these changes are regulated by lineage- and stage-specific activation of *cis* regulatory elements (promoters, enhancers, chromatin insulators, and others) that recruit transcription factors and structural proteins such as CTCF and cohesin that orchestrate changes which promote ordered recombination within each locus as outlined in Section 4. In this section, we focus on changes in nuclear organization that occur during rearrangement.

3.1 Subnuclear Localization

DNA FISH analyses revealed that there are links between the location of antigen receptor loci and their activation status. The first study to demonstrate that the proximity of a gene relative to the nuclear periphery is reflective of an inactive state focused on the *immunoglobulin loci* (Kosak et al., 2002). *Igh* alleles are located at the nuclear periphery in T lineage cells and they move inward just prior to the onset of recombination at the pro-B cell stage. Further detailed analysis revealed that in pre-pro-B cells and T lineage cells, the locus is anchored at the periphery through its 5′ V_H gene segments, while the 3′end, containing the $D_HJ_HC_H$ cluster of gene segments, is more centrally located. This orientation is compatible with D_HJ_H segments having access to euchromatically located recombinase enzymes and rearrangement being restricted to D_H and J_H gene segments in both these cell types (Fig. 6; Fuxa et al., 2004). Peripherally located V_H gene segments are refractory to RAG and do not recombine in these cells.

3.2 Locus Contraction

The antigen receptor loci all occupy large expanses of DNA ranging from 1 Mb (*Tcrb*) to over 3 Mb (*Igk*) that mostly encompass V gene segments. This presents a logistical problem since rearrangement requires the formation of a synapse between recombining regions that can be widely separated on the linear chromosome. The first 3D DNA FISH analyses of antigen receptor

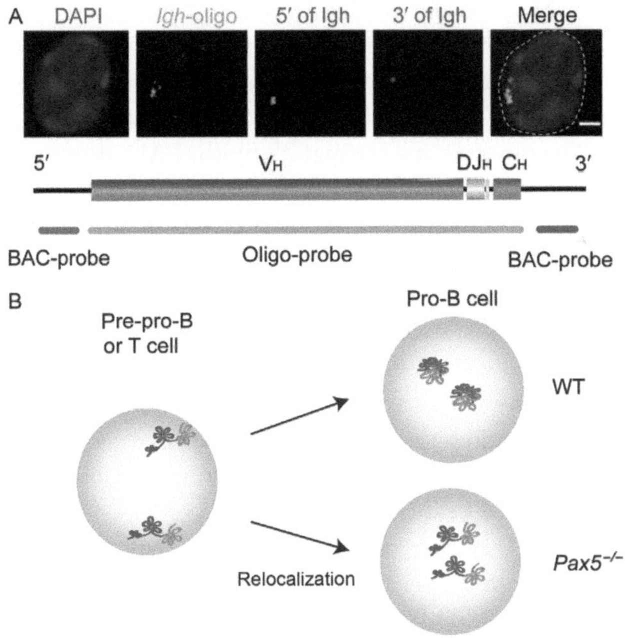

Figure 6 The *Igh* locus is activated by a two-step mechanism that involves relocation to the center of the nucleus and Pax5-mediated locus contraction at the time of recombination in pro-B cells. (A) 3D DNA FISH showing the location of the *Igh* locus and its orientation at the nuclear periphery when it is in a decontracted state. A probe scheme is shown below the FISH images identifying the location of 5′ and 3′ BAC probes relative to an oligonucleotide probe that covers the entire *Igh* locus. (B) Scheme showing the two-step activation of *Igh*. (See the color plate.)

loci in developing lymphocytes revealed that the *Igh* locus is in an extended position in lymphocyte progenitors and non-B cells and that in recombining cells, it is in a contracted conformation (Fuxa et al., 2004; Kosak et al., 2002). This is also the case for the *Igk*, *Tcrb*, and *Tcra* loci; however, unlike the other antigen receptor loci, *Igk* is contracted in pro-B cells, the stage prior to recombination, which may be a reflection of the fact that *Igk* can undergo low-level recombination in these cells (Roldan et al., 2005; Skok et al., 2007). In all cases, locus contraction brings widely dispersed V gene segments into contact with the proximal DJC domain through chromatin looping to provide every V gene an equal opportunity to rearrange. However, it should be noted that other factors, such as RSS sequence, chromatin status, and transcriptional activity, also influence which V gene is targeted for recombination.

Detailed FISH analysis investigating loop formation of the *Igh* locus used multiple small probes combined with mathematical modeling to reveal that the V_H genes of the *Igh* locus are folded into two 1 Mb rosette-like structures that are connected by linkers. The rosettes are separated from each other in pre-pro-B and T cells consistent with a decontracted state and they interact with each other in pro-B cells when the locus is contracted (Jhunjhunwala et al., 2008). The rosettes are compatible with TAD structures that were defined 4 years later as a result of Hi-C chromosome conformation analyses that provided insight into the interactome (Dixon et al., 2012; Nora et al., 2012; Sexton et al., 2012). TAD structures are highly self-interacting regions that are separated by distinct boundaries. 5C analyses delineate two distinct TADs for the proximal and distal V_H gene domains that are consistent with previous findings and with a separation of their regulation (Figs. 2B and 7). Contraction—or inter-TAD association—is a reversible process, and once recombination has taken place, the loci are once again found in a decontracted conformation. This is discussed in more detail in the context of allelic exclusion (see Section 5).

Most of the early work focusing on understanding the mechanisms underlying locus contraction centered on the *Igh* locus, in part because the B cell-specific transcription factor, Pax5, was the first factor to be identified as essential for altering locus conformation. Pax5 is upregulated at the pro-B cell stage after the initiation of D_H-to-J_H rearrangement, which starts at the earlier pre-pro-B cell stage (Fuxa et al., 2004). In the absence of Pax5, the 2 *Igh* alleles are found in an extended form in the center of the nucleus (Fuxa et al., 2004), and in this conformation, only the 4 most proximal V_H genes out of a total of nearly 200 undergo recombination, underlining the importance of

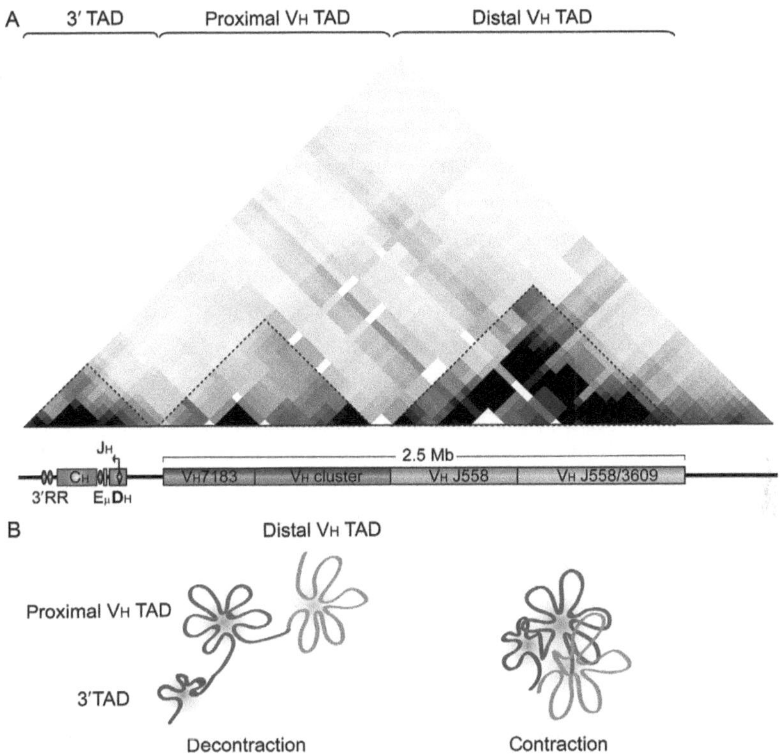

Figure 7 Locus contraction of the *Igh* locus involves interaction of two TADs or rosette structures that encompass the V_H gene region. (A) Scheme showing the various gene segments of the *Igh* locus relative to a 5C matrix of *Igh* interactions in DP cells (B.H. & J.A.S., unpublished). (B) Scheme showing rosette-like structure of the *Igh* locus as identified by DNA FISH analyses (Jhunjhunwala et al., 2008). (See the color plate.)

this process in generating antibody diversity (Roldan et al., 2005). Pax5 is essential for mediating contraction in B cells. However, ectopic expression of this factor in T cells cannot induce a change in locus conformation although it does have an impact on relocalizing the two alleles from the periphery to the center of the nucleus (Fig. 6). These findings indicate that another, yet unidentified, factor that is present in B cells but not T cells is required for locus contraction (Fuxa et al., 2004). Relocation to the center of the nucleus in T cells ectopically expressing Pax5 likely occurs as an indirect effect of Pax5 on upregulating another B cell-specific transcription factor, EBF.

Based on these observations, we put forward a two-step model for *Igh* activation (Fig. 6) (Roldan et al., 2005). In lymphoid progenitor cells, the *Igh* locus is found in an extended conformation and is anchored at the nuclear periphery via its 5′end. The central more accessible location of the D_HJ_H segments is compatible with D_H-to-J_H rearrangement occurring at low levels in T cells as previously reported (Chaumeil, Micsinai, Ntziachristos, Roth, et al., 2013; Kurosawa et al., 1981). Upregulation of EBF early in B cell development induces relocation of *Igh* to the center of the nucleus, which increases D_H-to-J_H rearrangement and allows V_H-to DJ_H recombination involving proximal V_H genes. Distal V_H gene rearrangement is not possible when the locus is in an extended conformation and thus, Pax5 expression at the pro-B cell stages is required for inclusion of these segments in the antibody repertoire.

No lineage- and stage-specific factors have been identified as essential for locus contraction of *Igk*, *Tcrb*, or *Tcra*. However, ubiquitously expressed YY1, which has been shown to be important for *Igh* locus contraction (Liu, Schmidt-Supprian, et al., 2007; Medvedovic et al., 2013), has also been identified as important for mediating *Igk* contraction (Liu, Schmidt-Supprian, et al., 2007; Pan et al., 2013). Furthermore, CCCTC-binding factor, CTCF, and its binding partner cohesin also play important roles, as reviewed in Chaumeil and Skok (2012) and discussed in more detail in Section 4.

3.3 Allelic Pairing and Pericentromeric Localization

Accessibility of antigen receptor loci has been linked to proximity to a second repressive compartment of the nucleus, pericentromeric heterochromatin (PCH). The initial reports revealed that transcriptionally inactive genes localize to PCH in developing T cells (Brown, Baxter, Graf, Merkenschlager, & Fisher, 1999) and subsequent studies led to the discovery that productively rearranged and nonproductively rearranged *Igh* alleles are found in distinct nuclear compartments in mature activated B cells (Skok et al., 2001). Differences in nuclear localization of highly expressed productively rearranged versus low-expressing nonproductively rearranged alleles led to the idea that repositioning to pericentric regions could play a role in allelic exclusion (Roldan et al., 2005; Skok et al., 2007). Not surprisingly this turns out to be just one aspect of control. In the case of the antigen receptor loci, pericentromeric localization is linked to homologous and heterologous antigen receptor allele pairing, which in turn is linked to control of allele and locus-specific accessibility ensuring that breaks are introduced asynchronously on one allele or locus at a time.

Briefly, it was shown that RAG proteins enhance association of recombining homologous and heterologous loci in euchromatic regions of the nucleus. Pairing is mediated by RAG-induced higher-order looping of one allele away from its respective chromosome territory. RAG-mediated cleavage is targeted to the looped out allele, and once the break is introduced, the DNA damage-sensing factor, ATM (ataxia telangiectasia mutated), is recruited in *cis* to the site to initiate repair. Both the C-terminal portions of RAG2 and ATM perform the same function and act in *trans* on other recombining alleles (homologues) or loci (heterologues, e.g., *Igh* and *Tcra*) repositioning them to PCH and inhibiting further higher-order loop formation (Fig. 8). This may be part of the mechanism that serves to prevent further cleavage until repair of the first break is completed (Chaumeil, Micsinai, Ntziachristos, Deriano, et al., 2013; Chaumeil, Micsinai, Ntziachristos, Roth, et al., 2013; Chaumeil & Skok, 2013; Hewitt et al., 2009). Together, these data support a model in which breaks occur on paired alleles; however, since allelic association is likely transient and cleavage a rapid event with repair occurring over several hours, it is not surprising we find repair foci on both paired and unpaired alleles. Live-imaging studies would provide further support for such a model.

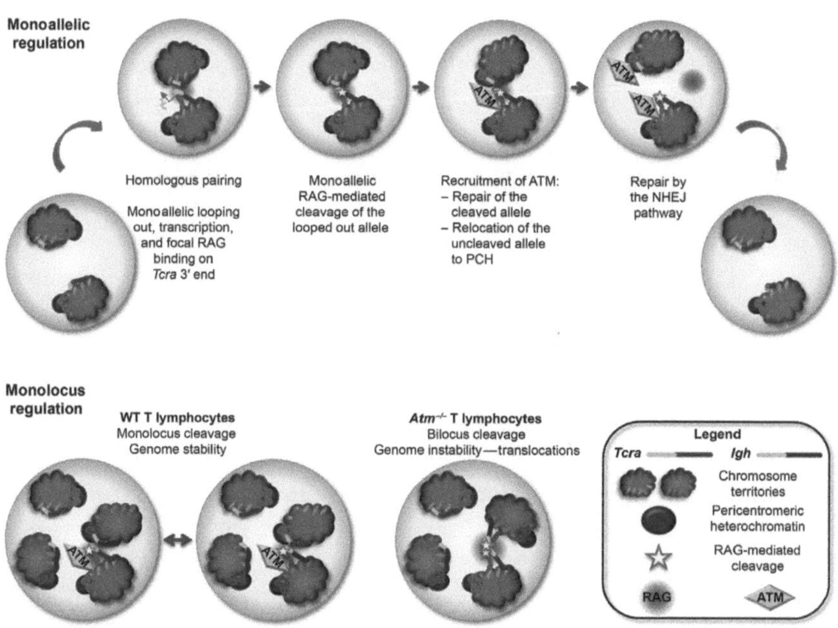

Figure 8 Model of ATM-mediated control of cleavage. Changes in nuclear accessibility of the antigen receptor loci are linked with monoallelic and monolocus rearrangement and maintenance of genome integrity. (See the color plate.)

Is pairing necessary for this level of control? We hypothesize that pairing is required for cleavage as well as feedback control of cleavage as close proximity could be important for coordinating *trans* control by ATM and the C-terminal portion of RAG2. Although RAG is found in abundance in euchromatic regions of the nucleus and is enriched at H3K4me3 marked chromatin (Ji et al., 2010), RAG-mediated cleavage is inherently inefficient as RAG-mediated breaks are only detected in around 20% or less of recombining cells as determined by immuno-FISH analyses (Chaumeil, Micsinai, Ntziachristos, Deriano, et al., 2013; Chaumeil, Micsinai, Ntziachristos, Roth, et al., 2013; Hewitt et al., 2009). It is thus not inconceivable that the chance of a break occurring improves as the local concentration of RAG increases. However, it could be argued that increasing the local concentration of RAG in recombination centers via association of RAG-bound genes (Chaumeil, Micsinai, Ntziachristos, Roth, et al., 2013; Chaumeil & Skok, 2013) would increase the risk of cleavage on closely associated loci. Not so, if feedback control of RAG cleavage via an ATM-mediated mechanism occurs in a localized fashion (Fig. 8).

Support for this idea comes from the feedback control of Spo-11p-mediated cleavage during meiosis: ATM appears to play a similar role in regulating the introduction of breaks in this process; however, it is clear that cleavage control occurs in a localized fashion in meiosis (Garcia, Gray, Allison, Cooper, & Neale, 2015; Lange et al., 2011). Since the feedback control of recombination appears to be conserved in meiosis and V(D)J recombination, it is conceivable that it shares common mechanistic features, the details of which have yet to be worked out. In both cases, feedback control of cleavage is important for maintenance of genome stability and in the case of V(D)J recombination, asynchronous cleavage provides a means of (i) maintaining genome stability and preventing the generation of translocations and (ii) initiating allelic exclusion (discussed in Section 5).

3.4 *Igk–Igh* Allelic Pairing and Its Impact on *Igh* Locus Contraction

A further example of pairing linked to PCH localization comes from a transient interaction between *Igh* and *Igk* that occurs at the pre-B cell stage of development (Hewitt et al., 2008). We found that association of one *Igk* allele with one *Igh* allele at PCH triggers (i) the repositioning of *Igh* to PCH and (ii) *Igh* locus decontraction. This serves to (i) reduce accessibility of partially recombined (DJ_H rearranged) *Igh* alleles that could otherwise go on being rearranged in pre-B cells and (ii) prevent ongoing mid and distal

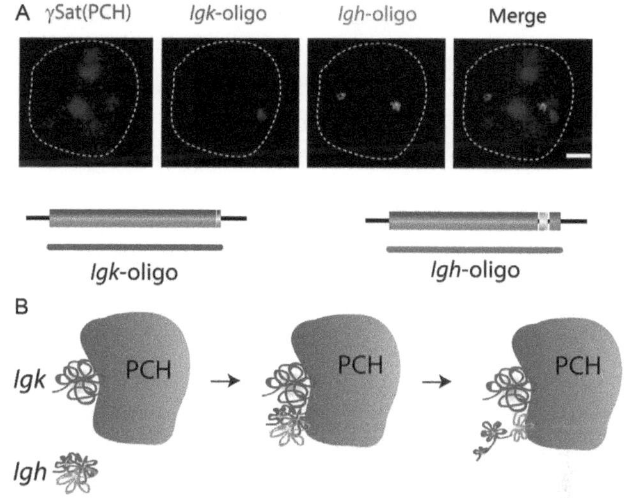

Figure 9 An interaction between *Igk* and *Igh* is responsible for *Igh* PCH association and *Igh* locus decontraction post-recombination, at the pre-B cell stage. (A) 3D DNA FISH showing *Igk–Igh* interaction at PCH in pre-B cells. Oligonucleotide probes encompassing the entire *Igk* and *Igh* loci were used for this analysis. (B) Scheme showing *Igk–Igh* interaction at PCH leading to *Igh* locus decontraction. (See the color plate.)

V_H rearrangement occurring during light chain rearrangement. Interlocus *Igk–Igh* pairing and *Igh* decontraction rely on the 3′Eκ *Igk* enhancer: in its absence, there is reduced *Igk–Igh* pairing and reduced *Igh* localization at PCH, and *Igh* remains in a contracted conformation increasing the level of mid and distal V_H rearrangement detected in pre-B cells (Fig. 9; Hewitt et al., 2008). Intriguingly, the intronic enhancer of *Igk*, MiEκ, has an antagonistic effect on *Igk–Igh* pairing, *Igh* localization at PCH, and decontraction which are all increased in its absence in association with reduced levels of mid and distal V_H rearrangement in pre-B cells (Hewitt et al., 2008).

4. *CIS*-ACTING ELEMENTS THAT CONTROL ACCESSIBILITY AND RECOMBINATION

In this section, we aim to highlight functions of *cis*-acting elements and their role in regulating accessibility, locus conformation, and ordered recombination. In particular, we will focus on the most recent work identifying regulatory elements that play an important role in all of these aspects of control. These elements are depicted in Figs. 2–5. In particular, in these figures we have focused on CTCF-binding elements (CBEs) as these are an important component of long-range interactions. For all loci, we have analyzed the orientation of the CBEs as a recent study demonstrates that loop bases involve a pair of CTCF sites in a head-to-head orientation (Rao et al., 2014). In addition, we analyzed the location of the closest CBE relative to the TSS of each V gene and have marked whether these are located upstream, downstream, or overlapping with it and whether there is any pattern to this organization on the individual loci. Segment annotations with coordinates, strand orientation, and functional status as well as coordinates for regulatory elements are provided in Table 1. Annotations were collected from NCBI (*Igh* gene ID: 111507; *Igk* gene ID: 243469; *Tcrb* gene ID: 21577; *Tcra* gene ID: 21473) and IMGT/LIGM-DB databases (Giudicelli et al., 2006) using the mm10 genome build that uses the C57BL/6 strain as the reference genome.

4.1 Important *cis*-Acting Elements and Their Function in *Igh* Rearrangement

4.1.1 The Intronic Enhancer Eμ

The intronic enhancer Eμ, located in the 700 kb region that separates the J_H and the C_H clusters, is a combination of a 220-bp core enhancer element (cEμ) and two 310–350 bp flanking matrix attachment regions (MARs). Deletion of Eμ has been shown to impair both D_H-to-J_H and V_H-to-DJ_H recombination. In Eμ knockout mouse models, sense μ0 (initiated at the D_HQ52 region), Iμ transcripts (which originate 3′ of the Eμ core (Lennon & Perry, 1985; Perlot, Alt, Bassing, Suh, & Pinaud, 2005)), and antisense transcripts in the J_H and D_H regions (Afshar, Pierce, Bolland, Corcoran, & Oltz, 2006) are severely impaired. However, despite the defect in V(D)J recombination and a partial block in B cell development at the pro-B cell stage, Eμ deletion (core or full length) does not severely affect germline sense or antisense transcription in the V_H region or V_H gene usage (Afshar et al., 2006; Perlot et al., 2005). However this enhancer appears to be important for efficient Igμ-chain expression and strong signaling through the pre-B cell receptor (pre-BCR) and BCR (Marquet et al., 2014).

4.1.2 The 3′ Regulatory Region

The 3′ regulatory region (3′RR), located 200 kb downstream of the C_H cluster, spans 30 kb and contains multiple enhancer elements with strict B-lineage specificity (HS3a, HS1–2, HS3b, and HS4) and a proposed insulator region containing CTCF-binding sites (HS5–8) which also bind cohesin and likely act as a *Igh* 3′ chromatin boundary (Degner et al., 2011;

Table 1 List of Segments and Regulatory Elements of *Igh, Igk, Tcrb,* and *Tcra-Tcrd*. Annotations correspond to the mm10/GRCm38 genome assembly which uses the C57BL/6 strain as genome reference and were collected from NCBI (*Igh* gene ID: 111507; *Igk* gene ID: 243469; *Tcrb* gene ID: 21577; *Tcra* gene ID: 21473) and IMGT/LIGMDB databases (Giudicelli et al., 2006).

Segment	Chr	Start	End	Strand	Size	Functional status	
HSCTCF	chr12	113215906	113224658	.	8752		3'RR
HS3	chr12	113233598	113235248	.	1650		
HS12	chr12	113243680	113245543	.	1863		
HS3a	chr12	113249772	113251171	.	1399		
Igha	chr12	113258768	113260236	–	1468	Functional	CONSTANT REGIONS
Ighe	chr12	113269263	113273248	–	3985	Functional	
Ighg2c	chr12	113287389	113288932	–	1543	Functional	
Ighg2b	chr12	113304314	113307933	–	3619	Functional	
Ighg1	chr12	113326544	113330523	–	3979	Functional	
Ighg3	chr12	113357442	113361232	–	3790	Functional	
Ighd	chr12	113407535	113416324	–	8789	Functional	
Ighm	chr12	113418826	113422730	–	3904	Functional	
Emu	chr12	113427403	113427623	.	220		ENHANCER
Ighj4	chr12	113428514	113428567	–	53	Functional	J GENES
Ighj3	chr12	113429085	113429132	–	47	Functional	
Ighj2	chr12	113429468	113429515	–	47	Functional	
Ighj1	chr12	113429781	113429833	–	52	Functional	
Ighd4-1	chr12	113430528	113430538	–	10	Functional	D GENES
Ighd3-2	chr12	113448214	113448229	–	15	Functional	
Ighd5-6	chr12	113449588	113449597	–	9	D ORF	
Ighd2-8	chr12	113450851	113450867	–	16	Functional	
Ighd5-5	chr12	113454942	113454951	–	9	D ORF	
Ighd2-7	chr12	113456720	113456736	–	16	Functional	
Ighd5-8	chr12	113459864	113459892	–	28	D ORF	
Ighd5-4	chr12	113460101	113460110	–	9	D ORF	
Ighd2-6	chr12	113461369	113461385	–	16	Functional	
Ighd5-7	chr12	113464524	113464552	–	28	D ORF	
Ighd5-3	chr12	113464761	113464770	–	9	D ORF	
Ighd2-5	chr12	113466027	113466043	–	16	Functional	
Ighd5-2	chr12	113469426	113469435	–	9	D ORF	
Ighd2-4	chr12	113470694	113470710	–	16	Functional	
Ighd6-2	chr12	113474875	113474903	–	28	Pseudo	
Ighd2-3	chr12	113475400	113475416	–	16	Functional	
Ighd6-1	chr12	113480143	113480171	–	28	Pseudo	
Ighd1-1	chr12	113482170	113482192	–	22	Functional	
IGCR1	chr12	113485095	113485503	.	408		INSULATOR
Ighd3-1	chr12	113525313	113525329	–	16	Functional	
Ighd5-1	chr12	113526800	113526809	–	9	Pseudo	
Ighv5-1	chr12	113572929	113573222	–	293	Pseudo	V GENES
Ighv2-1	chr12	113574247	113574540	–	293	Pseudo	
Ighv5-2	chr12	113578504	113578990	–	486	Functional	
Ighv2-2	chr12	113588267	113588700	–	433	Functional	
Ighv5-3	chr12	113589967	113590243	–	276	Pseudo	
Ighv5-4	chr12	113597448	113597741	–	293	Functional	
Ighv6-1	chr12	113603460	113603928	–	468	Pseudo	
Ighv2-3	chr12	113611184	113611476	–	292	Functional	
Ighv5-5	chr12	113612965	113613442	–	477	Pseudo	
Ighv5-6	chr12	113625508	113625801	–	293	Functional	
Ighv5-7	chr12	113634686	113634981	–	295	Pseudo	
Ighv2-4	chr12	113653291	113653583	–	292	Functional	
Ighv5-8	chr12	113654967	113655245	–	278	Pseudo	
Ighv5-9	chr12	113661771	113662228	–	457	Functional	
Ighv5-10	chr12	113669673	113669968	–	295	Pseudo	
Ighv2-5	chr12	113685482	113685774	–	292	Functional	
Ighv5-11	chr12	113687229	113687505	–	276	Pseudo	
Ighv5-12	chr12	113702126	113702419	–	293	Functional	
Ighv2-6	chr12	113716672	113716962	–	290	Functional	
Ighv5-9-1	chr12	113736113	113736406	–	293	Functional	

Continued

Table 1 *Igh*—cont'd

Segment	Chr	Start	End	Strand	Size	Functional status
Ighv5-12-4	chr12	113762251	113762544	–	293	Functional
Ighv2-9-1	chr12	113769857	113770142	–	285	Functional
Ighv5-13	chr12	113794063	113794358	–	295	Pseudo
Ighv2-6-8	chr12	113796138	113796430	–	292	Functional
Ighv2-7	chr12	113807314	113807747	–	433	Functional*
Ighv5-15	chr12	113826648	113826941	–	293	Functional
Ighv5-16	chr12	113838528	113838821	–	293	Functional
Ighv5-17	chr12	113859149	113859442	–	293	Functional
Ighv5-18	chr12	113875772	113876239	–	467	Pseudo
Ighv2-8	chr12	113877361	113877659	–	298	Pseudo
Ighv2-9	chr12	113879097	113879388	–	291	Functional
Ighv5-19	chr12	113884761	113885342	–	581	Pseudo
Ighv7-1	chr12	113896408	113896946	–	538	Functional*
Ighv7-2	chr12	113912025	113912483	–	458	Functional*
Ighv14-1	chr12	113931953	113932246	–	293	Functional
Ighv4-1	chr12	113948284	113948577	–	293	Functional
Ighv3-1	chr12	113964390	113964683	–	293	Functional
Ighv11-1	chr12	113981879	113982174	–	295	Functional
Ighv14-2	chr12	113994469	113994898	–	429	Functional*
Ighv4-2	chr12	114013144	114013439	–	295	Functional*
Ighv3-2	chr12	114033803	114034097	–	294	Functional*
Ighv11-2	chr12	114048241	114048704	–	463	Functional
Ighv14-3	chr12	114059845	114060273	–	428	Functional
Ighv16-1	chr12	114068828	114069126	–	298	Functional
Ighv6-2	chr12	114089387	114089681	–	294	Pseudo
Ighv9-1	chr12	114093928	114094221	–	293	Functional
Ighv12-1	chr12	114107259	114107563	–	304	Pseudo
Ighv9-2	chr12	114109001	114109430	–	429	Functional
Ighv12-2	chr12	114127533	114127984	–	451	Pseudo
Ighv9-3	chr12	114140692	114141122	–	430	Functional
Ighv7-3	chr12	114153180	114153644	–	464	Functional
Ighv15-1	chr12	114158025	114158315	–	290	Pseudo
Ighv14-4	chr12	114176438	114176867	–	429	Functional
Ighv3-3	chr12	114196439	114196875	–	436	Functional*
Ighv7-4	chr12	114222788	114223254	–	466	Functional
Ighv3-4	chr12	114253617	114253915	–	298	Functional
Ighv3-5	chr12	114262652	114262950	–	298	Functional
Ighv13-1	chr12	114267607	114267906	–	299	Functional*
Ighv3-6	chr12	114288152	114288447	–	295	Functional
Ighv9-4	chr12	114299961	114300254	–	293	Functional
Ighv3-7	chr12	114314186	114314479	–	293	Pseudo
Ighv5-21	chr12	114320030	114320324	–	294	Pseudo
Ighv3-8	chr12	114322376	114322666	–	290	Functional
Ighv8-1	chr12	114332796	114333086	–	290	Pseudo
Ighv1-1	chr12	114351705	114351995	–	290	Pseudo
Ighv13-2	chr12	114357761	114358060	–	299	Functional
Ighv12-3	chr12	114366526	114366819	–	293	Functional
Ighv6-3	chr12	114391712	114392010	–	298	Functional
Ighv6-4	chr12	114406474	114406773	–	299	Functional
Ighv6-5	chr12	114416596	114416895	–	299	Functional
Ighv6-6	chr12	114434788	114435087	–	299	Functional
Ighv6-7	chr12	114455626	114455925	–	299	Functional
Ighv8-2	chr12	114462290	114462589	–	299	V ORF
Ighv1-2	chr12	114469075	114469371	–	296	Pseudo
Ighv10-1	chr12	114479007	114479451	–	444	Functional
Ighv1-3	chr12	114481615	114481916	–	301	Pseudo
Ighv1-4	chr12	114487136	114487429	–	293	Functional
Ighv10-2	chr12	114496003	114496300	–	297	Pseudo

Table 1 *Igh*—cont'd

Segment	Chr	Start	End	Strand	Size	Functional status	
Ighv1-5	chr12	114513330	114513623	–	293	Functional	
Ighv10-3	chr12	114523441	114523887	–	446	Functional	
Ighv1-6	chr12	114532972	114533264	–	292	Pseudo	
Ighv1-7	chr12	114538495	114538788	–	293	Functional	
Ighv10-4	chr12	114547258	114547556	–	298	Pseudo	
Ighv1-8	chr12	114555610	114555902	–	292	Pseudo	
Ighv15-2	chr12	114564578	114564874	–	296	Functional	
Ighv1-9	chr12	114583569	114583862	–	293	Functional	
Ighv1-10	chr12	114597878	114598175	–	297	Pseudo	
Ighv1-11	chr12	114612244	114612536	–	292	Functional	
Ighv1-12	chr12	114615850	114616143	–	293	Functional	
Ighv1-13	chr12	114630680	114630973	–	293	Pseudo	
Ighv1-14	chr12	114646572	114646865	–	293	Functional	
Ighv1-15	chr12	114657353	114657646	–	293	Functional	
Ighv1-16	chr12	114665872	114666165	–	293	V ORF	
Ighv1-17-1	chr12	114671968	114672262	–	294	Functional	
Ighv1-17	chr12	114676749	114677042	–	293	Pseudo	
Ighv1-18	chr12	114682632	114682925	–	293	Functional	
Ighv1-19-1	chr12	114704147	114704440	–	293	Pseudo	
Ighv1-19	chr12	114708648	114708941	–	293	Functional	
Ighv1-20	chr12	114723772	114724065	–	293	Functional	
Ighv1-21-1	chr12	114736245	114736538	–	293	Pseudo	
Ighv1-21	chr12	114740383	114740840	–	457	Pseudo	
Ighv1-22	chr12	114746273	114746566	–	293	Functional	
Ighv1-23	chr12	114764450	114764743	–	293	V ORF	
Ighv1-24	chr12	114772928	114773221	–	293	V ORF	
Ighv1-25	chr12	114781409	114781702	–	293	Pseudo	
Ighv1-26	chr12	114788372	114788665	–	293	Functional	
Ighv1-27	chr12	114804616	114804909	–	293	Pseudo	
Ighv1-28	chr12	114809421	114809714	–	293	Pseudo	
Ighv1-29	chr12	114812321	114812605	–	284	Pseudo	
Ighv1-30	chr12	114817099	114817390	–	291	Pseudo	
Ighv1-31	chr12	114829264	114829557	–	293	Functional	
Ighv1-32	chr12	114835223	114835516	–	293	Pseudo	
Ighv1-33	chr12	114843693	114844083	–	390	Pseudo	
Ighv1-34	chr12	114851190	114851483	–	293	Functional	
Ighv1-35	chr12	114875302	114875597	–	295	Pseudo	
Ighv1-36	chr12	114879888	114880181	–	293	Functional	
Ighv1-37	chr12	114896238	114896531	–	293	Functional	
Ighv1-38	chr12	114910011	114910303	–	292	Pseudo	
Ighv1-39	chr12	114914599	114914892	–	293	Functional	
Ighv1-40	chr12	114924941	114925233	–	292	Pseudo	
Ighv1-41	chr12	114932371	114932664	–	293	Pseudo	
Ighv1-42	chr12	114937154	114937544	–	390	Functional*	
Ighv1-43	chr12	114945950	114946243	–	293	Functional	
Ighv1-44	chr12	114963380	114963674	–	294	Pseudo	
Ighv1-45	chr12	114970055	114970348	–	293	Pseudo	
Ighv1-46	chr12	114977693	114977986	–	293	Pseudo	
Ighv1-47	chr12	114991108	114991401	–	293	Functional	
Ighv8-3	chr12	115003418	115003712	–	294	Pseudo	
Ighv8-4	chr12	115024026	115024318	–	292	Functional	
Ighv1-48	chr12	115038227	115038520	–	293	Pseudo	
Ighv1-49	chr12	115055223	115055516	–	293	Functional	
Ighv8-5	chr12	115067560	115067860	–	300	Functional	
PAIR1	chr12	115070994	115072036	.	1042		PAIR1
Ighv1-50	chr12	115119748	115120041	–	293	Functional	
Ighv1-51	chr12	115131375	115131668	–	293	Pseudo	
Ighv1-52	chr12	115145487	115145780	–	293	Functional	
Ighv1-53	chr12	115158445	115158835	–	390	Functional	

Continued

Table 1 *Igh*—cont'd

Segment	Chr	Start	End	Strand	Size	Functional status	
Ighv8-6	chr12	115165777	115166077	–	300	Functional	
PAIR2	chr12	115171765	115172222	.	457		PAIR2
Ighv1-54	chr12	115193675	115193968	–	293	Functional	
Ighv1-55	chr12	115208137	115208527	–	390	Functional	
PAIR3	chr12	115222694	115223160	.	466		PAIR3
Ighv1-56	chr12	115242802	115243095	–	293	Functional	
Ighv8-7	chr12	115258139	115258432	–	293	Pseudo	
PAIR4	chr12	115262443	115262909	.	466		PAIR4
Ighv1-57	chr12	115268424	115268715	–	291	Pseudo	
Ighv8-8	chr12	115294066	115294362	–	296	Functional	
PAIR5	chr12	115297471	115298510	.	1039		PAIR5
Ighv1-58	chr12	115312166	115312597	–	431	Functional*	
Ighv1-59	chr12	115335057	115335513	–	456	Functional	
Ighv1-60	chr12	115344537	115344830	–	293	Pseudo	
Ighv1-61	chr12	115359140	115359433	–	293	Functional	
Ighv1-62	chr12	115371986	115372278	–	292	Pseudo	
Ighv1-62-1	chr12	115386707	115386988	–	281	Functional	
PAIR6	chr12	115407442	115407908	.	466		PAIR6
Ighv1-62-2	chr12	115446417	115446710	–	293	Functional	
Ighv1-62-3	chr12	115460999	115461431	–	432	Functional*	
Ighv8-9	chr12	115468331	115468632	–	301	Functional*	
PAIR7	chr12	115474319	115474776	.	457		PAIR7
Ighv1-63	chr12	115495625	115496058	–	433	Functional	
Ighv1-64	chr12	115507587	115507977	–	390	Functional	
Ighv8-10	chr12	115519340	115519640	–	300	Pseudo	
PAIR8	chr12	115524113	115524608	.	495		PAIR8
Ighv1-65	chr12	115532055	115532347	–	292	Pseudo	
Ighv8-11	chr12	115567149	115567449	–	300	Functional	
PAIR9	chr12	115574340	115574807	.	467		PAIR9
Ighv1-66	chr12	115593110	115593403	–	293	Functional	
Ighv1-67	chr12	115603940	115604233	–	293	Functional	
Ighv1-68	chr12	115611340	115611633	–	293	Pseudo	
Ighv1-69	chr12	115623203	115623595	–	392	Functional	
Ighv8-12	chr12	115647945	115648245	–	300	Functional	
PAIR10	chr12	115655807	115656267	.	460		PAIR10
Ighv1-70	chr12	115691889	115692182	–	293	Pseudo	
PAIR11	chr12	115703584	115704051	.	467		PAIR11
Ighv1-71	chr12	115742205	115742506	–	301	Functional	
Ighv1-72	chr12	115757984	115758277	–	293	Functional	
Ighv8-13	chr12	115765335	115765635	–	300	V ORF	
PAIR12	chr12	115771193	115771646	.	453		PAIR12
Ighv1-73	chr12	115790593	115790887	–	294	Pseudo	
Ighv1-74	chr12	115802690	115803080	–	390	Functional	
Ighv8-14	chr12	115808427	115808727	–	300	Pseudo	
PAIR13	chr12	115812742	115813210	.	468		PAIR13
PAIR14	chr12	115820652	115821118	.	466		PAIR14
Ighv1-75	chr12	115833950	115834243	–	293	Functional	
Ighv8-15	chr12	115844016	115844309	–	293	Pseudo	
Ighv1-76	chr12	115847881	115848174	–	293	Functional	
Ighv8-16	chr12	115858008	115858301	–	293	Pseudo	
Ighv1-77	chr12	115861867	115862160	–	293	Functional	
Ighv1-78	chr12	115868773	115869066	–	293	Functional	
Ighv1-79	chr12	115888096	115888389	–	293	Pseudo	
Ighv1-80	chr12	115912344	115912637	–	293	Functional	
Ighv1-81	chr12	115920279	115920572	–	293	Functional	
Ighv1-82	chr12	115952538	115952831	–	293	Functional	
Ighv1-83	chr12	115963817	115964109	–	292	Pseudo	
Ighv1-84	chr12	115980702	115981134	–	432	Functional	
Ighv1-85	chr12	116000028	116000321	–	293	Functional	
Ighv1-86	chr12	116009659	116009954	–	295	Pseudo	

Table 1 *Igk*—cont'd

Segment	Chr	Start	End	Strand	Size	Functional status	
Igkv2-137	chr6	67555636	67556216	+	580	Functional*	V GENES
Igkv1-136	chr6	67593876	67594308	+	432	Pseudo	
Igkv1-135	chr6	67609745	67610508	+	763	Functional	
Igkv14-134-1	chr6	67711174	67711628	–	454	Pseudo	
Igkv17-134	chr6	67720729	67721225	–	496	Pseudo	
Igkv1-133	chr6	67724914	67725661	+	747	Functional	
Igkv1-132	chr6	67759700	67760413	+	713	Functional	
Igkv1-131	chr6	67766036	67766772	–	736	V ORF	
Igkv14-130	chr6	67791046	67791511	+	465	Functional	
Igkv9-129	chr6	67839793	67840266	+	473	Functional	
Igkv9-128	chr6	67847409	67847874	+	465	Pseudo	
Igkv17-127	chr6	67861159	67861659	+	500	Functional	
Igkv14-126-1	chr6	67876011	67876477	+	466	Pseudo	
Igkv14-126	chr6	67896177	67896642	+	465	Pseudo	
Igkv11-125	chr6	67913573	67914052	+	479	Functional	
Igkv9-124	chr6	67942076	67942540	–	464	Functional	
Igkv9-123	chr6	67954230	67954690	–	460	Functional	
Igkv1-122	chr6	68016742	68017488	+	746	Functional	
Igkv17-121	chr6	68036818	68037318	+	500	Functional	
Igkv9-120	chr6	68049983	68050454	+	471	Functional	
Igkv9-119	chr6	68056345	68056810	+	465	Pseudo	
Igkv14-118-2	chr6	68066369	68066828	+	459	Pseudo	
Igkv14-118-1	chr6	68082547	68083145	+	598	Pseudo	
Igkv11-118	chr6	68105414	68105880	+	466	Pseudo	
Igkv1-117	chr6	68121091	68121825	+	734	Functional	
Igkv2-116	chr6	68151885	68152617	+	732	Pseudo	
Igkv1-115	chr6	68160797	68161716	+	919	Pseudo	
Igkv11-114	chr6	68164459	68164923	+	464	Pseudo	
Igkv2-113	chr6	68178801	68179564	+	763	Pseudo	
Igkv2-112	chr6	68219981	68220705	+	724	Functional	
Igkv14-111	chr6	68256404	68256869	+	465	Functional	
Igkv1-110	chr6	68270527	68271265	+	738	Functional	
Igkv2-109	chr6	68302439	68303158	+	719	Functional	
Igkv1-108	chr6	68312054	68312359	+	305	Pseudo	
Igkv2-107	chr6	68326098	68326752	+	654	Pseudo	
Igkv11-106	chr6	68339527	68339995	+	468	Pseudo	
Igkv2-105	chr6	68348661	68349441	+	780	Pseudo	
Igkv16-104	chr6	68425605	68426072	+	467	Functional	
Igkv15-103	chr6	68437468	68437925	+	457	V ORF	
Igkv15-102	chr6	68466126	68466590	–	464	Pseudo	
Igkv20-101-2	chr6	68474888	68475098	+	210	?	
Igkv15-101-1	chr6	68479082	68479573	+	491	Pseudo	
Igkv15-101	chr6	68481421	68481718	–	297	Pseudo	
Igkv14-100	chr6	68519012	68519477	+	465	Functional	
Igkv1-99	chr6	68541658	68542422	+	764	Functional	
Igkv12-98	chr6	68570763	68571235	+	472	Functional	
Igkv15-97	chr6	68591380	68591832	+	452	Pseudo	
Igkv10-96	chr6	68631965	68632430	–	465	Functional*	
Igkv2-95-2	chr6	68647999	68648531	–	532	Pseudo	
Igkv2-95-1	chr6	68670752	68671335	+	583	Pseudo	
Igkv10-95	chr6	68680379	68680848	+	469	Functional	
Igkv10-94	chr6	68704508	68704978	–	470	Functional	
Igkv2-93-1	chr6	68712925	68713508	–	583	Pseudo	
Igkv19-93	chr6	68736297	68736764	–	467	Functional	
Igkv4-92	chr6	68755038	68755593	–	555	Functional	
Igkv4-91	chr6	68768555	68769103	–	548	Functional	
Igkv4-90	chr6	68807180	68807708	–	528	Functional	

Continued

Table 1 *Igk*—cont'd

Segment	Chr	Start	End	Strand	Size	Functional status
Igkv13-89-1	chr6	68810495	68810543	+	48	Pseudo
Igkv12-89	chr6	68834846	68835307	–	461	Functional
Igkv1-88	chr6	68862265	68863031	–	766	Functional
Igkv13-87	chr6	68902801	68903266	+	465	Pseudo
Igkv4-86	chr6	68910411	68910938	–	527	Functional
Igkv13-85	chr6	68930269	68930734	–	465	Functional
Igkv13-84	chr6	68939602	68940067	+	465	Functional
Igkv4-83	chr6	68961378	68962019	–	641	Pseudo
Igkv13-82	chr6	68979198	68979663	+	465	Pseudo
Igkv4-81	chr6	68990758	68991294	–	536	Functional
Igkv13-80-1	chr6	69009143	69009513	+	370	Pseudo
Igkv4-80	chr6	69016558	69017080	–	522	Functional
Igkv4-79	chr6	69042972	69043505	–	533	Functional
Igkv13-78-1	chr6	69051444	69051794	+	350	Pseudo
Igkv4-78	chr6	69059690	69060224	–	534	Functional
Igkv4-77	chr6	69110904	69111432	–	528	Pseudo
Igkv13-76	chr6	69137866	69138332	+	466	Pseudo
Igkv4-75	chr6	69156118	69156659	–	541	Pseudo
Igkv13-74-1	chr6	69177416	69177787	+	371	Pseudo
Igkv4-74	chr6	69184826	69185361	–	535	Functional
Igkv13-73-1	chr6	69190073	69190282	+	209	Pseudo
Igkv4-73	chr6	69197583	69198116	–	533	Functional
Igkv4-72	chr6	69226854	69227383	–	529	Functional
Igkv13-71-1	chr6	69235927	69236158	+	231	Pseudo
Igkv4-71	chr6	69243160	69243688	–	528	Functional
Igkv4-70	chr6	69267888	69268412	–	524	Functional
Igkv4-69	chr6	69283794	69284319	–	525	Functional
Igkv4-68	chr6	69304834	69305363	–	529	Functional
Igkv12-67	chr6	69324177	69324646	–	469	Pseudo
Igkv12-66	chr6	69334604	69335076	–	472	Pseudo
Igkv4-65	chr6	69342472	69343004	–	532	Pseudo
Igkv13-64	chr6	69362722	69363187	+	465	Pseudo
Igkv4-63	chr6	69377944	69378472	–	528	Functional
Igkv13-62-1	chr6	69392439	69392815	+	376	Pseudo
Igkv4-62	chr6	69399812	69400344	–	532	V ORF
Igkv13-61-1	chr6	69409499	69409875	+	376	Pseudo
Igkv4-61	chr6	69416893	69417425	–	532	Functional
Igkv4-59	chr6	69438218	69438741	–	523	Functional
Igkv4-60	chr6	69463288	69463725	–	437	Pseudo
Igkv4-58	chr6	69500254	69500758	–	504	Functional
Igkv13-57-2	chr6	69523739	69524016	+	277	Pseudo
Igkv4-57-1	chr6	69544368	69544892	–	524	Functional
Igkv13-57-1	chr6	69569737	69570109	+	372	Pseudo
Igkv4-57	chr6	69575975	69576500	–	525	Functional
Igkv13-56-1	chr6	69581244	69581452	+	208	Pseudo
Igkv4-56	chr6	69587295	69587810	–	515	Pseudo
Igkv13-55-1	chr6	69599886	69600262	+	376	Pseudo
Igkv4-55	chr6	69607275	69607801	–	526	Functional
Igkv13-54-1	chr6	69617048	69617423	+	375	Pseudo
Igkv4-54	chr6	69631648	69632178	–	530	V ORF
Igkv4-53	chr6	69648824	69649354	–	530	Functional
Igkv4-51	chr6	69681406	69681939	–	533	Functional
Igkv4-50	chr6	69700767	69701287	–	520	Functional
Igkv12-49	chr6	69716498	69716956	–	458	Pseudo
Igkv5-48	chr6	69726573	69727125	–	552	Functional
Igkv12-47	chr6	69750854	69751310	–	456	Pseudo
Igkv12-46	chr6	69764523	69764992	–	469	Functional
Igkv5-45	chr6	69775750	69776306	–	556	Functional

Table 1 *Igk*—cont'd

Segment	Chr	Start	End	Strand	Size	Functional status
Igkv12-44	chr6	69814631	69815100	–	469	Functional
Igkv5-43	chr6	69823355	69823911	–	556	Functional
Igkv12-42	chr6	69834792	69835245	–	453	Pseudo
Igkv12-41	chr6	69858420	69858884	–	464	Functional
Igkv5-40-1	chr6	69868595	69869193	–	598	Pseudo
Igkv12-40	chr6	69879350	69879819	–	469	Pseudo
Igkv5-39	chr6	69900424	69900977	–	553	Functional
Igkv12-38	chr6	69943186	69943648	–	462	Functional
Igkv5-37	chr6	69963312	69963861	–	549	Functional
Igkv18-36	chr6	69992465	69992977	–	512	Functional
Igkv1-35	chr6	70010949	70011673	–	724	V ORF
Igkv8-34	chr6	70044112	70044678	–	566	Functional
Igkv7-33	chr6	70058632	70059199	–	567	Functional
Igkv6-32	chr6	70074024	70074584	–	560	Functional
Igkv8-31	chr6	70105860	70106179	–	319	Pseudo
Igkv8-30	chr6	70117061	70117617	–	556	Functional
Igkv6-29	chr6	70138462	70139082	–	620	Functional
Igkv8-28	chr6	70143593	70144161	–	568	Functional
Igkv8-27	chr6	70171809	70172270	–	461	Functional
Igkv8-26	chr6	70193228	70193797	+	569	V ORF
Igkv6-25	chr6	70215433	70215957	+	524	Functional
Igkv8-24	chr6	70216858	70217421	–	563	Functional
Igkv8-23-1	chr6	70240427	70240710	+	283	V ORF
Igkv6-23	chr6	70260409	70260933	–	524	Functional
Igkv8-22	chr6	70297554	70298119	–	565	Pseudo
Igkv8-21	chr6	70314895	70315452	–	557	Functional
Igkv6-20	chr6	70335841	70336479	–	638	Functional
Igkv8-19	chr6	70340876	70341448	–	572	Functional
Igkv8-18	chr6	70355877	70356445	+	568	V ORF
Igkv6-17	chr6	70371469	70371993	+	524	Functional
Igkv8-16	chr6	70386672	70387238	–	566	Functional
Igkv6-15	chr6	70406469	70406992	–	523	Functional
Igkv6-14	chr6	70434952	70435476	–	524	Functional
Igkv6-13	chr6	70457503	70458036	–	533	Functional
Igkv3-12-1	chr6	70497662	70498342	–	680	?
Igkv3-12	chr6	70518250	70518849	+	599	Functional
Igkv3-11	chr6	70553274	70554440	+	1166	Pseudo
Igkv3-10	chr6	70572633	70573230	+	597	Functional
Igkv3-9	chr6	70588189	70588777	+	588	Functional
Igkv3-8	chr6	70588871	70589452	+	581	Pseudo
Igkv3-7	chr6	70607437	70608036	+	599	Functional
Igkv3-6	chr6	70642405	70643565	+	1160	Pseudo
Igkv3-5	chr6	70663298	70663895	+	597	Functional
Igkv3-4	chr6	70671789	70672377	+	588	Functional
Igkv3-3	chr6	70686946	70687534	+	588	Functional
Igkv3-2	chr6	70698468	70699067	+	599	Functional
Igkv3-1	chr6	70703578	70704177	+	599	Functional
Cer	chr6	70709374	70710025	.	651	
Sis	chr6	70712108	70715840	.	3732	
Igkj1	chr6	70722562	70722599	+	37	Functional
Igkj2	chr6	70722916	70722954	+	38	Functional
Igkj3	chr6	70723223	70723260	+	37	Pseudo
Igkj4	chr6	70723548	70723585	+	37	Functional
Igkj5	chr6	70723886	70723923	+	37	Functional
MiEk	chr6	70725210	70725956	.	746	
Igkc	chr6	70726434	70726754	+	320	Functional
3Ek	chr6	70735257	70736522	.	1265	
Edk	chr6	70743710	70744950	.	1240	

Continued

Table 1 *Tcrb*—cont'd

Segment	Chr	Start	End	Strand	Size	Functional status	
Trbv1	chr6	40891296	40891885	+	589	Functional	V GENES
Trbv2	chr6	41047556	41047995	+	439	Functional	
Trbv3	chr6	41048394	41048824	+	430	Functional	
Trbv4	chr6	41059443	41059886	+	443	Functional	
Trbv5	chr6	41062359	41062803	+	444	Functional	
Trbv6	chr6	41066871	41067154	+	283	Pseudo	
Trbv7	chr6	41080253	41080654	+	401	Pseudo	
Trbv8	chr6	41084669	41085107	+	438	Pseudo	
Trbv9	chr6	41087991	41088288	+	297	Pseudo	
Trbv10	chr6	41091935	41092228	+	293	Pseudo	
Trbv11	chr6	41106805	41107303	+	498	Pseudo	
Trbv12-1	chr6	41113567	41114067	+	500	Functional	
Trbv13-1	chr6	41116036	41116468	+	432	Functional	
Trbv12-2	chr6	41118864	41119364	+	500	Functional	
Trbv13-2	chr6	41121396	41121832	+	436	Functional	
Trbv12-3	chr6	41127592	41128076	+	484	Pseudo	
Trbv13-3	chr6	41130147	41130585	+	438	Functional	
Trbv14	chr6	41135167	41135616	+	449	Functional	
Trbv15	chr6	41141211	41141658	+	447	Functional	
Trbv16	chr6	41151791	41152230	+	439	Functional	
Trbv17	chr6	41163092	41163556	+	464	Functional	
Trbv18	chr6	41175137	41175608	+	471	Pseudo	
Trbv19	chr6	41178589	41179048	+	459	Functional	
Trbv20	chr6	41188273	41188977	+	704	Functional	
Trbv21	chr6	41202587	41203044	+	457	V ORF	
Trbv22	chr6	41212075	41212534	+	459	Pseudo	
Trbv23	chr6	41216073	41216526	+	453	Functional	
Trbv24	chr6	41218091	41218557	+	466	Functional	
Trbv25	chr6	41221674	41222210	+	536	Pseudo	
Trbv26	chr6	41227525	41227913	+	388	Functional	
Trbv27	chr6	41258121	41258579	+	458	Pseudo	
Trbv28	chr6	41266697	41267150	+	453	Pseudo	
Trbv29	chr6	41271403	41271881	+	478	Functional	
Trbv30	chr6	41281376	41281990	+	614	Functional	
Trbd1	chr6	41533201	41533212	+	11	Functional	D GENE
Trbj1-1	chr6	41533864	41533911	+	47	Functional	J GENES
Trbj1-2	chr6	41534001	41534048	+	47	Functional	
Trbj1-3	chr6	41534323	41534372	+	49	Functional	
Trbj1-4	chr6	41534811	41534861	+	50	Functional	
Trbj1-5	chr6	41535084	41535133	+	49	Functional	
Trbj1-6	chr6	41535554	41535606	+	52	Pseudo	
Trbj1-7	chr6	41535642	41535687	+	45	Pseudo	
Trbc1	chr6	41538219	41539665	+	1446	Functional	CONSTANT REGION
Trbd2	chr6	41542163	41542176	+	13	Functional	D GENE
Trbj2-1	chr6	41542754	41542803	+	49	Functional	J GENES
Trbj2-2	chr6	41542957	41543007	+	50	Functional	
Trbj2-3	chr6	41543223	41543271	+	48	Functional	
Trbj2-4	chr6	41543362	41543410	+	48	Functional	
Trbj2-5	chr6	41543453	41543501	+	48	Functional	
Trbj2-6	chr6	41543598	41543645	+	47	Pseudo	
Trbj2-7	chr6	41543810	41543856	+	46	Functional	
Trbc2	chr6	41546730	41548181	+	1451	Functional	CONSTANT REGION
Eb	chr6	41554187	41554745	.	558		ENHANCER
Trbv31	chr6	41557693	41558371	.	678	Functional	

Table 1 *Tcra-Tcrd*—cont'd

Segment	Chr	Start	End	Strand	Size	Functional status	
Trav1	chr14	52427967	52428876	+	909	Functional	Vα GENES
Trav2	chr14	52567293	52568053	+	760	Functional	
Trav3-1	chr14	52580762	52581213	+	451	Functional	
Trav4-1	chr14	52607307	52607804	+	497	Pseudo	
Trav5-1	chr14	52622566	52623082	+	516	Functional	
Trav6-1	chr14	52638515	52638962	+	447	Functional	
Trav7-1	chr14	52654817	52655329	+	512	Functional	
Trav6-2	chr14	52667411	52667864	+	453	Functional	
Trav7d-2	chr14	52683964	52684472	+	508	Functional	
Trav4d-2	chr14	52710917	52711402	+	485	Pseudo	
Trav6d-3	chr14	52725308	52726919	+	1611	Functional	
Trav7d-3	chr14	52744314	52744838	+	524	Functional	
Trav6d-4	chr14	52753398	52753846	+	448	Functional	
Trav7d-4	chr14	52769863	52770392	+	529	Functional	
Trav8d-1	chr14	52778510	52778995	+	485	Functional	
Trav9d-1	chr14	52792316	52792776	+	460	Functional	
Trav6d-5	chr14	52795165	52795633	+	468	Functional	
Trav10d	chr14	52810997	52811496	+	499	Functional*	
Trav6d-6	chr14	52821922	52822359	+	437	Functional	
Trav11d	chr14	52824357	52824913	+	556	Functional	
Trav7d-5	chr14	52835861	52836387	+	526	Functional	
Trav12d-1	chr14	52844560	52845041	+	481	Functional	
Trav13d-1	chr14	52851355	52851866	+	511	Functional	
Trav14d-1	chr14	52855665	52856166	+	501	Functional*	
Trav15d-1-dv6d-1	chr14	52863365	52863933	+	568	Functional	
Trav3d-2	chr14	52887587	52888080	+	493	Pseudo	
Trav9d-2	chr14	52892186	52892618	+	432	Functional*	
Trav4d-3	chr14	52899972	52900470	+	498	Functional	
Trav5d-2	chr14	52914417	52914702	+	285	Pseudo	
Trav12d-2	chr14	52915915	52916426	+	511	Functional	
Trav9d-3	chr14	52920678	52921110	+	432	Functional	
Trav5d-3	chr14	52928320	52928532	+	212	Pseudo	
Trav12d-3	chr14	52929789	52930282	+	493	Functional*	
Trav13d-2	chr14	52942764	52943289	+	525	Functional	
Trav14d-2	chr14	52948668	52949169	+	501	Functional	
Trav15d-2-dv6d-2	chr14	52956828	52957400	+	572	Functional	
Trav3d-3	chr14	52972532	52973038	+	506	Functional	
Trav9d-4	chr14	52983456	52983770	+	314	Functional	
Trav4d-4	chr14	52990889	52991389	+	500	Functional	
Trav5d-4	chr14	53001713	53002231	+	518	Functional*	
Trav6d-7	chr14	53007632	53008098	+	466	Functional	
Trav7d-6	chr14	53020459	53020991	+	532	Functional	
Trav13d-3	chr14	53032902	53033419	+	517	Functional	
Trav8d-2	chr14	53042428	53042891	+	463	Functional	
Trav16d-dv11	chr14	53047287	53047821	+	534	Functional	
Trav13d-4	chr14	53072763	53073275	+	512	Functional	
Trav14d-3-dv8	chr14	53078547	53079045	+	498	Functional	
Trav15d-3	chr14	53082063	53082522	+	459	Pseudo	
Trav7n-4	chr14	53091346	53091875	+	529	Functional	
Trav9n-1	chr14	53101801	53102265	+	464	Pseudo	
Trav6n-5	chr14	53104871	53105344	+	473	Functional	
Trav10n	chr14	53122114	53122611	+	497	Functional	
Trav6n-6	chr14	53132700	53133137	+	437	Functional	
Trav11n	chr14	53135122	53135678	+	556	Pseudo	
Trav7n-5	chr14	53146922	53147448	+	526	Functional	
Trav12n-1	chr14	53155607	53156088	+	481	Functional	
Trav13n-1	chr14	53162399	53162910	+	511	Functional	
Trav14n-1	chr14	53166677	53167178	+	501	Functional	
Trav15n-1	chr14	53174325	53174896	+	571	Functional	
Trav3n-2	chr14	53198525	53199018	+	493	Pseudo	
Trav9n-2	chr14	53203122	53203544	+	422	Functional	
Trav4n-3	chr14	53210906	53211407	+	501	Functional	
Trav5n-2	chr14	53225393	53225669	+	276	Pseudo	
Trav12n-2	chr14	53226884	53227400	+	516	Functional	
Trav9n-3	chr14	53231668	53232102	+	434	Functional	
Trav5n-3	chr14	53239284	53239563	+	279	Pseudo	
Trav12n-3	chr14	53240751	53241244	+	493	Pseudo	
Trav13n-2	chr14	53253792	53254317	+	525	Functional	

Continued

Table 1 *Tcra-Tcrd*—cont'd

Segment	Chr	Start	End	Strand	Size	Functional status	
Trav14n-2	chr14	53259678	53260179	+	501	Functional	
Trav15n-2	chr14	53267963	53268535	+	572	Functional	
Trav3n-3	chr14	53283669	53284175	+	506	Functional	
Trav9n-4	chr14	53294596	53295030	+	434	Functional	
Trav4n-4	chr14	53302059	53302559	+	500	Functional	
Trav5n-4	chr14	53312904	53313422	+	518	Functional	
Trav6n-7	chr14	53318818	53319284	+	466	Functional	
Trav7n-6	chr14	53324676	53325208	+	532	Functional	
Trav13n-3	chr14	53337122	53337639	+	517	Functional	
Trav8n-2	chr14	53345961	53346424	+	463	Functional	
Trav16n	chr14	53351090	53351621	+	531	Functional	
Trav13n-4	chr14	53362368	53364106	+	1738	Functional	
Trav14n-3	chr14	53370077	53370575	+	498	Functional	
Trav15n-3	chr14	53374821	53375278	+	457	Pseudo	
Trav7-2	chr14	53390633	53391136	+	503	Functional	
Trav4-2	chr14	53418388	53418873	+	485	Functional	
Trav6-3	chr14	53428761	53430377	+	1616	Functional	
Trav7-3	chr14	53443314	53443839	+	525	Functional	
Trav6-4	chr14	53454327	53454784	+	457	Functional	
Trav7-4	chr14	53461209	53461738	+	529	Functional	
Trav8-1	chr14	53469756	53470231	+	475	Functional	
Trav9-1	chr14	53488106	53488567	+	461	Functional	
Trav6-5	chr14	53491152	53491622	+	470	Functional	
Trav10	chr14	53505790	53506286	+	496	Functional*	
Trav6-6	chr14	53516929	53517366	+	437	Functional	
Trav11	chr14	53519303	53519859	+	556	Functional*	
Trav7-5	chr14	53530786	53531313	+	527	Functional	
Trav12-1	chr14	53538266	53538738	+	472	Functional*	
Trav13-1	chr14	53545014	53545525	+	511	Functional	
Trav14-1	chr14	53554057	53554558	+	501	Functional	
Trav15-1-dv6-1	chr14	53559676	53560247	+	571	Functional	
Trav3-2	chr14	53586471	53586964	+	493	Pseudo	
Trav9-2	chr14	53591080	53591504	+	424	Functional	
Trav4-3	chr14	53598911	53599410	+	499	Functional	
Trav5-2	chr14	53614936	53615224	+	288	Pseudo	
Trav12-2	chr14	53616397	53616914	+	517	Functional	
Trav12-3	chr14	53621752	53622245	+	493	Functional	
Trav13-2	chr14	53634888	53635399	+	511	Functional	
Trav14-2	chr14	53640775	53641225	+	450	Functional	
Trav15-2-dv6-2	chr14	53649425	53649994	+	569	Functional	
Trav3-3	chr14	53666005	53666506	+	501	Functional	
Trav9-4	chr14	53676196	53676628	+	432	Functional	
Trav4-4-dv10	chr14	53683677	53684177	+	500	Functional	
Trav5-4	chr14	53704007	53704514	+	507	V ORF	
Trav6-7-dv9	chr14	53709943	53710419	+	476	Functional*	
Trav7-6	chr14	53716760	53717291	+	531	Functional	
Trav13-3	chr14	53729558	53730074	+	516	Functional*	
Trav8-2	chr14	53738375	53738838	+	463	Functional	
Trav16n	chr14	53743174	53743705	+	531	Functional	
Trav13-4-dv7	chr14	53757410	53757921	+	511	Functional	
Trav14-3	chr14	53763196	53763694	+	498	Functional	
Trav15-3	chr14	53766735	53767179	+	444	Pseudo	
Trav3-4	chr14	53777071	53777576	+	505	Functional	
Trav12-4	chr14	53778603	53778845	+	242	Pseudo	
Trav13-5	chr14	53795455	53795963	+	508	Functional	
Trav17	chr14	53806639	53807115	+	476	Functional	
Trav18	chr14	53831105	53831827	+	722	V ORF	
Trav19	chr14	53845322	53845827	+	505	Functional*	
Trav20	chr14	53863110	53863661	+	551	Pseudo	
Trav21/dv12	chr14	53876016	53876752	+	736	Functional	
Trdv1	chr14	53881612	53882217	+	605	Functional	Vδ GENE
Trav22	chr14	53926978	53927595	+	617	Pseudo	
Trdv2-1	chr14	53946073	53946660	+	587	Functional	Vδ GENE
Trdv2-2	chr14	53960993	53961602	+	609	Functional	Vδ GENE
Trav23	chr14	53977153	53977678	+	525	Pseudo	
Trdv3	chr14	54000948	54001253	+	305	Pseudo	Vδ GENE
Trdv4	chr14	54075004	54075515	+	511	Functional	Vδ GENE
Trdd1	chr14	54113468	54113476	+	8	Functional	Dδ GENES
Trdd2	chr14	54122226	54122241	+	15	Functional	

Table 1 *Tcra-Tcrd*—cont'd

Segment	Chr	Start	End	Strand	Size	Functional status	
Trdj1	chr14	54123138	54123188	+	50	Functional	Jδ GENES
Trdj2	chr14	54136779	54136837	+	58	Functional	
Ed	chr14	54138465	54138512	+	47		ENHANCER (δ)
Trdc	chr14	54142851	54146108	+	3257	Functional	CONSTANT REGION (δ)
Trdv5	chr14	54148662	54149201	–	539	Functional	Vδ GENE
TEAp	chr14	54152995	54153117	+	122		
Traj61	chr14	54155005	54155076	+	71	Pseudo	Jα GENES
Traj60	chr14	54155933	54155985	+	52	Pseudo	
Traj59	chr14	54156133	54156194	+	61	J ORF	
Traj58	chr14	54157280	54157342	+	62	Functional	
Traj57	chr14	54158507	54158569	+	62	Functional	
Traj56	chr14	54159263	54159325	+	62	Functional	
Traj55	chr14	54161439	54161501	+	62	Pseudo	
Traj54	chr14	54161990	54162043	+	53	Pseudo	
Traj53	chr14	54162644	54162709	+	65	Functional	
Traj52	chr14	54165316	54165381	+	65	Functional	
Traj51	chr14	54166239	54166281	+	42	Pseudo	
Traj50	chr14	54167590	54167652	+	62	Functional	
Traj49	chr14	54168686	54168744	+	58	Functional	
Traj48	chr14	54169808	54169868	+	60	Functional	
Traj47	chr14	54171802	54171856	+	54	J ORF	
Traj46	chr14	54172338	54172398	+	60	J ORF	
Traj45	chr14	54172831	54172890	+	59	Functional	
Traj44	chr14	54173690	54173750	+	60	J ORF	
Traj43	chr14	54174742	54174798	+	56	Functional	
Traj42	chr14	54175773	54175836	+	63	Functional	
Traj41	chr14	54176271	54176325	+	54	J ORF	
Traj40	chr14	54177921	54177981	+	60	Functional	
Traj39	chr14	54179962	54180024	+	62	Functional	
Traj38	chr14	54180574	54180635	+	61	Functional	
Traj37	chr14	54181518	54181577	+	59	Functional	
Traj36	chr14	54182402	54182464	+	62	Pseudo	
Traj35	chr14	54183774	54183838	+	64	Functional	
Traj34	chr14	54184699	54184756	+	57	Functional	
Traj33	chr14	54185358	54185414	+	56	Functional	
Traj32	chr14	54186101	54186166	+	65	Functional	
Traj31	chr14	54187895	54187951	+	56	Functional	
Traj30	chr14	54189866	54189924	+	58	Functional	
Traj29	chr14	54190946	54191005	+	59	J ORF	
Traj28	chr14	54191661	54191725	+	64	Functional	
Traj27	chr14	54192303	54192361	+	58	Functional	
Traj26	chr14	54194490	54194549	+	59	Pseudo	
Traj25	chr14	54194815	54194871	+	56	J ORF	
Traj24	chr14	54195645	54195700	+	55	Functional	
Traj23	chr14	54196081	54196140	+	59	Functional	
Traj22	chr14	54197248	54197307	+	59	Functional	
Traj21	chr14	54198790	54198846	+	56	Functional	
Traj20	chr14	54199441	54199498	+	57	J ORF	
Traj19	chr14	54200386	54200446	+	60	J ORF	
Traj18	chr14	54200777	54200842	+	65	Functional	
Traj17	chr14	54201775	54201837	+	62	Functional	
Traj16	chr14	54203134	54203194	+	60	Functional	
Traj15	chr14	54204422	54204481	+	59	Functional	
Traj14	chr14	54205111	54205144	+	33	Pseudo	
Traj13	chr14	54205741	54205797	+	56	Functional	
Traj12	chr14	54206552	54206610	+	58	Functional	
Traj11	chr14	54207139	54207197	+	58	Functional	
Traj9	chr14	54209393	54209450	+	57	Functional	
Traj8	chr14	54209918	54209954	+	36	Pseudo	
Traj7	chr14	54211470	54211528	+	58	J ORF	
Traj6	chr14	54212688	54212749	+	61	Functional	
Traj5	chr14	54213777	54213838	+	61	Functional	
Traj4	chr14	54216291	54216353	+	62	J ORF	
Traj3	chr14	54217292	54217357	+	65	J ORF	
Traj2	chr14	54217836	54217901	+	65	Functional	
Traj1	chr14	54218814	54218842	+	28	Pseudo	
Trac	chr14	54220521	54224198	+	3677	Functional	CONSTANT REGION
Ea	chr14	54227374	54227584	+	210		ENHANCER (α)

*Mark segments for which different annotations concerning the functional status coexist. Dark blue (dark gray in the print version) = enhancer regions and PAIR elements, light blue (light gray in the print version) = constant regions, yellow (white in the print version) = j genes, red (dark gray in the print version) = D genes, aqua (white in the print version) = insulators, green (light gray in the print version) = V genes.

Garrett et al., 2005). These hypersensitive sites mostly show occupancy by transcription factors in mature B cells as this enhancer is not implicated in V(D)J regulation but controls CSR and somatic hypermutation, which take place in mature germinal center B cells after encounter with antigen (Khamlichi, Pinaud, Decourt, Chauveau, & Cogne, 2000; Pinaud et al., 2011; Rouaud et al., 2013). In line with its role in these late events, the binding profile of Pax5 to the 3′RR is altered during CSR leading to enrichment on HS1–2, HS4, and HS7 (Chatterjee et al., 2011).

4.1.3 PAIR Elements

Fourteen PAIR (Pax5-activated intergenic repeats) elements were identified within the distal V_H region each of which contains functional CTCF-, E2A-, and Pax5-binding sites (Ebert et al., 2011). 11 of the 14 PAIR elements are found immediately upstream of V_H3609 genes interspersed within the distal V_HJ558 gene family. Detailed investigation of PAIRs 4, 6, and 7 demonstrates binding of Pax5, E2A, and CTCF in pro-B cells. In contrast, at the later pre-B cell stage, there is depletion of Pax5 occupancy at these sites (Ebert et al., 2011). Pax5 binding at the pro-B cell stage correlates with the presence of antisense transcripts that are distinct from those identified in other studies (Bolland et al., 2004). These PAIR elements are implicated in locus contraction since Pax5 binding at the pro-B cell stage coincides with contraction, and Pax5 depletion in pre-B cells corresponds to a decontracted state (Roldan et al., 2005). It is of interest that neither the *Igh* enhancers (Eμ and the 3′RR region) nor the intergenic control region 1 (IGCR1) insulator site have any impact on locus contraction, suggesting that only elements within the V_H cluster are required (Medvedovic et al., 2013). The involvement of PAIRs in *Igh* contraction will need to be confirmed with genetic approaches that target these elements.

4.1.4 The Intergenic Control Region 1

The IGCR1 is located within a 100-kb-long intergenic region, which separates the *Ig* V_H and D_H gene segments of *Igh*. It spans 4.3 kb and lies between V_H81x (Ighv5-1) and DFL16.1 (2.1 kb upstream of DFL16.1 also named Ighd1-1). IGCR1 consists of six hypersensitive (HS) sites (HS 1–6). Two conserved CTCF-binding sites, HS4/5 that exhibit enhancer blocking activity mark a sharp boundary of antisense transcription that stops at least 40 kb from the V_H genes (Featherstone, Wood, Bowen, & Corcoran, 2010). In T cells and early pre-pro-B cells undergoing D_H–J_H rearrangement, the two CTCF sites separate regions of active and inactive chromatin associated with the D_H and V_H regions, respectively. Antisense transcription, which occurs at high level within this region in these cells, is reduced in pro-B cells where V_H–DJ_H recombination takes place. Thus, it was proposed that the two CTCF sites act as an insulator preventing the spreading of chromatin activation and transcription into the V_H region during D_H–J_H rearrangement (Featherstone et al., 2010).

Deletion of the 4.1 kb fragment (named the IGCR1) encompassing both CBEs (CBE1/2) alongside potential binding sites for other regulators (YY1 and PU.1) demonstrated that mutant alleles in a RAG-deficient background were indeed associated with upregulation of proximal V_H7183 and V_HQ52 transcripts and an enrichment of active histone marks in pro-B cells. Increased accessibility/transcription of proximal V_H genes is linked to preferential rearrangement of V_H7183 and V_HQ52 at the expense of distal V_HJ558 gene rearrangement in recombination competent IGCR1 targeted cells. Furthermore, mutant CBE alleles can undergo V_H–D_H rearrangement prior to D_H–J_H rearrangement, indicating a role for these elements in regulating ordered rearrangement (Guo et al., 2011). Additionally, mutant *Igh* alleles can undergo V_H–DJ_H recombination in developing thymocytes, in contrast to wild-type counterparts which normally only undergo D_H–J_H rearrangement. Thus, the two CBE sites have a role in regulating lineage-specific, ordered rearrangement as reviewed in Chaumeil and Skok (2012).

In more recent studies, extended analysis of the CBE sites was performed after scrambling each element to separately assess their individual contributions to these processes (Lin et al., 2015). They demonstrate that scrambling of CBE1 but not CBE2 impacts allele expression such that in F1 mice harboring a 129 IgMa CBE1-mutated allele and a C57BL/6 IgMb WT allele, resulting B cells were found to express half as many IgMa as compared to IgMb alleles. However, this defect was much more severe in F1 mice if both CBEs were mutated simultaneously. In line with their previous findings (Guo et al., 2011), mutation of either CBE1 or CBE2 led to a decrease in distal V_H gene rearrangement but this defect was more pronounced in CBE1$^{-/-}$ mice. Double CBE1/CBE2$^{-/-}$ mice however had the most severe defect.

Mice carrying mutations in the CBE-binding sites also display defects in ordered rearrangement such that direct V_H-to-DJ_H joins were detected in CBE1$^{-/-}$ mice and variably in CBE2$^{-/-}$ mice, but again this was most pronounced in the double CBE1/CBE2$^{-/-}$ mutants. Finally, mutant CBE1 and 2 mice displayed low levels of proximal V_H-to-DJ_H rearrangements in T cells but lineage-inappropriate rearrangement was much more severe in the double CBE1/CBE2$^{-/-}$ mice. No defects in allelic inclusion were observed in any of the three mutant mice. However, in line with what was previously observed in the double CBE1/CBE2$^{-/-}$ mutant mice (Guo et al., 2011), the presence of a productively rearranged V1–8 knockin allele could not suppress proximal V_H-to-DJ_H rearrangement on endogenous alleles. These rearrangements were detected in spleen on CBE2$^{-/-}$ alleles but more so on CBE1$^{-/-}$ alleles; however, they were predominantly non-productive. No distal V_H gene rearrangements were observed perhaps because locus contraction was impaired as we previously showed in mice harboring a rearranged transgene that skip past the pro-B cell stage of development (Roldan et al., 2005).

Together, these studies indicate that CBE1 has a more pronounced effect on ordered rearrangement and feedback control than CBE2. Two distinct mechanisms have been proposed that underpin the distinct behaviors of the CBE elements. First, CBE1 contains binding sites for PU.1 and YY1 and the presence of these binding sites could impact these functions. Second, and more interesting, is the observation that the orientation of the CTCF site within CBE1 is in the opposite direction to the 60 V_H CTCF sites, while the CTCF site within CBE2 is in the opposite orientation to the 10 3′ CTCF sites (Fig. 2), with the implication that CBE2 interacts with the 3′RR region, promoting D_H-to-J_H rearrangement and inhibiting direct V_H-to-DJ_H joins. CBE1 on the other hand could interact with CBEs in V_H genes. It is of note that deletion of both CBE1/2 sites does not alter locus contraction as determined by 4C-seq from the Eμ viewpoint that measures interactions specifically with this region alone (Medvedovic et al., 2013). However, there has been no 4C-seq analyses from the viewpoint of CBE1/2, so it remains to be determined what impact combined or individual deletion of these CTCF insulator sites has on surrounding interactions, and whether the change in chromatin boundaries that accompanies their deletion is matched by a change in interaction boundaries, as shown in studies involving the *Hoxa* locus (Narendra et al., 2015). A detailed analysis of looping in wild-type versus mutant CBE1 and CBE2 B cells could help resolve these issues. Furthermore, it would be interesting to find out what effect reversing the orientation of the two CBE sites has on *Igh* locus assembly.

4.2 Important *cis*-Acting Elements and Their Function in *Igk* Rearrangement

4.2.1 Enhancers

Igk possesses three powerful B cell-specific transcriptional enhancers: the MAR and the intronic enhancer, iEκ (together known as the MiEκ), are located between the Jκ and Cκ gene segments, while two additional enhancers, 3′Eκ and Edκ, are found 8.5 and 15.5 kb downstream of the constant region (Liu et al., 2002; Meyer, Sharpe, Surani, & Neuberger, 1990; Zhou, Xiang, & Garrard, 2010). MiEκ and 3′Eκ are both important for rearrangement and deletion of either one leads to a reduction in the ratio of κ/λ expressing B cells, while the double mutant is sufficient to abrogate *Igk* recombination altogether (Inlay, Alt, Baltimore, & Xu, 2002; Inlay, Lin, Gao, & Xu, 2006). In contrast, an absence of both the 3′Eκ and Edκ leads to a dramatic reduction in germline and rearranged transcription, a reduction in active chromatin marks, increased DNA methylation, and reduced levels of rearrangement. Furthermore, in mature cells, IGλ is exclusively expressed on the cell surface despite functional rearrangement of *Igk*. This indicates that in the absence of both the 3′Eκ and Edκ, the intronic enhancer is incapable of triggering *Igk* transcription (Zhou et al., 2010). Conditional knockout of the 3′Eκ in mature cells with an Edκ deletion leads to complete silencing of the *Igk* locus (Zhou, Xiang, Ding, & Garrard, 2013). In these mice, the mature B cells partially dedifferentiate, inducing RAG1/2 expression along with other pro-B cell markers and redifferentiate after triggering *Igl* gene rearrangement. These findings demonstrate that 3′Eκ and Edκ are essential for both the establishment and maintenance of transcriptional activity of *Igk*.

4.2.2 Promoters That Influence Jκ Usage

Igk germline transcription is initiated from two promoters located 150 bp (proximal) and 3.5 kb (distal) upstream of Jκ1 that give rise to the κ^0 transcripts (Schlissel, 2004; Fig. 3). The κ^0 0.8 and κ^0 1.1 germline transcripts are initiated from the proximal and distal promoters, respectively, and spliced to the Cκ region (Engel, Rolink, & Weiss, 1999; Martin & van Ness, 1990). Germline transcription from these promoters is linked to rearrangement, and deletion of a 4 kb region encompassing the two promoters has a marked impact on rearrangement of the allele bearing the deletion (Cocea et al., 1999). More recent studies demonstrate a role for the proximal promoter in directing primary rearrangements to Jκ1, thereby ensuring the retention of other Jκ segments that can be used in subsequent rounds of recombination for receptor editing (Vettermann, Timblin, Lim, Lai, & Schlissel, 2015). It was shown that the distal but not the proximal promoter is active in both recombining and editing cells. Deletion of the proximal promoter leads to increased breaks on Jκ2 and decreased usage of Jκ1 and this correlates with an increase in H3K4me3 levels and transcription in the Jκ1 region. Thus, the proximal promoter acts as a suppressor of accessibility and secondary recombination. Since it is inactive in recombining B cells, it was suggested that this function could be a result of promoter interference that is also found in *Tcra* where the active TEA suppresses activity of downstream Jα promoters (Abarrategui & Krangel, 2006, 2007).

4.2.3 Sis and Cer

Two additional regulatory elements within *Igk*, Sis—hypersensitivity sites 3–6 (HS3–6) and Cer—hypersensitivity sites 1–2 (HS1–2), reside in the 18 kb intervening Vκ–Jκ sequence (Liu et al., 2002). Sis (silencer in the intervening sequence) is a recombination silencer and heterochromatin targeting element. It binds both Ikaros and CTCF and directs the repositioning of *Igk* to PCH in pre-B cells (Liu et al., 2006). Deletion of Sis leads to reduced distal Vκ and enhanced proximal Vκ usage (Ribeiro de Almeida et al., 2011; Xiang et al., 2011). The neighboring CTCF-binding site, Cer (contracting element for recombination), plays a role in *Igk* locus contraction (Xiang, Park, & Garrard, 2013). Like Sis, deletion of Cer increases proximal and diminishes distal Vκ usage although its deletion has no impact on germline transcription or chromatin structure. Additionally, an absence of Cer leads to rearrangement of *Igk* in T cells. This is somewhat surprising since, unlike *Igh*, there is no evidence for *Igk* activation in T cells. Double deletion of both Cer and Sis gives rise to increased transcription of proximal Vκ in both pre-B and splenic B cells (Xiang, Park, & Garrard, 2014). In this respect, Cer and Sis behave in a similar manner to the CBEs, CBE1/2 in the *Igh* locus, although mutation of the IGCR1 does not appear to impact *Igh* locus contraction as determined by 4C-seq (Medvedovic et al., 2013) (see Section 4.1.3), while in contrast DNA FISH analyses of *Igk* alleles with deleted Cer (or double-deleted Cer and Sis) demonstrate a dramatic effect on *Igk* contraction (Xiang et al., 2013, 2014). It is difficult to compare the effects of the 4C-seq and DNA FISH analyses in these two studies as neither give a complete picture of how interactions are altered across each locus in entirety. The 4C-seq analysis was performed from the Eμ viewpoint in IGCR1-mutated cells and this serves to highlight interactions from Eμ alone (Medvedovic et al., 2013), while the FISH analyses provide information on the distances separating three points on the *Igk* locus and offer no details of intralocus interactions (Xiang et al., 2013, 2014).

4.2.4 Pre-BCR Signaling and Its Impact on Long-Range Interactions

Functional rearrangement of one *Igh* allele in pro-B cells leads to cell surface expression of the pre-BCR, which is comprised of IGH paired with surrogate light chain. Pre-BCR signaling in large pre-B cells drives proliferation and subsequent differentiation to the small pre-B cell stage where cells exit cell cycle and *Igk* rearrangement is initiated. To examine changes in *Igk* locus conformation by 4C-seq during the transition from the pro- to pre-B cell stage, pre-BCR signaling mutants of increasing severity were used (mice lacking *Btk*, *Slp65*, or both together) on a RAG-deficient background (Stadhouders et al., 2014). These analyses revealed that pre-BCR signaling reduces interactions of the three enhancers with *Igk* flanking sequences and increases interactions of the 3′Eκ with the Vκ regions, without altering Vκ interactions with the MiEκ (these are already in close contact at the pro-B cell stage). It is of note that in all cases, the enhancers interact more frequently with functional versus nonfunctional Vκs in pre-B cells. The Sis element also displays an altered interaction pattern within the *Igk* locus in pro-B and pre-B cells, interacting more frequently with the proximal domain (JκCκ) in pro-B cells compared to pre-B cells. In the latter, the interaction profile spreads to the Vκ gene region, which may be a reflection of a change in transcriptional activity although transcriptional profiles in pro-B cells are not shown in this study (Stadhouders et al., 2014). Vκ interactions correlate strongly with binding of E2A and Ikaros that are frequently found close to promoters and bind to the 3′Eκ, MiEκ, and Sis regulatory elements (Bossen, Mansson, & Murre, 2012; Kil et al., 2012; Ribeiro de Almeida et al., 2011, 2012). Furthermore, interactions occur preferentially if both E2A and Ikaros are present together versus Ikaros alone and the presence of both factors is linked to frequency of Vκ gene usage.

4.3 Important *cis*-Acting Elements and Their Function in *Tcrb* Rearrangement

4.3.1 The Eβ Enhancer and Promoters

Eβ is the sole known enhancer of *Tcrb*. It spans 550 bp and is located 6 kb downstream of the Cβ2 region and about 3 kb upstream Vβ31 (Fig. 4). Eβ facilitates activation of promoters flanking each of the two Dβ segments (McMillan & Sikes, 2008; Sikes, Gomez, Song, & Oltz, 1998). PDβ1, positioned immediately 5′ of the Dβ1 12-RSS, was the first germline *Tcrb* promoter discovered. It uses a TATA element situated in the RSS spacer to initiate transcription at Dβ1. PDβ1 is bound by T cell-restricted transcription factors including SP1, GATA-3, and members of the ETS, RUNX, and bHLH families. Most of these factors also bind Eβ (Doty, Xia, Nguyen, Hathaway, & Willerford, 1999; Sikes et al., 1998; Tripathi et al., 2000). Deletion of Eβ or PDβ1 dramatically affects T cell development in the following way. An Eβ deficiency gives rise to a similar phenotype as RAG deficiency in terms of *Tcrb* rearrangement. A total absence of germline transcription in the proximal DJCβ1–DJCβ2 domain leads to a failure of *Tcrb* rearrangement and a block in T cell development at the DN3 stage (Bories, Demengeot, Davidson, & Alt, 1996; Bouvier et al., 1996). In contrast, targeted deletion of PDβ1 specifically attenuates DJβ1 rearrangement without affecting DJβ2 and V–DJβ2 rearrangements (Whitehurst, Chattopadhyay, & Chen, 1999). These phenotypes demonstrate the importance of Eβ in modulating chromatin accessibility across both DJCβ regions, while the contribution of each promoter is specifically directed toward their associated DJβ clusters.

The DJβ2 region is comprised of two promoters, one upstream (5′PDβ2) and one downstream (3′PDβ2) of the Dβ2 segment. Analogous to PDβ1, 5′PDβ2 is located immediately 5′ to Dβ2 and binds GATA-3, RUNX1, and E47 (McMillan & Sikes, 2009). However, PDβ2 is inactive prior to Dβ2–Jβ2 recombination. Germline transcription at the DJβ2 cluster is driven by the NFκB-dependent promoter 3′PDβ2, located several hundred base pairs downstream of Dβ2 (McMillan & Sikes, 2008). Repression of 5′PDβ2 is ensured by USF-1, a constitutively expressed bHLH protein that binds in the spacer of the Dβ2 12-RSS. It has recently been shown that the introduction of DNA DSBs relieves the USF-mediated repression of Dβ2 (Stone et al., 2012). Following DJβ recombination, 5′PDβ2 is activated and both DJβ clusters are transcribed and can rearrange with distant Vβ genes.

Little is known about the developmental regulation of Vβ promoters. They are responsible for germline as well as rearranged transcription of Vβ elements. Similarly, they are involved in regulating recombinase accessibility as deletion of the Vβ31 promoter leads to a 10-fold decrease in Vβ31 rearrangement (Ryu et al., 2004). Unlike proximal promoters, Vβ promoters do not appear to require Eβ for their transcriptional activation in DN3 (Mathieu, Hempel, Spicuglia, Verthuy, & Ferrier, 2000). However, this enhancer can increase expression of the most highly transcribed subset of *Tcrb* Vβ segments in DN thymocytes.

4.3.2 Long-Range Interactions

Recent studies have investigated the role of enhancers and an insulator in shaping the interaction landscape of *Tcrb* that is so important for ensuring diversification of the *Tcrb* repertoire (Majumder et al., 2015). Using 3C (one-to-one interaction analysis) with an anchor on Eβ, they show that an absence of this enhancer leads to reduced interactions with the rest of the locus; however, interactions from this viewpoint to the mid Vβ gene region (Vβ12–13, Vβ14, Vβ16, and Vβ20) are maintained, although the 3C signal is lower than controls (Majumder et al., 2015). In line with contraction analyses of other enhancer-deficient antigen receptor loci, it was demonstrated that deletion or inactivation of Eβ (through introduction of mutations in RUNX-binding sites) does not disrupt the interaction between Dβ clusters and the Vβ gene segments despite ablation of germline transcription and reduced H3K4me3 levels in the region. This indicates that transcription is dispensable for long-range interactions between Vβ, Dβ, and Jβ gene segments. Nonetheless, Eβ could still have an impact on Vβ gene repertoire as it alters germline transcription of a subset of Vβ genes, but this is not testable because Eβ is essential for activation and recombination of the DβJβ region.

These studies demonstrated that the promoter PDβ1 is important for interactions between the Dβ2 region and distal Vβ genes because a 3.5 kb deletion impacts the 3C signal on these Vβ genes when Dβ2 is used as an anchor. This deletion also reduces CTCF levels at distal Vβ genes without impacting transcription or cohesin levels. Reduced CTCF binding may explain alterations in Vβ interaction frequency with the proximal DβJβ domain. It is of note that the PDβ1 deletion does not alter interactions from the distal Vβ5 viewpoint. In addition, proximal Vβ gene interactions with the DβJβ domain do not require the PDβ1; however, interactions between proximal and distal Vβ genes do. In contrast, a minimal deletion of the PDβ1 promoter does not impact long-range Vβ to DβJβ interactions, despite its impact on Dβ1 transcription. Interactions between the distal Vβs and the DβJβ thus rely on a 3 kb region located just upstream of the PDβ1 promoter. However, distal V gene interactions occur most robustly with the 5′*Prss2*-CTCF (5′PC) site, which is intact in the 3.5 kb deleted promoter allele and furthermore CTCF binding is not altered at the 5′PC if this region is deleted. The 5′PC can be distinguished by a 5′ repetitive tract (which contains a viral LTR that is expressed at low levels in DN cells that harbors insulator properties) and a pair of CTCF/RAD21-binding sites. The *Prss2* gene is normally inactive in WT, minimal PDβ1 deletion, or Eβ mutated alleles. However, it is activated if the entire 3.5 kb promoter is deleted and the genomic region associated with the promoter is enriched for H3K4me3. This mark spreads from the PDβ1 and PDβ2 region all the way up to the 5′PC in the PDβ1 mutant, suggesting that a chromatin boundary has been disrupted. This chromatin barrier appears to be required for mediating interactions between the distal Vβ gene segments (where, in contrast to the proximal domain, there is robust CTCF binding) and the PDβ region. Future studies will be required to identify the transcription factors that are involved in this interaction. Certainly, it is clear that these insulator sites are a common feature of antigen receptor loci as they are found in *Tcrb*, *Igk*, and *Igh* loci. In the case of *Igk* and *Igh* loci, deletion of these elements increases transcription and recombination in the proximal domain, perhaps disrupting interactions with distal V gene segments (see Section 6.2).

4.4 Important *cis*-Acting Elements and Their Function in *Tcra/d* Rearrangement

4.4.1 Enhancers and Promoters

There are two enhancers in the *Tcra/d* locus: Eδ and Eα regulate *Tcrd* and *Tcra* rearrangement, respectively (Fig. 5; Krangel, 2009). The Eδ, which is located between the Jδ and Cδ gene segments, regulates germline transcription of the promoter pDδ in the DJδ region. The Eδ region functions locally in adult DN cells and deletion of Eδ reduces *Tcrd* rearrangement by 10-fold although it does not alter accessibility of the Vδ genes (Hao & Krangel, 2011). In addition, Eδ appears to be dispensable for *Tcrd* expression after rearrangement (Monroe et al., 1999).

Eα is essential for *Tcra* rearrangement and it also regulates expression from rearranged *Tcrd* alleles (Krangel, 2009; Monroe et al., 1999). Deletion of Eα blocks *Tcra* rearrangement and T cell development (Sleckman, Bardon, Ferrini, Davidson, & Alt, 1997). Eα regulates a "T early α" promoter (TEAp) located just upstream of the Jα array, and this in turn activates the Jα array. Specifically, the TEA promoter targets primary rearrangement at the extreme 5′ of the Jα array by opening up the RSSs of these genes (Hawwari, Bock, & Krangel, 2005). This maximizes the use of Jα segments during secondary rearrangement as the use of the 3′ Jαs in the first round of rearrangement could result in deletion of intervening segments leaving few substrates for subsequent recombination events. The Eα RR also controls germline transcription and accessibility of the proximal Vα genes via long-range interactions (over 500 kb) (Hawwari et al., 2005). In addition, it mediates interactions between the proximal Vα and Jα gene segments, which are essential for synapsis and rearrangement (Shih et al., 2012).

As with the other loci, long-distance interactions between *cis* elements are essential for *Tcra* and *Tcrd* rearrangement. Chromatin organizers like cohesin and CTCF play an important role in mediating long-range interactions (Seitan et al., 2011; Shih et al., 2012; and reviewed in Chaumeil & Skok, 2012). Cohesin binds to the *Tcra* locus control region, the Eα enhancer, the Jα49 promoter, the TEAp, sites located between *Tcrd* and the first Vα segments and to Vα gene promoters in DP cells. Deletion of Rad21, one of the cohesin complex components, impairs the interaction between Eα and TEAp, which in turn impacts activation of distal 3′ Jα genes and impairs secondary *Tcra* rearrangements (Seitan et al., 2011). CTCF also mediates interaction between the Eα and TEAp and binds to the proximal Vα gene promoters, which may assemble into a rosette-like topology with Vα, Eα, and Jα (Shih et al., 2012).

5. ALLELIC EXCLUSION

Antigen receptors are expressed from only one allele in individual lymphocytes to ensure unique receptor specificity. This is fundamental to the proper functioning of the adaptive immune response, which relies on clonal expansion of lymphocytes expressing receptors that specifically recognize an invading pathogen. Elucidating the mechanisms underlying monospecific receptor expression—allelic exclusion—has proved to be a challenging puzzle, likely because the process involves multiple levels of control. However, allelic exclusion is not infallible as dual *Tcr* or *Ig* receptor expressing T and B cells are found at low frequency in the periphery (Fournier et al., 2012; Pelanda, 2014). Although tolerance mechanisms exist to restrain dual receptor cells with a self-reactive receptor (Fournier et al., 2012), these cells can become activated and cause autoimmune disease if the nonself-reactive receptor recognizes a pathogenic antigen (Ercolini & Miller, 2009; Flodstrom-Tullberg, 2003; Ji et al., 2010; Pelanda, 2014). Nonetheless, the effects of these dual receptor cells are not altogether negative as their presence can be beneficial in counteracting infection because allelically included cells expand the receptor repertoire and may in some instances be important for combatting an invading pathogen (He et al., 2002). Thus, evolution may tolerate a certain frequency of allelically included dual receptor cells, balancing an autoimmune outcome with that of counteracting infection (Fig. 10).

All the antigen receptor loci are regulated in a unique manner and in particular they are all subjected to different controls when it comes to allelic inclusion. The *Igh* locus is subject to stringent allelic exclusion and only 2–4% of mouse spleen B cells contain two in-frame rearrangements with 0.01% expressing dual receptors. *Igk* and *Tcrb* are also subject to fairly stringent controls and allelic inclusion occurs at a frequency of 1–7% and 1–3%, respectively (Brady, Steinel, & Bassing, 2010). *Tcra*, on the other hand, can

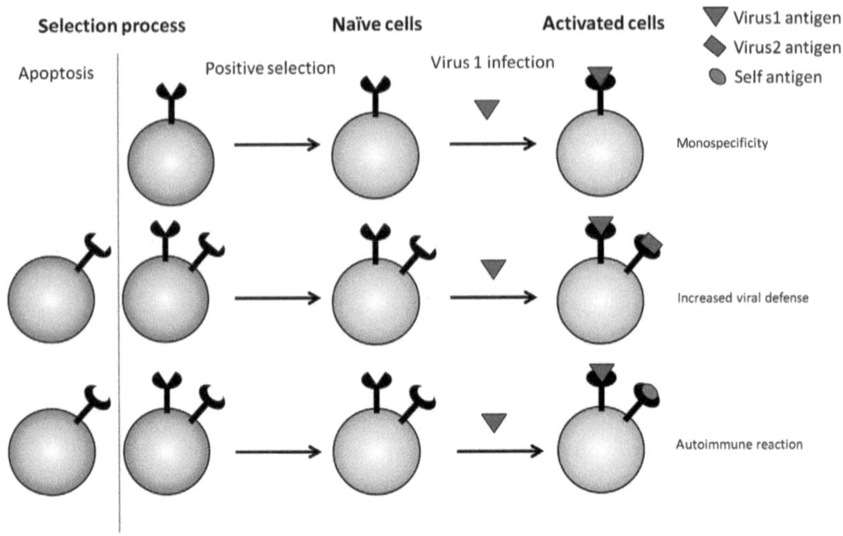

Figure 10 Evolution may tolerate a certain frequency of allelically included dual receptor cells, balancing an autoimmune outcome with that of counteracting infection.

rearrange both alleles prior to differentiation, but the frequency of allelic inclusion on the cell surface is only 10%, which may be due to post-translation control (Brady et al., 2010).

As mentioned above, allelic exclusion ensures the expression of only one productively rearranged allele (Jung, Giallourakis, Mostoslavsky, & Alt, 2006; Vettermann & Schlissel, 2010). The other allele can be nonfunctional for one of three reasons: (i) it remains in germline configuration (*Igk* or *Tcra*) or is partially recombined having undergone D-to-J but not V-to-DJ rearrangement (*Tcrb* or *Igh*), (ii) the allele has an out-of-frame rearrangement and the mRNA is degraded by the nonsense-mediated decay (NMD) pathway, or (iii) the allele encodes a protein that cannot pair with its partner (i.e., *Igh* with *Igk* or *Tcrb* with *Tcra*) and thus a receptor cannot be assembled on the surface. In this way, allelic exclusion is very different from other well-known monoallelically expressed genes such as olfactory receptors or those resulting from X inactivation or parental imprinting.

As a general rule, allelic exclusion is enforced during the process of V(D)J rearrangement (Fig. 11). However, in some cells with dual rearrangements, the product of only one allele is expressed at the cell surface as a result of posttranslational silencing, and in this case allelic exclusion is enforced by a later event (Alam & Gascoigne, 1998).

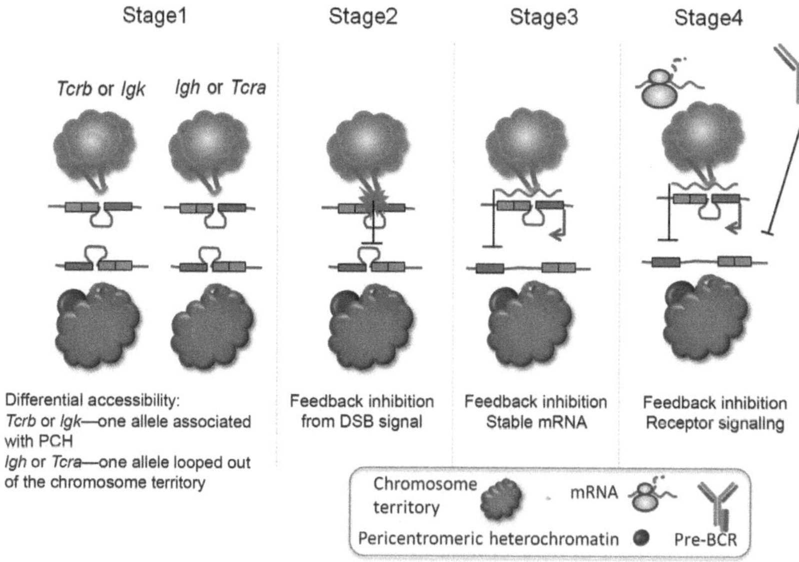

Figure 11 Allelic exclusion is enforced at multiple stages of development.

5.1 Asynchronous Rearrangement

Early models proposed that asynchronous recombination occurred as a result of low-efficiency recombination (RAG breaks are introduced in around 20% or less cells at any one time) which reduces the chances of rearrangement occurring on the two alleles at the same time. Added to this, the imprecise nature of junctions results in a high failure rate of rearrangements (two out of three will be nonproductive) and this in itself will contribute to the initiation of allelic exclusion. While these facts are indisputable, it is also now well established that breaks are introduced in a regulated asynchronous manner on all antigen receptor alleles analyzed (*Igh*, *Igk*, and *Tcra*) (Fig. 8; Chaumeil, Micsinai, Ntziachristos, Deriano, et al., 2013; Chaumeil & Skok, 2013; Hewitt et al., 2009). Rearrangement on one allele at a time involves regulation *in trans* and allelic communication, which may or may not be reliant on pairing (discussed in Section 3.3). As described in Section 3.3, the introduction of RAG-mediated cleavage on one allele recruits ATM to the site of the break and this acts in *trans* on the other allele preventing the introduction of further breaks by a mechanism that involves repositioning of the other allele to repressive PCH and curtailment of higher-order looping (Fig. 8).

DNA FISH analyses have revealed that in most cells, the two *Igh* and *Tcra* alleles are both located in euchromatic regions of the nucleus prior to the onset of recombination, and thus by this means of assessment homologues appear to be equivalently accessible to RAG. Despite this observation, it has recently been suggested that monoallelic targeting of *Igh* and *Tcra* occurs preferentially on highly transcribed alleles that are looped outside of their respective chromosome territories and that for both these loci, loop formation occurs on only one allele at a time (Chaumeil, Micsinai, Ntziachristos, Deriano, et al., 2013; Chaumeil, Micsinai, Ntziachristos, Roth, et al., 2013). It remains to be determined whether RAG proteins are properly targeted across the *Igk* and *Tcrb* loci in the context of higher-order loop formation; however, we do know that in contrast to *Igh* and *Tcra*, prior to recombination, one *Igk* and *Tcrb* allele are associated with repressive PCH and/or the nuclear lamina in pre-B and DN T cells, respectively, while the other alleles are found in a euchromatic location where RAG targeting occurs (Roldan et al., 2005; Schlimgen, Reddy, Singh, & Krangel, 2008; Skok et al., 2007). Thus, differential accessibility of the two alleles may play a role in determining which allele is targeted (Fig. 12). However, these studies provide no information about whether differential positioning of the two alleles is heritably transmitted or whether the two alleles are equally likely to find themselves in opposite locations in the same or a subsequent cell cycle.

5.1.1 Replication Timing

Early studies support a deterministic model of accessibility that relies on the observation that homologous antigen receptor alleles are asynchronously replicated (Mostoslavsky et al., 2001). As a general principle, early replicating loci are more active than late-replicating loci. Thus, based on this premise, differences in replication timing likely reflect differences in activation status of homologues and differences in accessibility that may predispose one allele to recombine before the other. According to these observations, allelic choice is a random process that mirrors the process of X inactivation. Through lineage tracing experiments, it was determined that allelic choice (which correlates with differences in replication timing) could be imposed early on in lymphoid development at the common lymphoid progenitor (CLP) stage; a subgroup of single CLP cells gives rise to mature B cells that all express *Igk* from the same allele (Farago et al., 2012). Thus, commitment occurs prior to rearrangement, but once allelic differences are imposed, they are heritably transmitted throughout development.

5.1.2 The Impact of Nonproductive Rearrangements

A recent study presents an alternative view (Alves-Pereira et al., 2014). Clonal analysis of reconstituted single IgMa/IgMb heterozygous

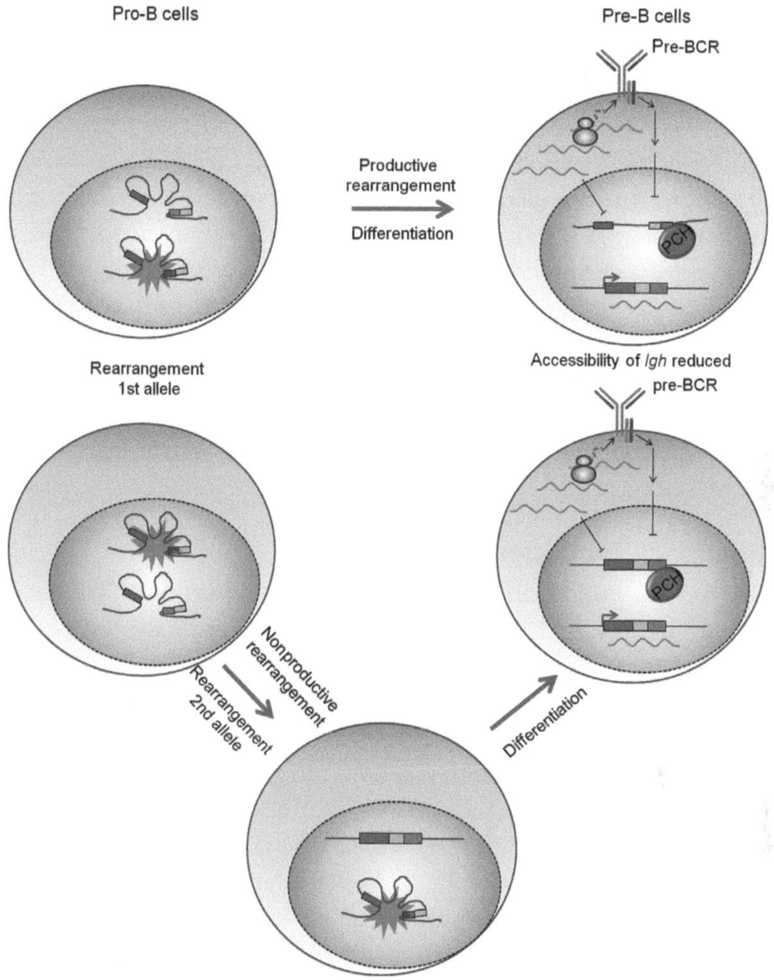

Figure 12 Reduced accessibility of *Igh* downstream of pre-BCR signaling occurs as a result of differentiation and an alteration in signaling pathways that do not support continued accessibility of this locus.

hematopoietic stem cells (HSCs) in irradiated RAG-deficient recipients consistently generated equal numbers of IgMa- and IgMb-expressing B cells in each animal. Moreover, PCR analysis showed the expected differences in the retention of a V_H–D_H intergenic fragment (60% in the VDJ_H/DJ_H and 40% in VDJ_H/VDJ_H configuration). In contrast, CLP-derived clones were completely skewed to either IgMa- or IgMb-expressing cells and highly skewed clones were found more frequently in Ly6d+ B cell progenitors compared to the more uncommitted Ly6d− cells. Furthermore,

Ly6d+ cells and pro-B cells have a similar capability to generate a skewed B cell population. Analyses of *Igh* rearrangement status on both the productive and silent alleles in skewed clones indicate that the bias to rearrange one allele can be explained by the impact of nonproductive rearrangement. Thus, their findings support the idea that the two *Igh* alleles are synchronously competent to undergo rearrangement. Furthermore, in contrast to previous studies, they demonstrate that the bias observed for *Igh* is not reproduced for *Igk*, suggesting that this locus is not precommitted in the CLP stage of B cell development. Taken together, these data suggest that allelic exclusion of *Ig* loci differs from X-chromosome inactivation as no stable epigenetic mark is propagated until pro-B cells start rearranging. The differences observed in both studies can be explained by the fact that in the latter study, recombined silent alleles were included in the analysis.

5.2 Feedback Inhibition

5.2.1 Feedback Inhibition Through the Introduction of a DSB Break

As discussed in Section 3.3, feedback inhibition occurs at the level of breaks (Figs. 8 and 11). A break in one allele or locus inhibits further breaks during repair as a result of ATM- and RAG2-mediated control. In the absence of the C-terminus of RAG2 and ATM, biallelic and bilocus breaks are introduced and this can lead to the generation of intralocus translocations (Chaumeil, Micsinai, Ntziachristos, Deriano, et al., 2013; Chaumeil, Micsinai, Ntziachristos, Roth, et al., 2013; Chaumeil & Skok, 2013; Deriano et al., 2011; Hewitt et al., 2009) which are a hallmark of ATM deficiency. Clearly controlling the number of breaks that are introduced per cell at any one time is important for maintenance of genome stability and thwarting the occurrence of translocations. However, feedback control also contributes to the initiation of allelic exclusion by preventing the simultaneous rearrangement of homologues that could lead to allelic inclusion. Support for this comes from the observation that ATM-deficient mice have increased allelic inclusion of *Igh*, *Igk*, and *Tcrb* (Steinel, Fisher, Yang-Iott, & Bassing, 2014; Steinel et al., 2013). The C-terminus of RAG2 and ATM also inhibit the introduction of biallelic breaks on *Tcra* (Chaumeil, Micsinai, Ntziachristos, Deriano, et al., 2013) even though this locus is not subjected to stringent enforcement of allelic exclusion. Thus, ATM-mediated control of cleavage appears to be a common mechanism that is shared by different loci in recombining lymphocytes as well as in cells undergoing meiosis (see Section 3.3) (Lange et al., 2011).

5.2.2 Control of Recombination via Regulation of RAG Expression—Implications for Allelic Exclusion

It is clearly critical to have mechanisms in place to control RAG activity to ensure that cleavage does not occur across cell cycle as this could lead to genome instability. Productive rearrangement of an *Igh* or *Tcrb* allele in pro-B or DN cells, respectively, leads to cell surface expression of the pre-BCR or pre-TCR. Signaling through these two receptors results in a proliferative burst and subsequent differentiation to the pre-B or DP cell stage where *Igk*, *Igl*, or *Tcra* are recombined. There are two known mechanisms that have evolved to prevent the introduction of breaks during cell cycle. The first involves degradation of RAG2 protein (Lee & Desiderio, 1999) and the second involves control of *Rag1* expression (Johnson et al., 2012). Both mechanisms are also important for preventing the introduction of further breaks on the second allele, which could violate allelic exclusion.

5.2.3 Feedback Inhibition by Productive mRNA

Two-thirds of transcripts generated by rearrangement are out of frame. In contrast to mRNAs from productively rearranged alleles, these are degraded by the NMD pathway, which selectively degrades transcripts harboring premature termination codons (Fig. 11) (Weischenfeldt et al., 2008). Recent studies suggest that mRNA from productively rearranged alleles can have a role in suppressing rearrangement and initiating allelic exclusion. The evidence for this comes from two complementary mouse models. The first harbors a dominant-negative mutation of Rent1/hUpf1, an essential *trans*-effector of the NMD pathway. This mutation induces premature shut-off of *Tcrb* rearrangement causing a block in T cell development at the DN stage. This defect can be rescued with a productively rearranged *Tcrb* transgene, suggesting that mRNA has a function in V(D)J recombination independent of its protein product (Frischmeyer-Guerrerio et al., 2011).

Further support for a negative regulatory role for mRNA in recombination comes from a mouse model that has the endogenous DQ52J_H cluster replaced by a V_HB1–8 VDJ exon rendered nonproductive by the introduction of a termination codon at position 5 on one allele (Ter5 allele) (Lutz et al., 2011). Transcription of the targeted *Igh* allele is driven by its physiological *Igh* chain promoter in one mouse line (Ter5hi), while in the other transcription is driven by a weak, truncated DQ52 promoter (Ter5lo). This results in the production of Ter5 high and low amounts of stable *Igh* transcripts, respectively, that do not encode protein. Thus, stable Igμ mRNA is separated from translation into IGH protein. The presence of stable Ter5hi transcripts leads to a severe block in B cell differentiation at the pro-B cell stage and a corresponding decrease in the pre-B cell compartment. Importantly, the block in development is linked to a decreased frequency of recombined *Igh* alleles, while RAG expression remains unaffected (Lutz et al., 2011) (Fig. 11). Recombination of the wild-type allele is inhibited in the Ter5hi heterozygous cells, preventing the generation of a productively rearranged *Igh* allele that could drive development to the pre-B cell stage. In contrast, there is a significantly higher frequency of *Igh* rearrangement on the nontargeted allele in the heterozygous Ter5lo mice. Thus, it appears that the difference in mRNA stability allows pro-B cells to distinguish between productive and nonproductive *Ig* gene rearrangements and that the presence of stable Igμ transcripts contributes to *Igh* chain allelic exclusion.

5.2.4 Feedback Inhibition at the Level of Protein

It has been known for some time that cell surface expression of a productively rearranged IGH prevents ongoing rearrangement on remaining DJ_H rearranged *Igh* alleles (Fig. 12). This is in large part because productive rearrangement drives development forward and any antigen receptor locus moving to a new developmental compartment will be subject to changes in signaling pathways and transcriptional profiles that do not support its continued accessibility and ongoing rearrangement. This is exemplified in mice that express a prerearranged *Igh* (knockin or transgenic) allele, which drives B cell development forward skipping out the pro-B cell stage where *Igh* recombination normally takes place. As a result, B cells reach the pre-B cell stage without an opportunity to fully open up V_H genes via IL-7/STAT5 signaling (Bertolino et al., 2005; Chowdhury & Sen, 2003) and without undergoing locus contraction (Roldan et al., 2005). At the pre-B cell stage, there is a reduction in IL-7/STAT5 signaling and Pax5 binding at PAIRs is reduced to an extent that may impact the ability of the *Igh* locus to contract (Ebert et al., 2011). Indeed, rearrangement on decontracted endogenous *Igh* loci in these mice is limited to the four most proximal V_H gene segments (Roldan et al., 2005).

Further evidence for this idea comes from mice with mutations in components of the pre-BCR and pre-TCR signaling pathways that result in a partial or total block in B and T cell development that forces cells to remain in the pro-B or DN cell stages, respectively. In this situation, the presence of mRNA or protein originating from productive *Igh* or *Tcrb* rearrangement is not sufficient to prevent ongoing rearrangement on accessible alleles that can be targeted by RAG leading to a violation of exclusion. To some extent, staying in the same compartment indefinitely may help to overcome constraints that are not 100% efficient at preventing recombination on the second allele.

5.3 Maintenance of Allelic Exclusion

As summarized above, there are multiple levels at which allelic exclusion is enforced starting with differential accessibility of the two alleles (which may or may not be heritably predetermined), and moving on to feedback control at the level of (i) ATM- and RAG2-mediated regulation of asynchronous cleavage, (ii) stable mRNA production, and (iii) protein production. Protein expression of a functionally rearranged *Igh* or *Tcrb* drives differentiation forward, and at the subsequent stages of development, these loci are subjected to different signaling pathways and transcriptional factor profiles that alter their accessibility to maintain allelic exclusion. It is particularly important for the *Igh* or *Tcrb* loci that accessibility is reduced in pre-B and DP cells as RAG expression is once again upregulated and this could potentially target unrearranged alleles during *Igk*/*Igl* and *Tcra* recombination, respectively, that could then lead to allelic inclusion. There are many transcription factors involved in regulating rearrangement but only a few (for example E2A) which are reported to function in allelic exclusion (Hauser, Grundstrom, & Grundstrom, 2014). In DN thymocytes, E2A binds to the *Tcrb* DJβ region, Eβ, and some Vβ gene promoters and activates germline transcription (Belle & Zhuang, 2014). The E2A inhibitor, ID3, is upregulated downstream of pre-TCR signaling and E2A binding to *Tcrb* is reduced, which in turn reduces accessibility. In contrast, enforced expression of E2A in thymocytes overrides allelic exclusion in mice expressing a rearrarranged *Tcrb* transgene (Agata et al., 2007). In addition, IL-7/STAT5 signaling has been implicated in regulating allelic exclusion of the *Igh* locus. Thus, in mice expressing constitutively active STAT5a, accessibility of the *Igh* locus is maintained in pre-B cells supporting ongoing rearrangement (Hewitt et al., 2009).

In addition to changes in accessibility, *Igh* and *Tcrb* loci both undergo decontraction in pre-B and DP cells, respectively, but this is not sufficient to block rearrangement on proximal V genes. The lineage-specific factors that induce locus contraction and decontraction of *Tcrb* have not been identified. In contrast, this is much better understood for the *Igh* locus (see Section 3.4) although there are details that need further clarification.

Previous studies have demonstrated that a 3′Eκ-mediated interaction between *Igh* and *Igk* loci at the pre-B cell stage is important for relocating unrearranged *Igh* alleles to repressive PCH and for inducing locus decontraction (Fig. 9). Going forward, it will be important to determine if the 3′Eκ enhancer is important for reducing Pax5 binding to PAIR elements at the pre-B cell stage. Currently, it is not known why Pax5 binding to PAIR elements is altered in cells where Pax5 expression remains at high levels.

In conclusion, V(D)J recombination is tightly regulated at multiple levels in order to limit the possible hazards associated with the introduction of DSBs. As discussed in this section, many of the mechanisms controlling accessibility and cleavage may also contribute to enforcement of the allelic exclusion process. Finally, although great progress has been made in recent years, we still have some way to go before we fully understand the process.

6. FUTURE DIRECTIONS

6.1 Long-Range Genomic Interactions in V(D)J Recombination

Long-range genomic interactions appear to be an important component of V(D)J recombination (Skok, 2010) and studies that have focused on this aspect of control have contributed to our understanding of gene regulation as a whole. Indeed, recombination involves (i) intralocus interactions that are important for the generation of repertoire diversity, (ii) homologue and heterologue pairing which is linked to feedback control of cleavage via a mechanism that involves ATM and RAG2, and (iii) enhancer-mediated interloci interactions that are important for reducing accessibility and inducing locus decontraction after productive rearrangement (between *Igh* and *Igk* for example, see Section 3.4). For now, these are the ones we know about, but it seems likely that future studies will reveal additional factors and mechanisms that mediate long-range genomic interactions.

6.2 A Holistic Approach for Analyzing Interactions

Among the ensemble of long-range genomic interactions, the intralocus interactions are the most well studied as demonstrated by the multitude of papers that have been published on this topic. Nonetheless, we still do not have a holistic understanding of the structure of each locus and the impact of *cis-* and *trans-*acting factors on looping. This is because most studies reported here have analyzed locus conformation at the molecular level using 3C or 4C-seq. These approaches provide limited information about the structure of the locus as a whole because interactions are only examined from a particular viewpoint and this gives no information on TAD structure. In addition, most of the available Hi-C data are too low in resolution and do not provide sufficient information on looping. For a complete picture, a Capture C or High-resolution Hi-C (Rao et al., 2014) approach will be required to generate a high-resolution matrix of inter- and intra-TAD interactions. Next-generation approaches that are currently emerging (Hughes et al., 2013; Kolovos et al., 2014) and constantly being improved upon will be useful for providing the information we are lacking. Furthermore, use of these approaches will allow us to compare one locus with another and to determine the impact of regulatory elements on structure and regulation using gene-targeted mouse models.

6.3 The Impact of RAG on Organization of Antigen Receptor Loci

Finally, the RAG proteins, which are such a fundamental component of recombination, are largely eliminated in 3C and 4C analyses because of the impact of recombination on locus structure. However, we cannot rule out that the RAG proteins themselves have no influence on locus conformation as we already know from our studies that the RAG proteins have a significant impact on the organization of each locus. For example, we know that the RAG proteins are required for pairing of homologous and heterologous antigen receptor alleles and for bringing RAG-bound loci together in the nucleus (Brandt, Hewitt, & Skok, 2010; Chaumeil, Micsinai, Ntziachristos, Deriano, et al., 2013; Chaumeil, Micsinai, Ntziachristos, Roth, et al., 2013; Chaumeil & Skok, 2013; Hewitt et al., 2009). In addition, we have shown that the presence of RAG is important for inducing the formation of higher-order loops that separate the 3′end of the locus from the chromosome territory which is linked to enrichment of RAG and active histone modifications as well as directed cleavage in this region. Thus, future studies should not ignore the impact of this important player.

ACKNOWLEDGMENTS

J.A.S. is a Leukemia & Lymphoma Society (LLS) scholar and is supported by NIH Grants R01 GM086852, NIAID R56 A1099111, R21 AI107069 and R01GM112192. JC is supported by the ARC foundation for cancer research (France).

REFERENCES

Abarrategui, I., & Krangel, M. S. (2006). Regulation of T cell receptor-alpha gene recombination by transcription. *Nature Immunology*, *7*, 1109–1115.

Abarrategui, I., & Krangel, M. S. (2007). Noncoding transcription controls downstream promoters to regulate T-cell receptor alpha recombination. *The EMBO Journal*, *26*, 4380–4390.

Afshar, R., Pierce, S., Bolland, D. J., Corcoran, A., & Oltz, E. M. (2006). Regulation of IgH gene assembly: Role of the intronic enhancer and 5′DQ52 region in targeting DHJH recombination. *Journal of Immunology*, *176*, 2439–2447.

Agata, Y., Tamaki, N., Sakamoto, S., Ikawa, T., Masuda, K., Kawamoto, H., et al. (2007). Regulation of T cell receptor beta gene rearrangements and allelic exclusion by the helix-loop-helix protein, E47. *Immunity*, *27*, 871–884.

Alam, S. M., & Gascoigne, N. R. (1998). Posttranslational regulation of TCR Valpha allelic exclusion during T cell differentiation. *Journal of Immunology*, *160*, 3883–3890.

Alves-Pereira, C. F., de Freitas, R., Lopes, T., Gardner, R., Marta, F., Vieira, P., et al. (2014). Independent recruitment of Igh alleles in V(D)J recombination. *Nature Communications*, *5*, 5623.

Arakawa, H., Shimizu, T., & Takeda, S. (1996). Re-evaluation of the probabilities for productive arrangements on the kappa and lambda loci. *International Immunology*, *8*, 91–99.

Belle, I., & Zhuang, Y. (2014). E proteins in lymphocyte development and lymphoid diseases. *Current Topics in Developmental Biology*, *110*, 153–187.

Bertolino, E., Reddy, K., Medina, K. L., Parganas, E., Ihle, J., & Singh, H. (2005). Regulation of interleukin 7-dependent immunoglobulin heavy-chain variable gene rearrangements by transcription factor STAT5. *Nature Immunology*, *6*, 836–843.

Bolland, D. J., Wood, A. L., Johnston, C. M., Bunting, S. F., Morgan, G., Chakalova, L., et al. (2004). Antisense intergenic transcription in V(D)J recombination. *Nature Immunology*, *5*, 630–637.

Bories, J. C., Demengeot, J., Davidson, L., & Alt, F. W. (1996). Gene-targeted deletion and replacement mutations of the T-cell receptor beta-chain enhancer: The role of enhancer elements in controlling V(D)J recombination accessibility. *Proceedings of the National Academy of Sciences of the United States of America*, *93*, 7871–7876.

Bossen, C., Mansson, R., & Murre, C. (2012). Chromatin topology and the regulation of antigen receptor assembly. *Annual Review of Immunology*, *30*, 337–356.

Boudil, A., Matei, I. R., Shih, H. Y., Bogdanoski, G., Yuan, J. S., Chang, S. G., et al. (2015). IL-7 coordinates proliferation, differentiation and Tcra recombination during thymocyte beta-selection. *Nature Immunology*, *16*(4), 397–405.

Bouvier, G., Watrin, F., Naspetti, M., Verthuy, C., Naquet, P., & Ferrier, P. (1996). Deletion of the mouse T-cell receptor beta gene enhancer blocks alphabeta T-cell development. *Proceedings of the National Academy of Sciences of the United States of America*, *93*, 7877–7881.

Brady, B. L., Steinel, N. C., & Bassing, C. H. (2010). Antigen receptor allelic exclusion: An update and reappraisal. *Journal of Immunology*, *185*, 3801–3808.

Brandt, V. L., Hewitt, S. L., & Skok, J. A. (2010). It takes two: Communication between homologous alleles preserves genomic stability during V(D)J recombination. *Nucleus*, *1*, 23–29.

Brown, K. E., Baxter, J., Graf, D., Merkenschlager, M., & Fisher, A. G. (1999). Dynamic repositioning of genes in the nucleus of lymphocytes preparing for cell division. *Molecular Cell*, *3*, 207–217.

Chatterjee, S., Ju, Z., Hassan, R., Volpi, S. A., Emelyanov, A. V., & Birshtein, B. K. (2011). Dynamic changes in binding of immunoglobulin heavy chain 3′ regulatory region to protein factors during class switching. *The Journal of Biological Chemistry*, *286*, 29303–29312.

Chaumeil, J., Micsinai, M., Ntziachristos, P., Deriano, L., Wang, J. M., Ji, Y., et al. (2013). Higher-order looping and nuclear organization of Tcra facilitate targeted RAG cleavage and regulated rearrangement in recombination centers. *Cell Reports*, *3*(2), 359–370.

Chaumeil, J., Micsinai, M., Ntziachristos, P., Roth, D. B., Aifantis, I., Kluger, Y., et al. (2013). The RAG2 C-terminus and ATM protect genome integrity by controlling antigen receptor gene cleavage. *Nature Communications*, *4*, 2231.

Chaumeil, J., & Skok, J. A. (2012). The role of CTCF in regulating V(D)J recombination. *Current Opinion in Immunology*, *24*, 153–159.

Chaumeil, J., & Skok, J. A. (2013). A new take on V(D)J recombination: Transcription driven nuclear and chromatin reorganization in rag-mediated cleavage. *Frontiers in Immunology*, *4*, 423.

Chowdhury, D., & Sen, R. (2003). Transient IL-7/IL-7R signaling provides a mechanism for feedback inhibition of immunoglobulin heavy chain gene rearrangements. *Immunity*, *18*, 229–241.

Ciofani, M., & Zuniga-Pflucker, J. C. (2010). Determining gammadelta versus alphass T cell development. *Nature Reviews. Immunology*, *10*, 657–663.

Cocea, L., De Smet, A., Saghatchian, M., Fillatreau, S., Ferradini, L., Schurmans, S., et al. (1999). A targeted deletion of a region upstream from the Jkappa cluster impairs kappa chain rearrangement in cis in mice and in the 103/bcl2 cell line. *The Journal of Experimental Medicine*, *189*, 1443–1450.

Degner, S. C., Verma-Gaur, J., Wong, T. P., Bossen, C., Iverson, G. M., Torkamani, A., et al. (2011). CCCTC-binding factor (CTCF) and cohesin influence the genomic architecture of the Igh locus and antisense transcription in pro-B cells. *Proceedings of the National Academy of Sciences of the United States of America*, *108*, 9566–9571.

Deriano, L., Chaumeil, J., Coussens, M., Multani, A., Chou, Y., Alekseyenko, A. V., et al. (2011). The RAG2 C terminus suppresses genomic instability and lymphomagenesis. *Nature*, *471*, 119–123.

Dixon, J. R., Selvaraj, S., Yue, F., Kim, A., Li, Y., Shen, Y., et al. (2012). Topological domains in mammalian genomes identified by analysis of chromatin interactions. *Nature*, *485*, 376–380.

Doty, R. T., Xia, D., Nguyen, S. P., Hathaway, T. R., & Willerford, D. M. (1999). Promoter element for transcription of unrearranged T-cell receptor beta-chain gene in pro-T cells. *Blood*, *93*, 3017–3025.

Ebert, A., McManus, S., Tagoh, H., Medvedovic, J., Salvagiotto, G., Novatchkova, M., et al. (2011). The distal V(H) gene cluster of the Igh locus contains distinct regulatory elements with Pax5 transcription factor-dependent activity in pro-B cells. *Immunity*, *34*, 175–187.

Engel, H., Rolink, A., & Weiss, S. (1999). B cells are programmed to activate kappa and lambda for rearrangement at consecutive developmental stages. *European Journal of Immunology*, *29*, 2167–2176.

Ercolini, A. M., & Miller, S. D. (2009). The role of infections in autoimmune disease. *Clinical and Experimental Immunology*, *155*, 1–15.

Farago, M., Rosenbluh, C., Tevlin, M., Fraenkel, S., Schlesinger, S., Masika, H., et al. (2012). Clonal allelic predetermination of immunoglobulin-kappa rearrangement. *Nature*, *490*, 561–565.

Featherstone, K., Wood, A. L., Bowen, A. J., & Corcoran, A. E. (2010). The mouse immunoglobulin heavy chain V-D intergenic sequence contains insulators that may regulate ordered V(D)J recombination. *The Journal of Biological Chemistry*, *285*, 9327–9338.

Feddersen, R. M., Martin, D. J., & Van Ness, B. G. (1990). Novel recombinations of the IG kappa-locus that result in allelic exclusion. *Journal of Immunology*, *145*, 745–750.

Flodstrom-Tullberg, M. (2003). Viral infections: Their elusive role in regulating susceptibility to autoimmune disease. *Microbes and Infection/Institut Pasteur, 5*, 911–921.

Fournier, E. M., Velez, M. G., Leahy, K., Swanson, C. L., Rubtsov, A. V., Torres, R. M., et al. (2012). Dual-reactive B cells are autoreactive and highly enriched in the plasmablast and memory B cell subsets of autoimmune mice. *The Journal of Experimental Medicine, 209*, 1797–1812.

Frischmeyer-Guerrerio, P. A., Montgomery, R. A., Warren, D. S., Cooke, S. K., Lutz, J., Sonnenday, C. J., et al. (2011). Perturbation of thymocyte development in nonsense-mediated decay (NMD)-deficient mice. *Proceedings of the National Academy of Sciences of the United States of America, 108*, 10638–10643.

Fuxa, M., Skok, J., Souabni, A., Salvagiotto, G., Roldan, E., & Busslinger, M. (2004). Pax5 induces V-to-DJ rearrangements and locus contraction of the immunoglobulin heavy-chain gene. *Genes & Development, 18*, 411–422.

Garcia, V., Gray, S., Allison, R. M., Cooper, T. J., & Neale, M. J. (2015). Tel1-mediated interference suppresses clustered meiotic double-strand-break formation. *Nature, 520*(7545), 114–118.

Garrett, F. E., Emelyanov, A. V., Sepulveda, M. A., Flanagan, P., Volpi, S., Li, F., et al. (2005). Chromatin architecture near a potential 3′ end of the igh locus involves modular regulation of histone modifications during B-cell development and in vivo occupancy at CTCF sites. *Molecular and Cellular Biology, 25*, 1511–1525.

Giudicelli, V., Duroux, P., Ginestoux, C., Folch, G., Jabado-Michaloud, J., Chaume, D., et al. (2006). IMGT/LIGM-DB, the IMGT comprehensive database of immunoglobulin and T cell receptor nucleotide sequences. *Nucleic Acids Research, 34*, D781–D784.

Guo, C., Yoon, H. S., Franklin, A., Jain, S., Ebert, A., Cheng, H. L., et al. (2011). CTCF-binding elements mediate control of V(D)J recombination. *Nature, 477*, 424–430.

Halverson, R., Torres, R. M., & Pelanda, R. (2004). Receptor editing is the main mechanism of B cell tolerance toward membrane antigens. *Nature Immunology, 5*, 645–650.

Hao, B., & Krangel, M. S. (2011). Long-distance regulation of fetal V(delta) gene segment TRDV4 by the Tcrd enhancer. *Journal of Immunology, 187*, 2484–2491.

Hauser, J., Grundstrom, C., & Grundstrom, T. (2014). Allelic exclusion of IgH through inhibition of E2A in a VDJ recombination complex. *Journal of Immunology, 192*, 2460–2470.

Hawwari, A., Bock, C., & Krangel, M. S. (2005). Regulation of T cell receptor alpha gene assembly by a complex hierarchy of germline Jalpha promoters. *Nature Immunology, 6*, 481–489.

Hawwari, A., & Krangel, M. S. (2005). Regulation of TCR delta and alpha repertoires by local and long-distance control of variable gene segment chromatin structure. *The Journal of Experimental Medicine, 202*, 467–472.

He, X., Janeway, C. A., Jr., Levine, M., Robinson, E., Preston-Hurlburt, P., Viret, C., et al. (2002). Dual receptor T cells extend the immune repertoire for foreign antigens. *Nature Immunology, 3*, 127–134.

Helmink, B. A., & Sleckman, B. P. (2012). The response to and repair of RAG-mediated DNA double-strand breaks. *Annual Review of Immunology, 30*, 175–202.

Hewitt, S. L., Chaumeil, J., & Skok, J. A. (2010). Chromosome dynamics and the regulation of V(D)J recombination. *Immunological Reviews, 237*, 43–54.

Hewitt, S. L., Farmer, D., Marszalek, K., Cadera, E., Liang, H. E., Xu, Y., et al. (2008). Association between the Igk and Igh immunoglobulin loci mediated by the 3′Igk enhancer induces 'decontraction' of the Igh locus in pre-B cells. *Nature Immunology, 9*, 396–404.

Hewitt, S. L., Yin, B., Ji, Y., Chaumeil, J., Marszalek, K., Tenthorey, J., et al. (2009). RAG-1 and ATM coordinate monoallelic recombination and nuclear positioning of immunoglobulin loci. *Nature Immunology, 10*, 655–664.

Hughes, J. R., Lower, K. M., Dunham, I., Taylor, S., De Gobbi, M., Sloane-Stanley, J. A., et al. (2013). High-resolution analysis of cis-acting regulatory networks at the alpha-globin locus. *Philosophical Transactions of the Royal Society of London. Series B, Biological Sciences*, *368*, 20120361.

Inlay, M., Alt, F. W., Baltimore, D., & Xu, Y. (2002). Essential roles of the kappa light chain intronic enhancer and 3′ enhancer in kappa rearrangement and demethylation. *Nature Immunology*, *3*, 463–468.

Inlay, M. A., Lin, T., Gao, H. H., & Xu, Y. (2006). Critical roles of the immunoglobulin intronic enhancers in maintaining the sequential rearrangement of IgH and Igk loci. *The Journal of Experimental Medicine*, *203*, 1721–1732.

Jhunjhunwala, S., van Zelm, M. C., Peak, M. M., Cutchin, S., Riblet, R., van Dongen, J. J., et al. (2008). The 3D structure of the immunoglobulin heavy-chain locus: Implications for long-range genomic interactions. *Cell*, *133*, 265–279.

Ji, Y., Resch, W., Corbett, E., Yamane, A., Casellas, R., & Schatz, D. G. (2010). The in vivo pattern of binding of RAG1 and RAG2 to antigen receptor loci. *Cell*, *141*, 419–431.

Johnson, K., Chaumeil, J., Micsinai, M., Wang, J. M., Ramsey, L. B., Baracho, G. V., et al. (2012). IL-7 functionally segregates the pro-B cell stage by regulating transcription of recombination mediators across cell cycle. *Journal of Immunology*, *188*, 6084–6092.

Johnson, K., Chaumeil, J., & Skok, J. A. (2010). Epigenetic regulation of V(D)J recombination. *Essays in Biochemistry*, *48*, 221–243.

Jung, D., Giallourakis, C., Mostoslavsky, R., & Alt, F. W. (2006). Mechanism and control of V(D)J recombination at the immunoglobulin heavy chain locus. *Annual Review of Immunology*, *24*, 541–570.

Khamlichi, A. A., Pinaud, E., Decourt, C., Chauveau, C., & Cogne, M. (2000). The 3′ IgH regulatory region: A complex structure in a search for a function. *Advances in Immunology*, *75*, 317–345.

Kil, L. P., de Bruijn, M. J., van Nimwegen, M., Corneth, O. B., van Hamburg, J. P., Dingjan, G. M., et al. (2012). Btk levels set the threshold for B-cell activation and negative selection of autoreactive B cells in mice. *Blood*, *119*, 3744–3756.

Kim, D. R., Dai, Y., Mundy, C. L., Yang, W., & Oettinger, M. A. (1999). Mutations of acidic residues in RAG1 define the active site of the V(D)J recombinase. *Genes & Development*, *13*, 3070–3080.

Kim, M. S., Lapkouski, M., Yang, W., & Gellert, M. (2015). Crystal structure of the V(D)J recombinase RAG1-RAG2. *Nature*, *518*, 507–511.

Kolovos, P., van de Werken, H. J., Kepper, N., Zuin, J., Brouwer, R. W., Kockx, C. E., et al. (2014). Targeted Chromatin Capture (T2C): A novel high resolution high throughput method to detect genomic interactions and regulatory elements. *Epigenetics & Chromatin*, 7, 10.

Kosak, S. T., Skok, J. A., Medina, K. L., Riblet, R., Le Beau, M. M., Fisher, A. G., et al. (2002). Subnuclear compartmentalization of immunoglobulin loci during lymphocyte development. *Science*, *296*, 158–162.

Krangel, M. S. (2009). Mechanics of T cell receptor gene rearrangement. *Current Opinion in Immunology*, *21*, 133–139.

Kurosawa, Y., von Boehmer, H., Haas, W., Sakano, H., Trauneker, A., & Tonegawa, S. (1981). Identification of D segments of immunoglobulin heavy-chain genes and their rearrangement in T lymphocytes. *Nature*, *290*, 565–570.

Landree, M. A., Wibbenmeyer, J. A., & Roth, D. B. (1999). Mutational analysis of RAG1 and RAG2 identifies three catalytic amino acids in RAG1 critical for both cleavage steps of V(D)J recombination. *Genes & Development*, *13*, 3059–3069.

Lange, J., Pan, J., Cole, F., Thelen, M. P., Jasin, M., & Keeney, S. (2011). ATM controls meiotic double-strand-break formation. *Nature*, *479*(7372), 237–240.

Lee, J., & Desiderio, S. (1999). Cyclin A/CDK2 regulates V(D)J recombination by coordinating RAG-2 accumulation and DNA repair. *Immunity, 11*, 771–781.

Lee, G. S., Neiditch, M. B., Salus, S. S., & Roth, D. B. (2004). RAG proteins shepherd double-strand breaks to a specific pathway, suppressing error-prone repair, but RAG nicking initiates homologous recombination. *Cell, 117*, 171–184.

Lennon, G. G., & Perry, R. P. (1985). C mu-containing transcripts initiate heterogeneously within the IgH enhancer region and contain a novel 5′-nontranslatable exon. *Nature, 318*, 475–478.

Lin, S. G., Guo, C., Su, A., Zhang, Y., & Alt, F. W. (2015). CTCF-binding elements 1 and 2 in the Igh intergenic control region cooperatively regulate V(D)J recombination. *Proceedings of the National Academy of Sciences of the United States of America, 112*(6), 1815–1820.

Liu, Z. M., George-Raizen, J. B., Li, S., Meyers, K. C., Chang, M. Y., & Garrard, W. T. (2002). Chromatin structural analyses of the mouse Igkappa gene locus reveal new hypersensitive sites specifying a transcriptional silencer and enhancer. *The Journal of Biological Chemistry, 277*, 32640–32649.

Liu, H., Schmidt-Supprian, M., Shi, Y., Hobeika, E., Barteneva, N., Jumaa, H., et al. (2007). Yin Yang 1 is a critical regulator of B-cell development. *Genes & Development, 21*, 1179–1189.

Liu, Y., Subrahmanyam, R., Chakraborty, T., Sen, R., & Desiderio, S. (2007). A plant homeodomain in RAG-2 that binds hypermethylated lysine 4 of histone H3 is necessary for efficient antigen-receptor-gene rearrangement. *Immunity, 27*, 561–571.

Liu, Z., Widlak, P., Zou, Y., Xiao, F., Oh, M., Li, S., et al. (2006). A recombination silencer that specifies heterochromatin positioning and Ikaros association in the immunoglobulin kappa locus. *Immunity, 24*, 405–415.

Livak, F., Tourigny, M., Schatz, D. G., & Petrie, H. T. (1999). Characterization of TCR gene rearrangements during adult murine T cell development. *The Journal of Immunology: Official Journal of the American Association of Immunologists, 162*, 2575–2580.

Lutz, J., Heideman, M. R., Roth, E., van den Berk, P., Muller, W., Raman, C., et al. (2011). Pro-B cells sense productive immunoglobulin heavy chain rearrangement irrespective of polypeptide production. *Proceedings of the National Academy of Sciences of the United States of America, 108*, 10644–10649.

Majumder, K., Koues, O. I., Chan, E. A., Kyle, K. E., Horowitz, J. E., Yang-Iott, K., et al. (2015). Lineage-specific compaction of Tcrb requires a chromatin barrier to protect the function of a long-range tethering element. *The Journal of Experimental Medicine, 212*, 107–120.

Marquet, M., Garot, A., Bender, S., Carrion, C., Rouaud, P., Lecardeur, S., et al. (2014). The Emu enhancer region influences H chain expression and B cell fate without impacting IgVH repertoire and immune response in vivo. *Journal of Immunology, 193*, 1171–1183.

Martin, D. J., & van Ness, B. G. (1990). Initiation and processing of two kappa immunoglobulin germ line transcripts in mouse B cells. *Molecular and Cellular Biology, 10*, 1950–1958.

Mathieu, N., Hempel, W. M., Spicuglia, S., Verthuy, C., & Ferrier, P. (2000). Chromatin remodeling by the T cell receptor (TCR)-beta gene enhancer during early T cell development: Implications for the control of TCR-beta locus recombination. *The Journal of Experimental Medicine, 192*, 625–636.

Matthews, A. G., Kuo, A. J., Ramon-Maiques, S., Han, S., Champagne, K. S., Ivanov, D., et al. (2007). RAG2 PHD finger couples histone H3 lysine 4 trimethylation with V(D)J recombination. *Nature, 450*, 1106–1110.

McMillan, R. E., & Sikes, M. L. (2008). Differential activation of dual promoters alters Dbeta2 germline transcription during thymocyte development. *Journal of Immunology, 180*, 3218–3228.

McMillan, R. E., & Sikes, M. L. (2009). Promoter activity 5′ of Dbeta2 is coordinated by E47, Runx1, and GATA-3. *Molecular Immunology*, *46*, 3009–3017.

Medvedovic, J., Ebert, A., Tagoh, H., Tamir, I. M., Schwickert, T. A., Novatchkova, M., et al. (2013). Flexible long-range loops in the VH gene region of the Igh locus facilitate the generation of a diverse antibody repertoire. *Immunity*, *39*, 229–244.

Mendes, R. D., Sarmento, L. M., Cante-Barrett, K., Zuurbier, L., Buijs-Gladdines, J. G., Povoa, V., et al. (2014). PTEN microdeletions in T-cell acute lymphoblastic leukemia are caused by illegitimate RAG-mediated recombination events. *Blood*, *124*, 567–578.

Meyer, K. B., Sharpe, M. J., Surani, M. A., & Neuberger, M. S. (1990). The importance of the 3′-enhancer region in immunoglobulin kappa gene expression. *Nucleic Acids Research*, *18*, 5609–5615.

Mombaerts, P., Iacomini, J., Johnson, R. S., Herrup, K., Tonegawa, S., & Papaioannou, V. E. (1992). RAG-1-deficient mice have no mature B and T lymphocytes. *Cell*, *68*, 869–877.

Monroe, R. J., Sleckman, B. P., Monroe, B. C., Khor, B., Claypool, S., Ferrini, R., et al. (1999). Developmental regulation of TCR delta locus accessibility and expression by the TCR delta enhancer. *Immunity*, *10*, 503–513.

Mostoslavsky, R., Singh, N., Tenzen, T., Goldmit, M., Gabay, C., Elizur, S., et al. (2001). Asynchronous replication and allelic exclusion in the immune system. *Nature*, *414*, 221–225.

Mullighan, C. G., Miller, C. B., Radtke, I., Phillips, L. A., Dalton, J., Ma, J., et al. (2008). BCR-ABL1 lymphoblastic leukaemia is characterized by the deletion of Ikaros. *Nature*, *453*, 110–114.

Nadel, B., Tang, A., Escuro, G., Lugo, G., & Feeney, A. J. (1998). Sequence of the spacer in the recombination signal sequence affects V(D)J rearrangement frequency and correlates with nonrandom Vkappa usage in vivo. *The Journal of Experimental Medicine*, *187*, 1495–1503.

Naik, A. K., Hawwari, A., & Krangel, M. S. (2015). Specification of Vdelta and Valpha usage by Tcra/Tcrd locus V gene segment promoters. *Journal of Immunology*, *194*, 790–794.

Narendra, V., Rocha, P. P., An, D., Raviram, R., Skok, J. A., Mazzoni, E. O., et al. (2015). Transcription. CTCF establishes discrete functional chromatin domains at the Hox clusters during differentiation. *Science*, *347*, 1017–1021.

Nora, E. P., Lajoie, B. R., Schulz, E. G., Giorgetti, L., Okamoto, I., Servant, N., et al. (2012). Spatial partitioning of the regulatory landscape of the X-inactivation centre. *Nature*, *485*, 381–385.

Onozawa, M., & Aplan, P. D. (2012). Illegitimate V(D)J recombination involving non-antigen receptor loci in lymphoid malignancy. *Genes, Chromosomes & Cancer*, *51*, 525–535.

Pan, X., Papasani, M., Hao, Y., Calamito, M., Wei, F., Quinn Iii, W. J., et al. (2013). YY1 controls Igkappa repertoire and B-cell development, and localizes with condensin on the Igkappa locus. *The EMBO Journal*, *32*, 1168–1182.

Papaemmanuil, E., Rapado, I., Li, Y., Potter, N. E., Wedge, D. C., Tubio, J., et al. (2014). RAG-mediated recombination is the predominant driver of oncogenic rearrangement in ETV6-RUNX1 acute lymphoblastic leukemia. *Nature Genetics*, *46*, 116–125.

Pelanda, R. (2014). Dual immunoglobulin light chain B cells: Trojan horses of autoimmunity? *Current Opinion in Immunology*, *27*, 53–59.

Pelanda, R., Schwers, S., Sonoda, E., Torres, R. M., Nemazee, D., & Rajewsky, K. (1997). Receptor editing in a transgenic mouse model: Site, efficiency, and role in B cell tolerance and antibody diversification. *Immunity*, *7*, 765–775.

Perlot, T., Alt, F. W., Bassing, C. H., Suh, H., & Pinaud, E. (2005). Elucidation of IgH intronic enhancer functions via germ-line deletion. *Proceedings of the National Academy of Sciences of the United States of America*, *102*, 14362–14367.

Pinaud, E., Marquet, M., Fiancette, R., Peron, S., Vincent-Fabert, C., Denizot, Y., et al. (2011). The IgH locus 3′ regulatory region: Pulling the strings from behind. *Advances in Immunology, 110*, 27–70.

Prak, E. L., & Weigert, M. (1995). Light chain replacement: A new model for antibody gene rearrangement. *The Journal of Experimental Medicine, 182*, 541–548.

Rao, S. S., Huntley, M. H., Durand, N. C., Stamenova, E. K., Bochkov, I. D., Robinson, J. T., et al. (2014). A 3D map of the human genome at kilobase resolution reveals principles of chromatin looping. *Cell, 159*, 1665–1680.

Ribeiro de Almeida, C., Stadhouders, R., de Bruijn, M. J., Bergen, I. M., Thongjuea, S., Lenhard, B., et al. (2011). The DNA-binding protein CTCF limits proximal Vkappa recombination and restricts kappa enhancer interactions to the immunoglobulin kappa light chain locus. *Immunity, 35*, 501–513.

Ribeiro de Almeida, C., Stadhouders, R., Thongjuea, S., Soler, E., & Hendriks, R. W. (2012). DNA-binding factor CTCF and long-range gene interactions in V(D)J recombination and oncogene activation. *Blood, 119*, 6209–6218.

Roldan, E., Fuxa, M., Chong, W., Martinez, D., Novatchkova, M., Busslinger, M., et al. (2005). Locus 'decontraction' and centromeric recruitment contribute to allelic exclusion of the immunoglobulin heavy-chain gene. *Nature Immunology, 6*, 31–41.

Rouaud, P., Vincent-Fabert, C., Saintamand, A., Fiancette, R., Marquet, M., Robert, I., et al. (2013). The IgH 3′ regulatory region controls somatic hypermutation in germinal center B cells. *The Journal of Experimental Medicine, 210*, 1501–1507.

Ryu, C. J., Haines, B. B., Lee, H. R., Kang, Y. H., Draganov, D. D., Lee, M., et al. (2004). The T-cell receptor beta variable gene promoter is required for efficient V beta rearrangement but not allelic exclusion. *Molecular and Cellular Biology, 24*, 7015–7023.

Schatz, D. G., & Swanson, P. C. (2011). V(D)J recombination: Mechanisms of initiation. *Annual Review of Genetics, 45*, 167–202.

Schlimgen, R. J., Reddy, K. L., Singh, H., & Krangel, M. S. (2008). Initiation of allelic exclusion by stochastic interaction of Tcrb alleles with repressive nuclear compartments. *Nature Immunology, 9*, 802–809.

Schlissel, M. S. (2004). Regulation of activation and recombination of the murine Igkappa locus. *Immunological Reviews, 200*, 215–223.

Seitan, V. C., Hao, B., Tachibana-Konwalski, K., Lavagnolli, T., Mira-Bontenbal, H., Brown, K. E., et al. (2011). A role for cohesin in T-cell-receptor rearrangement and thymocyte differentiation. *Nature, 476*, 467–471.

Sexton, T., Yaffe, E., Kenigsberg, E., Bantignies, F., Leblanc, B., Hoichman, M., et al. (2012). Three-dimensional folding and functional organization principles of the Drosophila genome. *Cell, 148*, 458–472.

Shih, H. Y., Verma-Gaur, J., Torkamani, A., Feeney, A. J., Galjart, N., & Krangel, M. S. (2012). Tcra gene recombination is supported by a Tcra enhancer- and CTCF-dependent chromatin hub. *Proceedings of the National Academy of Sciences of the United States of America, 109*, E3493–E3502.

Shinkai, Y., Rathbun, G., Lam, K.-P., Oltz, E. M., Stewart, V., Mendelsohn, M., et al. (1992). RAG-2-deficient mice lack mature lymphocytes owing to inability to initiate V(D)J rearrangement. *Cell, 68*, 855–867.

Sikes, M. L., Gomez, R. J., Song, J., & Oltz, E. M. (1998). A developmental stage-specific promoter directs germline transcription of D beta J beta gene segments in precursor T lymphocytes. *Journal of Immunology, 161*, 1399–1405.

Skok, J. (2010). V(D)J recombination: A paradigm for studying chromosome interactions in mammalian cells. *Epigenomics, 2*, 176–177.

Skok, J. A., Brown, K. E., Azuara, V., Caparros, M. L., Baxter, J., Takacs, K., et al. (2001). Nonequivalent nuclear location of immunoglobulin alleles in B lymphocytes. *Nature Immunology, 2*, 848–854.

Skok, J. A., Gisler, R., Novatchkova, M., Farmer, D., de Laat, W., & Busslinger, M. (2007). Reversible contraction by looping of the Tcra and Tcrb loci in rearranging thymocytes. *Nature Immunology, 8*, 378–387.

Sleckman, B. P., Bardon, C. G., Ferrini, R., Davidson, L., & Alt, F. W. (1997). Function of the TCR alpha enhancer in alphabeta and gammadelta T cells. *Immunity, 7*, 505–515.

Spanopoulou, E., Roman, C. A. J., Corcoran, L., Schlissel, M., Silver, D. P., Nemazee, D., et al. (1994). Functional immunoglobulin transgenes guide ordered B cell differentiation in RAG-1-deficient mice. *Genes & Development, 8*, 1030–1042.

Stadhouders, R., de Bruijn, M. J., Rother, M. B., Yuvaraj, S., Ribeiro de Almeida, C., Kolovos, P., et al. (2014). Pre-B cell receptor signaling induces immunoglobulin kappa locus accessibility by functional redistribution of enhancer-mediated chromatin interactions. *PLoS Biology, 12*, e1001791.

Steinel, N. C., Fisher, M. R., Yang-Iott, K. S., & Bassing, C. H. (2014). The ataxia telangiectasia mutated and cyclin D3 proteins cooperate to help enforce TCRbeta and IgH allelic exclusion. *Journal of Immunology, 193*, 2881–2890.

Steinel, N. C., Lee, B. S., Tubbs, A. T., Bednarski, J. J., Schulte, E., Yang-Iott, K. S., et al. (2013). The ataxia telangiectasia mutated kinase controls Igkappa allelic exclusion by inhibiting secondary Vkappa-to-Jkappa rearrangements. *The Journal of Experimental Medicine, 210*, 233–239.

Stone, J. L., McMillan, R. E., Skaar, D. A., Bradshaw, J. M., Jirtle, R. L., & Sikes, M. L. (2012). DNA double-strand breaks relieve USF-mediated repression of Dbeta2 germline transcription in developing thymocytes. *Journal of Immunology, 188*, 2266–2275.

Tiegs, S. L., Russell, D. M., & Nemazee, D. (1993). Receptor editing in self-reactive bone marrow B cells. *The Journal of Experimental Medicine, 177*, 1009–1020.

Tonegawa, S. (1983). Somatic generation of antibody diversity. *Nature, 302*, 575–581.

Tripathi, R. K., Mathieu, N., Spicuglia, S., Payet, D., Verthuy, C., Bouvier, G., et al. (2000). Definition of a T-cell receptor beta gene core enhancer of V(D)J recombination by transgenic mapping. *Molecular and Cellular Biology, 20*, 42–53.

Vettermann, C., & Schlissel, M. S. (2010). Allelic exclusion of immunoglobulin genes: Models and mechanisms. *Immunological Reviews, 237*, 22–42.

Vettermann, C., Timblin, G. A., Lim, V., Lai, E. C., & Schlissel, M. S. (2015). The proximal J kappa germline-transcript promoter facilitates receptor editing through control of ordered recombination. *PLoS One, 10*, e0113824.

Weischenfeldt, J., Damgaard, I., Bryder, D., Theilgaard-Monch, K., Thoren, L. A., Nielsen, F. C., et al. (2008). NMD is essential for hematopoietic stem and progenitor cells and for eliminating by-products of programmed DNA rearrangements. *Genes & Development, 22*, 1381–1396.

Whitehurst, C. E., Chattopadhyay, S., & Chen, J. (1999). Control of V(D)J recombinational accessibility of the D beta 1 gene segment at the TCR beta locus by a germline promoter. *Immunity, 10*, 313–322.

Xiang, Y., Park, S. K., & Garrard, W. T. (2013). Vkappa gene repertoire and locus contraction are specified by critical DNase I hypersensitive sites within the Vkappa-Jkappa intervening region. *Journal of Immunology, 190*, 1819–1826.

Xiang, Y., Park, S. K., & Garrard, W. T. (2014). A major deletion in the Vkappa-Jkappa intervening region results in hyperelevated transcription of proximal Vkappa genes and a severely restricted repertoire. *Journal of Immunology, 193*, 3746–3754.

Xiang, Y., Zhou, X., Hewitt, S. L., Skok, J. A., & Garrard, W. T. (2011). A multifunctional element in the mouse Igkappa locus that specifies repertoire and Ig loci subnuclear location. *Journal of Immunology, 186*, 5356–5366.

Yamagami, T., ten Boekel, E., Andersson, J., Rolink, A., & Melchers, F. (1999). Frequencies of multiple IgL chain gene rearrangements in single normal or kappaL chain-deficient B lineage cells. *Immunity, 11*, 317–327.

Zhou, X., Xiang, Y., Ding, X., & Garrard, W. T. (2013). Loss of an Igkappa gene enhancer in mature B cells results in rapid gene silencing and partial reversible dedifferentiation. *Molecular and Cellular Biology*, *33*, 2091–2101.

Zhou, X., Xiang, Y., & Garrard, W. T. (2010). The Igkappa gene enhancers, E3′ and Ed, are essential for triggering transcription. *Journal of Immunology*, *185*, 7544–7552.

CHAPTER FIVE

Dynamic Control of Long-Range Genomic Interactions at the Immunoglobulin κ Light-Chain Locus

Claudia Ribeiro de Almeida*, Rudi W. Hendriks†,1, Ralph Stadhouders‡,1

*Sir William Dunn School of Pathology, University of Oxford, Oxford, United Kingdom
†Department of Pulmonary Medicine, Erasmus MC, Rotterdam, The Netherlands
‡Gene Regulation, Stem Cells and Cancer Programme, Centre for Genomic Regulation (CRG), Barcelona, Spain
[1]Corresponding authors: e-mail address: r.hendriks@erasmusmc.nl; ralph.stadhouders@crg.eu

Contents

Advances in Immunology, Volume 128
ISSN 0065-2776
http://dx.doi.org/10.1016/bs.ai.2015.07.004
© 2015 Elsevier Inc.
All rights reserved.

Abstract

The *Igκ* locus, which is spread over 3 Mb of genomic DNA and contains >100 variable (V) genes, serves as an important model system to study long-range chromatin interactions. Here, we will discuss how in developing B cells in the bone marrow the accessibility of individual V_κ segments is controlled by many lineage-specific and ubiquitously expressed transcription factors that act on various *cis*-regulatory elements, including promoters, enhancers, and insulators. This dynamic control furthermore involves changes in subnuclear localization, histone modification, DNA demethylation, and three-dimensional locus compaction. In pro-B cells, the *Igκ* locus adopts a poised conformation as full contraction has been achieved and many key transcription factors already occupy the locus. Subsequently, the combined activation of pre-B cell antigen receptor signaling pathways and attenuation of IL-7R signaling in small resting pre-B cells dramatically modifies the transcription factor landscape, supporting the induction of monoallelic *Igκ* gene rearrangements. Hereby, the intronic and 3′ *Igκ* enhancer elements coordinately focus their activities in the V_κ region toward frequently used V_κ genes. Recent work has drawn attention to the intriguing role of the CTCF-associated regulatory elements Cer and Sis, which are located in the V_κ–J_κ intervening region and control *Igκ* locus contraction and V_κ repertoire diversity. This involves CTCF-mediated locus insulation, restricting enhancer activity to the V_κ region and suppressing the preferential recombination to proximal V_κ genes. A picture emerges in which the dynamic control of long-range genomic interactions ensures correct timing of *Igκ* locus recombination and provides appropriate opportunities for individual V_κ gene segments to engage in V_κ–J_κ rearrangement.

1. INTRODUCTION

The B cell antigen receptor (BCR) is expressed on the surface of B lymphocytes and provides our immune system with a virtually unlimited capacity to detect foreign invaders. The BCR and its secreted form, called antibody, are composed of identical pairs of immunoglobulin (Ig) heavy (H) and light (L) chains, both with an N-terminal highly variable region and a C-terminal constant region. Whereas the variable regions specifically recognize the antigen, the constant region of the IgH chain determines antibody effector function. The generation of a BCR repertoire with a wide range of antigen-binding specificities is an astonishing evolutionary accomplishment. Exactly how a mammalian genome is able to carry the genetic information required to produce an almost infinite BCR repertoire was a matter of controversy and fierce debate among leading immunologists, until it was demonstrated that somatic recombination of gene segments is the molecular mechanism underlying antibody diversity (Hozumi & Tonegawa, 1976; Weigert, Gatmaitan, Loh, Schilling, & Hood, 1978).

The *Ig* genes are encoded within the *IgH* and *IgL* chain loci, which are among the largest multigene family loci present in our genome. They contain large collections of variable (V), diversity (D), joining (J), and constant (C) gene segments that are recombined through a process called V(D)J recombination (reviewed in Jung & Alt, 2004; Schatz & Swanson, 2011). The lymphocyte-specific endonucleases recombination activation gene-1 (RAG1) and RAG2 cooperate to induce double-strand DNA breaks at specific recombination signal sequences that flank the V, D, and J gene segments (Oettinger, Schatz, Gorka, & Baltimore, 1990). Subsequently, these gene segments are joined by ubiquitously expressed components of the nonhomologous end-joining DNA repair machinery.

The process of B cell development is characterized by strictly ordered V(D)J recombination events in B cell precursors. Recombination first takes place between the D_H and J_H genes of the *IgH* locus in pro-B cells and is followed by V_H-to-DJ_H recombination. Productive *IgH* chain rearrangement is monitored by deposition of the IgH μ chain protein on the cell surface, together with the preexisting surrogate light-chain (SLC) proteins, as the pre-BCR complex. Expression of the pre-BCR serves as a checkpoint that monitors for functional *IgH* chain gene rearrangement. It triggers the proliferative expansion of *IgH* μ chain protein expressing pre-B cells. Subsequently, pre-B cells exit the cell cycle and undergo a transition from large cycling into small resting pre-B cells, after which V(D)J recombinase activity is redirected to the *Igκ* and *Igλ L* chain loci (Fig. 1). A single-step V-to-J recombination event is first invoked at one of the *Igκ* loci. The lack of D elements in the *IgL* chain loci ensures that autoreactive BCR specificities can be replaced by ongoing *IgL* chain V-to-J gene recombination, a process that is referred to as "receptor editing" (Nemazee, 2006). If V-to-J recombination is not successful at one of the *Igκ* loci, rearrangement at the *Igλ* locus often follows. The majority of the IgL chains synthesized by human and mouse B lymphocytes are of the κ subtype, although the κ:λ ratio in mice (10:1) is significantly more skewed than in humans (1.7:1 on average) (Ghia et al., 1995).

Parallel to *IgH* and *IgL* chain rearrangement in B cells, differentiation of T lymphocytes depends on rearrangement of T cell receptor (TCR) gene

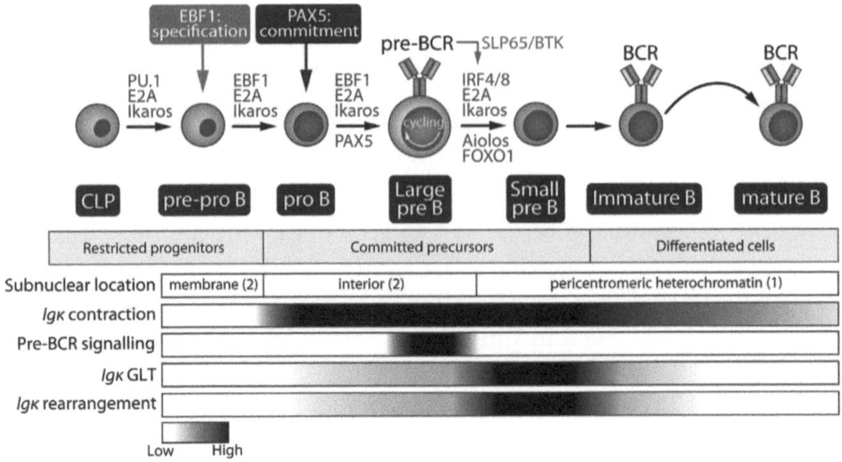

Figure 1 Overview and characteristics of murine B cell development in the bone marrow. The sequential stages of early mouse B cell development and the key transcription factors involved in their progression are shown. Critical events include (i) B cell fate specification by EBF1 at the pre-pro-B cell stage, (ii) B cell fate commitment by PAX5 at the pro-B cell stage, (iii) pre-BCR signaling, (iv) large pre-B cell proliferation, and (v) differentiation into small resting pre-B cells in which *Igκ* light-chain GLT and rearrangement are induced. Note that *Igκ* locus contraction already occurs at the pro-B cell stage and that the subnuclear localization of the *Igκ* locus dynamically changes during differentiation (the number of alleles present at a specific location during development is denoted in parentheses). BCR, B cell receptor; CLP, common lymphoid progenitor; GLT, germline transcription.

segments in four loci, TCRα, -β, -γ, and -δ, mediated by the same recombinase machinery. In most progenitor T-lineage cells in the thymus, a TCRβ chain is assembled from V_β, D_β, and J_β elements, and—as a next step—the TCRα chain is generated from the joining of V_α and J_α elements. TCRγδ$^+$ T cells represent only a small fraction of all T cells in lymphoid organs.

Because the V, D, and J gene segments within the *Ig* and *TCR* loci are spread over several megabases of the genome, it is obvious that locus compaction and folding are essential for the site-specific DNA recombination reactions that create a diverse repertoire of Ig and TCRs. The question how exactly V(D)J recombination is regulated over great genomic distances, providing appropriate recombination opportunities for individual gene segments, is still largely unknown and currently subjected to intense study. Given that all *Ig* and *TCR* loci are rearranged by a common recombination machinery (Yancopoulos, Blackwell, Suh, Hood, & Alt, 1986), an essential level of regulation of the V(D)J recombination process relies on lineage and developmental stage-specific changes in the accessibility of antigen receptor loci for recombination. This "accessibility hypothesis" was proposed by Alt and coworkers 30 years ago based on the initial observation that rearranging gene segments are concomitantly undergoing germline transcription (Yancopoulos & Alt, 1985).

In this review, we focus on the *Igκ* locus, which, since the original identification of the intronic *Igκ* enhancer element (Emorine, Kuehl, Weir, Leder, & Max, 1983; Picard & Schaffner, 1984; Queen & Baltimore, 1983), served as an important model system to study enhancer function, both in the control of *Ig* gene rearrangement and in the long-distance regulation of *Ig* gene transcription. Research on the *Igκ* locus has been facilitated by the relatively large numbers of pre-B cells present in the bone marrow that are in the process of rearranging their *IgL* genes, in contrast to the comparatively low numbers of pro-B cells. Moreover, *Igκ* locus recombination can be studied *in vitro* in interleukin-7 (IL-7)-driven bone marrow cultures, in which upon IL-7 withdrawal proliferating pre-B cells stop cycling and initiate *Igκ* locus recombination (Rolink, Kudo, Karasuyama, Kikuchi, & Melchers, 1991), as well as in Abelson virus (AMuLV) transformed pre-B cell lines, in which both *Igκ* germline transcription and rearrangement can be induced by LPS or by pharmacological inhibition of the transforming viral kinase v-Abl (Muljo & Schlissel, 2003). Finally, transient transfection experiments in the nonlymphoid human embryonic cell line BOSC23 have proven to be valuable for investigating the effects of transcription factors (TFs) on the activation of the *Igκ* locus for gene recombination (Romanow et al., 2000).

It has now become clear that accessibility of individual gene segments in the *Igκ* locus is tightly controlled by many *cis*-regulatory elements, including promoters, enhancers, and silencers, and involves subnuclear relocation, histone acetylation and methylation, DNA demethylation, and locus compaction. In particular, recent work has drawn attention to the intriguing role of various regulatory elements present in the V_κ–J_κ intervening region that control *Igκ* locus contraction and specify V_κ repertoire. In this review, we will focus on the function of *cis*-regulatory elements across the *Igκ* locus in the regulation of V_κ gene accessibility and how the generation of a diverse and balanced V_κ repertoire is dependent on the activity of various lineage-specific and ubiquitously expressed nuclear factors that cooperate in the dynamic control of long-range genomic interaction at the *Igκ* locus.

2. THE IMMUNOGLOBULIN κ LOCUS: GENE SEGMENTS AND REGULATORY ELEMENTS

In this section, we will review our current knowledge of the genetic architecture of the human and mouse *Igκ* loci, placing special emphasis on the key regulatory elements required for proper rearrangement of the *Igκ* locus.

2.1 Coding Gene Segments of the *Igκ* Locus

The *Igκ* loci in humans and mice are organized in a broadly similar fashion (see Fig. 2 for a general outline). They consist of a large V_κ region spanning 1.4 and 3.2 Mb in humans and mice, respectively, followed by a ~30 kb J_κ–C_κ region that also harbors the κ regulatory elements. The murine *Igκ* locus on chromosome 6 accommodates over 100 functional V_κ genes, 60 V_κ pseudogenes, 4 functional J_κ genes, 1 J pseudogene, and 1 C_κ gene. All *Ig* (pseudo)gene numbers and genomic size measurements in this chapter were obtained from the latest GENCODE (Harrow et al., 2006) and IMGT (Lefranc et al., 2005) databases. In the human locus on chromosome 2, a fairly recent and human-specific genetic duplication (Sitnikova & Nei, 1998) has organized the 44 functional V_κ genes and >30 V_κ pseudogenes into two clusters separated by an ~800 kb gene desert. The human *Igκ* locus further contains five functional J_κ genes and a single C_κ gene. The large numbers of V_κ genes in the mouse and human *Igκ* loci have been further divided into subgroups based on substantial sequence similarity (>75% for the IMGT-defined subgroups (Giudicelli & Lefranc, 1999)). These

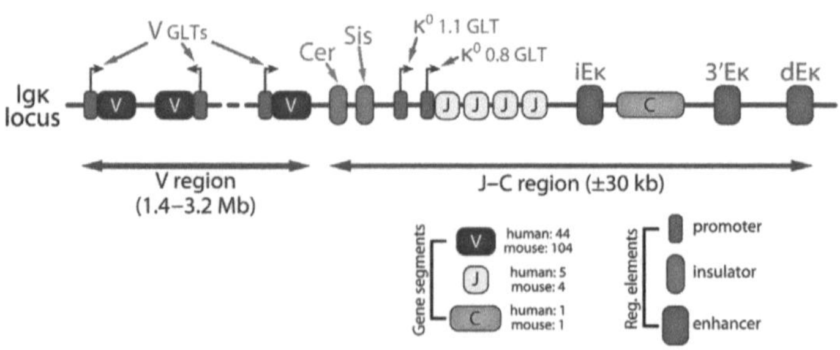

Figure 2 *Igκ* locus overview. Schematic overview of the mouse/human *Igκ* locus depicting the relative position and the number of different functional Variable (V), Joining (J), and Constant (C) gene segments. The key regulatory elements involved in the regulation of recombination are also shown. See text for a complete description of the different features.

subgroups occur in a semiclustered fashion with a fairly high degree of interspersion. V_κ genes almost always consist of two exons separated by a ~100–400 bp intron, in which the first small "leader" exon encodes a signal peptide that guides the produced L chains through the endoplasmic reticulum. This signal peptide is not incorporated into the antibody, as it is cleaved before assembly of the final Ig molecule. Perhaps not surprisingly, this basic mechanism of protein transport across membranes via an N-terminal signal peptide was first described for an IgL chain protein (Blobel & Dobberstein, 1975). Located 3′ of the last V_κ gene exon and 5′ of the J_κ exons are specific sequences known as recombination signal sequence (RSS) elements, which are recognized by the RAG1 and RAG2 proteins that form the core of the V(D)J recombination machinery (Oettinger et al., 1990). In both *IgL* loci, the orientation of the RSS of V_L and J_L genes allows direct V_L-to-J_L rearrangement, thus enabling multiple rounds of V_L-to-J_L rearrangements at the *Igκ* and *Igλ* loci. This is in contrast to the *IgH* locus, where the orientation of the V_H and J_H RSS elements is not directly compatible and requires the 5′ and 3′ D_H gene RSS elements to achieve $V_HD_HJ_H$ joining. Because the first VDJ recombination event at the *IgH* locus results in a deletion of all remaining D genes, subsequent rearrangement is not possible (Jung & Alt, 2004). Rearrangement of the *Igκ* locus generates a single fused V_κ–J_κ exon, which is then later joined to the C_κ gene by splicing out the J_κ–C_κ intervening region after transcription.

Although V(D)J recombination is widely regarded as a random mechanism (Tonegawa, 1983), it quickly became apparent that certain *Ig* genes are used more frequently than others (Brezinschek, Brezinschek, & Lipsky, 1995; Kalled & Brodeur, 1991; Wood & Coleclough, 1984). The recent application of high-throughput sequencing technology to analyze the *IgH* and *Igκ* V gene repertoires in pro-B and pre-B cells, respectively, has revealed significant locus-wide differences in primary V gene usage (Aoki-Ota, Torkamani, Ota, Schork, & Nemazee, 2012; Choi et al., 2013; Pan et al., 2013). Remarkably, it was shown that only one-third of the V_κ genes accounted for >85% of the V_κ genes used by C57BL/6 murine B cells (Aoki-Ota et al., 2012). It is currently unclear why these biases in V_κ gene usage have evolved, but the simplest explanation would be that particular V(D)J combinations give rise to effective pathogen-neutralizing antibodies and are thus favored by natural selection. Alternatively, as proposed by Aoki-Ota et al. (2012), a rearrangement bias at the *Igκ* locus could represent a sequential order of recombination that facilitates secondary rearrangements by ensuring that the maximum number of V_κ genes is retained after primary rearrangement (as discussed below). The molecular mechanisms underpinning unequal *Ig* gene rearrangement frequencies are still poorly understood. Several factors are thought to contribute to this phenomenon, including gene-specific differences in RSS strength, chromatin structure or accessibility, and binding of sequence-specific TFs (Feeney, 2009; Stadhouders et al., 2014). Sections 3–6 provide an in-depth discussion of the roles played by chromatin structure and TFs in regulating *Igκ* locus rearrangement, including their potential influence on V_κ usage.

The organizational layout of the murine and human *Igκ* locus is similar and exhibits several unique features, when compared to the other *Ig* loci (Fig. 2). First, the *Igκ* locus encodes only a single C_κ gene as opposed to the multiple C genes found in the *IgH* and *Igλ* loci. Second, the *Igκ* locus contains many V_κ genes that are in an opposite transcriptional orientation to the J_κ genes (exemplified by the second V_κ gene in Fig. 2). Recombination of these V_κ genes occurs via an inversion that does not lead to a deletion of the V_κ–J_κ intervening sequence, thus retaining the other V_κ genes for possible secondary rearrangements (Bassing, Swat, & Alt, 2002). In the *IgH* and *Igλ* loci, all V genes follow the transcriptional orientation of the J–C cluster.

Finally, the *Igκ* locus harbors an RSS-containing genetic element ~25 kb downstream of C_κ called the recombination sequence (RS) or kappa deleting element (Kde) (Durdik, Moore, & Selsing, 1984; Siminovitch, Bakhshi, Goldman, & Korsmeyer, 1985). Recombination of this element to V_κ genes or specific sites within the J_κ–C_κ intron results in a deletion of the J_κ–C_κ region and its regulatory elements. This mechanism prevents continued *Igκ* rearrangement or the undesired transcription of a nonfunctional V_κ–J_κ exon and plays an important role in receptor editing (Vela, Ait-Azzouzene, Duong, Ota, & Nemazee, 2008). In fact, almost all (>75%) human and mouse B cells expressing *Igλ* light chains carry such RS recombinations (Brauninger, Goossens, Rajewsky, & Kuppers, 2001; Nadel, Cazenave, & Sanchez, 1990).

Recombination to the RS element is common, as it is present in ~30% of Ig κ^+ and in almost all λ^+ mature human B lymphocytes. The chromosomal DNA excised from the genome is circularized into stable episomal structures, which are diluted twofold after each cell division. Therefore, the replication history of mouse and human B cell subsets can be quantified by calculating the ratio between genomic coding joints and signals joints on κ-deleting recombination excision circles of the *Igκ*-deleting rearrangement (van Zelm, Szczepanski, van der Burg, & van Dongen, 2007). The relative position of Kde is conserved in mammals, including platypus (Das, Nikolaidis, & Nei, 2009).

2.2 Regulatory Elements of the *Igκ* Locus

The intricate and highly regulated events accompanying *Ig* locus rearrangement are controlled by numerous powerful regulatory elements. The different regulatory elements of the *Igκ* locus are depicted in Fig. 2 and are among the best-studied gene regulatory elements known. The *Ig* loci have been fruitful model systems for studying gene regulation. The first cellular eukaryotic enhancers were identified in the 1980s in the *IgH* (Banerji, Olson, & Schaffner, 1983) and *Igκ* loci (Queen & Baltimore, 1983) and were found to stimulate the expression of *Ig* reporter genes. Similarly, long noncoding RNA transcripts—now widely considered as functional molecules with numerous important biological functions (Mercer, Dinger, & Mattick, 2009)—were first discovered in the *Igκ* (Van Ness et al., 1981) and *IgH* loci (Nelson, Haimovich, & Perry, 1983). Molecular studies of the *Ig* loci have also significantly contributed to our understanding of tissue-specific transcriptional control (Staudt & Lenardo, 1991). Today, its large size and well-characterized elements still make the *Igκ* locus an extremely suitable model system to study enhancer and insulator actions over long distances, including the role of noncoding transcription, higher-order chromatin structure, and DNA methylation. In this paragraph, we will discuss the different κ regulatory elements, including their identification, discovery, and function.

2.2.1 Igκ Promoters and Germline Transcription

Directly 5′ of the first exon of every functional V_κ gene segment lies a short stretch (<200 bp) of promoter DNA. RNA polymerase II and other basic transcriptional machinery components are recruited to this region, which contains the transcription start site. At the V_κ promoter, incoming gene regulatory signals are integrated into the transcription initiation rate (Lenhard, Sandelin, & Carninci, 2012). After V_κ-to-J_κ recombination, the promoter of the V_κ gene involved will drive transcription of the new $V_\kappa J_\kappa$ exon, which will be spliced to the C_κ region to produce an *Igκ* L chain mRNA. V_κ promoters contain several important sequence elements, but only two universal elements have been identified (Staudt & Lenardo, 1991). The first element is a non-cell-type-specific TATA box that serves as a recruitment site for the transcription apparatus (Maston, Evans, & Green, 2006) and is required for high-level *Ig* gene transcription. The second element is a conserved ATGCAAAT octamer motif (Falkner & Zachau, 1984) that not only stimulates high-level transcription but also confers tissue-specificity to *Ig* gene expression (Grosschedl & Baltimore, 1985; Mason, Williams, & Neuberger, 1985; Staudt & Lenardo, 1991). The strict B cell-specific expression pattern of *Ig* genes is not only achieved at the promoter level, but it also relies on regulatory signals from enhancers and insulators (see Sections 2.2.2 and 2.2.3). TFs operate at regulatory elements by binding to specific DNA motifs (Maston et al., 2006). The proper activation of *Ig* gene transcription (or any other gene expressed specifically in B cells) demands the presence of a unique ensemble of TFs only found in B cells (Nutt & Kee, 2007; Staudt & Lenardo, 1991). A detailed description of the TFs involved in *Igκ* locus biology is presented in Sections 4 and 5.

The above-mentioned transcription of a rearranged *Ig* gene generates coding mRNA molecules. However, preceding this coding transcription is the developmentally timed production of noncoding RNA transcripts—a signature feature of *Ig* and *TCR* loci (Abarrategui & Krangel, 2009). In order to distinguish them from coding V(D)J transcription, these noncoding transcripts are referred to as sterile or germline transcripts (GLTs). Like their coding counterparts, GLTs are produced from V gene promoters (Yancopoulos & Alt, 1985). GLTs were first discovered originating upstream of the first J_κ gene (Van Ness et al., 1981) and noncoding transcription was later detected over the J_H region (Lennon & Perry, 1985) and the *Igλ* locus (Engel, Rolink, & Weiss, 1999). The biological relevance of germline transcription (GLT) is still poorly defined. Yancopoulos and Alt (1985) observed that the production of GLTs from *Ig* gene segments was correlated with their recombination in a strict cell-type and differentiation stage-specific manner. Their findings were later confirmed for virtually all other antigen receptor loci (Abarrategui & Krangel, 2009) and formed the basis for the accessibility model of V(D)J recombination. This model states that developmentally timed GLT "opens" the chromatin environment of gene segments, making them accessible to the V(D)J recombination machinery. Although the accessibility model has gained broad support, the causal relationship between GLT and V(D)J recombination has been difficult to directly examine. Through the insertion of a transcriptional terminator that interferes with GLT, it was shown that transcription can indeed contribute to TCR J_α gene accessibility (Abarrategui & Krangel, 2006). Nevertheless, many questions concerning the mechanism by which GLT promotes recombination—including the potential importance of GLTs themselves—remain to be answered.

Besides the V_κ gene promoters from which GLTs are produced (Schlissel & Baltimore, 1989), the mouse *Igκ* locus contains two germline promoters referred to as κ^0 promoters (Engel et al., 1999; Martin & van Ness, 1990; Van Ness et al., 1981). The proximal κ^0 promoter is located just upstream of the $J_\kappa 1$ gene, yielding a 0.8 kb GLT ($\kappa^0 0.8$) spliced to the C_κ region. The distal κ^0 promoter initiates ~3.5 kb upstream of $J_\kappa 1$ at a region with homology to the Ig leader sequence, yielding a 1.1 kb GLT ($\kappa^0 1.1$) (Fig. 2). These promoters regulate the production of several long noncoding RNA transcripts (Amin et al., 2009; Taylor et al., 2009). The κ^0 promoters carry an octamer motif similar to the one found in V_κ gene promoters, and promoter location, structure, and transcript processing appear largely conserved between mice and humans (Thompson et al., 1992). GLT over the J_κ–C_κ region is initiated at the pre-B cell stage (Schlissel & Baltimore, 1989; Van Ness et al., 1981) and is important for *Igκ* recombination, as deletion of both the proximal and distal κ^0 promoters resulted in impaired V_κ-to-J_κ rearrangement (Cocea et al., 1999). The usage of κ^0 promoters appears to be developmentally regulated (Taylor et al., 2009), and the vast majority of κ^0 GLTs are produced from the distal κ^0 promoter (Amin et al., 2009). The relative contributions of both κ^0 promoters to *Igκ* recombination have been controversial (Amin et al., 2009; Taylor et al., 2009) and are still incompletely understood. Analysis of knock-in mice that report transcriptional activity of the κ^0 promoters revealed that GLT is biallelic and occurs in all pre-B cells (Amin et al., 2009). However, the generation of mice deficient for the proximal κ^0 promoter revealed that this promoter is not required for *Igκ* recombination and is even inactive in pre-B cells prior to rearrangement (Vettermann, Timblin, Lim, Lai, & Schlissel, 2015). Deletion of the proximal κ^0 promoter increased activating H3K4me3 histone marks in the J_κ region, suggesting that this promoter acts as a suppressor of recombination by limiting chromatin accessibility (because RAG2 specifically binds to H3K4me3, as discussed below). Despite this lack of activity, the proximal κ^0 promoter turned out to influence J_κ gene choice by inducing "$J_\kappa 1$ skipping" (Vettermann et al., 2015). Normally, primary *Igκ* V_κ–J_κ rearrangements are strongly biased toward $J_\kappa 1$ (Yamagami, ten Boekel, Andersson, Rolink, & Melchers, 1999), leaving the other three functional J_κ genes available for secondary rearrangements during receptor editing. Indeed, pre-B cells from proximal κ^0 knockout mice exhibited reduced potential for secondary rearrangements (Vettermann et al., 2015).

In summary, *Igκ* promoter elements drive transcription of the rearranged $V_\kappa J_\kappa$ exon after recombination, and κ^0 promoters are involved in the developmental regulation of V_κ-to-J_κ rearrangement and appear involved in J_κ gene choice.

2.2.2 Three *Igκ* Locus Enhancer Elements

The *Igκ* locus contains three defined transcriptional enhancers located in the J_κ–C_κ region (Fig. 2). Enhancers are small (~100–1500 bp) *cis*-acting genomic elements that are critical for the activation of their target genes (Maston et al., 2006). They can be located >1 Mb upstream or downstream from their target promoters and act in a direction-independent fashion (de Laat & Duboule, 2013). Deploying their regulatory function in a highly tissue-specific manner, enhancers are the prime determinants of tissue-specific gene expression and therefore key components in controlling development and differentiation (Ong & Corces, 2011). Enhancers are operated by the combinatorial action of TFs that recognize specific DNA motifs clustered within the enhancer elements. In order to bring the recruited regulatory proteins in the vicinity of a target gene, enhancers are thought to engage in long-range chromatin interactions (Ong & Corces, 2011). Although the mechanisms underlying such "chromatin looping" remain poorly understood, accumulating evidence indicates that three-dimensional (3D) genome topology plays a key role in regulating gene expression (de Laat & Duboule, 2013). The three κ enhancers are highly conserved among mammals, including their relative positions and orientation (Das et al., 2009), and play pivotal roles in orchestrating V_κ-to-J_κ recombination and tissue-specific *Igκ* gene expression.

A conserved sequence in the J_κ–C_κ region was discovered as the intronic κ enhancer element (iEκ) using a plasmid containing a rearranged *Igκ* gene (Emorine et al., 1983; Picard & Schaffner, 1984; Queen & Baltimore, 1983). Deleting parts of the J_κ–C_κ intron resulted in a loss of Igκ transcription when these constructs were transfected into myeloma cells (Queen & Baltimore, 1983; Queen & Stafford, 1984). Although these and other *in vitro* experiments all pointed toward an important tissue-specific role for iEκ in the *Igκ* locus (Staudt & Lenardo, 1991), a critical *in vivo* assessment of iEκ function was lacking until gene-targeting approaches in mice finally confirmed the importance of iEκ for B cell development (Inlay, Alt, Baltimore, & Xu, 2002; Inlay, Lin, Gao, & Xu, 2006; Inlay, Tian, Lin, & Xu, 2004; Xu, Davidson, Alt, & Baltimore, 1996). In contrast to several of the early *in vitro* studies mentioned above, iEκ deletion in mice did not result in reduced expression of the rearranged *Igκ* allele. However, mice lacking iEκ showed decreased levels of GLT, *Igκ* rearrangement, and reduced κ:λ ratios. The iEκ element was further proposed to be important for activation of *Igκ* DNA demethylation and maintenance of monoallelic *Igκ* demethylation (Inlay et al., 2002; Lichtenstein, Keini, Cedar, & Bergman, 1994). The latter process has been implicated in the establishment of allelic exclusion (Mostoslavsky et al., 1998), the mechanism that ensures that every individual B cell only expresses one *Igκ* allele (Bergman & Cedar, 2004; Mostoslavsky, Alt, & Rajewsky, 2004). The iEκ element closely resembles its *IgH* counterpart, the *IgH* intronic enhancer designated Eμ, both in terms of location and because of the presence of binding sites for similar B cell-specific TF combinations (discussed below and in Section 4). Nevertheless, their activities are strictly stage specific, as was elegantly illustrated by a study in which iEκ was replaced by Eμ *in vivo*. The presence of Eμ in the *Igκ* J_κ–C_κ intron prematurely induced GLT, DNA methylation, and recombination of the *Igκ* locus in pro-B cells (Inlay, Lin, et al., 2006), corresponding to the developmental stage at which Eμ promotes *IgH* rearrangement (Chen et al., 1993; Serwe & Sablitzky, 1993). However, Eμ was unable to evoke *Igκ* rearrangements in pre-B cells, indicating that despite their similar nature Eμ and iEκ induce V(D)J recombination only within a very specific developmental window (Inlay, Lin, et al., 2006).

Not long after the discovery of the intronic *IgH* and *Igκ* enhancers, several studies reported the expression of rearranged *Ig* genes apparently independent from intronic enhancer activity, suggesting the existence of additional enhancer elements (Atchison & Perry, 1987, 1988; Klein, Sablitzky, & Radbruch, 1984; Wabl & Burrows, 1984). For the *Igκ* locus, this notion was confirmed through the identification of an enhancer located ~9 kb 3′ of the C_κ gene, designated 3′Eκ (Fig. 2) (Meyer & Neuberger, 1989). The 3′Eκ element is B cell-specific and approximately sevenfold stronger *in vitro* than the fairly weak iEκ element (Meyer & Neuberger, 1989). Similar to findings in iEκ-deficient mice, targeted deletion of 3′Eκ *in vivo* resulted in a reduced κ:λ ratio and diminished levels of κ^0 GLT, locus demethylation, and recombination (Gorman et al., 1996; Inlay et al., 2002). However, in contrast to iEκ, the 3′Eκ element appeared to be an important mediator of high-level *Igκ* expression in mature B cells (Gorman et al., 1996; Inlay, Lin, et al., 2006) and is continuously required for *Igκ* mRNA expression (Zhou, Xiang, Ding, & Garrard, 2013). The 3′Eκ element has also been implicated in regulating affinity maturation of antibodies through somatic hypermutation (SHM) (Odegard & Schatz, 2006), although its role in this process has been controversial and appeared to be only quantitative (Inlay, Gao, et al., 2006; van der Stoep, Gorman, & Alt, 1998). Finally, 3′Eκ has been implicated in directing the nuclear position of the *Igκ* and *Igh* loci (Hewitt et al., 2008; Park, Xiang, Feng, & Garrard, 2014) and in regulating allelic exclusion (Zhou, Xiang, Ding, & Garrard, 2012).

Both iEκ and 3′Eκ are quantitatively important for *Igκ* recombination, but they are individually not absolutely required for rearrangement (Gorman et al., 1996; Xu et al., 1996). However, when both elements are deleted, *Igκ* rearrangement is essentially abolished (Inlay et al., 2002), demonstrating that the two enhancers share redundant but essential functions in promoting *Igκ* locus recombination. Nevertheless, they also play distinct roles in *Igκ* locus biology. Both enhancers might contribute to κ locus demethylation in partly different ways, because specifically iEκ appeared to mediate the maintenance of monoallelic *Igκ* demethylation (Inlay et al., 2002). As already stated above, only 3′Eκ is important for *Igκ* expression in mature B cells and has been implicated in SHM. Accumulating evidence indicates that iEκ and 3′Eκ are sequentially activated during B cell development and might regulate distinct phases of *Igκ* rearrangement. The iEκ element—but not 3′Eκ—is necessary for initial V_κ gene accessibility (Inlay, Lin, et al., 2006), which solely depends on binding of the E2A early B cell TF (Inlay, Lin, et al., 2006; Inlay et al., 2004). Moreover, iEκ is already engaged in long-range chromatin interactions with the V_κ region in pro-B cells (Lin et al., 2012) and shows preferential interactions with V_κ genes at this stage (Stadhouders et al., 2014)—events that are considered important for *Igκ* rearrangement (discussed in detail in Section 6). In contrast, the 3′Eκ element appears to rely on the activity of the TF interferon regulatory factor-4 (IRF4; see Section 4.2), which is highly induced at the pre-B cell stage (Johnson et al., 2008; Lazorchak, Schlissel, & Zhuang, 2006). Unlike for iEκ, chromatin interactions mediated by 3′Eκ only start to exhibit specificity for V_κ genes at the pre-B cell stage (Stadhouders et al., 2014). Sections 4–6 further detail the differential roles of the iEκ and 3′Eκ elements—including the various TFs that bind and operate them—in orchestrating the molecular events that culminate into rearrangement of the *Igκ* locus.

A third *Igκ* enhancer was discovered and characterized by the Garrard laboratory (Liu et al., 2002; Xiang & Garrard, 2008). This distal κ enhancer (dEκ)—identified as DNAse I hypersensitive site (HS) 9 in a screen for potential regulatory elements in the J_κ–C_κ region (Liu et al., 2002)—is located ~8 kb downstream of 3′Eκ (Fig. 2). The dEκ does not play a role in regulating *Igκ* recombination (Xiang & Garrard, 2008), but mice lacking dEκ show a mild phenotype composed of (i) reduced κ:λ ratio, (ii) impaired *Igκ* mRNA expression in splenic B cells, and (iii) reduced levels of SHM, similar to those seen in 3′Eκ-deficient B cells (Xiang & Garrard, 2008). Hence, dEκ mainly functions at later stages of B cell maturation, although follow-up studies utilizing 3′Eκ-dEκ double-deficient mice revealed a compensatory role for dEκ in 3′Eκ-deficient pre-B cells. Whereas dEκ-deficient pre-B cells exhibit no *Igκ* rearrangement defects (Xiang & Garrard, 2008), 3′Eκ-dEκ double-deficient pre-B cells showed reductions in *Igκ* GLT, DNA demethylation, and recombination that exceeded those seen in 3′Eκ single-deficient pre-B cells (Zhou, Xiang, & Garrard, 2010). These experiments suggest that 3′Eκ and dEκ are partially redundant and that 3′Eκ-deficient pre-B cells might prematurely activate dEκ to compensate for this loss (Zhou et al., 2010).

In conclusion, the general contributions of the three κ enhancers to *Igκ* activity can be summarized as follows: (i) the iEκ, 3′Eκ, and dEκ elements promote locus accessibility, DNA demethylation, and subsequent V_{κ}-to-J_{κ} rearrangement in a synergistic and partly redundant manner; and (ii) 3′Eκ and dEκ—but not iEκ—drive and maintain high-level *Igκ* mRNA expression and stimulate SHM of a functionally rearranged *Igκ* gene. Thus, the three core *Igκ* transcriptional enhancers control virtually all aspects of *Igκ* rearrangement and expression through a cooperative and synergistic mode of action.

2.2.3 The Sis and Cer Elements Upstream of $J_{\kappa}1$

A third class of regulatory elements that plays an important role in *Igκ* locus biology involves the insulator. Traditionally, insulators are thought to have two characteristic properties: (i) they "block" enhancer–promoter interactions to confine the action of enhancers to specific regions, and (ii) they act as barrier elements to separate chromatin domains, e.g., to prevent active chromatin from spreading into a heterochromatin region or vice versa (Gaszner & Felsenfeld, 2006). Insulator-binding proteins, in mammals mainly represented by the CCCTC-binding factor (CTCF) (see Section 5.1), have also been implicated in mediating 3D genome folding, allowing them to spatially influence gene regulatory processes (Ong & Corces, 2014). The powerful transcriptional enhancers present in *Ig* and *TCR* loci are involved in long-range chromatin interactions that may extend over >1 Mb (Guo, Gerasimova, et al., 2011; Ribeiro de Almeida et al., 2011). Such large and complex regulatory landscapes demand the presence of insulator elements to control enhancer activity and to prevent inappropriate intermingling of chromatin domains. Indeed, insulator elements have currently been identified for many *Ig* and *TCR* loci and their presence is essential for generating a diverse V(D)J rearrangement repertoire (Guo, Yoon, et al., 2011; Majumder et al., 2015; Xiang, Park, & Garrard, 2013).

Much of the pioneering work on characterizing *Igκ* locus insulator elements came from the Garrard laboratory. The same screen for DNAse I HSs in the J_κ–C_κ region that identified dEκ (as described above) also revealed two clusters of HSs within a ~5.5 kb region located ~12–6.5 kb upstream of $J_\kappa 1$ (see Fig. 2) (Liu et al., 2002). The proximal HS cluster was studied first and shown to confer transcriptional silencing onto a minimal V_κ promoter in cell line reporter assays (Liu et al., 2002). As such, this region was dubbed "Sis," which stands for "Silencer in the Intervening Sequence." A subsequent study employed a yeast artificial chromosome-based transgenic approach that allowed conditional deletion of Sis from an *Igκ* mini-locus transgene *in vivo* (Liu et al., 2006). Sis deletion resulted in a marked increase in V_κ–J_κ joining and a reduced colocalization of the transgene with centromeric heterochromatin, again advocating a negative role for this element in *Igκ* rearrangement (Liu et al., 2006). Adding to a potential insulator function, Sis was found to be occupied by CTCF in early B cells (Degner, Wong, Jankevicius, & Feeney, 2009). Next, mice were generated and characterized that carry a deletion across the Sis region (Xiang, Zhou, Hewitt, Skok, & Garrard, 2011). This study confirmed earlier results using a transgenic approach (Liu et al., 2006), but additionally revealed a surprising requirement for Sis in ensuring the generation of a diverse V_κ repertoire. Importantly, in Sis-deficient B cells, usage of the proximal V_κ genes was greatly favored over distal genes (Xiang et al., 2011)—an observation that showed a striking resemblance to the severely skewed rearrangement patterns observed in CTCF-deficient pre-B cells (Ribeiro de Almeida et al., 2011) (discussed in Section 5.1). Thus, it was postulated that Sis represents a barrier element that prevents the *Igκ* enhancers from preferentially targeting the proximal V_κ genes.

Recently, mice were generated in which the distal HS cluster was deleted (Xiang et al., 2013), as well as animals that lack both distal and proximal clusters through a ~6.3 kb deletion encompassing all six HSs (Xiang, Park, & Garrard, 2014). B cells lacking the distal HS cluster showed "decontraction" of their *Igκ* loci; hence the region was named "Contracting element for recombination" or Cer (Xiang et al., 2013). Locus contraction is a 3D topological state in which the proximal and distal ends of an *Ig* or *TCR* locus are spatially brought together (Jhunjhunwala, van Zelm, Peak, & Murre, 2009). Rearranging *Ig* loci are contracted, and locus contraction provides an attractive model for explaining how appropriate sampling of all V genes in a Mb-sized region can be achieved (also see Section 3.5). In agreement with this notion, locus decontraction leads to biased rearrangement repertoires (Fuxa et al., 2004; Liu et al., 2007; Reynaud et al., 2008), as also observed in the Cer-deficient mice (Xiang et al., 2013, 2014). Cer is also bound by CTCF and remains the only element that has directly been implicated in regulating *Igκ* locus contraction, as individual deletion of the iEκ, 3′Eκ, and Sis elements *in vivo* had no effect on locus contraction (Hewitt et al., 2008; Xiang et al., 2011). Its deletion also affected J_κ choice (Xiang et al., 2013). Analysis of Sis/Cer double knockout mice further revealed that neither of the two elements is involved in *Igκ* allelic exclusion or total *Igκ* rearrangement levels. However, V_κ usage patterns were dramatically skewed, as recombination with middle or distal V genes was essentially eliminated in Sis/Cer-deficient pre-B cells (Xiang et al., 2014). Furthermore, proximal V_κ gene accessibility as measured by GLT levels was greatly increased, suggesting that Cer and Sis synergistically repress GLT of the proximal V_κ region (Xiang et al., 2014).

Developmental timing of V_κ-to-J_κ rearrangement was not affected by deletion of the Sis or Cer regions (Liu et al., 2006; Xiang et al., 2013, 2011); nevertheless, Cer-deficient mice—but not Sis-deficient mice—did manifest *Igκ* gene rearrangements in $CD4^+CD8^+$ thymocytes. In this respect, the Cer region resembles the InterGenic Control Region 1 (IGCR1) that lies between the V_H and D_H clusters in the *IgH* locus, which suppresses V_H-to-DJ_H joining in thymocytes (Guo, Yoon, et al., 2011).

In summary, the CTCF-bound *Igκ* insulator elements Sis and Cer are localized to a small region upstream of $J_\kappa1$ (Fig. 2). Here, they appear to "insulate" the proximal V_κ genes from the activity of the downstream *Igκ* enhancer elements to limit the accessibility of proximal V_κ genes for recombination. As a result, a diverse and effective antibody repertoire can be readily established. Surprisingly, Sis and Cer are not conserved between mice and humans. This implies that humans have either evolved different ways to restrict proximal V_κ gene usage, or that the human equivalents of Sis and Cer are located at different positions in the V_κ–J_κ intervening region. Since insulator elements have not been described for the human *Igκ* locus, future efforts will have to reveal whether they play a similar role in regulating human antibody diversity.

2.2.4 Additional Functional Noncoding Elements in the Igκ Locus

Besides the well-characterized trio of *Igκ* enhancers discussed above, several other (potential) *Igκ* enhancer elements have been identified to date.

Zhou et al. (2012) described the identification of HS10, a conserved region of open chromatin ~2 kb downstream of the *Igκ* flanking gene ribose

5-phosphate isomerase (*Rpia*). HS10 deletion in mice did not affect *Igκ* expression and recombination in developing B cells (Zhou et al., 2012). Similar to dEκ, HS10 is important though for maximal *Igκ* expression in plasma cells and plays a minor role in promoting SHM (Zhou et al., 2012). An unbiased genomics approach was recently used to scan the entire *Igκ* locus for regions displaying enhancer-associated chromatin signatures, including combinatorial TF binding and the presence of specific histone modifications (Predeus et al., 2014). One potential enhancer element identified, named κRE1, was selected by the authors and exhibited B cell-specific enhancer activity *in vitro* (Predeus et al., 2014). Unlike the other κ enhancer elements, κRE1 is located within the V_κ region. One possible function of this enhancer could be the local stimulation of GLT to increase the accessibility of a subset of V_κ genes, but follow-up studies *in vivo* should validate whether κRE1 is indeed a bona fide *Igκ* enhancer.

The presence of a matrix association region (MAR) directly adjacent to the iEκ was reported (Cockerill & Garrard, 1986). MARs are regions in the genome that interact with the nuclear matrix and might influence gene expression (Kumar et al., 2007; Laemmli, Kas, Poljak, & Adachi, 1992). The sequence of the *Igκ* MAR is conserved between mouse and human (Whitehurst et al., 1992). *In vivo* deletion of the *Igκ* MAR resulted in only very minor phenotypes, the most striking being a modest reduction in SHM of rearranged *Igκ* genes in germinal center B cells (Yi, Wu, Trevorrow, Claflin, & Garrard, 1999).

Weaver and Baltimore noted the presence of two conserved binding sites for the B cell-specific Paired box TF PAX5 immediately upstream of the proximal κ^0 promoter, which they called KI and KII (Weaver & Baltimore, 1987). Intriguingly, deletion of the KI–KII sites *in vivo* appeared to diminish V_κ–J_κ rearrangement efficiency without altering GLT (Ferradini et al., 1996). The PAX5 TF (discussed in Section 4.4) has been implicated in mediating KI–KII function (Sato, Wang, & Kudo, 2001; Tian, Okabe, Miyazaki, Takeshita, & Kudo, 1997).

3. REGULATION OF *Igκ* LOCUS RECOMBINATION DURING B CELL DEVELOPMENT

3.1 Induction of *Igκ* Locus Recombination by Pre-BCR Signaling

The initiation of gene rearrangement events in the *Igκ* locus in pre-B cells requires the activity of the RAG1 and RAG2 proteins and the attenuation of the cell cycle, as well as accessibility of V_κ and J_κ gene segments, all of which are thought to be largely dependent on pre-BCR signaling.

The pre-BCR is transiently expressed on the cell surface of pre-B cells in the bone marrow and serves to monitor successful *IgH* gene rearrangement. The IgH μ protein is deposited on the cell surface, together with the SLC proteins and the signaling subunits Ig-α and Ig-β, as the pre-BCR complex (Melchers, 2005). The SLC components are two germline-encoded invariant proteins, termed λ5 (encoded by *Igll1* and resembling the constant region of λ L chain) and VpreB (either VpreB1 or VpreB2 (Shimizu, Mundt, Licence, Melchers, & Martensson, 2002)), which form a heterodimer and substitute for the not yet rearranged IgL chain. It is currently not entirely clear how pre-BCR signaling is initiated. It was shown (Ohnishi & Melchers, 2003) that the non-Ig-like tail of λ5 has the capacity to induce cell-autonomous signaling, suggesting ligand independence. It now appears that functional pre-BCR formation and autonomous signaling require the N-linked glycosylation site N46 in the CH1 domain of IgH μ chain (Ubelhart et al., 2010). In contrast, the finding that the pre-BCR checkpoint shapes the V_H repertoire at the pro-B to pre-B cell transition suggested ligand-mediated pre-BCR activation (ten Boekel, Yamagami, Andersson, Rolink, & Melchers, 1999; Wardemann et al., 2003). In support of such a model, interactions have been identified between the non-Ig tail of λ5 of the pre-BCR to galectin-1 or to stromal-cell-associated heparin sulfate (Bradl, Wittmann, Milius, Vettermann, & Jack, 2003; Gauthier, Rossi, Roux, Termine, & Schiff, 2002). Furthermore, the pre-BCR is capable of recognizing multiple (self-)antigens, including DNA, lipopolysaccharide (LPS), and insulin, via the non-Ig part of λ5 (Kohler et al., 2008). It has been hypothesized that pre-BCR autoreactivity serves to clonally expand pre-B cells in the absence of foreign antigens, which would explain why the majority of antibodies expressed in early immature B cell compartments display self-reactivity, including poly-reactive and antinuclear specificities (Wardemann et al., 2003). A pre-B cell selection model based on autorecognition would be supported by the finding that in SLC-deficient mice autoreactive pre-B cells are preferentially selected (Keenan et al., 2008).

The primary functions of the pre-BCR include (i) prevention of further rearrangement of the *IgH* allelic counterpart, a process referred to as allelic exclusion; (ii) triggering of proliferative expansion of IgH μ-expressing large cycling pre-B cells; and (iii) subsequent induction of developmental progression of large cycling into small resting pre-B cells in which the recombination machinery is reactivated for *IgL* chain rearrangement (Fig. 1). In the mouse, but not in human, pre-BCR signals synergize with IL-7R signals, and as a result, pre-B cells acquire the capacity to respond to low concentrations of IL-7 which drives the proliferative expansion of the IgH μ^+ large pre-B cell pool (Hendriks & Middendorp, 2004; Melchers, 2005). Through activation of Janus kinase-1 (JAK1) and JAK3, signal transducer and activator of transcription-5 (STAT5) is recruited and activated, which drives the expression of the cell cycle-promoting cyclin D3 (encoded by the *Ccnd3* gene) (Cooper et al., 2006; Mandal et al., 2009). In cycling pre-B cells, transcription of the *Rag* genes is low and RAG2 protein is unstable due to cell cycle-dependent, phosphorylation-mediated degradation (Li, Dordai, Lee, & Desiderio, 1996). For developmental progression, pre-B cells must cease dividing. This is thought to be essential to prevent the initiation of RAG-mediated double-strand DNA breaks during DNA replication, which would compromise genomic integrity in dividing lymphocytes, leading to chromosomal aberrations. Pre-B cells in the bone marrow exit the cell cycle when they migrate away from IL-7-expressing stromal cells (Tokoyoda, Egawa, Sugiyama, Choi, & Nagasawa, 2004).

Signaling through the pre-BCR downstream adapter molecule SLP65 (SH2-domain-containing leukocyte protein of 65 kDa, also known as BLNK or BASH) switches cell fate from proliferation to differentiation. This is evident from the phenotype of SLP65-deficient mice, which is characterized by an arrest of B cell development at the large cycling pre-B cell stage (Jumaa et al., 1999). SLP65-mediated pre-BCR signaling (i) induces expression of the IRF4, Ikaros, and Aiolos TFs that downregulate expression of the SLC component λ5 (Ma, Pathak, Trinh, & Lu, 2008; Thompson et al., 2007); (ii) inhibits proliferation by directly suppressing c-Myc expression (Ma et al., 2010); and (iii) reduces inhibitory phosphorylation of Forkhead box type O (FOXO) TFs (reviewed in Herzog, Reth, & Jumaa, 2009). Like in SLP65-mutant mice, B cell development in IRF4/IRF8-double-deficient mice is also arrested at the large cycling pre-B cell stage, whereby pre-B cells fail to rearrange *IgL* loci and express high levels of SLC, thus maintaining high surface expression of the pre-BCR (Lu, Medina, Lancki, & Singh, 2003). Nevertheless, SLC silencing is not critical for the limitation of pre-B cell expansion, because constitutive SLC expression is not associated with increased numbers of pre-B cells or hampered *Igκ* rearrangement (van Loo, Dingjan, Maas, & Hendriks, 2007). In response to pre-BCR signaling, IL-7Rα expression is downregulated and SLP65 binds JAK3 to interfere with IL-7R signaling

(Marshall, Fleming, Wu, & Paige, 1998; Nakayama et al., 2009; Schebesta, Pfeffer, & Busslinger, 2002). All these changes result in attenuation of the cell cycle and thus RAG protein stabilization. Moreover, *Rag1* and *Rag2* gene transcription is induced by FOXO proteins (Amin & Schlissel, 2008).

In recent studies, the effects of single and double-deficiency of SLP65 and Bruton's tyrosine kinase (BTK), representing a gradient of progressively diminishing pre-BCR signaling, were analyzed (Stadhouders et al., 2014). BTK-deficient mice manifest a specific developmental delay within the small pre-B cell compartment when compared with wild-type mice (Middendorp, Dingjan, & Hendriks, 2002). A synergistic role of SLP65 and BTK in B cell development was demonstrated by the almost complete block in B cell development at the large cycling pre-B cell stage in BTK/SLP65 double-deficient mice (Jumaa, Mitterer, Reth, & Nielsen, 2001). The levels of κ^0 0.8 and 1.1 GLTs are apparently normal in BTK-deficient pre-B cells, modestly reduced in SLP65-deficient pre-B cells and almost absent in BTK/SLP65 double-deficient pre-B cells. GLT across the V_κ region showed a similar pattern of sensitivity to this gradient of diminishing pre-BCR signaling (Kersseboom et al., 2006; Stadhouders et al., 2014). Interestingly, BTK or SLP65 single-deficient pre-B cells have severely reduced GLT over the J_λ and V_λ segments and BTK-deficient mice show a ~50% reduction in the frequency of Igλ chain-expressing B cells (Dingjan et al., 2001). Therefore, BTK-mediated pre-BCR signaling is particularly required for the induction of *Igλ* gene rearrangement.

Genome-wide expression profiling in BTK and SLP65 single- and double-deficient pre-B cells confirmed that pre-BCR signaling decreases the expression of *Dntt* (encoding terminal deoxynucleotidyl transferase) and the SLC components *Vpreb* and *Igll1* and increases expression of the cell surface markers CD2, CD22, CD25 (IL-2R), and MHC class II (Stadhouders et al., 2014). We also found that pre-BCR signaling upregulates the expression of various genes implicated in *IgL* recombination, such as *Ikzf3* (Aiolos), *Ikzf1* (Ikaros), *Irf4*, *Spib*, *Pou2f2* (OCT2), polymerase-μ, and *Hivep1* (MBP-1), some of which were already known to be induced by pre-BCR signaling. The function of these nuclear proteins will be discussed in detail below. Moreover, this analysis identified genes encoding DNA-binding or modifying factors that were not previously associated with *IgL* recombination, including *Lmo4*, *Zfp710*, *Arid1a/3a/3b*, the lysine-specific demethylases *Aof1* and *Phf2*, *Prdm2* (a H3K9 methyltransferase), *Sik1* (encoding a histone deacetylase (HDAC) kinase), the DNA repair protein gene *Rev1*, and several *Hdac* genes. Future experiments may reveal a role for these nuclear proteins in the control of *Igκ* locus accessibility or recombination. Although we did not find significant differences in the expression levels of several other TFs implicated in *Ig* gene recombination, e.g., OBF1/OCA-B, PAX5, E2A, and IRF8, pre-BCR signaling might control their activity via posttranslational modifications or through modulation of the expression of their respective regulators. For example, the activity of E2A is restricted by basic helix–loop–helix protein inhibitors of DNA-binding ID2 and ID3, which lack a DNA-binding domain but are capable of forming nonfunctional heterodimers with E2A (Kee, Quong, & Murre, 2000). Since differentially expressed genes (comparing wild-type, BTK and SLP65 single- and double-deficient pre-B cells) included a wide array of genes—involved in, e.g., signal transduction, cell–cell interaction, and survival—we conclude that pre-BCR signaling may additionally affect the transcription of many genes in an indirect fashion or through a complex regulatory network.

3.2 Repression of *Igκ* Locus Recombination by IL-7R Signaling

It has been shown that *Igκ* locus rearrangement can occur independently of IgH μ chain expression, suggesting that pre-BCR signaling does not play a role in the induction of *IgL* gene rearrangement. For example, *Igκ* transcription and rearrangements have been detected in IgH μ chain-negative pro-B cells in mouse bone marrow (Novobrantseva et al., 1999) and can be induced by LPS in transformed pro-B cell lines via activation of nuclear factor κB (NF-κB) (Sen & Baltimore, 1986a). Moreover, removal of IL-7 from *in vitro* cultured (i) pro-B cells from mice with a targeted deletion of J_H, (ii) pro-B cells harboring a D_H-to-J_H rearrangement on both *IgH* alleles from Bcl2-Tg mice, or (iii) pre-B cells from λ5-deficient mice that cannot express the pre-BCR on the cell surface, all resulted in the differentiation into IgL chain- expressing B cells with apparently normal frequencies and kinetics (Grawunder, Haasner, Melchers, & Rolink, 1993; Melchers et al., 2000; Rolink et al., 1993). Finally, expression of constitutively active RAS induced *IgL* gene rearrangement in J_H-deficient pro-B cells in the absence of IgH μ chain expression, thereby maintaining an extent of *Igκ* rearrangement and a preference for *κ* over λ recombination similar to that of wild-type B cells (Shaw, Swat, Davidson, & Alt, 1999).

On the other hand, several lines of evidence indicate that pre-BCR signaling is important for the induction of *IgL* gene rearrangement. First, surface IgH μ expression correlates with GLT and rearrangement of the *Igκ* locus (reviewed in Schlissel, 2004). Second, in the absence of SLP65, κ^0 GLT is reduced (Kersseboom et al., 2006). Third, BTK-deficient mice show a ~3-h delay in the progression of pre-B cells into both surface Ig κ^+ and surface Ig λ^+ B cells (Middendorp et al., 2002) and manifest reduced *Igλ* L chain GLT and *Igλ* usage (Dingjan et al., 2001; Kersseboom et al., 2006). Fourth, transgenic expression of the constitutively active E41K-BTK mutant in IgH μ chain-negative pro-B cells induced premature *Igκ* rearrangement and protein expression (Kersseboom et al., 2006).

A picture is emerging in which pre-BCR and IL-7R signaling coordinately regulate pre-B cell proliferation and subsequent induction of *Igκ* gene rearrangement. It was shown that pre-BCR-mediated activation of the RAS–MEK–ERK pathway silences transcription of *Ccnd3* to induce cell cycle exit and concomitantly augments *Igκ* locus accessibility by down-regulating ID3, which increases nuclear E2A-binding activity (Mandal et al., 2009). The capacity of pre-BCR signaling to diminish IL-7 responsiveness plays a crucial role in this process, as this stabilizes RAG2 levels and enhances *Igκ* locus accessibility: STAT5 activation, induced by IL-7R signaling, directly represses *IgL* chain rearrangement in pro-B cells by binding to iEκ and recruiting the Polycomb repressive complex-2 (PRC2) that confers transcriptional repression through deposition of histone 3 lysine 27 trimethylation (H3K27me3) marks (Malin et al., 2010; Mandal et al., 2011, 2009). When IL-7R signaling ceases in pre-B cells, STAT5-mediated repression is released and E2A can efficiently access iEκ to establish an accessible chromatin structure at the *Igκ* locus through the induction of activating epigenetics marks (see Section 4.8).

Taken together, these findings indicate that the low level of *Igκ* gene rearrangement in pro-B cells is greatly enhanced in pre-B cells by pre-BCR signaling and the attenuation of IL-7R signaling, which cooperate to increase *Igκ* locus accessibility, RAG activity, and the expression of TFs crucially involved in *Igκ* rearrangement. In this bimodular model, the relative binding to iEκ of the two key TFs STAT5 and E2A (see Sections 4.5 and 4.8) crucially regulates *Igκ* GLT and rearrangement in pre-B cells.

3.3 *Igκ* Locus Recombination and Receptor Editing

Whereas most Igκ L chain-expressing cells retain their *Igλ* loci in germline configuration, Igλ expressing cells generally have nonfunctional *Igκ* gene rearrangements on both alleles (Alt, Enea, Bothwell, & Baltimore, 1980;

Coleclough, Perry, Karjalainen, & Weigert, 1981; Hieter, Korsmeyer, Waldmann, & Leder, 1981). Yet, *Igλ* gene rearrangements are independent of *Igκ* gene rearrangements, because Igλ-expressing cells are also generated in targeted mice that cannot rearrange their *Igκ* loci (Takeda et al., 1993; Zou, Takeda, & Rajewsky, 1993). In labeling experiments, using BrdU—which is mainly incorporated in large cycling pre-B cells that subsequently differentiate into small pre-B cells that rearrange their *IgL* loci—it was shown that the first Ig κ^+ or λ^+ B cells appear after ~8 or ~24 h, respectively (Arakawa, Shimizu, & Takeda, 1996).

Further differentiation of surface IgM^+ B cells upon productive *IgL* chain rearrangement is dependent on the quality of IgL chains produced. The IgL chain must have the capacity to efficiently pair with the IgH μ chain (Yamagami et al., 1999) to enable IgM surface expression, and this newly formed BCR should not recognize self-antigens. In situations where either (i) rearrangement is not in frame, (ii) pairing is not successful, or (iii) the resulting BCR is autoreactive, immature B cells can perform secondary rearrangements. This process has been examined in a number of autoreactive transgenic mouse models, showing that immature B cells with an autoreactive BCR die by apoptosis (clonal deletion), become nonresponsive (anergic), or revise the IgL chains of their BCR to eliminate self-reactivity, a process referred to as "receptor editing" (Gay, Saunders, Camper, & Weigert, 1993; Goodnow, Crosbie, Jorgensen, Brink, & Basten, 1989; Nemazee & Burki, 1989; Radic & Zouali, 1996; Tiegs, Russell, & Nemazee, 1993). Not only the strength of the interaction between BCR and autoantigen, but also the developmental stage of the immature B cell regulates the choice of either one of these options: surface IgM^{low} B cells advancing in their development lose their ability to undergo receptor editing when they become surface IgM^{high} and then become more susceptible to antigen-mediated apoptosis (Melamed, Benschop, Cambier, & Nemazee, 1998). Receptor editing can involve the rearranged *Igκ* allele through rearrangement of a more upstream V_κ to a more downstream J_κ element, thereby deleting a preexisting V_κ–J_κ junction. In this process, *Igκ* alleles are often deleted by recombination involving the RS element. BrdU-labeling experiments in mice have shown that the receptor editing process—including the detection of autoreactivity, reinduction of *IgL* chain recombination, and surface BCR expression—delays B cell development by only ~2 h (Casellas et al., 2001). When editing is unsuccessful on the already rearranged *Igκ* allele, rearrangement events may proceed to the other, germline, *Igκ* allele or to the *Igλ* loci.

Receptor editing is not limited to transgenic mice expressing an autoreactive BCR, but is a frequent event during normal B cell development as deduced from the quantification of the in-frame *Igκ* genes in Ig λ^+ B cells (Retter & Nemazee, 1998). Moreover, using knock-in mice in which one of the two C_κ regions was replaced by the human counterpart, the extent to which editing occurs in developing B cells was measured: ~25% of all antibody molecules are produced by gene replacement (Casellas et al., 2001). Recently, next-generation sequencing was used to characterize the primary *Igκ* repertoire from bone marrow, splenic, and lymph node B cells of C57BL/6 mice (Aoki-Ota et al., 2012). Interestingly, consistent with an important role for receptor editing in shaping the *Igκ* repertoire, a recombination bias was found for inversional rearrangements to $J_\kappa1$ or deletional rearrangements of proximal V_κ segments to $J_\kappa1$, which would retain the maximum numbers of V_κ genes available for subsequent rearrangements. The extend of receptor editing is limited by the life span of pre-B cells, since overexpression of the antiapoptotic protein BCL-2 increased the relative frequency of editing and λ usage (Lang et al., 1997). In this context, an *in vitro* system with purified nuclear substrates was used to show that added recombinant RAG proteins could access and cleave J_κ RSSs even in mature B lymphocyte chromatin (Stanhope-Baker, Hudson, Shaffer, Constantinescu, & Schlissel, 1996). Therefore, cessation of *Igκ* locus recombination *in vivo* cannot be explained by a reduction of *Igκ* locus accessibility and is primarily due to the termination of RAG expression, consistent with the efficient downregulation of RAG levels in mature peripheral B cells (Nagaoka, Gonzalez-Aseguinolaza, Tsuji, & Nussenzweig, 2000).

Both in humans and in mice, a considerable proportion of mature B cells (~10%) express both κ and λ L chains or two different κ L chains, a phenomenon called isotypic or allelic L chain inclusion, respectively (Giachino, Padovan, & Lanzavecchia, 1995; Huang, Kearney, Grusby, Benoist, & Mathis, 2006; Liu et al., 2005). The finding that transgenic B cells expressing a preexisting anti-DNA autoreactive BCR are especially subject to *IgL* allelic inclusion would support the notion that the expression of two different IgL chains is an alternative way to limit autoreactivity of B cells (Witsch & Bettelheim, 2008). BrdU-labeling experiments have shown that allelically included B cells take longer to develop than allelically excluded B cells, supporting the notion that dual receptor B cells arise by receptor editing (Casellas et al., 2007). Moreover, simultaneous rearrangement of the two allelic *Igκ* loci is thought to be a rare event, because one of the *Igκ* alleles is epigenetically marked and thereby predetermined for rearrangement, precluding the concomitant accessibility of both *Igκ* alleles for recombination in a single cell (Farago et al., 2012; Goldmit et al., 2005).

3.4 Nuclear Localization of the *Igκ* Locus and Locus Accessibility

During the subsequent stages of B cell development, the spatial organization of the *Igκ* locus is actively remodeled. Various different epigenetic mechanisms ensure that *Igκ* rearrangement is favored in pre-B cells and suppressed in pro-B cells and T cells. These mechanisms include *Igκ* locus contraction, subnuclear relocation, DNA methylation, chromatin remodeling, histone modification, and GLT.

Interestingly, the *Igκ* allele that undergoes rearrangement first in pre-B cells is already predetermined prior to the pre-B cell stage. Lineage tracing experiments have shown that in hematopoietic stem cells or multipotent progenitors, allelic choice is still plastic (Farago et al., 2012). At the common lymphoid progenitor (CLP) stage, cells become committed to the choice of a single allele and this decision is then stably maintained in a clonal fashion during the subsequent stages (in pre-pro-B, pro-B, and pre-B cells) up to the rearrangement process (Farago et al., 2012). It has furthermore been demonstrated that the allelic choice at the CLP stage is not subject to feedback inhibition. In mice carrying two productively pre-rearranged *Igκ* loci, most B cells were able to express both alleles simultaneously, but they still underwent monoallelic demethylation at the *Igκ* locus and the unmethylated allele was still the preferred substrate for SHM in mature B cells (Fraenkel et al., 2007).

One of the mechanisms that mark the allelic choice is asynchronous replication timing. The genome is divided into well-defined zones that replicate in a programmed manner throughout the S phase of the cell cycle. Early-replicating zones contain a large proportion of housekeeping genes and tissue-specific genes that are actively expressed, while repressed genes are mostly associated with late replicating zones. The rearranged *Igκ* alleles in B cells are always early replicating and recombination in pre-B cells almost always occurs on the early-replicating *Igκ* allele (Mostoslavsky et al., 2001). The choice which of the two *Igκ* alleles will be early replicating is established at the late CLP stage and clonally maintained (Farago et al., 2012). Thus, asynchronous replication is an early marker of the clonal monoallelic predetermination for *Igκ* rearrangement. An asynchronous replication profile is also found in regions flanking the *Igκ* locus, e.g., the *Rpia* gene ~40 kb downstream. It has been shown that the control of asynchronous replication is not affected by targeted deletion of iEκ or 3′Eκ (Mostoslavsky et al., 2001), but it remains unclear how early replication is translated into preferential accessibility for rearrangement.

Another parameter that is important for the regulation of *Igκ* locus accessibility is its nuclear localization. At the time of *Igκ* gene recombination in pre-B cells, one allele is generally located at the nuclear periphery within pericentromeric heterochromatin, which represents a repressive and inaccessible chromatin environment. The other allele is positioned away from the periphery and localized to more euchromatic regions of the nucleus to allow rearrangement. Expression of RAG1 protein, but not its cleavage activity, is required for monoallelic *Igκ* locus association with pericentromeric heterochromatin in pre-B cells (Hewitt et al., 2009). In the absence of the DNA repair checkpoint protein Ataxia telangiectasia mutated (ATM), monoallelic recruitment of *Igκ* to heterochromatin was reduced and many cells contained RAG-mediated double-strand DNA breaks on both alleles (Hewitt et al., 2009).

Taken together, in pre-B cells, there is one *Igκ* allele destined to undergo preferential rearrangement, characterized by early replication, activating H3K4me3 activation marks, and located away from pericentromeric heterochromatin. However, these allelic differences appear to have little impact on GLT over the J_{κ} region, initiated from the two κ^0 promoters, which has been shown to be biallelic (Amin et al., 2009; Singh, Bergman, Cedar, & Chess, 2003). Nevertheless, the association with heterochromatin will sequester one allele away from the recombinase. It is possible that GLT over the V_{κ} region is monoallelic, thereby contributing to monoallelic rearrangement, but this remains unexplored.

In mature splenic B cells, one *Igκ* allele was still found to be associated with heterochromatin (Skok et al., 2001) and also monoallelic demethylation, SHM, and asynchronous replication were maintained—independently of rearrangement (Fraenkel et al., 2007). Thus, epigenetic marks at the *Igκ* locus not only direct allele-specific V_{κ}-to-J_{κ} rearrangement in pre-B cells but also direct allele specificity during later events such as SHM. In this respect, the *Igκ* locus shows marked differences with the *IgH* locus. Although it was previously reported that *in situ* hybridization analyses showed nonequivalent nuclear location of the *IgH* alleles in activated mature B cells (Roldan et al., 2005; Skok et al., 2001), recent microscopy measurements did not uncover pronounced differences in pericentromeric heterochromatin proximity between productive and nonproductive *IgH* alleles in cycling splenic B cells (Holwerda et al., 2013). Furthermore, using an allele-specific 4C-seq strategy, we found that the two physically different *IgH* alleles in splenic B cells occupy the same chromatin compartments. The overall chromatin environment of *IgH* was very similar in B and T cells and in resting and activated lymphocytes and both *IgH* alleles were localized to an active compartment (Holwerda et al., 2013). These findings are consistent with earlier identified similarities between productive and nonproductive *IgH* alleles including chromatin marks, transcription, and frequencies of SHM (Holwerda et al., 2013). Thus, it can be concluded that only for the *Igκ* locus—and not for the *IgH* locus—monoallelic epigenetic marks are maintained in mature B cells.

The *Igκ* locus is relocated during the sequential stages of early B cell development (Fig. 1). In embryonic stem cells, multipotent progenitors and pre-pro-B cells from IL-7R-deficient mice (Kosak et al., 2002) or E2A-deficient mice (Stadhouders et al., 2014), both *Igκ* alleles interact with the nuclear membrane. Remarkably, in committed pro-B cells, both *Igκ* loci move away from the nuclear membrane to a more central region, despite that one of the two alleles is already epigenetically predetermined for rearrangement. Nuclear repositioning is independent of RAG2 and does not result in an appreciable GLT of *Igκ* in pro-B cells (Kosak et al., 2002). Next, in pre-B cells, one of the two *Igκ* alleles becomes associated with the Ikaros and heterochromatin protein 1γ (HP1γ) proteins in the repressive environment of pericentromeric heterochromatin, thereby favoring rearrangement of the centrally positioned other allele (Goldmit et al., 2005). At the pre-B cell stage, the heterochromatin-associated *Igκ* and *IgH* alleles colocalize, resulting in a decontraction of the *IgH* allele involved. Intriguingly, both phenomena are mediated by the 3′Eκ (Hewitt et al., 2008). It is possible that Ikaros—acting as a repressor—is first recruited to the excluded *Igκ* allele and the 3′Eκ at the pre-B cell stage, after which the *Igκ* locus relocates to pericentromeric heterochromatin. Subsequent Ikaros recruitment to the allelically excluded *IgH* locus, possibly via close spatial proximity with the Ikaros-bound *Igκ* allele, could then induce *IgH* locus decontraction and contribute to allelic exclusion. Although current ChIP-seq methodology cannot easily distinguish between individual alleles, such a model is supported by published data showing that Ikaros binding is observed across the *IgH* locus in pre-B cells (Ferreiros-Vidal et al., 2013). In support of such a model, the Ikaros-bound Sis region of the *Igκ* locus was found to be largely responsible for targeting *IgH* loci to heterochromatin domains (Xiang et al., 2011). Interestingly, in the absence of Sis, the *Igκ* and *IgH loci* still interact in pre-B cells (as this is mediated by 3′Eκ and not by Sis), but these alleles are no longer recruited to heterochromatin. Thus, from these findings, it can be concluded that Sis is required in *trans* for the recruitment of the *IgH* locus to pericentromeric heterochromatin by a "piggyback" mechanism.

There appears to be a parallel with antibody secreting plasma cells, in which the functionally rearranged *IgH* and *Igκ* loci preferentially colocalize in transcription factories, often near the nuclear periphery (Park et al., 2014). However, in plasma cells *Ig* genes are located in close proximity to nuclear pores, favorable for the efficient export of *Ig* transcripts into the cytoplasm (as opposed to the Ikaros-mediated association of silenced genes with pericentromeric heterochromatin and the nuclear lamina). Interestingly, plasma cells from 3′Eκ-deficient mice exhibited reduced transcriptional activity, colocalization in transcription factories, and positioning near the nuclear periphery. 3C experiments confirmed the importance of the 3′Eκ element in mediating *trans*-chromosomal interactions between *Igκ* and the Eμ and 3′RR enhancer elements in the *IgH* locus. It has been hypothesized that this nuclear organization maximizes coordinated transcriptional and trafficking control to potentiate optimal antibody assembly (Park et al., 2014).

3.5 *Igκ* Locus Contraction, Epigenetic Modifications, and Rearrangement

Both in the *IgH* and *Igκ* loci, the V genes are spread over a large genomic region, which is thought to impede efficient recombination of distantly located V gene segments. To facilitate long-range V(D)J recombination, *Ig* loci undergo compaction (Kosak et al., 2002). Overall, locus contraction is achieved through long-range chromatin interactions between proximal and distal regions within *Ig* loci. This process brings distal V genes in close proximity to the D or J regions to which the RAG proteins bind. *IgH* locus contraction is dependent on PAX5, which induces both large-scale contraction and distal V_H-to-DJ_H rearrangements (Fuxa et al., 2004). However, no factor engaged in a parallel function for *Igκ* locus contraction has been identified to date. Initial FISH studies in FACS-sorted B cell precursors showed that V_κ and C_κ genes were often separated by large distances in pro-B cells and large cycling pre-B cells, but were particularly colocalized in most small pre-B and early immature B cells (Roldan et al., 2005). In contrast, in 103X7 pre-B cells, *Igκ* alleles are predominantly contracted before induction of recombination (Fitzsimmons, Bernstein, Max, Skok, & Shapiro, 2007). Moreover, recent Hi-C and 3C-seq studies provided evidence for substantial lineage-specific interactions between the iEκ or 3′Eκ elements and the distal V_κ region in pro-B cells, indicating that the *Igκ* locus is already contracted early in B cell development (Lin et al., 2012; Ribeiro de Almeida et al., 2011; Stadhouders et al., 2014). However, E2A-deficient pre-pro-B cells manifested very few interactions between iEκ and the distal V_κ region (Lin et al., 2012). Based on these findings—together with our 3D-FISH analyses showing that the *Igκ* locus adopted a decontracted topology in E2A-deficient pre-pro-B cells but not in RAG1-deficient pro-B cells (Stadhouders et al., 2014)—we conclude that *Igκ* locus contraction is achieved in pro-B cells, whereby E2A likely has a critical role in controlling *Igκ* locus topology (Fig. 1).

Beyond subnuclear localization and locus contraction, the two *Igκ* alleles display additional developmentally regulated epigenetic features. In pro-B cells, the *Igκ* allele predetermined for gene recombination is preferentially decorated with activating histone methylation and acetylation marks such as H3K4me3 and H3Ac (Farago et al., 2012; Goldmit et al., 2005). These histone modifications most likely support allele-specific gene recombination, as it was shown that RAG2 has the capacity to efficiently bind to chromatin decorated with H3K4me3 and RAG2 preferentially binds to the acetylated allele in pre-B cell clones (Farago et al., 2012; Ji et al., 2010).

Finally, for many genes, DNA methylation plays an important role in the regulation of transcription and chromatin structure. In most cell types, the *Igκ* locus is heavily methylated, but in mature B cells, critical sites are unmethylated on the rearranged allele and remain methylated on the non-rearranged allele (Mostoslavsky et al., 1998). This pattern is generated by demethylation, which is initiated on a single *Igκ* allele prior to rearrangement in pre-B cells and requires the iEκ and 3′Eκ elements. It is thought that demethylation enhances rearrangement, because (i) in pre-B cells double-strand DNA breaks in the *Igκ* locus are preferentially formed on unmethylated alleles and (ii) methylated J_κ alleles are poor substrates for the rearrangement machinery in *in vitro* RAG-mediated cleavage assays (Goldmit, Schlissel, Cedar, & Bergman, 2002).

Taken together, a picture is emerging in which asynchronous replication of the *Igκ* locus is the first epigenetic mark that signifies the allelic choice for gene recombination at the late CLP stage. Next, at the pro-B cell stage, histone modifications follow and both *Igκ* alleles move away from the nuclear periphery as they undergo large-scale compaction to facilitate V_κ-to-J_κ rearrangement. As a result of *Igκ* locus contraction, the *Igκ* enhancers display high interaction frequencies with the distal V_κ region and even beyond the boundaries of the *Igκ* locus (Stadhouders et al., 2014). Pre-BCR signaling induces significant changes in the expression levels of most TFs involved in *Igκ* recombination, as well as a functional redistribution of enhancer-mediated long-range interactions (see Section 6). Whereas one *Igκ* allele becomes strongly demethylated and marked with active histone modifications—both supporting RAG-mediated recombination—the other *Igκ* allele remains methylated and is recruited to pericentromeric heterochromatin. How the silenced *Igκ* allele becomes reactivated in the context of receptor editing events is currently unknown.

4. LINEAGE-SPECIFIC TFs INVOLVED IN *Igκ* LOCUS RECOMBINATION

The driving force behind the complex cascade of events that culminates in balanced V_{κ}-to-J_{κ} recombination is a stage and tissue-specific combination of key TFs. These proteins are recruited to many regions within the *Igκ* region and, most importantly, to the different κ regulatory elements. Given the key role of critically involved TFs in B cell development and V(D)J recombination, numerous laboratories are actively investigating where and when they occupy the B cell genome. The advent of high-throughput sequencing technology in particular has greatly increased the feasibility of genome-wide studies into TF binding. A wealth of genome-wide TF-binding data from pro- and pre-B cells is currently available, including *in vivo* binding profiles for the *Igκ* locus. Several of these studies have already shown that TFs can display striking *Igκ* locus binding patterns that are suggestive of their possible mode of action (Bossen, Mansson, & Murre, 2012; Lucas, Bossen, & Murre, 2011). In order to facilitate hypothesis generation, we took advantage of available genome-wide datasets from pro-B and pre-B cells, the majority of which were generated by chromatin immune-precipitation combined with genome-wide DNA sequencing (ChIP-seq) experiments. We compiled these datasets to provide an overview of TF binding and the chromatin state at the *Igκ* locus in developing B cells. Table 1 summarizes the binding/enrichment profiles of several key TFs and chromatin modifications relevant to *Igκ* recombination, including references to the original reports. Genome browser snapshots showing the measured binding profiles at selected regions within the *Igκ* locus are provided in Figs. 3 and 4.

4.1 Ikaros and Aiolos

The Ikaros family of TFs comprises Ikaros (encoded by *Ikzf1*), Aiolos (*Ikzf3*), Helios (*Ikzf2*), Eos (*Ikzf4*), and Pegasus (*Ikzf5*) and is characterized

Table 1 Transcription Factor Occupancy and Chromatin Features of the *Igκ* Locus in Precursor B Cells

Dataset[a] (Cell Type)	V_κ Region	V_κ Prom.[b]	κRE1	Cer	Sis	κ^0 (1.1)	iEκ	3′Eκ	dEκ	References
Chromatin features										
DHS (pro-B)	+	+	+	+	+	–	+	+	+	Revilla et al. (2012)
DHS (pre-B)	+	+	+	+	+	+	+	+	+	Unpublished observations
H3K4Me2 (pro-B)	+	+	+	+	+	+	+	+	+	Lin et al. (2010)
H3K4Me3 (pro-B)	–	–	–	–	–	–	–	–	–	Lin et al. (2010)
Pro-B cell only										
MED1 (pro-B)[c,d]	+ (center)	+	+	+	–	–	–	+	–	Whyte et al. (2013)
EBF1 (pro-B)[c]	+ (distal)	+/–	+	–	–	–	–	–	–	Schwickert et al. (2014)
IRF4 (pro-B)[c]	+	+/–	+	+	+	–	+	+	+	Schwickert et al. (2014)
PU.1 (pro-B)[c]	+	+	+	+	+	–	+	+	+	Schwickert et al. (2014)
P300 (pro-B)[c]	+	–	+	–	–	–	+	+	+	Lin et al. (2012)
Pro- and pre-B cell										
PAX5 (pro-B)	+	+/–	+	–	+	–	–	+	–	Revilla et al. (2012)
PAX5 (pre-B)	+	+	+	+	+	–	+	+	+	Ochiai et al. (2012)
FOXO1 (pro-B)	–	–	–	–	–	–	–	–	–	Lin et al. (2010)
FOXO1 (pre-B)	+	+	+	+	+	–	+	+	–	Ochiai et al. (2012)

Ikaros (pro-B)	–	–	+	–	+	–	–	–	–	Schwickert et al. (2014)
Ikaros (pre-B)	+	+	+	+	+	+	+	+	+	Ferreiros-Vidal et al. (2013)
YY1 (pro-B)	+	–	+	+	–	–	+	+	+	Medvedovic et al. (2013)
YY1 (pre-B)	+	+/–	ND	ND	ND	ND	ND	+	ND	Pan et al. (2013)
E2A (pro-B)	+	+	+	+	+	–	+	+	+	Lin et al. (2010)
E2A (pre-B)	+	+	ND	ND	ND	ND	+	+	ND	Stadhouders et al. (2014)
CTCF (pro-B)	+	–	–	+	+	–	–	–	–	Choi et al. (2013)
CTCF (pre-B)	+	–	+	+	+	–	–	–	–	Ribeiro de Almeida et al. (2011)
RAD21 (pro-B)	+	–	–	+	+	–	–	–	–	Choi et al. (2013)
RAD21 (pre-B)	+	ND	ND	ND	+	ND	ND	ND	ND	Degner et al. (2009)

[a]Based on published (genome-wide) DNase I hypersensitivity (DHS) profiling and chromatin immunoprecipitation (ChIP) assays.
[b]+ indicates association of the TF or chromatin state with a significant number (>±20) of individual V_κ promoters; +/– refers to only sporadic promoter binding.
[c]There is currently (almost) no pre-B cell binding data available for these factors. IRF4 (Lu et al., 2003) and PU.1 (Pongubala et al., 1992) have been shown to bind the 3′Eκ in pre-B cells; P300 has been shown to bind the V_κ region and selected V_κ promoters (Sakamoto et al., 2012).
[d]MED1 ChIP-Seq data from Whyte et al. (2013) were generated using the 38B9 pro-B cell line.
ND, not determined. "+" indicates significant binding/enrichment; "–" indicates no or very little binding/enrichment.

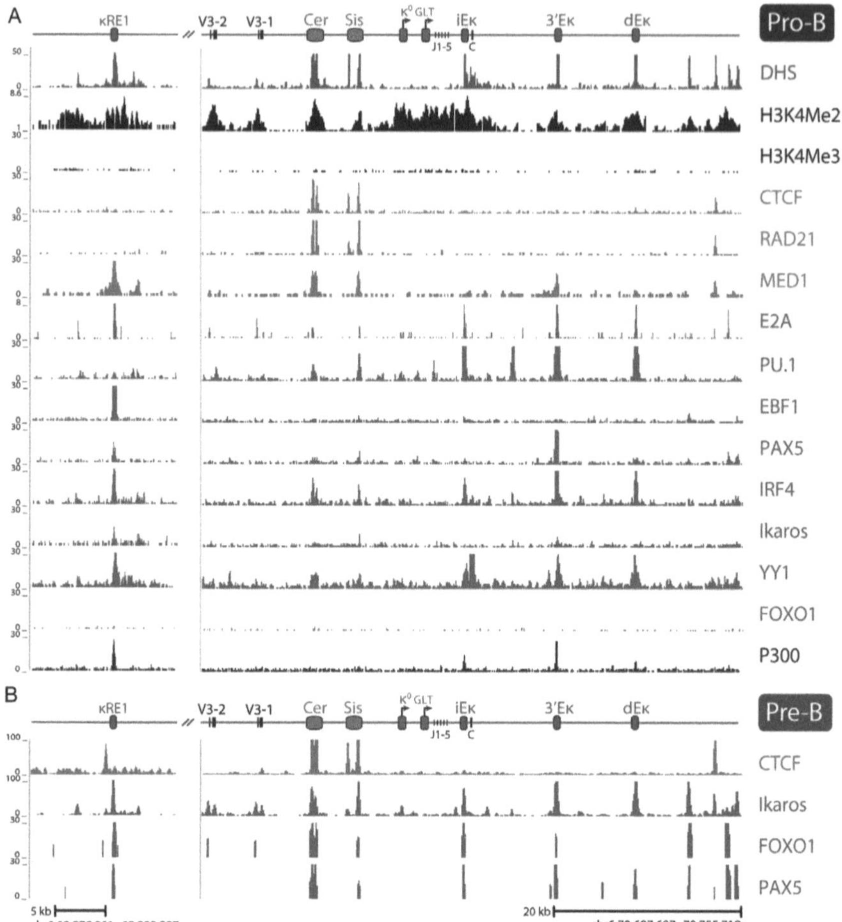

Figure 3 Transcription factor binding and chromatin features of the murine *Igκ* locus. (A) Pro-B cell enrichment profiles for various chromatin features and transcription factors (TFs) as determined by DNase I-seq and ChIP-seq experiments. DNase hypersensitivity (DHS) is shown in gray, histone modifications in black, insulator/architectural proteins in light blue, key B cell TFs in green, and chromatin-modifying factors in dark blue. Relevant gene segments and regulatory elements are indicated on the top. The murine κRE1 putative enhancer element (*left*) and the $J_κ$–$C_κ$ regulatory region (*right*) are shown. Genome-wide datasets were downloaded from public databases (references are included in Table 1) and aligned to the mm10 mouse genome build for visualization. (B) As in (A), but showing pre-B cell profiles. Scale bar and genomic coordinates of the regions displayed are shown at the bottom of the figure. (See the color plate.)

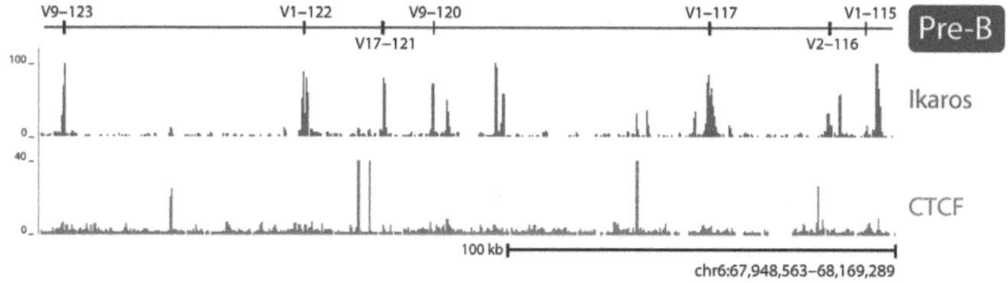

Figure 4 Differential binding of Ikaros and CTCF to the V_κ region. Depicted are Ikaros and CTCF ChIP-seq profiles obtained from pre-B cells, showing a representative small section of the V_κ region. Note how Ikaros consistently occupies both V_κ promoters and intergenic sites, while CTCF only targets intergenic sites. V_κ gene segments are indicated on the top. Scale bar and genomic coordinates of the regions displayed are shown at the bottom of the figure. (See the color plate.)

by the presence of two sets of highly conserved C_2H_2-type zinc finger motifs (reviewed in Cobb & Smale, 2005; John & Ward, 2011). A set of four N-terminal zinc fingers mediate DNA binding, and both Ikaros and Aiolos recognize the canonical sequence GGAA. Two C-terminal zinc fingers enable these proteins to oligomerize with themselves, with other Ikaros family members or with other TFs. Ikaros proteins have been shown to interact with chromatin-remodeling complexes such as the Nucleosome Remodeling Histone Deacetylase (NuRD) and the SWItch/Sucrose Non-Fermentable (SWI/SNF) complexes, allowing them to function as both repressors and activators of gene transcription (Kim et al., 1999).

4.1.1 *Ikaros in Pre-B Cell Differentiation*

Expressed in virtually all murine and human hematopoietic cell types, the Ikaros TF occupies a central role in the regulation of blood cell formation (Georgopoulos, 2002). Addressing the function of Ikaros in the differentiation of early committed B cell progenitors and V(D)J recombination has been complicated by the absolute requirement of a functional Ikaros allele for the development of the earliest B cell progenitors (Sellars, Kastner, & Chan, 2011). In fact, in the absence of Ikaros, multipotent progenitors are unable to differentiate into CLPs. To circumvent this issue, Ikaros hypomorphic mice were generated. B cell development in these animals exhibited several intrinsic defects: a marked reduction in the pre-pro-B cell population, a partial block in pro-B to pre-B cell differentiation, defective IL-7-dependent proliferation of pre-B cells, and impaired differentiation of IgM^- to IgM^+ B cells (Kirstetter, Thomas, Dierich, Kastner, & Chan, 2002). Results from additional *in vitro* studies indicated that Ikaros, together with

Aiolos, repressed the expression of the pre-BCR SLC components λ5 and VpreB, as well as several cell cycle genes to facilitate the transition from the pre-BCR-driven large cycling into the small resting pre-B cell stage (Ma et al., 2010, 2008; Mandal et al., 2009; Thompson et al., 2007). Moreover, Ikaros (and Aiolos) expression levels increase during B cell progenitor differentiation and are strongly induced by pre-BCR signaling (Morgan et al., 1997; Stadhouders et al., 2014). These studies thus implicated Ikaros as a key regulator of early B cell development in. However, they provided only limited insight into the full complement of Ikaros target genes in pro- and pre-B cells. Importantly, conditional loss-of-function mouse models would be required to probe Ikaros function during individual stages of B cell development *in vivo*.

Several important recent studies have addressed these gaps in our knowledge of Ikaros function, providing an extensive catalog of Ikaros target genes and a wealth of novel insight into the physiological function of Ikaros in pro/pre-B cell differentiation (Ferreiros-Vidal et al., 2013; Heizmann, Kastner, & Chan, 2013; Joshi et al., 2014; Schwickert et al., 2014). One study described global Ikaros occupancy across the pre-B cell genome (Ferreiros-Vidal et al., 2013). Ikaros-controlled genes were strongly represented in virtually all key aspects of pre-B cell biology, including pre-BCR signaling, cell cycle regulation, *Ig* rearrangement, and developmental maturation. The Chan/Kastner and Busslinger laboratories used conditional inactivation of Ikaros in early pro-B cells using an Mb1-Cre transgene (Hobeika et al., 2006) to analyze Ikaros function *in vivo* (Heizmann et al., 2013; Schwickert et al., 2014). Georgopoulos and coworkers generated a conditional Ikaros allele in which loxP sites flanked the two DNA-binding zinc fingers in exon 5 of *Ikzf1* (Joshi et al., 2014). Using a Cd2-Cre transgene, exon 5 deletion was achieved in early pro-B cell stages, resulting in the expression of a dominant-negative protein. All three groups reported that Ikaros loss of function essentially caused an early developmental arrest at the pre-B cell stage. In an *in vitro* rescue culture system, it was shown that Ikaros—in cooperation with IL-7 withdrawal—induces cell cycle exit, *Igκ* GLT, and *IgL* gene rearrangement and controls many developmentally regulated genes (Heizmann et al., 2013). Schwickert et al. crossed their conditional Ikaros-deficient mice on a $Rag2^{-/-}$ background and evaluated gene expression differences between control and Ikaros-deficient pro-B cells that were derived from short-term IL-7-driven *in vitro* cultures (Schwickert et al., 2014). These experiments suggest that Ikaros enables pre-BCR signaling through a combined activation and repression of respectively positive and negative regulators of pre-BCR signaling. For example, Ikaros was found to play an important role in the activation of the pre-BCR downstream phosphoinositide-3-kinase, mitogen-activated protein kinase, and calcium signaling pathways, in the migration of pro-B cells toward the chemokine CXCL12 (SDF-1α), and in preventing pro-B cell adhesion through the $\alpha_4\beta_1$ integrin (Joshi et al., 2014; Schwickert et al., 2014).

4.1.2 The Activities of Ikaros and Aiolos Are Partially Nonoverlapping

The closely related Ikaros and Aiolos TFs share highly conserved zinc finger domains that confer DNA-binding specificity (Cobb & Smale, 2005), and ChIP-seq experiments in T cells revealed numerous shared binding sites (Zhang et al., 2012). Both are upregulated at the pro-to-pre-B cell transition and *in vitro* experiments have suggested that Ikaros and Aiolos cooperate to silence pre-BCR genes and promote *IgL* gene rearrangement (Ma et al., 2010, 2008; Mandal et al., 2009; Thompson et al., 2007). Moreover, comparison of deregulated genes in Aiolos and Ikaros-deficient pre-B cells revealed a ~50% overlap (Ferreiros-Vidal et al., 2013). Taken together, these findings suggest a synergistic regulation of early B cell development by Ikaros and Aiolos. However, in stark contrast to the complete lack of B cell development in Ikaros-deficient animals described above, Aiolos knockout mice exhibited a phenotype characterized by only a twofold increase in pre-B cells and constitutive activation of mature B cells (Wang et al., 1998). Pro-B cells in which Ikaros was conditionally deleted manifested an upregulation of Aiolos expression, but this did not rescue the developmental defect associated with Ikaros deficiency (Schwickert et al., 2014), indicating that Aiolos is unable to compensate for a loss of Ikaros. This would also be supported by the finding that Ikaros-null mice and the Ikaros dominant-negative mutant mice—in which also Aiolos is thought to be nonfunctional (Joshi et al., 2014)—have a largely similar phenotype with respect to the developmental arrest at the pre-B cell stage (as described above). However, in the Ikaros dominant-negative mutant mice, the arrested pre-B cells were highly proliferative and induced pre-B acute lymphoblastic leukemia upon transfer in immunodeficient recipient mice (Joshi et al., 2014). Because this phenotype was not found in the conditional Ikaros-null mice (Heizmann et al., 2013; Schwickert et al., 2014), it is conceivable that the proliferative phenotype is caused by the dominant-negative nature of the Ikaros mutation, possibly by interfering with Aiolos function. This notion is supported by the finding that a homozygous deletion of Aiolos in a heterozygous Ikaros dominant-negative background resulted in a similar pre-B cell phenotype (Joshi et al., 2014). The presence of Aiolos in conditional Ikaros-null pre-B cells may prevent the aberrant cellular proliferation observed in Ikaros dominant-negative mutant pre-B cells in which the activity of both Aiolos and Ikaros is compromised.

Thus, although a substantial body of evidence supports Ikaros and Aiolos fulfilling redundant functions in early B cells, they also appear to play non-overlapping roles in regulating transcription and chromatin.

4.1.3 Ikaros as a Regulator of Ig Gene Rearrangement

Although the absence of Ikaros prohibits B cell development, the generation of pro-B cells could be restored by the introduction of another TF, early B cell factor (EBF1; see Section 4.6) in Ikaros-deficient hematopoietic precursors cells, allowing the analysis of the early B cell compartment in an Ikaros-deficient background *in vitro* (Reynaud et al., 2008). It appeared that Ikaros is essential for *IgH* gene rearrangement: Ikaros-deficient pro-B cells (i) failed to upregulate *Rag* expression, (ii) showed greatly reduced *IgH* locus accessibility and recombination, and (iii) displayed *IgH* locus decontraction (Reynaud et al., 2008). However, in contrast to these *in vitro* studies, VDJ rearrangements of the *IgH* locus appeared unaffected in all three conditional Ikaros mutant mice, as IgH μ protein expression in pre-B cells was similar to controls (Heizmann et al., 2013; Joshi et al., 2014; Schwickert et al., 2014). Moreover, D_H-to-J_H or proximal and distal V_H-to-D_H–J_H recombination events at the *IgH* locus were detected at similar frequencies in conditional dominant-negative Ikaros knockout and control pre-B cells (Joshi et al., 2014). These apparent phenotypic discrepancies may be attributed to a relatively late onset of excision by the Mb1-Cre or Cd2-Cre strains used, allowing significant numbers of pro-B cells to complete *IgH* gene rearrangement before Ikaros protein levels have significantly diminished. Alternatively, Ikaros may not have a significant impact on *IgH* locus rearrangement. In this scenario, the reported reduction of *IgH* accessibility and locus compaction in Ikaros-deficient EBF1-rescued pro-B cells (Reynaud et al., 2008) can be explained by the very early developmental arrest at the multipotent progenitor stage associated with Ikaros deficiency: although the EBF1-rescued Ikaros-deficient pro-B cells expressed PAX5 and YY1, they were still not fully committed to the B cell fate.

Nevertheless, Ikaros-null mice—as well as Ikaros dominant-negative mutant mice—showed a phenotype essentially characterized by an arrest at the pre-B cell stage before the induction of *Igκ* recombination. A lack of *Igκ* recombination alone did not cause this pre-B cell maturation defect, since the introduction of a prearranged *Igκ* gene did not rescue the phenotype (Joshi et al., 2014). Instead, transcriptome analyses as well as functional experiments in Ikaros-deficient pro-B or pre-B cells demonstrated defective pre-BCR signaling, as well as enhanced integrin signaling and adhesion.

In addition to its role in differentiation of pre-B cells, we postulate that Ikaros is likely to function in pre-B cells as an essential regulator of V_κ-to-J_κ rearrangement. How Ikaros achieves this at the molecular level is still unclear. We recently noticed that, using published ChIP-Seq data (Ferreiros-Vidal et al., 2013), in primary pre-B cells Ikaros is recruited to virtually every V_κ gene and every known regulatory element in the *Igκ* locus (Stadhouders et al., 2014) (Table 1; Figs. 3 and 4). In strong contrast, in pro-B cells the *Igκ* locus is almost completely devoid of Ikaros binding (Goldmit et al., 2005; Schwickert et al., 2014) (Table 1; Fig. 3). Massive Ikaros recruitment to the *Igκ* locus specifically in pre-B cells correlates with the robust induction of Ikaros expression at that stage and the onset of V_κ-to-J_κ rearrangement. Almost all V_κ regions harbored Ikaros-binding sites, but also many intergenic Ikaros-binding sites were identified in the *Igκ* locus in pre-B cells (Fig. 4). These observations imply that Ikaros is directly involved in *Igκ* gene recombination by controlling the regulatory elements that orchestrate this process. Ikaros does not appear required for contraction of the *Igκ* locus, as *Igκ* contraction is already achieved in pro-B cells prior to Ikaros recruitment to the *Igκ* locus (Stadhouders et al., 2014). In association with HP1γ and NuRD complexes, Ikaros has the capacity to recruit genes to pericentromeric heterochromatin in developing lymphocytes (Brown et al., 1997; Kim et al., 1999). Indeed, at the time when monoallelic association of the *Igκ* locus with pericentromeric heterochromatin is initiated, increased Ikaros and HP1γ binding to the J_κ–C_κ region was observed (Goldmit et al., 2005). Moreover, Ikaros associates with the Sis regulatory region that specifies heterochromatin positioning to germline mouse *Igκ* transgenes in pre-B cells (Liu et al., 2006) and is involved in recruiting endogenous *Igκ* loci to heterochromatin (Xiang et al., 2011). Monoallelic colocalization of *Ig* alleles with regions of pericentromeric heterochromatin is associated with transcriptional silencing and the establishment of allelic exclusion (Skok et al., 2001). Nevertheless, in mice with a targeted deletion of the Sis region, monoallelic heterochromatin localization of *Igκ* alleles is reduced, but despite their euchromatic location, allelic, and isotype exclusion remained intact (Xiang et al., 2011).

Ikaros certainly does not only play a role as a repressor protein, because rescue experiments in IL-7-driven pre-B cell cultures showed that Ikaros induced *Igκ* GLTs (within ~4 h), *Rag1* and *Rag2* transcription, and *IgL* gene rearrangement (Heizmann et al., 2013). Given the strong upregulation of Aiolos at the pro-B to pre-B cell transition, it will be of interest to perform similar ChIP-seq experiments for Aiolos to assess its binding dynamics at the *Igκ* locus during early B cell development.

4.2 Interferon Regulatory Factors IRF4 and IRF8

IRF4 and IRF8 are highly homologous TFs belonging to the IRF TF family. IRF4 and IRF8 are selectively expressed in lymphoid and myeloid cells (Taniguchi, Ogasawara, Takaoka, & Tanaka, 2001) and have long been implicated in early B cell development and *Igκ* rearrangement (Lu, 2008). IRF4 was first described as part of a B cell-specific heterodimer that is recruited to the 3′Eκ (Eisenbeis, Singh, & Storb, 1995). In fact, IRF4,8 bind quite weakly to DNA and their recruitment relies on interactions with other TFs, in particular, the Ets-family TFs PU.1 and Spi-B (see Section 4.3) (Brass, Zhu, & Singh, 1999; Eisenbeis et al., 1995; Escalante et al., 2002; Pongubala et al., 1992).

Indeed, IRF4 and IRF8 are required for early B cell development as homozygous compound-mutant mice exhibit a B cell developmental arrest at the pre-B cell stage (Lu et al., 2003). IRF4,8-deficient pre-B cells are highly proliferative, fail to downregulate the SLC components λ5 and VpreB, and display a severe impairment of *Igκ* rearrangement. Subsequent studies showed that IRF4 and IRF8 function redundantly in promoting *Igκ* locus accessibility and recombination: expression of either IRF4 or IRF8 in IRF4,8-deficient pre-B cells induces *Igκ* GLTs and rearrangement (Johnson et al., 2008; Ma, Turetsky, Trinh, & Lu, 2006). The defect in IRF4,8-deficient pre-B cells could not be rescued by Bcl2-mediated survival signals and forced cell cycle exit of proliferating IRF4,8-deficient pre-B cells (induced by IL-7 withdrawal) did not evoke *Igκ* recombination.

It is conceivable that IRF4,8 indirectly regulate pre-B cell maturation and *Igκ* recombination through the upregulation of Ikaros/Aiolos, and several lines of evidence would support this. First, the phenotype observed in IRF4,8-deficient mice closely resembles that of Ikaros mutant mice (as described in Section 4.1). Second, both IRF4 and IRF8 are able to induce Ikaros/Aiolos expression (Ma et al., 2008). Third, both Ikaros and IRF4 have the capacity to induce *Igκ* GLTs and rearrangement (Heizmann et al., 2013; Ma et al., 2006). Fourth, both Ikaros and IRF4 regulate the expression of chemokine receptor CXCR4 and promote the migration of pre-B cells in response to the chemokine ligand CXCL12 (Johnson et al., 2008; Joshi et al., 2014; Schwickert et al., 2014). However, in contrast to Ikaros, which recruits one of the *Igκ* alleles to pericentromeric heterochromatin via binding to the Sis element, IRF4 promotes monoallelic positioning of the *Igκ* locus away from pericentromeric heterochromatin (Johnson et al., 2008). IRF4,8 bind to 3′Eκ and regulate its activity, as histone acetylation at these enhancers was severely compromised in IRF4,8-deficient pre-B cells. Although it was speculated that IRF4 might associate with an RNA polymerase II factory, the mechanism by which IRF4 counteracts the association of an *Igκ* allele with pericentromeric heterochromatin is currently unknown (Johnson et al., 2008).

The role of IRF4,8 in the induction of *Igκ* recombination therefore remains incompletely understood at the molecular level. Lazorchak et al. reported that IRF4 and E2A cooperate to promote *Igκ* locus accessibility (Lazorchak et al., 2006). They proposed that IRF4 binding stimulates E2A co-recruitment to 3′Eκ, after which IRF4 and E2A synergistically activate the *Igκ* locus via 3′Eκ. As IRF4,8 act at the pre-B cell stage and are strongly induced by pre-BCR signaling, full activation of 3′Eκ was hypothesized to occur by pre-BCR signaling—in contrast to iEκ activation, which was found to be more pre-BCR signaling-independent (also see Sections 2.2 and 6) (Lazorchak et al., 2006; Stadhouders et al., 2014). Despite the very low levels of IRF4 expression in pro-B cells (Stadhouders et al., 2014), ChIP-Seq experiments in RAG2-deficient pro-B cells revealed significant binding of IRF4 to the *Igκ* locus (Schwickert et al., 2014)—including the three core *Igκ* enhancers (Table 1; Fig. 3). Although it is currently unclear what the relevance of this early IRF4 recruitment to the *Igκ* locus in pro-B cells could be, it is consistent with the finding that a small fraction of pro-B cells (~4–7%) rearranges the *Igκ* locus before the assembly of a productive $V_HD_HJ_H$ joint (Novobrantseva et al., 1999).

4.3 The ETS-Family Members PU.1, Spi-B, and Spi-C

The ETS-family PU.1 TF is required for the development of a broad range of hematopoietic cells, including B lymphocytes (Scott, Simon, Anastasi, & Singh, 1994). Interestingly, the first PU.1 target described in mouse myeloma cells was the 3′Eκ (Pongubala et al., 1992), and PU.1 binding was later observed at V_κ promoters in B cell lines (Schwarzenbach, Newell, & Matthias, 1995). These findings suggested that PU.1 plays a role in pre-B cell development and *Igκ* recombination. However, hematopoietic defects in PU.1-deficient mice already occur at the multipotent progenitor cell stage (Scott et al., 1997). For example, PU.1 is required for IL-7R expression (DeKoter, Lee, & Singh, 2002), which is essential for the survival and maturation of lymphoid progenitors *in vivo* (Peschon et al., 1994). PU.1 was suggested to act directly upstream of the B cell specification factor EBF1, because ectopic expression of EBF1 overcomes the B cell lineage arrest in PU.1-deficient progenitors (Medina et al., 2004). These early maturation defects precluded a direct dissection of PU.1 function in early B cell development *in vivo* using conventional knockout mice.

The generation of a conditional PU.1 allele by several laboratories provided the first *in vivo* models to study PU.1 function in pro- and pre-B cells (Iwasaki et al., 2005; Polli et al., 2005; Ye, Ermakova, & Graf, 2005). Surprisingly, deleting PU.1 at the late pro-B cell stage using a Cd19-Cre transgene allowed for relatively normal B cell development, although one study did observe a reduction in pro/pre-B cells (Ye et al., 2005). Additionally, cultured PU.1 knockout fetal liver hematopoietic progenitors could give rise to B cells with rearranged *IgH* loci that produced μ heavy-chain protein, albeit slow and inefficient (Ye et al., 2005). Alternative genetic models employing a hypomorphic PU.1 allele did report developmental blockades at the early pro-B cell stage (Houston, Kamath, Schweitzer, Chlon, & DeKoter, 2007; Rosenbauer et al., 2006).

The importance of PU.1 for early B cell development thus remains controversial. *In vivo* studies have clearly shown that PU.1 plays a critical role in the specification and survival of early lymphoid progenitors. However, after lymphoid progenitors have committed to the B cell fate, the importance of PU.1 seems to diminish—suggesting that PU.1 does not play a leading role in rearrangement of the *Igκ* locus. Nevertheless, evidence from *in vitro* studies continues to implicate PU.1 as an activator of *Igκ* recombination via 3′Eκ (Hodawadekar, Park, Farrar, & Atchison, 2012; McDevit, Perkins, Atchison, & Nikolajczyk, 2005; Schweitzer & DeKoter, 2004). When examining recently published PU.1 ChIP-Seq data from pro-B cells (Schwickert et al., 2014; Whyte et al., 2013), we indeed observed widespread binding of PU.1 to the *Igκ* locus, including all *Igκ* regulatory elements and many V_{κ} promoters (Table 1; Fig. 3). Such dedicated recruitment immediately implies a functional role for PU.1 in *Igκ* locus biology—at least at the pro-B cell stage. Locus-wide binding data from pre-B cells are currently not available, but ChIP assays in primary B cells indicated that PU.1 binding to 3′Eκ increases during the pro-B to pre-B cell transition and *Igκ* locus activation (McDevit et al., 2005). A possible confounding factor in dissecting PU.1 function in early B cell progenitors is the role of the closely related ETS-family factors Spi-B and Spi-C (Bemark, Martensson, Liberg, & Leanderson, 1999; Ray et al., 1992). Functional redundancy between PU.1 and Spi-B has been suggested, although they also perform distinct functions (Dahl, Ramirez-Bergeron, Rao, & Simon, 2002; Garrett-Sinha, Dahl, Rao, Barton, & Simon, 2001; Hu et al., 2001). Importantly, both PU.1 and Spi-B have the capacity to recruit IRF4 and IRF8 to ETS-IRF composite elements present in 3′Eκ and in the *Igλ* enhancers (Eisenbeis et al., 1995; Escalante et al., 2002). Moreover, PU.1 binding to 3′Eκ was also shown to increase binding of other TFs, including E2A and PAX5 (Hodawadekar et al., 2012). Detailed molecular studies of PU.1 function at the pro- and pre-B cell stage, preferentially involving Spi-B and Spi-C, are therefore required to unravel the significance of PU.1 binding at the *Igκ* locus.

4.4 The B Cell Commitment Factor PAX5

Paired box protein PAX5 is a key TF involved in B cell development and expressed from the onset of B lymphopoiesis up to mature B cells (Adams et al., 1992; Barberis, Widenhorn, Vitelli, & Busslinger, 1990; Fuxa & Busslinger, 2007). In PAX5-deficient mice, B cell development is arrested at an early pro-B cell stage, characterized by the expression of B220 together with the early markers c-kit, CD43, λ5, VpreB, and HSA, and the absence of the later markers CD25 and BP-1 (Nutt, Urbanek, Rolink, & Busslinger, 1997; Urbanek, Wang, Fetka, Wagner, & Busslinger, 1994). In PAX5-deficient embryos however, B lymphopoiesis is arrested even before the appearance of $B220^{+}$ progenitors in the fetal liver, indicating a differential requirement of PAX5 in fetal and adult B cell development (Nutt et al., 1997). Although PAX5-deficient pro-B cells express several B cell-specific genes, they are not yet restricted in their lineage fate (Heavey, Charalambous, Cobaleda, & Busslinger, 2003; Hoflinger et al., 2004; Nutt, Heavey, Rolink, & Busslinger, 1999; Nutt, Morrison, Dorfler, Rolink, & Busslinger, 1998; Rolink, Nutt, Melchers, & Busslinger, 1999; Schaniel, Bruno, Melchers, & Rolink, 2002; Schaniel, Gottar, Roosnek, Melchers, & Rolink, 2002). The ability of PAX5-deficient hematopoietic progenitors to differentiate into various lineages demonstrates the critical role of PAX5 in B-cell lineage commitment (Fig. 1). This is accomplished through activating the transcription of B cell-specific genes while simultaneously repressing transcription of lineage-inappropriate genes (Delogu et al., 2006; Holmes, Carotta, Corcoran, & Nutt, 2006; Nutt et al., 1999, 1998; Schebesta et al., 2007; Souabni, Cobaleda, Schebesta, & Busslinger, 2002).

PAX5 was first implicated in the control of V(D)J rearrangement at the *IgH* locus, as PAX5-deficient pro-B cells showed a ~100-fold lower recombination efficiency of the distal V_HJ558 gene family compared to wild-type pro-B cells, despite normal D_H-to-J_H recombination (Nutt et al., 1997). Subsequent analyses of $PAX5^{-/-}$ pro-B cells revealed efficient V_H-to-D_HJ_H recombination of the most proximal V_H7183 genes, as well as normal accessibility and GLT of proximal and distal V_H genes (Hesslein et al., 2003). 3D-FISH studies showed that in PAX5-deficient pro-B cells, the *IgH* locus is in an extended configuration, thus preventing the distal V_H genes from physically contacting and recombining with the proximal D_HJ_H domain (Fuxa et al., 2004). Although these studies demonstrate that PAX5 promotes *IgH* locus contraction, only more recently a mechanism was provided. Fourteen PAX5-activated intergenic repeat (PAIR) elements were identified in the distal V_H gene region by active chromatin mark profiling (Ebert et al., 2011). PAIR elements bound by PAX5 and E2A, the ubiquitous proteins YY1, and the architectural CTCF/cohesin complexes carry PAX5-dependent active chromatin and give rise to PAX5-regulated noncoding antisense RNA exclusively in pro-B cells (Ebert et al., 2011; Medvedovic et al., 2013). These PAIR elements have been suggested to participate in the looping of distal V_H genes in pro-B cells (Ebert et al., 2011), which has to some extent been supported by 3C and 4C-seq experiments (Medvedovic et al., 2013; Verma-Gaur et al., 2012). In addition, PAX5 has also been implicated in other aspects of *IgH* chromatin regulation, such as the removal of repressive H3K9me2 modifications (Johnson et al., 2004) or recruitment of the RAG complex to V_H genes (Zhang et al., 2006).

Taken together, these studies have provided extensive knowledge on the role of PAX5 in *IgH* gene rearrangement, but only few reports have focused on how PAX5 regulates *Igκ* gene rearrangement. As described in Section 2.2, two tandem PAX5-binding sites upstream of the proximal κ^0 promoter named KI and KII play a crucial role in *Igκ* locus rearrangement (Ferradini et al., 1996; Tian et al., 1997; Weaver & Baltimore, 1987). Specific mutations of the KI and KII sites reduced the frequency of V_κ-to-J_κ rearrangement in mice to 1/10 of the normal level, although *Igκ* GLT was not affected. Binding sites for PAX5 have also been identified in 3′Eκ (Maitra & Atchison, 2000; Roque, Smith, & Blasquez, 1996; Shaffer, Peng, & Schlissel, 1997) (Table 1; Fig. 3). Interestingly, PAX5 dissociated from the KI, KII, and 3′Eκ during *in vitro* differentiation from pre-B to immature B cells induced by IL-7 withdrawal (Sato et al., 2001; Shaffer et al., 1997). Because PAX5 is released from its binding sites when the *Igκ* locus is activated, it may be concluded PAX5 binding to the *Igκ* locus might play a repressive role (Maitra & Atchison, 2000; Sato et al., 2001). Nevertheless, PAX5 is clearly necessary for the activation of the *Igκ* locus: PAX5-deficient IgH μ^+ pre-B cells fail to initiate *Igκ* GLT and gene rearrangements upon IL-7 withdrawal (Sato, Saito-Ohara, Inazawa, & Kudo, 2004). Furthermore, 3D-FISH analyses showed that in the absence of PAX5, the *Igκ* locus is preferentially located at the nuclear periphery (Sato et al., 2004). These findings do not necessarily imply a direct role for PAX5 in *Igκ* locus activation for gene rearrangement, because it is very well possible that these defects in *Igκ* recombination result from a role for PAX5 in the transcriptional regulation of pre-BCR components (Nutt et al., 1998; Revilla et al., 2012; Schebesta et al., 2002; Thevenin, Nutt, & Busslinger, 1998) and the activation of the key pre-BCR adaptor protein SLP65 (Ochiai et al., 2012) and TFs induced by pre-BCR signaling, including Aiolos, IRF4, and IRF8 (Pridans et al., 2008; Schebesta et al., 2007). Recently, a more direct role for PAX5 in *Igκ* locus activation was proposed, because PAX5 was found to bind a region upstream of the iEκ/MAR in pre-B cells (Levin-Klein, Kirillov, Rosenbluh, Cedar, & Bergman, 2014). This region has been denominated Dm for "demethylation element" and is necessary for DNA demethylation of a pre-rearranged *Igκ* transgene. However, deletion of the Dm element in the endogenous *Igκ* locus did not affect the demethylation process, possibly due to redundancy with other enhancers in the *Igκ* locus. Additional evidence for a direct role for PAX5 in mediating *Igκ* rearrangement comes from recent ChIP-seq studies, which have shown that PAX5 is recruited to the *Igκ* in primary pro-B cells and especially in IRF4,8-deficient cycling pre-B cells, where robust PAX5 binding was observed at all known insulator and enhancer elements (Table 1; Fig. 3). However, as IRF4,8-deficient pre-B cells cannot initiate *Igκ* rearrangement (Lu et al., 2003), the presence of PAX5 by itself is not sufficient to induce recombination—perhaps because of the inability to activate the IRF factors (Pridans et al., 2008; Schebesta et al., 2007).

In conclusion, PAX5 functions as a master regulator of B cell lineage specification and as a crucial factor for *IgH* locus rearrangement. Thus far however, its involvement in almost every aspect of B cell development has complicated unraveling the *in vivo* mechanistic role of PAX5 in *Igκ* locus recombination. Additional analysis involving, for example, the mutagenesis of specific PAX5-binding *cis*-regulatory elements will be crucial to elucidate the molecular mechanism of PAX5 action at the *Igκ* locus.

4.5 The Basic Helix–Loop–Helix Factor E2A

The E2A proteins comprise two highly related basic helix–loop–helix TFs, named E12 and E47, which differ in their DNA-binding region. E12 and E47 arise through alternative splicing of the *Tcfe2a* gene. E2A is a major transcriptional determinant that acts in concert with the B cell specification and commitment factors EBF1 (see Section 4.6) and PAX5, as part of a global network of TFs necessary for the initiation of the B cell lineage transcriptional program at the earliest stages of B lymphopoiesis (Lin et al., 2010; O'Riordan & Grosschedl, 1999) (Fig. 1). Ablation of the *Tcfe2a* gene in mice completely blocks B cell development at the pre-pro-B cell stage before the onset of D_H–J_H rearrangement in the *IgH* locus (Bain et al., 1994; Zhuang, Soriano, & Weintraub, 1994). *IgH* gene rearrangements and *Cd19*, *Mb-1*, *Igll1*, and *Pax5* transcripts are not detectable in bone marrow cells derived from E2A-deficient mice (Bain et al., 1997).

E2A is a key regulator of *Igκ* gene rearrangements, consistent with the original identification of E2A as a DNA-binding entity interacting with iEκ (Murre, McCaw, & Baltimore, 1989). In fact, already in pro-B cells, E2A binds to the three *Igκ* enhancers, as well as to the κRE1 element and many $V_κ$ genes (Table 1; Fig. 3). Targeted mutation of the two E2A-binding sites in iEκ affected *Igκ* gene rearrangement as severely as deletion of the entire enhancer, providing direct functional evidence for the critical importance of E2A proteins in mediating iEκ activity (Inlay et al., 2004). Moreover, E2A proteins have also been shown to directly bind to E-box sites located in 3′Eκ (Greenbaum & Zhuang, 2002) and to $V_κ$ gene promoters that contain E-boxes in close proximity to RSSs (Bemark, Liberg, & Leanderson, 1998; Brekke & Garrard, 2004; Schable & Zachau, 1993). In fact, the distribution of E2A occupancy across the $V_κ$ region is not random and most E2A occupancy is observed close to the Vκ gene segments (particular within ~200 bp), either at their 5′- or 3′-end (Bossen et al., 2012; Brekke & Garrard, 2004; Lin et al., 2010). In fact, a significant proportion (~37%) of all functional $V_κ$ genes display E2A occupancy in close proximity. We found that the presence of an E2A-binding site strongly correlated with the usage of $V_κ$ genes in the B cell repertoire and with the frequency of long-range chromatin interactions with the iEκ and 3′Eκ elements (Stadhouders et al., 2014). Ectopic expression of E2A proteins in conjunction with RAG1 and RAG2 readily promotes *Igκ* GLT and V_{κ}-to-J_{κ} gene rearrangement in nonlymphoid cells (Romanow et al., 2000). Furthermore, in E2A-deficient pre-B cells, the ability to undergo V_{κ}-to-J_{κ} gene rearrangement is completely abolished (Goebel et al., 2001; Kwon et al., 2008; Lazorchak et al., 2006). Interestingly, expression of E47 is sufficient to rescue *Igκ* GLT and rearrangement, suggesting that E47 is the critical component in promoting *Igκ* recombination (Beck, Peak, Ota, Nemazee, & Murre, 2009; Lazorchak et al., 2006). Finally, secondary rearrangements and receptor editing are severely impaired in immature B cells of E2A-heterozygous mice, which is consistent with E2A functioning to promote *Igκ* locus activation (Quong et al., 2004). E2A may play a role in regulating *Igκ* locus transcription and chromatin accessibility to recombination by promoting histone acetylation through direct recruitment of histone acetyltransferase complexes (Bradney et al., 2003; Lazorchak et al., 2006; Massari et al., 1999; Sakamoto et al., 2012). Moreover, proper activation of the *Igκ* locus is in part dependent on the synergistic interaction between IRF4 and E2A, which is best exemplified by IRF4-mediated recruitment of E2A to 3′Eκ in pre-B cells (Lazorchak et al., 2006; Nagulapalli & Atchison, 1998).

In summary, E2A proteins clearly play key roles in controlling *Igκ* gene rearrangement, apparently by regulating chromatin accessibility and *Igκ* locus 3D chromatin topology. Although prime candidates, it remains to be determined whether E2A proteins are directly involved in instigating *Igκ* locus contraction already at the pro-B cell stage, either alone or in combination with other TFs such as IRF4, CTCF, or PU.1.

4.6 Early B Cell Factor 1

EBF1 is one of the key factors determining B-cell lineage fate, together with E2A and PAX5 (Fig. 1). EBF1-deficient mice display a complete block of B lymphopoiesis at the pre-pro-B cell stage (Lin & Grosschedl, 1995), a similar phenotype to the one observed in E2A-deficient mice (Bain et al., 1994; Zhuang et al., 1994). Ectopic expression of EBF1 can rescue the developmental block in E2A-deficient hematopoietic progenitor cells to a large extent, suggesting that one of the main functions of E2A is to initiate *Ebf1* transcription in early lymphoid progenitors (Seet, Brumbaugh, & Kee, 2004). In addition to E2A, other early B cell factors such as Ikaros, PU.1, PAX5, and the RUNX1/CBF-β complex are also involved in the activation of the *Ebf1* promoter (Dias, Silva, Cumano, & Vieira, 2005; Kikuchi, Lai, Hsu, & Kondo, 2005; Roessler et al., 2007; Smith, Gisler, & Sigvardsson, 2002; Timblin & Schlissel, 2013). Once established, *Ebf1* expression is enforced by a reciprocal positive feedback loop with E2A, FOXO1, and IL-7R signaling as central players (Dengler et al., 2008; Kee & Murre, 1998; Mansson et al., 2012; Welinder et al., 2011). In addition, EBF1 can also positively regulate its own transcription through direct binding to its own promoter or by activation of *Pax5* via binding to an intergenic enhancer in the *Pax5* locus (Decker et al., 2009; Roessler et al., 2007).

Recent genome-wide ChIP-Seq studies in pro-B cells identified several EBF1 target genes, mainly involved in determining B cell identity or in the (pre-)BCR signaling cascade (Lin et al., 2010; Treiber et al., 2010). Interestingly, many EBF1-regulated genes also contain binding sites for E2A, a synergy that has been observed before for the *Igll1, VpreB*, and *Cd79a* genes (Gisler & Sigvardsson, 2002; O'Riordan & Grosschedl, 1999; Sigvardsson, 2000; Sigvardsson, O'Riordan, & Grosschedl, 1997). Besides its role in B cell specification, EBF1 has also been implicated in B-cell commitment, mainly because *Ebf1*-deficient lymphoid progenitors can give rise to several other hematopoietic lineages (Pongubala et al., 2008). Importantly, EBF1 restricts alternative lineage potential independently of its target gene *Pax5* (Banerjee, Northrup, Boukarabila, Jacobsen, & Allman, 2013; Pongubala et al., 2008) and is also essential for the maintenance of B-cell identity in committed pro-B cells (Nechanitzky et al., 2013). EBF1 initiates gene activation or repression by inducing a diverse array of epigenetic changes, e.g., including the activating histone mark H3K4me2 (Gyory et al., 2012; Treiber et al., 2010), DNA demethylation at the *Cd79a* promoter (Maier et al., 2004), and the repressive histone mark H3K27me3 at non-B lineage genes (Banerjee et al., 2013; Treiber et al., 2010).

As for so many other early B cell TFs, studying the role of EBF1 in *Ig* rearrangements has been hampered by the pronounced developmental block present in conventional knockout mice. EBF1-deficient lymphoid progenitors arrest at the pre-pro-B cell stage before the onset of D_H-to-J_H recombination (Lin & Grosschedl, 1995). EBF1-deficient cells cultured *in vitro* show D_H-to-J_H but not V_H-to-D_HJ_H recombination (Pongubala et al., 2008). Like for E2A, ectopic expression of EBF1 in conjunction with RAG1 and RAG2 can induce D_H-to-J_H and V_κ-J_κ rearrangements in non-lymphoid cells (Goebel et al., 2001; Romanow et al., 2000).

Recent ChIP-seq studies have revealed that EBF1 already binds the *Igκ* locus in pro-B cells, although binding was not observed at *Igκ* regulatory elements and was mainly focused to the distal part of the V_κ region (Table 1; Fig. 3). This preference for the 5′-distal part is intriguing and could perhaps indicate that EBF1 is involved—together with E2A—in activating *Igκ* locus contraction in pro-B cells. It is currently unclear to which extent EBF1 binds the *Igκ* locus in pre-B cells. Further evidence for a role of EBF1 in *Igκ* recombination came from the identification of potential EBF1-binding sites within or near V_κ genes, in the introns of $V_\kappa4$ genes, and in a subset of V_κ promoter regions (Bemark et al., 1998; Brekke & Garrard, 2004; Hagman, Belanger, Travis, Turck, & Grosschedl, 1993; Sigvardsson, Akerblad, & Leanderson, 1996; Sigvardsson, Bemark, & Leanderson, 1995). EBF1-binding sites at V_κ promoters have been linked to increased transcription activity (Sigvardsson et al., 1996, 1995) or V_κ GLT in nonlymphoid cells upon ectopic EBF1 expression (Goebel et al., 2001).

At present, it is difficult to ascertain the exact role of EBF1 in V(D)J recombination, because it also controls—probably in a dose-dependent fashion (Lukin et al., 2011)—the expression of important modulators of *Ig* gene rearrangement, including, e.g., RAG1, RAG2, and FOXO1 (Zandi et al., 2008). It has been suggested that low levels of EBF1 at early stages of B cell development drive B cell specification, activate *Foxo1* and *Rag* gene expression, and initiate *IgH* rearrangement. Enhanced IL-7R signaling and concomitant proliferation at the late pro-B or large pre-B cell stage significantly upregulate the expression of EBF1. High EBF1 levels, together with c-Myb, strongly repress *Rag* transcription in cycling cells through negative regulation of FOXO1 and positive regulation of the TF GFI-1B (Timblin & Schlissel, 2013). GFI-1B is both a negative regulator of *Rag* transcription and a FOXO1 antagonist at the *Rag* locus (Schulz et al., 2012). Attenuation of IL-7R signaling and differentiation to the small pre-B cell stage is accompanied by reduction of EBF1 expression and an upregulation of RAG levels (Timblin & Schlissel, 2013).

It is likely that EBF1 is involved in *Igκ* locus rearrangement through direct binding to the V_κ region, as ectopic expression experiments in nonlymphoid cells clearly demonstrate that EBF has the capacity to induce *Igκ* locus rearrangement. Hereby, the V_κ region was not opened globally, but some V_κ genes showed high levels of recombination, other V_κ families were not induced to rearrange (Goebel et al., 2001). As no correlation between recombination levels and the presence of EBF1-binding sites in V_κ promoters was observed, EBF1-binding does not appear to act by local induction of GLTs, but may rather act by promoting chromatin remodeling. In line with this notion, it has been shown that EBF1 interacts, e.g., with the coactivator p300/CBP, which acts as a bridge between TFs and the SWI/SNF chromatin-remodeling complex (Agalioti et al., 2000; Gao et al., 2009; Zhao, McCarrick-Walmsley, Akerblad, Sigvardsson, & Kadesch, 2003).

4.7 FOXO Factors

TFs of the FOXO subfamily are the ubiquitously expressed downstream targets of the PI3K–AKT/PKB signaling pathway (Zhang, Tang, Hadden, & Rishi, 2011). The AKT kinase phosphorylates FOXO TFs such as FOXO1 and FOXO3a, which as a consequence are transported out of the nucleus to be retained or degraded in the cytoplasm (Amin & Schlissel, 2008; Biggs, Meisenhelder, Hunter, Cavenee, & Arden, 1999; Brunet et al., 1999). As such, high levels of AKT activity prevent transcriptional regulation by FOXO TFs. During B cell development, IL-7R signaling activates the PI3K–AKT pathway to stimulate cell proliferation (Corcoran et al., 1996; Venkitaraman & Cowling, 1994). Attenuation of IL-7R signaling during pre-B cell differentiation diminishes PI3K–AKT pathway activity, leading to cell cycle exit, and the induction of *IgL* rearrangements (reviewed in Clark, Mandal, Ochiai, & Singh, 2014; Herzog et al., 2009). Because of the important function of PI3K–AKT signaling, FOXO TFs were expected to play important roles during B cell differentiation.

Conventional FOXO1 knockout mice die during midgestation because of vascular defects (Hosaka et al., 2004). However, conditional deletion strategies to specifically deplete FOXO1 in hematopoietic progenitor or B cells revealed a critical role in specifying and enforcing B cell lineage commitment and early B cell differentiation (Dengler et al., 2008; Mansson et al., 2012). FOXO1 was shown to activate the expression of the B cell specification factor EBF1 in CLPs to synergistically induce and maintain the pro-B cell transcriptional program together with EBF1 and E2A (Lin et al., 2010; Mansson et al., 2012). *Foxo1* gene expression is strongly upregulated at the early pro-B cell stage (Rumfelt, Zhou, Rowley, Shinton, & Hardy, 2006) and ablation of *Foxo1* at this stage essentially prevents differentiation to the small pre-B cell stage (Dengler et al., 2008). FOXO1-deficient pro-B cells exhibited severe proliferation defects, which were attributed to a failure to upregulate IL-7R expression (Dengler et al., 2008). Importantly, FOXO1 was found to directly activate *Rag* gene transcription and FOXO1-deficient pro-B cells expressed aberrantly low levels of RAG proteins, leading to impaired *IgH* rearrangements (Amin & Schlissel, 2008; Dengler et al.,

2008). Later deletion of *Foxo1* resulted in a developmental block at the pre-B cell stage, likely through reduced *Rag* gene expression (Dengler et al., 2008; Ochiai et al., 2012). Importantly, FOXO1, along with PAX5, is required for the activation of the critical pre-BCR signaling component SLP65 (Ochiai et al., 2012), which in turn activates FOXO TFs in pre-B cells by inactivating AKT (Herzog et al., 2008). Of note, several other FOXO TFs have also been implicated in regulating early B cell development (i.e., FOXP1 (Hu et al., 2006) and FOXO3a (Herzog et al., 2008; Ochiai et al., 2012)), and a possible functional redundancy between FOXO1 and FOXO3a has been previously suggested (Herzog et al., 2008; Ochiai et al., 2012).

In summary, FOXO1 plays an essential role in early B cell development as a downstream target of PI3K–AKT signaling. As in pre-B cells IL-7-driven proliferation ceases, the PI3K–AKT pathway is inactivated, allowing FOXO1 to accumulate in the nucleus, which (i) drives *Rag* transcription, (ii) stimulates SLP65-mediated pre-BCR signaling, and (iii) induces *IgL* chain recombination (Clark et al., 2014). However, whether FOXO1 directly acts on the *Igκ* locus to regulate recombination is not clear. Transduction of FOXO1 cDNA into IRF4,8-deficient cycling pre-B cells activates *Rag* transcription but not κ^0 GLT (Ochiai et al., 2012), suggesting that FOXO1 alone cannot induce *Igκ* locus accessibility in the absence of IRF4. Nevertheless, examining published FOXO1 ChIP-Seq data from pro- and pre-B cell populations does implicate FOXO1 in mediating *Igκ* rearrangements in a direct manner (Table 1; Fig. 3). FOXO1 binding across the *Igκ* locus was only observed in pre-B cells, where most *Igκ* regulatory elements and many Vκ promoters were occupied. Additional experiments are necessary to elucidate a potential direct role for FOXO1 in mediating *Igκ* recombination. As FOXO1 and FOXO3a binding to the pre-B cell genome largely overlap (Ochiai et al., 2012), such studies will have to take possible overlapping functions by these FOXO family members into account.

4.8 Other Factors Implicated in *Igκ* Recombination

The TFs discussed above and displayed in Table 1 and Fig. 3 do not represent an exhaustive list. Several other factors have been implicated in directly regulating *Igκ* recombination, which we will briefly discuss below.

4.8.1 Nuclear Factor κB

NF-κB has long since been associated with the regulating of *Igκ* expression. In fact, NF-κB was discovered as a TF binding to a short sequence (called the κB site) in iEκ (Sen & Baltimore, 1986b). Further supporting the notion that NF-κB is an essential regulator of the *Igκ* locus were the observations that the induction of NF-κB DNA binding correlated with *Igκ* transcription and rearrangement, and that NF-κB inactivation blocked both processes (reviewed by Sen, 2004). However, the importance of NF-κB for the activation of V_κ-to-J_κ recombination was challenged by the generation of a targeted mouse specifically lacking the κB site in the iEκ element (Inlay et al., 2004). *Igκ* locus transcription and recombination were unaffected in these κB-deficient mice. Considering that NF-κB binds specifically to iEκ and that E2A binding to this enhancer is responsible for its full activation (Inlay et al., 2004), these findings show that NF-κB is not essential for *Igκ* locus recombination. Although *Igλ* gene rearrangements are shown to depend on NF-κB signals (Derudder et al., 2009), this requirement could be rescued by Bcl-2 overexpression, supporting a role of NF-κB in survival rather than *Igλ* locus accessibility. Similar to NF-κB, the HIVEP1 (also called MBP-1) protein is able to bind the iEκ κB motif *in vitro* (Baldwin, LeClair, Singh, & Sharp, 1990). Little is known about the cellular function of the HIVEP TF family (Hicar, Liu, Allen, & Wu, 2001) and no function for HIVEP1 in B cell development has been reported. Interestingly, *Hivep1* expression is induced by pre-BCR signaling (Stadhouders et al., 2014), making it a promising candidate to follow up on.

4.8.2 OCA-B/OCT Factors

OCA-B (also known as OBF1 or BOB1, encoded by the *Pou2af1* gene) is a transcriptional coactivator predominantly expressed in B cells. It forms a ternary complex with the TFs OCT1 and OCT2, enhancing their DNA-binding affinity to conserved octamer motifs present in *Ig* gene promoters and enhancers, as well as other target genes (Gstaiger, Knoepfel, Georgiev, Schaffner, & Hovens, 1995; Luo, Fujii, Gerster, & Roeder, 1992; Pierani, Heguy, Fujii, & Roeder, 1990; Strubin, Newell, & Matthias, 1995). It has been proposed that OCA-B, together with ubiquitously expressed OCT1 and B-cell-restricted OCT2 proteins, regulates *Ig* gene transcription levels by specific interactions with the TATA box-associated basal transcription machinery. However, mice deficient for OCA-B showed normal B cell development to the IgM^+ immature B cell stage, indicating that *Ig* gene transcription is essentially unaffected (Kim et al., 1996; Nielsen, Georgiev, Lorenz, & Schaffner, 1996; Schubart, Rolink, Kosco-Vilbois, Botteri, & Matthias, 1996). Nevertheless, *Ocab*$^{-/-}$ mice have reduced numbers of immature and mature B cell subsets and an absence of germinal center formation. Similar phenotypes were observed in OCT2-deficient (Corcoran & Karvelas, 1994; Corcoran et al., 1993) or OCA-B/OCT2-double-deficient mice (Schubart et al., 2001). In contrast, OCT1 is dispensable for B cell development and *Ig* transcription (Wang, Tantin, Chen, & Sharp, 2004).

Subsequent studies demonstrated that the role of OCA-B in *Ig* gene expression is more specific than initially anticipated. It was shown that not all octamer motifs in *Ig* gene promoters and enhancers depend on OCA-B and that OCT factors can dimerize on octamer elements in a way that permits or precludes OCA-B interaction (Remenyi et al., 2001; Tomilin et al., 2000). Octamer motifs in V_H promoters are refractory to OCA-B and $OCA\text{-}B^{-/-}$ mice express a normal V_H repertoire (Schubart, Rolink, Schubart, & Matthias, 2000). However, OCA-B deficiency has an impact on the transcription of IgH class switched isotypes, at least in part as a result of its activity at the 3′RR enhancer of the *IgH* locus (Stevens, Ong, Kim, Eckhardt, & Roeder, 2000).

In contrast to V_H promoters, OCA-B interacts with all V_κ octamers, but the extent to which V_κ promoters depend on OCA-B varies between promoters. As a consequence of OCA-B requirement for rearrangement of a subset of V_κ genes, $Ocab^{-/-}$ mice show a skewed V_κ repertoire (Casellas et al., 2002). In OCA-B-deficient pre-B cells, the *Igκ* locus chromatin domain is accessible, as measured by DNA demethylation and histone acetylation, and OCT1 and the basal transcriptional machinery are recruited. *Igκ* genes are transcribed, albeit at suboptimal levels, suggesting that OCA-B controls antibody gene recombination by regulating the efficiency of V_κ gene expression. Therefore, in the absence of OCA-B, weak V_κ promoters cannot acquire enough transcriptional activity to ensure efficient recombination, in comparison to stronger V_κ promoters occupied by additional transcription regulators.

Finally, at the pre-B cell stage, OCA-B is also involved in regulating the expression of other crucial B cell genes. In particular, OCA-B together with Aiolos is required for the downregulation of the SLC genes *Igll1* and *VpreB* (Karnowski et al., 2008) and OCA-B overexpression results in upregulation of the E2A inhibitor ID2 in pre-B cells (Bordon et al., 2008).

4.8.3 STAT5

The two highly related isoforms STAT5A and STAT5B—commonly referred to as STAT5—critically control the survival of pro-B cells (Malin et al., 2010). Complete STAT5 inactivation causes severely impaired lymphoid development with cells accumulating at the pre-pro-B cell stage (Yao et al., 2006). Upon IL-7R signaling in pro-B cells, cytoplasmic STAT5 is phosphorylated, resulting in the formation of STAT5 dimers with TF activity. Importantly, activation of STAT5 in pro-B cells inhibits *Igκ* locus transcription and hampers E2A binding to iEκ (Mandal et al., 2011). In particular, STAT5 binds as a tetramer to iEκ and recruits the PRC2 protein enhancer of zeste homolog 2 (EZH2), which then decorates the iEκ element and the J_{κ} cluster with repressive H3K27me3 marks. As a result, *Igκ* locus GLT is silenced during IL-7R-mediated pro-B and pre-B cell proliferation. Subsequently, pre-BCR signaling results in the downregulation of IL-7R signaling and STAT5 activity, leading to a loss of H3K27me3 marks and the concomitant activation of *Igκ* locus rearrangement. It has been shown that a loss of STAT5 and IL-7R signaling is sufficient to induce GLTs and *Igκ* gene rearrangement in pro-B cells in which survival is rescued by BCL-2 (Malin et al., 2010). Attenuation of IL-7R signaling results in increased binding of E2A to the iEκ enhancer (Johnson et al., 2008), accounting for an active state of iEκ. In addition, an approximately fourfold increase in histone H4 acetylation at iEκ was found upon attenuation of IL-7R signaling. The STAT5-repressive mechanism extends to 3′Eκ, because STAT5 was also found to inhibit 3′Eκ activity through the same motif recognized by PU.1 (Hodawadekar et al., 2012). STAT5 has the capacity to displace PU.1 at 3′Eκ, whereby loss of PU.1 also results in loss of IRF4 occupancy to its adjacent site. However, strong IRF4 expression can override the repressive activity of STAT5.

Taken together, these findings demonstrate that STAT5 acts as a molecular switch that controls TF occupancy at two *Igκ* enhancers: in pro-B cells STAT5 reduces E2A and PU.1/IRF4 binding to iEκ and 3′Eκ, respectively. Subsequent pre-BCR signaling reduces STAT5 activity, promoting the activation of both enhancer elements (Hodawadekar et al., 2012; Mandal et al., 2011).

4.8.4 DNA Polymerases

The template-dependent DNA polymerases are enzymes involved in processing *Ig* junctions during V(D)J recombination and contribute to antibody diversity (Weill & Reynaud, 2008). Interestingly, both *IgH* and *IgL* loci employ a specific template-dependent polymerase: gene-targeting studies in mice revealed that *IgH* rearrangement involves polymerase λ (pol λ), while *IgL* rearrangement requires polymerase μ (pol μ) (Bertocci, De Smet, Berek, Weill, & Reynaud, 2003; Bertocci, De Smet, Weill, & Reynaud,

2006). In agreement with its stage-specific function, pol μ expression is strongly upregulated at the pro- to pre-B cell transition (Bertocci et al., 2006), and gene expression analysis of pre-BCR signaling-regulated transcription revealed a significant induction of pol μ levels by signals mediated by the pre-BCR (Stadhouders et al., 2014). Future investigations should address which factors mediate the tissue-specific expression patterns of pol μ and pol λ, as this is currently unknown.

5. GENERAL NUCLEAR FACTORS INVOLVED IN *Igκ* LOCUS TOPOLOGY

5.1 The Polycomb Group Protein YY1

Yin Yang 1 (YY1) is a ubiquitously expressed TF that has been shown to contribute to long-range chromatin interactions at the *IgH* locus (Liu et al., 2007; Medvedovic et al., 2013; Verma-Gaur et al., 2012). Conditional deletion of *Yy1* at the onset of B cell development using Mb1-Cre resulted in impaired distal V_H–DJ_H recombination and defective *IgH* locus contraction (Liu et al., 2007). It has been suggested that YY1 controls *IgH* locus contraction and V_H-DJ_H recombination through direct interaction with the Eμ enhancer (Liu et al., 2007), where it strongly binds the *IgH* locus (Church, Ephrussi, Gilbert, & Tonegawa, 1985; Ephrussi, Church, Tonegawa, & Gilbert, 1985; Liu et al., 2007). More recent studies have shown that YY1 also binds across the V_H gene region, mostly in the distal V_HJ558 region (Medvedovic et al., 2013; Verma-Gaur et al., 2012) and occasionally in association with PAX5, E2A, CTCF and Cohesin at PAIR elements (Ebert et al., 2011). Importantly, the noncoding antisense transcripts detected at PAIR elements depend on YY1 expression, and looping between the Eμ and PAIR elements or the 3′CBE region is severely impaired in YY1-deficient pro-B cells (Medvedovic et al., 2013; Verma-Gaur et al., 2012). Taken together, these studies indicate that YY1 has a direct role in *IgH* locus rearrangement by controlling antisense transcription and long-range interactions.

An interesting aspect of YY1 is that it functions as a Polycomb Group (PcG) protein and can recruit other PcG proteins to DNA, including the histone methyltransferase EZH2 (Atchison, Ghias, Wilkinson, Bonini, & Atchison, 2003; Caretti, Di Padova, Micales, Lyons, & Sartorelli, 2004; Satijn, Hamer, den Blaauwen, & Otte, 2001; Wilkinson, Park, & Atchison, 2006). Similar to YY1, EZH2 has been shown to have a critical role in early B cell development and loss of EZH2 results in impaired distal V_H–DJ_H

rearrangement and *IgH* locus contraction (Su et al., 2003). The PcG function of YY1 resides in a short 25 amino acid motif, which was called the REPO domain for its role in recruitment of Polycomb (Wilkinson et al., 2006). Deletion of this motif selectively ablates YY1 PcG function without affecting all other known YY1 functions, including DNA binding, transcriptional regulation, and interaction with HDAC proteins (Wilkinson et al., 2006). Recently, reconstitution experiments have been used to show that expression of the YY1 REPO domain mutant protein (YY1ΔREPO) in YY1-null B cells failed to rescue B-cell development (Pan et al., 2013). Interestingly, YY1 PcG function appears to be largely dispensable for *IgH* rearrangements. In contrast, YY1ΔREPO pre-B cells showed a dramatic skewing of the V_κ repertoire, suggesting that YY1 PcG function may control *Igκ* rearrangement. In YY1ΔREPO pre-B cells, only a small fraction of the V_κ genes were found to be rearranged and >40% of rearrangements occurred with the two most distal V_κ genes (V_κ2–137 and V_κ1–135). Other rearrangements were more or less spread evenly over de V_κ region. Many of the V_κ genes that are most frequently rearranged in wild-type mice were not detected in YY1ΔREPO cells and significant recombination occurred to V_κ genes that are normally underrepresented (Pan et al., 2013). These findings suggest that when the YY1-REPO domain is mutated, the 3D topology of the *Igκ* locus is dramatically different and most of the DNA loops required for rearrangement of a normal repertoire are lost. Nevertheless, recombination to the most distal V_κ genes remained possible and was even preferred, indicating that some long-range interactions were independent of the YY1 REPO domain and that locus contraction was maintained. These YY1-REPO independent loops may require other TFs, such as E2A, PAX5, or CTCF. In this context, it is interesting to note that in the absence of CTCF, many long-range interactions in the *Igκ* locus were preserved, but particularly interactions with the most distal part the V_κ region, containing V_κ2–137 and V_κ1–135, were decreased (see Section 5.2) (Ribeiro de Almeida et al., 2011). These two V_κ genes are often used in normal B cells, but not in CTCF-deficient B cells, despite that the levels of V_κ2–137 and V_κ1–135 GLTs were found to be unafffected (Ribeiro de Almeida et al., 2011). Thus, it is conceivable that in the absence of the YY1-REPO domain *Igκ* locus topology mainly relies on CTCF-mediated interactions. These findings would be consistent with a model in which CTCF and YY1 play complementary roles in mediating long-range chromatin interactions.

A role for YY1 past the pro-B cell stage has been anticipated from the observation that in Mb1-Cre *Yy1*$^{fl/fl}$ mice very few mature B cells are generated despite that proximal V_H–DJ_H recombination occurs (Liu et al., 2007). Similarly, the expression of a rearranged *IgH* only partially rescued the pro-B to pre-B block caused by loss of YY1 (Liu et al., 2007). At the *Igκ* locus, the YY1 REPO domain seems to be involved in the recruitment of EZH2 and Condensin proteins to numerous sites across the $V_κ$ gene region where YY1, EZH2, and Condensin proteins colocalize (Pan et al., 2013). In addition, YY1 has also been shown to bind to the 3′Eκ enhancer (Park & Atchison, 1991; Roque et al., 1996).

Whether YY1 is required for *Igκ* locus contraction similar to what has been observed at the *IgH* locus remains to be determined. However, YY1 has been proposed to have a direct effect on *Igκ* locus structure through the nucleation of PcG and Condensin proteins across the $V_κ$ gene region, thereby promoting—in a presently unclear fashion—the usage of a highly diverse $V_κ$ gene repertoire (Pan et al., 2013). Interestingly, both PcG and Condensin proteins have been implicated in regulating chromosomal interactions (Haeusler, Pratt-Hyatt, Good, Gipson, & Engelke, 2008; Tiwari et al., 2008). Nevertheless, the fact that expression of YY1ΔREPO does not result in changes in proximal versus distal $V_κ$ gene rearrangement might point toward a role for YY1 at the *Igκ* locus that is intrinsically distinct from that observed for the *IgH* locus.

5.2 CCCTC-Binding Factor

5.2.1 CTCF and Igκ Long-Range Interactions, the Sis and Cer Regions

CTCF is a ubiquitously expressed and highly conserved 11-zinc finger protein that regulates gene expression at developmentally regulated loci. Classically, CTCF has been regarded as the archetypal protein that binds insulator sequences. Insulators are genomic elements that affect gene expression by regulating the communication between neighboring regulatory elements, including enhancer and promoters, thereby acting as enhancer blockers or chromatin barriers (also see Section 2.2.3). CTCF is believed to do so through the formation of chromatin loop domains (reviewed in Ong & Corces, 2014; Ribeiro de Almeida, Stadhouders, Thongjuea, Soler, & Hendriks, 2012). One of the best-studied examples is the imprinted *H19/Igf2* locus, where CTCF-mediated chromatin loops provide enhancer-blocking insulation and, consequently, parent-of-origin-specific *H19* and *Igf2* gene expression patterns (Kurukuti et al., 2006; Murrell, Heeson, & Reik, 2004). In the mouse *β-globin* locus, CTCF-based intrachromosomal interactions have also been extensively reported (Palstra et al., 2003; Splinter et al., 2006; Tolhuis, Palstra, Splinter, Grosveld, &

de Laat, 2002), but the functional significance of such high-order chromatin structures remains unclear (Bender et al., 2006; Bender et al., 1998). Likewise, CTCF also appears to be directly involved in promoter–enhancer interactions. For instance, CTCF activates the transcription of the human *MHCII* locus and the mouse *IFNγ* locus in T-helper 1 cells by bringing promoters and enhancers into close proximity while cooperating with cell-type-specific TFs (Majumder, Gomez, Chadwick, & Boss, 2008; Sekimata et al., 2009). In addition to these well-studied loci, genome-wide studies have uncovered an extensive number of CTCF-binding sites in the human and mouse genomes (Barski et al., 2007; Chen et al., 2008; Kim et al., 2007). The discovery that a high proportion of CTCF-binding sites are shared with Cohesin proteins provided a mechanistic explanation for CTCF-mediated chromatin loop formation (Hadjur et al., 2009; Ong & Corces, 2014; Parelho et al., 2008; Rubio et al., 2008; Wendt et al., 2008).

A putative role for CTCF in *Ig* loci long-range interactions was initially highlighted by mapping CTCF- and cohesin-binding sites across the *IgH* and *Igκ* loci (Degner et al., 2009; Ribeiro de Almeida et al., 2011). In contrast to the *IgH* locus where CTCF occupancy appears similar in thymocytes, pro- and pre-B cells, the *Igκ* locus recruits CTCF in an apparent lineage and developmental stage-specific manner (Degner et al., 2009). Multiple sites for CTCF and cohesin were identified throughout the V_H and V_κ regions (Degner et al., 2009) implicating CTCF in *Ig* chromatin loop formation and effective distal V–(D)J recombination. In the proximal V_H region, CTCF-binding sites are located within close genomic proximity to V_H genes (Bossen et al., 2012; Lucas et al., 2011). In the distal part of the *IgH* locus, some CTCF-binding sites localize to the aforementioned PAIR elements, upstream of V_H3609 genes, and interspersed among V_HJ558 genes (Ebert et al., 2011). Immediately upstream of the CTCF-binding sites at PAIR elements, noncoding antisense transcription is initiated exclusively in pro-B cells and reduced CTCF levels lead to increased antisense transcription (Degner et al., 2009). It has been suggested that PAIR elements mediate looping of distal V_H genes in pro-B cells, which is supported by 3C and 4C-seq experiments to some extent (Medvedovic et al., 2013; Verma-Gaur et al., 2012). In addition, anti-CTCF ChIP-loop assays have provided evidence for CTCF-mediated multiloop domain formation in the 5′ V_HJ558/V_H3609 region containing the PAIR elements (Guo, Gerasimova, et al., 2011). However, functional characterization of the PAIR elements to address whether they are directly involved in *IgH* locus contraction requires additional experiments, which should focus on the deletion of these proposed regulatory elements.

Within the V_κ region, CTCF occupancy is located between rather than adjacent to V_κ gene segments (Fig. 4), implying that CTCF function in the *IgH* and *Igκ* loci might be functionally different (see below).

Interestingly, a recently described putative *Igκ cis*-regulatory element designated κRE1 with stage- and cell type-specific enhancer characteristics (Predeus et al., 2014) is bound by CTCF in pre-B cells (Fig. 3). It has been suggested that κRE1 might have a role in controlling the recombination potential of some mouse V_κ gene segments. However, κRE1 is located in a relatively low V_κ-usage region—the average usage of the five nearest V_κ genes up- and downstream is ~0.25% (Aoki-Ota et al., 2012). However, it remains possible that κRE1 acts over a distance. The concomitant binding of the κRE1 element by PAX5, E2A, YY1, and CTCF (Fig. 3), its location within the V_κ region, and the strong long-range interactions that we detected between both iEκ/3′Eκ and κRE1 using 3C-Seq analysis (Stadhouders et al., 2014; R.S., C.R.d.A., & R.W.H., unpublished findings) are reminiscent of the characteristics of the PAIR elements present in the *IgH* locus (Ebert et al., 2011). Whether the κRE1 element is involved in locus contraction or plays a role in the expression of GLTs initiating from more distal V_κ genes remains to be experimentally tested.

Using 3C-based experiments in B cells where CTCF protein levels were downregulated by shRNA-mediated CTCF depletion or Mb1-Cre-mediated conditional deletion of the *Ctcf* gene *in vivo*, we and others have shown that in the absence of CTCF *IgH* locus contraction is only modestly affected (Degner et al., 2009) and *Igκ* long-range chromatin interactions are largely kept (Ribeiro de Almeida et al., 2011). Importantly, detailed analysis of *Igκ* interactions in pre-B cells revealed a correlation between V_κ usage, the local density of CTCF-binding sites, and the frequencies of contacts between the V_κ region and the regulatory elements Sis, iEκ, and 3′Eκ (Ribeiro de Almeida et al., 2011). Loss of CTCF resulted in significantly increased interactions with the most proximal V_κ genes containing the $V_\kappa3$ family. In line with these findings, CTCF-deficient pre-B cells showed significantly increased $V_\kappa3$ GLTs and recombination (Ribeiro de Almeida et al., 2011). On the other hand, loss of CTCF resulted in reduced contacts with the two most distal V_κ genes, $V_\kappa2$–137 and $V_\kappa1$–135, which are often rearranged in control but not in CTCF-deficient pre-B cells. Similar findings were obtained when two CTCF-binding elements within the V_κ–J_κ intergenic region were deleted, either individually or in combination (Xiang et al., 2013, 2014, 2011). The first of these CTCF-binding elements was identified as a cluster of four DNase I-hypersensitive sites (DHS3–6) corresponding to the Sis region (see Section 2.2.3; Liu et al., 2002,

2006). Interestingly, deletion of Sis resulted in decreased distal and enhanced proximal V_κ gene usage, despite normal *Igκ* locus contraction (Xiang et al., 2011). More recently, another two DNase I-hypersensitive sites (DHS1–2) corresponding to the Cer element were identified in close proximity to the Sis region. Deletion of the Cer element in the V_κ–J_κ intergenic resulted in (i) a ~50% reduction in overall locus contraction, (ii) markedly increased proximal and greatly diminished upstream V_κ gene usage, and (iii) nonlineage-specific *Igκ* rearrangement in T cells (Xiang et al., 2013). Combined deletion of both Sis and Cer elements recapitulated the phenotypes of individual deletions and additionally resulted in a striking upregulation of proximal V_κ gene GLTs and, consequently, more pronounced usage of proximal V_κ genes in comparison with deletion of either Sis or Cer alone (Xiang et al., 2014).

5.2.2 CTCF in the Sis and Cer Regions: Parallels with IgH and TCRβ Loci

Remarkably, in the *IgH* locus two CTCF-binding sites in the V_H–D_H IGCR1 may share functional features with Cer and Sis in the *Igκ* locus. These CTCF-binding sites, named CBE1 and CBE2, show a cooperative, potentially synergistic, effect on *IgH* locus rearrangement, although not fully functionally equivalent (Lin, Guo, Su, Zhang, & Alt, 2015). Deletion of the IGCR1 or mutation of both CTCF-binding sites causes upregulated proximal V_H gene GLT and a marked increase in proximal V_H gene rearrangements, as well as nonlineage-specific $V_HD_HJ_H$ rearrangements in thymocytes (Giallourakis et al., 2010; Guo, Yoon, et al., 2011). In contrast to Cer or DHS1–6 deletion in the *Igκ* locus, IGCR1 deletion did not affect global *IgH* long-range interactions or locus contraction (Medvedovic et al., 2013). Nevertheless, the CTCF-bound IGCR1 region interacts with other CTCF-binding sites in the 3′-end of the *IgH* locus and in some V_H genes (Guo, Yoon, et al., 2011; Medvedovic et al., 2013). Based on these findings—and together with the evidence for enhancer-blocking activity of the V_H–D_H intergenic region (Featherstone, Wood, Bowen, & Corcoran, 2010)—it has been proposed that CTCF binding to the IGCR1 is involved in the formation of chromatin loop domains that control ordered and lineage-specific V(D)J recombination. Such loops would exclude transcriptional activation and rearrangement of proximal V_H genes by separating the V_H region from the downstream D_H–J_H–C_μ portion of the *IgH* locus, which contains strong transcriptional enhancers.

Likewise, in the *Igκ* locus, CTCF binding at the Cer and/or Sis regions may serve an enhancer-blocking insulator function to prevent the downstream enhancers from conferring active chromatin conformations to the proximal V_{κ} genes. In line with this view, and despite extensive CTCF binding throughout the *Igκ* locus, conditional *Ctcf* deletion in pre-B cells results in increased GLT and rearrangement of proximal V_{κ} genes, similar to Sis or Cer deletion. In contrast, *Igκ* locus contraction is not impaired in Mb1-Cre $Ctcf^{lox/lox}$ pre-B cells (Ribeiro de Almeida et al., 2011). This suggests that other Cer-binding proteins than CTCF are involved in *Igκ* locus contraction. In this regard, Cer also recruits several other TFs, including E2A, PU.1, YY1, and IRF4, according to published pro-B cell ChIP-Seq data (Fig. 3). Alternatively, it cannot be excluded that Mb1-Cre-mediated *Ctcf* gene deletion is not complete when Cer-mediated *Igκ* locus contraction is established, which is thought to occur at the pro-B cell stage (Lin et al., 2010; Stadhouders et al., 2014).

Interestingly, recent 3C and 3D-FISH studies of the *TCRβ* locus have shed light on the possible mechanistic role of CTCF in orchestrating locus conformation—more specifically in the context of its insulator function (Majumder et al., 2015). Majumder et al. have shown that *TCRβ* chromatin conformation depends on an intervening region separating the V_{β} gene segments from the D_{β}–J_{β} clusters where the TCRβ enhancer is located. This region harbors chromatin barrier activity and consists of a proximal region with relatively weak CTCF binding, and a nearby distal high-affinity CTCF-binding site which functions as the major contact point for V_{β} genes segments. Importantly, loss of the V_{β}–D_{β} intervening barrier region activates this anchor CTCF site to become a new chromatin barrier, disarming it as the major contact point for distal V_{β} gene segments. These findings demonstrate for the first time a separation of function between two different CTCF-binding sites within the V_{β}–D_{β} intervening region. Although the two CTCF-binding sites have different roles in the *TCRβ* locus, they nevertheless have interdependent functions. The long-range contact point needs to be insulated from the hyperactive landscape of the *TCRβ* enhancer to serve as a tether for distal V_{β} genes. However, the insulation function is somewhat dominant over V_{β} tethering as disruption of the barrier decommissions the nearest upstream CTCF site from its tethering function to become an insulator. It is possible that, analogous to the *TCRβ* locus, Cer and Sis at the *Igκ* locus work as a bifunctional insulator-tethering element. However, specific topological requirements in different antigen receptor loci must also be considered to explain why Sis deletion at the *Igκ* locus does not result in impaired Cer-dependent *Igκ* locus contraction.

In conclusion, whether CTCF has a role in *Igκ* locus contraction is still an open question, although there is strong evidence implicating CTCF in specifying *Igκ* repertoire and appropriate $V_κ$ usage. Future studies focused on mutating individual CTCF sites are absolutely essential to dissect the role of CTCF at Cer and Sis elements, as well as to assess the relevance of CTCF-binding sites in the $V_κ$ region—both near frequently used $V_κ$ gene segments and near the putative κRE1 enhancer. Whether CTCF binding to Cer and Sis contributes to the same function in different extents or has different functions that are partially redundant remains to be elucidated. More generally, a common theme for the role of CTCF at antigen receptor loci appears to be emerging, in which a division of labor between structural/tethering and insulator sites should be taken into consideration. Certainly, the role of CTCF at these loci is very complex and still only partially understood to date.

6. A GENERAL MODEL FOR DYNAMIC CONTROL OF LONG-RANGE INTERACTIONS AT THE *Igκ* LOCUS

In small pre-B cells, *Igκ* rearrangement is initiated by two key interdependent phenomena: the occurrence of pre-BCR signaling and the cessation of IL-7-induced proliferation. The convergence of these events in pre-B cells results in increased *Igκ* locus accessibility to the RAG recombinase machinery and ultimately to $V_κ$–$J_κ$ rearrangement.

Igκ locus accessibility is a multifaceted concept comprised of subnuclear localization, epigenetic modification, TF binding, long-range chromatin interactions, and GLT. Mining of publicly available genome-wide datasets for key TFs and chromatin modifications relevant to *Igκ* recombination unraveled extensive TF binding and the presence of active chromatin marks at *Igκ* regulatory elements and $V_κ$ genes—surprisingly often already at the pro-B cell stage (Table 1; Fig. 3). Hi-C experiments (Lin et al., 2012), 3C-seq, and DNA FISH studies (Stadhouders et al., 2014) have shown that *Igκ* locus contraction is already apparent in pro-B cells. Taken together, these observations support the notion that the occurrence of several key events essential for the accessibility and recombination of the *Igκ* locus precedes the developmental stage during which *Igκ* locus rearrangement occurs.

Locus-wide 3C experiments show that despite extensive *Igκ* locus contraction at the pro-B cell stage, developmental progression to the pre-B cell stage is accompanied by focused and coordinated contacts between *Igκ* regulatory elements and the $V_κ$ region (Stadhouders et al., 2014). This regulatory focusing in small pre-B cells entails increased interaction frequency of both the iEκ and 3′Eκ enhancers with highly used $V_κ$ genes, which are often marked by E2A binding (Stadhouders et al., 2014). Interestingly, long-range interactions of iEκ were already partially directed to highly used $V_κ$ genes in RAG-deficient pro-B cells (a "prefocused" state), while 3′Eκ interactions strictly depended on pre-BCR signaling to induce preferential interactions with highly used $V_κ$ genes. As both enhancers induce $V_κ$ accessibility in a synergistic fashion (Inlay et al., 2002) and reach full activation at the small pre-B cell stage, we expected pre-BCR signaling to evoke coordination at the level of long-range chromatin interactions between the two individual enhancers. Indeed, correlations between the $V_κ$ interaction frequencies of the iEκ and 3′Eκ elements were dramatically enhanced during pro-B to pre-B cell differentiation, in particular for the most frequently used $V_κ$ genes (Fig. 5; Stadhouders et al., 2014). Therefore,

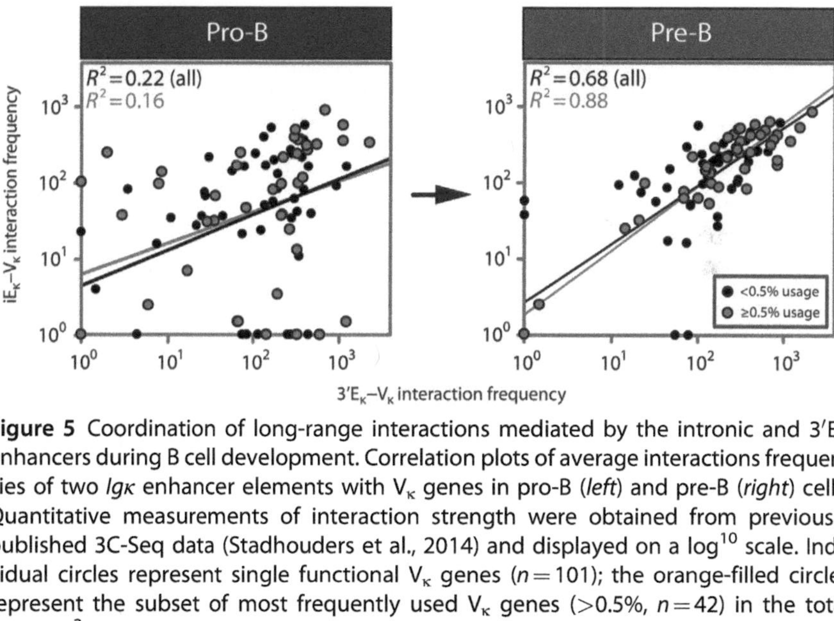

Figure 5 Coordination of long-range interactions mediated by the intronic and 3′Eκ enhancers during B cell development. Correlation plots of average interactions frequencies of two *Igκ* enhancer elements with $V_κ$ genes in pro-B (*left*) and pre-B (*right*) cells. Quantitative measurements of interaction strength were obtained from previously published 3C-Seq data (Stadhouders et al., 2014) and displayed on a $\log^{10}$ scale. Individual circles represent single functional $V_κ$ genes ($n=101$); the orange-filled circles represent the subset of most frequently used $V_κ$ genes (>0.5%, $n=42$) in the total set. An R^2 correlation coefficient was calculated for all genes (black) and for highly used genes (orange). Note the weak interaction correlations observed at the pro-B cell stage (FACS-sorted *Rag1*$^{-/-}$ pro-B cells), which significantly increase as cells mature into pre-B cells (FACS-sorted V_H81X μ^+ *Rag1*$^{-/-}$ pre-B cells) ready to rearrange the *Igκ* locus—in particular for highly used $V_κ$ genes. We interpret this as enhanced coordination between the activity of the iEκ and 3′Eκ enhancers—induced by pre-BCR signaling—to facilitate optimal locus accessibility and $V_κ$ gene usage. (See the color plate.)

we conclude that pre-BCR signaling induces a functional redistribution of enhancer-mediated chromatin interaction: interactions with V_κ genes are stimulated and focused according to their usage pattern, while promiscuous interactions with regions flanking the *Igκ* locus were abolished (Stadhouders et al., 2014).

Pre-BCR signaling does not significantly modulate long-range interactions between iEκ and V_κ genes, which are already robust in pro-B cells. Supporting this notion, active histone modifications and E2A were already detected at the *Igκ* enhancers and V_κ genes at the pro-B cell stage (Goldmit et al., 2005; Lin et al., 2010), whereby E2A was frequently found at the base of long-range chromatin interactions together with CTCF and PU.1, possibly acting as "anchors" to organize genome topology (Lin et al., 2012). The observed correlation between E2A binding, V_κ gene usage, and iEκ proximity in pro-B cells (Stadhouders et al., 2014) further strengthens an early critical role for E2A in regulating *Igκ* locus topology, V_κ gene accessibility, and recombination. Taken together, these findings support a model in which the initial increase in *Igκ* locus accessibility depends on E2A binding to iEκ (Inlay, Lin, et al., 2006; Inlay et al., 2004; Lazorchak et al., 2006). In contrast to the prefocused iEκ, the 3′Eκ element depends on pre-BCR signals to focus its long-range interactions. This is consistent with the requirement of 3′Eκ to recruit the pre-BCR signaling-induced IRF4 TF to promote locus accessibility (Johnson et al., 2008; Lazorchak et al., 2006). Further recruitment of E2A to both *Igκ* enhancers and (highly used) V_κ genes (Lazorchak et al., 2006; Sakamoto et al., 2012) then allows the synergistic promotion of *Igκ* recombination by both enhancers, which now both operate in a highly focused and coordinated fashion (Fig. 6).

Thus, the low levels of *Igκ* locus rearrangement in pro-B cells are greatly enhanced in pre-B cells by three main mechanisms triggered by the onset of pre-BCR signaling and the attenuation of IL-7R signaling: (i) induction of the expression, activity, and/or binding of key TFs to *Igκ* regulatory elements (e.g., Aiolos, Ikaros, IRF4, FOXO1, E2A), together with a loss of STAT5/EZH2 occupancy; (ii) upregulation of GLT over V_κ genes and the J_κ–C_κ region; and (iii) initiation of a focused and coordinated set of long-range chromatin interactions between the regulatory elements and the V_κ region, particularly with highly used V_κ genes (Fig. 6).

Of equal importance is the emerging evidence for a division of labor between the *Igκ* regulatory elements Cer and Sis as a bifunctional CTCF-controlled insulator-anchoring region—similar to what was recently observed for the *TCRβ* locus (Majumder et al., 2015). In this model, Cer

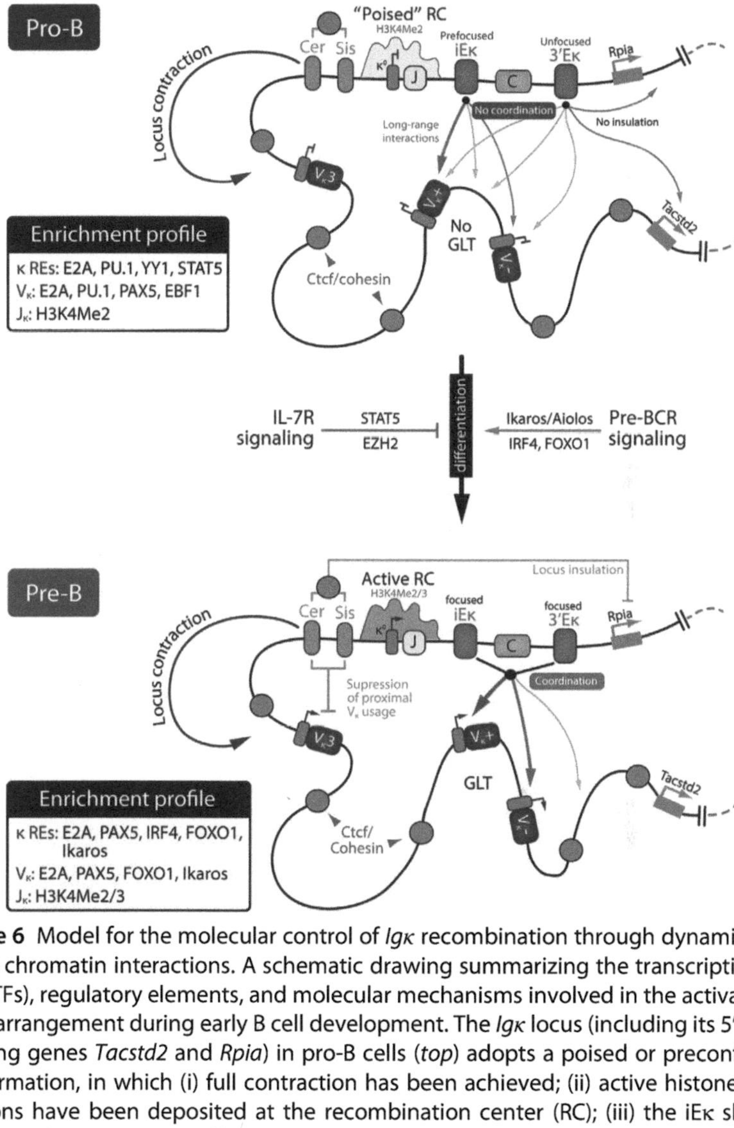

Figure 6 Model for the molecular control of *Igκ* recombination through dynamic long-range chromatin interactions. A schematic drawing summarizing the transcription factors (TFs), regulatory elements, and molecular mechanisms involved in the activation of *Igκ* rearrangement during early B cell development. The *Igκ* locus (including its 5′ and 3′ flanking genes *Tacstd2* and *Rpia*) in pro-B cells (*top*) adopts a poised or preconfigured conformation, in which (i) full contraction has been achieved; (ii) active histone modifications have been deposited at the recombination center (RC); (iii) the iEκ shows a prefocused interaction profile; and (iv) many key B cell TFs already occupy the locus. The combined activation of pre-BCR signaling and attenuation of IL-7R signaling lead to a modified TF landscape compatible with the induction of *Igκ* recombination through (i) full RC activation and RAG expression; (ii) induction of locus accessibility via germline transcription (GLT); (iii) CTCF-mediated locus insulation to restrict enhancer activity to the V_κ region and to suppress preferential recombination to proximal V_κ genes; (iv) focusing of the iEκ and 3′Eκ toward V_κ genes, in particular highly used (V_κ+) genes over less frequently used (V_κ−) genes; and (v) enhanced coordination between the V_κ interactions of both enhancers. These events converge to allow efficient V_κ-to-J_κ

(Continued)

would function as an anchor point that promotes looping of the distal V_κ regions to the J_κ region to induce locus contraction, while Sis operates as an insulator element that prevents the spreading of active chromatin associated with the J_κ region into Cer (Fig. 6). However, this hypothesis needs to be directly tested, for example, through mutation of CTCF-binding sites in Cer and Sis both individually and in combination.

Most interestingly, recent high-resolution Hi-C assays were used to investigate human genome architecture at kilobase resolution to reveal the principles of chromatin looping (Rao et al., 2014). This study revealed that the vast majority of chromatin loops has a striking property: they are in >90% of the cases anchored at a pair of convergent CTCF motifs occupied by CTCF and Cohesin. This observation suggests that the orientation of CTCF-binding site motifs at the *Igκ* locus defines the directionality of CTCF-mediated long-range interactions. Experiments in which the presence or the orientation of CTCF sites across an antigen receptor locus is altered could possibly enable the engineering of chromatin loops, domains, and other structures. Such approaches may prove to be particularly relevant for the understanding of *Igκ* locus topology, where rearrangement of V_κ genes embedded in reverse orientation results in an inversion of the V_κ–J_κ intervening sequence, allowing for secondary rearrangements with retained V_κ genes.

7. CONCLUDING REMARKS

Thirty years after the accessibility hypothesis was proposed based on the observation that rearranging gene segments are concomitantly undergoing GLT (Yancopoulos & Alt, 1985), it still continues to garner experimental support. Dynamic changes in *Igκ* locus chromatin structure and topology have been identified throughout B cell development, using an ever-increasing array of powerful new technologies including ChIP and high-resolution 3C-based methodologies. Although many studies have addressed

Figure 6—Cont'd rearrangement and the generation of a diverse antibody repertoire. The dominant TFs binding the *Igκ* locus at its regulatory elements (REs) and V_κ genes are listed on the left for each stage (enrichment profile); widespread CTCF/cohesin binding is common between both stages and shown as blue circles. Arrows depict long-range chromatin interactions between *Igκ* enhancers and V_κ genes, with thick red arrows indicating high interaction frequencies and thin orange arrows low interaction frequencies. See text for a more in-depth discussion. (See the color plate.)

the role of both lineage-specific and ubiquitously expressed key TFs in *Igκ* accessibility and rearrangement, our current understanding of the extremely complex and dynamic mechanisms involved in shaping long-range interactions at the *Igκ* locus remains limited.

The initial 3D FISH analyses demonstrating global changes in *Igκ* locus contraction in the absence of key TFs or after deletion of important κ regulatory elements can now be dissected in more detail using 3C-based studies. Despite allowing high-resolution analyses, current 3C-based studies can only be routinely performed at the cell-population level, precluding the detection of cell-to-cell differences. Thus, it remains elusive how exactly the signals and molecular events that allow the initiation of V_κ-to-J_κ recombination are integrated at the single pre-B cell level. A single cell Hi-C variant has recently been developed (Nagano et al., 2013), but is currently still very challenging to employ for analyzing genomic loci at high resolution (<10 kb). Thus, it remains elusive how exactly the signals and molecular events that allow the initiation of V_κ-to-J_κ recombination are integrated at the single pre-B cell level. In addition, it would be very informative to analyze long-range chromatin interactions and TF binding at the *Igκ* locus in an allele-specific manner. At a given moment in an individual small pre-B cell, only one allele is actively recombined, while the other allele is positioned in a repressive chromatin environment to suppress recombination. Although κ^0 GLTs are produced from both alleles (Amin et al., 2009; Singh et al., 2003), locus accessibility, TF-binding profiles, and long-range chromatin interactions are expected to widely differ between the active and inactive *Igκ* alleles. Allele-specific 4C and ChIP-seq strategies are feasible (Furey, 2012; Holwerda et al., 2013), and they have the potential to provide unprecedented insight into the molecular and biochemical differences between the two types of *Igκ* alleles. For example, it is expected that Sis-mediated interactions greatly differ between the two *Igκ* alleles in small pre-B cells, as this element plays a crucial role in monoallelic recruitment of *Igκ* to heterochromatin. Likewise, allele-specific Ikaros ChIP-Seq analysis would reveal whether the reported Ikaros-binding profiles in pre-B cells (Ferreiros-Vidal et al., 2013) are a reflection of Ikaros recruitment to the recombining allele, the pericentromeric heterochromatin-associated allele, or a combination of both. Many of the TFs operating at *Igκ* regulatory elements are likely to function in a cooperative fashion. In addition, binding sites for the same TF across the *Igκ* locus—and in particular the vast V_κ region—may show some level of functional redundancy. Hence, unraveling the individual contributions of single TFs or TF-binding sites is not a trivial task. Mining of TF-binding profiles relevant to *Igκ* locus recombination should give clues into their modes of action and relationships. This approach should be helpful when defining experimental approaches to specifically address TF function at particular regions of the *Igκ* locus.

Taken together, 30 years of research have clearly established that a concerted action between various TFs and GLT forms the basis of the dynamic changes in *Igκ* locus topology during B cell development. In this regard, it is intriguing to note that CTCF was recently demonstrated to bind RNA with high affinity at multiple genomic loci (Kung et al., 2015). Together with other recent studies implicating noncoding RNAs in mediating chromatin looping (Lai et al., 2013; Li et al., 2013), it is tempting to contemplate the existence of a cooperative role between CTCF (and possibly other TFs) together with noncoding GLTs in shaping 3D genome topology at the *Igκ* locus. Quite surprisingly, RNA–protein interactions involving GLTs have not been investigated to date—providing yet another interesting topic for future study.

We believe that future studies of the *Igκ* locus will continue to unravel the fundamental gene regulatory processes involved in the generation of a diverse antibody repertoire, with a special emphasis on the complex relationships between long-range chromatin interactions, TF binding, and noncoding transcription.

ACKNOWLEDGMENTS

These studies were partially supported by EMBO Long-term Fellowships to C.R.d.A. (ALTF 1351-2011) and R.S. (ALTF 1201-2014), and a Marie Sklodowska-Curie Actions Individual Fellowship (proposal number 654933) (R.S.). The authors would like to thank Antonio Gomez Moruno (CRG, Barcelona) and Dr. Markus Jaritz (IMP, Vienna) for bioinformatic support and helpful discussions.

Pre-B cell CTCF ChIP-Seq was deposited in the Sequence Read Archive (SRA), run SRR397837.

REFERENCES

Abarrategui, I., & Krangel, M. S. (2006). Regulation of T cell receptor-alpha gene recombination by transcription. *Nature Immunology*, *7*, 1109.

Abarrategui, I., & Krangel, M. S. (2009). Germline transcription: A key regulator of accessibility and recombination. *Advances in Experimental Medicine and Biology*, *650*, 93.

Adams, B., Dorfler, P., Aguzzi, A., Kozmik, Z., Urbanek, P., Maurer-Fogy, I., et al. (1992). Pax-5 encodes the transcription factor BSAP and is expressed in B lymphocytes, the developing CNS, and adult testis. *Genes & Development*, *6*, 1589.

Agalioti, T., Lomvardas, S., Parekh, B., Yie, J., Maniatis, T., & Thanos, D. (2000). Ordered recruitment of chromatin modifying and general transcription factors to the IFN-beta promoter. *Cell*, *103*, 667.

Alt, F. W., Enea, V., Bothwell, A. L., & Baltimore, D. (1980). Activity of multiple light chain genes in murine myeloma cells producing a single, functional light chain. *Cell*, *21*, 1.

Amin, R. H., Cado, D., Nolla, H., Huang, D., Shinton, S. A., Zhou, Y., et al. (2009). Biallelic, ubiquitous transcription from the distal germline Ig{kappa} locus promoter during B cell development. *Proceedings of the National Academy of Sciences of the United States of America*, *106*, 522.

Amin, R. H., & Schlissel, M. S. (2008). Foxo1 directly regulates the transcription of recombination-activating genes during B cell development. *Nature Immunology*, *9*, 613.

Aoki-Ota, M., Torkamani, A., Ota, T., Schork, N., & Nemazee, D. (2012). Skewed primary Igkappa repertoire and V-J joining in C57BL/6 mice: Implications for recombination accessibility and receptor editing. *Journal of Immunology*, *188*, 2305.

Arakawa, H., Shimizu, T., & Takeda, S. (1996). Re-evaluation of the probabilities for productive arrangements on the kappa and lambda loci. *International Immunology*, *8*, 91.

Atchison, L., Ghias, A., Wilkinson, F., Bonini, N., & Atchison, M. L. (2003). Transcription factor YY1 functions as a PcG protein in vivo. *The EMBO Journal*, *22*, 1347.

Atchison, M. L., & Perry, R. P. (1987). The role of the kappa enhancer and its binding factor NF-kappa B in the developmental regulation of kappa gene transcription. *Cell*, *48*, 121.

Atchison, M. L., & Perry, R. P. (1988). Complementation between two cell lines lacking kappa enhancer activity: Implications for the developmental control of immunoglobulin transcription. *The EMBO Journal*, *7*, 4213.

Bain, G., Maandag, E. C., Izon, D. J., Amsen, D., Kruisbeek, A. M., Weintraub, B. C., et al. (1994). E2A proteins are required for proper B cell development and initiation of immunoglobulin gene rearrangements. *Cell*, *79*, 885.

Bain, G., Robanus Maandag, E. C., te Riele, H. P., Feeney, A. J., Sheehy, A., Schlissel, M., et al. (1997). Both E12 and E47 allow commitment to the B cell lineage. *Immunity*, *6*, 145.

Baldwin, A. S., Jr., LeClair, K. P., Singh, H., & Sharp, P. A. (1990). A large protein containing zinc finger domains binds to related sequence elements in the enhancers of the class I major histocompatibility complex and kappa immunoglobulin genes. *Molecular and Cellular Biology*, *10*, 1406.

Banerjee, A., Northrup, D., Boukarabila, H., Jacobsen, S. E., & Allman, D. (2013). Transcriptional repression of Gata3 is essential for early B cell commitment. *Immunity*, *38*, 930.

Banerji, J., Olson, L., & Schaffner, W. (1983). A lymphocyte-specific cellular enhancer is located downstream of the joining region in immunoglobulin heavy chain genes. *Cell*, *33*, 729.

Barberis, A., Widenhorn, K., Vitelli, L., & Busslinger, M. (1990). A novel B-cell lineage-specific transcription factor present at early but not late stages of differentiation. *Genes & Development*, *4*, 849.

Barski, A., Cuddapah, S., Cui, K., Roh, T. Y., Schones, D. E., Wang, Z., et al. (2007). High-resolution profiling of histone methylations in the human genome. *Cell*, *129*, 823.

Bassing, C. H., Swat, W., & Alt, F. W. (2002). The mechanism and regulation of chromosomal V(D)J recombination. *Cell*, *109*(Suppl.), S45.

Beck, K., Peak, M. M., Ota, T., Nemazee, D., & Murre, C. (2009). Distinct roles for E12 and E47 in B cell specification and the sequential rearrangement of immunoglobulin light chain loci. *The Journal of Experimental Medicine*, *206*, 2271.

Bemark, M., Liberg, D., & Leanderson, T. (1998). Conserved sequence elements in K promoters from mice and humans: Implications for transcriptional regulation and repertoire expression. *Immunogenetics*, *47*, 183.

Bemark, M., Martensson, A., Liberg, D., & Leanderson, T. (1999). Spi-C, a novel Ets protein that is temporally regulated during B lymphocyte development. *The Journal of Biological Chemistry*, *274*, 10259.

Bender, M. A., Byron, R., Ragoczy, T., Telling, A., Bulger, M., & Groudine, M. (2006). Flanking HS-62.5 and 3′ HS1, and regions upstream of the LCR, are not required for beta-globin transcription. *Blood*, *108*, 1395.

Bender, M. A., Reik, A., Close, J., Telling, A., Epner, E., Fiering, S., et al. (1998). Description and targeted deletion of 5′ hypersensitive site 5 and 6 of the mouse beta-globin locus control region. *Blood*, *92*, 4394.

Bergman, Y., & Cedar, H. (2004). A stepwise epigenetic process controls immunoglobulin allelic exclusion. *Nature Reviews. Immunology*, *4*, 753.

Bertocci, B., De Smet, A., Berek, C., Weill, J. C., & Reynaud, C. A. (2003). Immunoglobulin kappa light chain gene rearrangement is impaired in mice deficient for DNA polymerase mu. *Immunity*, *19*, 203.

Bertocci, B., De Smet, A., Weill, J. C., & Reynaud, C. A. (2006). Nonoverlapping functions of DNA polymerases mu, lambda, and terminal deoxynucleotidyltransferase during immunoglobulin V(D)J recombination in vivo. *Immunity*, *25*, 31.

Biggs, W. H., 3rd, Meisenhelder, J., Hunter, T., Cavenee, W. K., & Arden, K. C. (1999). Protein kinase B/Akt-mediated phosphorylation promotes nuclear exclusion of the winged helix transcription factor FKHR1. *Proceedings of the National Academy of Sciences of the United States of America*, *96*, 7421.

Blobel, G., & Dobberstein, B. (1975). Transfer of proteins across membranes. I. Presence of proteolytically processed and unprocessed nascent immunoglobulin light chains on membrane-bound ribosomes of murine myeloma. *The Journal of Cell Biology*, *67*, 835.

Bordon, A., Bosco, N., Du Roure, C., Bartholdy, B., Kohler, H., Matthias, G., et al. (2008). Enforced expression of the transcriptional coactivator OBF1 impairs B cell differentiation at the earliest stage of development. *PLoS One*, *3*, e4007.

Bossen, C., Mansson, R., & Murre, C. (2012). Chromatin topology and the regulation of antigen receptor assembly. *Annual Review of Immunology*, *30*, 337.

Bradl, H., Wittmann, J., Milius, D., Vettermann, C., & Jack, H. M. (2003). Interaction of murine precursor B cell receptor with stroma cells is controlled by the unique tail of lambda 5 and stroma cell-associated heparan sulfate. *Journal of Immunology*, *171*, 2338.

Bradney, C., Hjelmeland, M., Komatsu, Y., Yoshida, M., Yao, T. P., & Zhuang, Y. (2003). Regulation of E2A activities by histone acetyltransferases in B lymphocyte development. *The Journal of Biological Chemistry*, *278*, 2370.

Brass, A. L., Zhu, A. Q., & Singh, H. (1999). Assembly requirements of PU.1-Pip (IRF-4) activator complexes: Inhibiting function in vivo using fused dimers. *The EMBO Journal*, *18*, 977.

Brauninger, A., Goossens, T., Rajewsky, K., & Kuppers, R. (2001). Regulation of immunoglobulin light chain gene rearrangements during early B cell development in the human. *European Journal of Immunology*, *31*, 3631.

Brekke, K. M., & Garrard, W. T. (2004). Assembly and analysis of the mouse immunoglobulin kappa gene sequence. *Immunogenetics*, *56*, 490.

Brezinschek, H. P., Brezinschek, R. I., & Lipsky, P. E. (1995). Analysis of the heavy chain repertoire of human peripheral B cells using single-cell polymerase chain reaction. *Journal of Immunology*, *155*, 190.

Brown, K. E., Guest, S. S., Smale, S. T., Hahm, K., Merkenschlager, M., & Fisher, A. G. (1997). Association of transcriptionally silent genes with Ikaros complexes at centromeric heterochromatin. *Cell*, *91*, 845.

Brunet, A., Bonni, A., Zigmond, M. J., Lin, M. Z., Juo, P., Hu, L. S., et al. (1999). Akt promotes cell survival by phosphorylating and inhibiting a Forkhead transcription factor. *Cell*, *96*, 857.

Caretti, G., Di Padova, M., Micales, B., Lyons, G. E., & Sartorelli, V. (2004). The Polycomb Ezh2 methyltransferase regulates muscle gene expression and skeletal muscle differentiation. *Genes & Development*, *18*, 2627.

Casellas, R., Jankovic, M., Meyer, G., Gazumyan, A., Luo, Y., Roeder, R., et al. (2002). OcaB is required for normal transcription and V(D)J recombination of a subset of immunoglobulin kappa genes. *Cell, 110*, 575.

Casellas, R., Shih, T. A., Kleinewietfeld, M., Rakonjac, J., Nemazee, D., Rajewsky, K., et al. (2001). Contribution of receptor editing to the antibody repertoire. *Science, 291*, 1541.

Casellas, R., Zhang, Q., Zheng, N. Y., Mathias, M. D., Smith, K., & Wilson, P. C. (2007). Igkappa allelic inclusion is a consequence of receptor editing. *The Journal of Experimental Medicine, 204*, 153.

Chen, X., Xu, H., Yuan, P., Fang, F., Huss, M., Vega, V. B., et al. (2008). Integration of external signaling pathways with the core transcriptional network in embryonic stem cells. *Cell, 133*, 1106.

Chen, J., Young, F., Bottaro, A., Stewart, V., Smith, R. K., & Alt, F. W. (1993). Mutations of the intronic IgH enhancer and its flanking sequences differentially affect accessibility of the JH locus. *The EMBO Journal, 12*, 4635.

Choi, N. M., Loguercio, S., Verma-Gaur, J., Degner, S. C., Torkamani, A., Su, A. I., et al. (2013). Deep sequencing of the murine IgH repertoire reveals complex regulation of nonrandom V gene rearrangement frequencies. *Journal of Immunology, 191*, 2393.

Church, G. M., Ephrussi, A., Gilbert, W., & Tonegawa, S. (1985). Cell-type-specific contacts to immunoglobulin enhancers in nuclei. *Nature, 313*, 798.

Clark, M. R., Mandal, M., Ochiai, K., & Singh, H. (2014). Orchestrating B cell lymphopoiesis through interplay of IL-7 receptor and pre-B cell receptor signalling. *Nature Reviews. Immunology, 14*, 69.

Cobb, B. S., & Smale, S. T. (2005). Ikaros-family proteins: In search of molecular functions during lymphocyte development. *Current Topics in Microbiology and Immunology, 290*, 29.

Cocea, L., De Smet, A., Saghatchian, M., Fillatreau, S., Ferradini, L., Schurmans, S., et al. (1999). A targeted deletion of a region upstream from the Jkappa cluster impairs kappa chain rearrangement in cis in mice and in the 103/bcl2 cell line. *The Journal of Experimental Medicine, 189*, 1443.

Cockerill, P. N., & Garrard, W. T. (1986). Chromosomal loop anchorage of the kappa immunoglobulin gene occurs next to the enhancer in a region containing topoisomerase II sites. *Cell, 44*, 273.

Coleclough, C., Perry, R. P., Karjalainen, K., & Weigert, M. (1981). Aberrant rearrangements contribute significantly to the allelic exclusion of immunoglobulin gene expression. *Nature, 290*, 372.

Cooper, A. B., Sawai, C. M., Sicinska, E., Powers, S. E., Sicinski, P., Clark, M. R., et al. (2006). A unique function for cyclin D3 in early B cell development. *Nature Immunology, 7*, 489.

Corcoran, L. M., & Karvelas, M. (1994). Oct-2 is required early in T cell-independent B cell activation for G1 progression and for proliferation. *Immunity, 1*, 635.

Corcoran, L. M., Karvelas, M., Nossal, G. J., Ye, Z. S., Jacks, T., & Baltimore, D. (1993). Oct-2, although not required for early B-cell development, is critical for later B-cell maturation and for postnatal survival. *Genes & Development, 7*, 570.

Corcoran, A. E., Smart, F. M., Cowling, R. J., Crompton, T., Owen, M. J., & Venkitaraman, A. R. (1996). The interleukin-7 receptor alpha chain transmits distinct signals for proliferation and differentiation during B lymphopoiesis. *The EMBO Journal, 15*, 1924.

Dahl, R., Ramirez-Bergeron, D. L., Rao, S., & Simon, M. C. (2002). Spi-B can functionally replace PU.1 in myeloid but not lymphoid development. *The EMBO Journal, 21*, 2220.

Das, S., Nikolaidis, N., & Nei, M. (2009). Genomic organization and evolution of immunoglobulin kappa gene enhancers and kappa deleting element in mammals. *Molecular Immunology, 46*, 3171.

de Laat, W., & Duboule, D. (2013). Topology of mammalian developmental enhancers and their regulatory landscapes. *Nature*, *502*, 499.

Decker, T., Pasca di Magliano, M., McManus, S., Sun, Q., Bonifer, C., Tagoh, H., et al. (2009). Stepwise activation of enhancer and promoter regions of the B cell commitment gene Pax5 in early lymphopoiesis. *Immunity*, *30*, 508.

Degner, S. C., Wong, T. P., Jankevicius, G., & Feeney, A. J. (2009). Cutting edge: Developmental stage-specific recruitment of cohesin to CTCF sites throughout immunoglobulin loci during B lymphocyte development. *Journal of Immunology*, *182*, 44.

DeKoter, R. P., Lee, H. J., & Singh, H. (2002). PU.1 regulates expression of the interleukin-7 receptor in lymphoid progenitors. *Immunity*, *16*, 297.

Delogu, A., Schebesta, A., Sun, Q., Aschenbrenner, K., Perlot, T., & Busslinger, M. (2006). Gene repression by Pax5 in B cells is essential for blood cell homeostasis and is reversed in plasma cells. *Immunity*, *24*, 269.

Dengler, H. S., Baracho, G. V., Omori, S. A., Bruckner, S., Arden, K. C., Castrillon, D. H., et al. (2008). Distinct functions for the transcription factor Foxo1 at various stages of B cell differentiation. *Nature Immunology*, *9*, 1388.

Derudder, E., Cadera, E. J., Vahl, J. C., Wang, J., Fox, C. J., Zha, S., et al. (2009). Development of immunoglobulin lambda-chain-positive B cells, but not editing of immunoglobulin kappa-chain, depends on NF-kappaB signals. *Nature Immunology*, *10*, 647.

Dias, S., Silva, H., Jr., Cumano, A., & Vieira, P. (2005). Interleukin-7 is necessary to maintain the B cell potential in common lymphoid progenitors. *The Journal of Experimental Medicine*, *201*, 971.

Dingjan, G. M., Middendorp, S., Dahlenborg, K., Maas, A., Grosveld, F., & Hendriks, R. W. (2001). Bruton's tyrosine kinase regulates the activation of gene rearrangements at the lambda light chain locus in precursor B cells in the mouse. *The Journal of Experimental Medicine*, *193*, 1169.

Durdik, J., Moore, M. W., & Selsing, E. (1984). Novel kappa light-chain gene rearrangements in mouse lambda light chain-producing B lymphocytes. *Nature*, *307*, 749.

Ebert, A., McManus, S., Tagoh, H., Medvedovic, J., Salvagiotto, G., Novatchkova, M., et al. (2011). The distal V(H) gene cluster of the Igh locus contains distinct regulatory elements with Pax5 transcription factor-dependent activity in pro-B cells. *Immunity*, *34*, 175.

Eisenbeis, C. F., Singh, H., & Storb, U. (1995). Pip, a novel IRF family member, is a lymphoid-specific, PU.1-dependent transcriptional activator. *Genes & Development*, *9*, 1377.

Emorine, L., Kuehl, M., Weir, L., Leder, P., & Max, E. E. (1983). A conserved sequence in the immunoglobulin J kappa-C kappa intron: Possible enhancer element. *Nature*, *304*, 447.

Engel, H., Rolink, A., & Weiss, S. (1999). B cells are programmed to activate kappa and lambda for rearrangement at consecutive developmental stages. *European Journal of Immunology*, *29*, 2167.

Ephrussi, A., Church, G. M., Tonegawa, S., & Gilbert, W. (1985). B lineage-specific interactions of an immunoglobulin enhancer with cellular factors in vivo. *Science*, *227*, 134.

Escalante, C. R., Brass, A. L., Pongubala, J. M., Shatova, E., Shen, L., Singh, H., et al. (2002). Crystal structure of PU.1/IRF-4/DNA ternary complex. *Molecular Cell*, *10*, 1097.

Falkner, F. G., & Zachau, H. G. (1984). Correct transcription of an immunoglobulin kappa gene requires an upstream fragment containing conserved sequence elements. *Nature*, *310*, 71.

Farago, M., Rosenbluh, C., Tevlin, M., Fraenkel, S., Schlesinger, S., Masika, H., et al. (2012). Clonal allelic predetermination of immunoglobulin-kappa rearrangement. *Nature*, *490*, 561.

Featherstone, K., Wood, A. L., Bowen, A. J., & Corcoran, A. E. (2010). The mouse immunoglobulin heavy chain V-D intergenic sequence contains insulators that may regulate ordered V(D)J recombination. *The Journal of Biological Chemistry*, *285*, 9327.

Feeney, A. J. (2009). Genetic and epigenetic control of V gene rearrangement frequency. *Advances in Experimental Medicine and Biology*, *650*, 73.

Ferradini, L., Gu, H., De Smet, A., Rajewsky, K., Reynaud, C. A., & Weill, J. C. (1996). Rearrangement-enhancing element upstream of the mouse immunoglobulin kappa chain J cluster. *Science*, *271*, 1416.

Ferreiros-Vidal, I., Carroll, T., Taylor, B., Terry, A., Liang, Z., Bruno, L., et al. (2013). Genome-wide identification of Ikaros targets elucidates its contribution to mouse B-cell lineage specification and pre-B-cell differentiation. *Blood*, *121*, 1769.

Fitzsimmons, S. P., Bernstein, R. M., Max, E. E., Skok, J. A., & Shapiro, M. A. (2007). Dynamic changes in accessibility, nuclear positioning, recombination, and transcription at the Ig kappa locus. *Journal of Immunology*, *179*, 5264.

Fraenkel, S., Mostoslavsky, R., Novobrantseva, T. I., Pelanda, R., Chaudhuri, J., Esposito, G., et al. (2007). Allelic 'choice' governs somatic hypermutation in vivo at the immunoglobulin kappa-chain locus. *Nature Immunology*, *8*, 715.

Furey, T. S. (2012). ChIP-seq and beyond: New and improved methodologies to detect and characterize protein-DNA interactions. *Nature Reviews. Genetics*, *13*, 840.

Fuxa, M., & Busslinger, M. (2007). Reporter gene insertions reveal a strictly B lymphoid-specific expression pattern of Pax5 in support of its B cell identity function. *Journal of Immunology*, *178*, 8222.

Fuxa, M., Skok, J., Souabni, A., Salvagiotto, G., Roldan, E., & Busslinger, M. (2004). Pax5 induces V-to-DJ rearrangements and locus contraction of the immunoglobulin heavy-chain gene. *Genes & Development*, *18*, 411.

Gao, H., Lukin, K., Ramirez, J., Fields, S., Lopez, D., & Hagman, J. (2009). Opposing effects of SWI/SNF and Mi-2/NuRD chromatin remodeling complexes on epigenetic reprogramming by EBF and Pax5. *Proceedings of the National Academy of Sciences of the United States of America*, *106*, 11258.

Garrett-Sinha, L. A., Dahl, R., Rao, S., Barton, K. P., & Simon, M. C. (2001). PU.1 exhibits partial functional redundancy with Spi-B, but not with Ets-1 or Elf-1. *Blood*, *97*, 2908.

Gaszner, M., & Felsenfeld, G. (2006). Insulators: Exploiting transcriptional and epigenetic mechanisms. *Nature Reviews Genetics*, *7*, 703.

Gauthier, L., Rossi, B., Roux, F., Termine, E., & Schiff, C. (2002). Galectin-1 is a stromal cell ligand of the pre-B cell receptor (BCR) implicated in synapse formation between pre-B and stromal cells and in pre-BCR triggering. *Proceedings of the National Academy of Sciences of the United States of America*, *99*, 13014.

Gay, D., Saunders, T., Camper, S., & Weigert, M. (1993). Receptor editing: An approach by autoreactive B cells to escape tolerance. *The Journal of Experimental Medicine*, *177*, 999.

Georgopoulos, K. (2002). Haematopoietic cell-fate decisions, chromatin regulation and ikaros. *Nature Reviews. Immunology*, *2*, 162.

Ghia, P., Gratwohl, A., Signer, E., Winkler, T. H., Melchers, F., & Rolink, A. G. (1995). Immature B cells from human and mouse bone marrow can change their surface light chain expression. *European Journal of Immunology*, *25*, 3108.

Giachino, C., Padovan, E., & Lanzavecchia, A. (1995). kappa+lambda+ dual receptor B cells are present in the human peripheral repertoire. *The Journal of Experimental Medicine*, *181*, 1245.

Giallourakis, C. C., Franklin, A., Guo, C., Cheng, H. L., Yoon, H. S., Gallagher, M., et al. (2010). Elements between the IgH variable (V) and diversity (D) clusters influence antisense transcription and lineage-specific V(D)J recombination. *Proceedings of the National Academy of Sciences of the United States of America*, *107*, 22207.

Gisler, R., & Sigvardsson, M. (2002). The human V-preB promoter is a target for coordinated activation by early B cell factor and E47. *Journal of Immunology, 168*, 5130.

Giudicelli, V., & Lefranc, M. P. (1999). Ontology for immunogenetics: The IMGT-ONTOLOGY. *Bioinformatics, 15*, 1047.

Goebel, P., Janney, N., Valenzuela, J. R., Romanow, W. J., Murre, C., & Feeney, A. J. (2001). Localized gene-specific induction of accessibility to V(D)J recombination induced by E2A and early B cell factor in nonlymphoid cells. *The Journal of Experimental Medicine, 194*, 645.

Goldmit, M., Ji, Y., Skok, J., Roldan, E., Jung, S., Cedar, H., et al. (2005). Epigenetic ontogeny of the Igk locus during B cell development. *Nature Immunology, 6*, 198.

Goldmit, M., Schlissel, M., Cedar, H., & Bergman, Y. (2002). Differential accessibility at the kappa chain locus plays a role in allelic exclusion. *The EMBO Journal, 21*, 5255.

Goodnow, C. C., Crosbie, J., Jorgensen, H., Brink, R. A., & Basten, A. (1989). Induction of self-tolerance in mature peripheral B lymphocytes. *Nature, 342*, 385.

Gorman, J. R., van der Stoep, N., Monroe, R., Cogne, M., Davidson, L., & Alt, F. W. (1996). The Ig(kappa) enhancer influences the ratio of Ig(kappa) versus Ig(lambda) B lymphocytes. *Immunity, 5*, 241.

Grawunder, U., Haasner, D., Melchers, F., & Rolink, A. (1993). Rearrangement and expression of kappa light chain genes can occur without mu heavy chain expression during differentiation of pre-B cells. *International Immunology, 5*, 1609.

Greenbaum, S., & Zhuang, Y. (2002). Identification of E2A target genes in B lymphocyte development by using a gene tagging-based chromatin immunoprecipitation system. *Proceedings of the National Academy of Sciences of the United States of America, 99*, 15030.

Grosschedl, R., & Baltimore, D. (1985). Cell-type specificity of immunoglobulin gene expression is regulated by at least three DNA sequence elements. *Cell, 41*, 885.

Gstaiger, M., Knoepfel, L., Georgiev, O., Schaffner, W., & Hovens, C. M. (1995). A B-cell coactivator of octamer-binding transcription factors. *Nature, 373*, 360.

Guo, C., Gerasimova, T., Hao, H., Ivanova, I., Chakraborty, T., Selimyan, R., et al. (2011). Two forms of loops generate the chromatin conformation of the immunoglobulin heavy-chain gene locus. *Cell, 147*, 332.

Guo, C., Yoon, H. S., Franklin, A., Jain, S., Ebert, A., Cheng, H. L., et al. (2011). CTCF-binding elements mediate control of V(D)J recombination. *Nature, 477*, 424.

Gyory, I., Boller, S., Nechanitzky, R., Mandel, E., Pott, S., Liu, E., et al. (2012). Transcription factor Ebf1 regulates differentiation stage-specific signaling, proliferation, and survival of B cells. *Genes & Development, 26*, 668.

Hadjur, S., Williams, L. M., Ryan, N. K., Cobb, B. S., Sexton, T., Fraser, P., et al. (2009). Cohesins form chromosomal cis-interactions at the developmentally regulated IFNG locus. *Nature, 460*, 410.

Haeusler, R. A., Pratt-Hyatt, M., Good, P. D., Gipson, T. A., & Engelke, D. R. (2008). Clustering of yeast tRNA genes is mediated by specific association of condensin with tRNA gene transcription complexes. *Genes & Development, 22*, 2204.

Hagman, J., Belanger, C., Travis, A., Turck, C. W., & Grosschedl, R. (1993). Cloning and functional characterization of early B-cell factor, a regulator of lymphocyte-specific gene expression. *Genes & Development, 7*, 760.

Harrow, J., Denoeud, F., Frankish, A., Reymond, A., Chen, C. K., Chrast, J., et al. (2006). GENCODE: Producing a reference annotation for ENCODE. *Genome Biology*, 7(Suppl. 1), S4.1.

Heavey, B., Charalambous, C., Cobaleda, C., & Busslinger, M. (2003). Myeloid lineage switch of Pax5 mutant but not wild-type B cell progenitors by C/EBPalpha and GATA factors. *The EMBO Journal, 22*, 3887.

Heizmann, B., Kastner, P., & Chan, S. (2013). Ikaros is absolutely required for pre-B cell differentiation by attenuating IL-7 signals. *The Journal of Experimental Medicine, 210*, 2823.

Hendriks, R. W., & Middendorp, S. (2004). The pre-BCR checkpoint as a cell-autonomous proliferation switch. *Trends in Immunology, 25*, 249.

Herzog, S., Hug, E., Meixlsperger, S., Paik, J. H., DePinho, R. A., Reth, M., et al. (2008). SLP-65 regulates immunoglobulin light chain gene recombination through the PI(3)K-PKB-Foxo pathway. *Nature Immunology, 9*, 623.

Herzog, S., Reth, M., & Jumaa, H. (2009). Regulation of B-cell proliferation and differentiation by pre-B-cell receptor signalling. *Nature Reviews. Immunology, 9*, 195.

Hesslein, D. G., Pflugh, D. L., Chowdhury, D., Bothwell, A. L., Sen, R., & Schatz, D. G. (2003). Pax5 is required for recombination of transcribed, acetylated, 5′ IgH V gene segments. *Genes & Development, 17*, 37.

Hewitt, S. L., Farmer, D., Marszalek, K., Cadera, E., Liang, H. E., Xu, Y., et al. (2008). Association between the Igk and Igh immunoglobulin loci mediated by the 3′ Igk enhancer induces 'decontraction' of the Igh locus in pre-B cells. *Nature Immunology, 9*, 396.

Hewitt, S. L., Yin, B., Ji, Y., Chaumeil, J., Marszalek, K., Tenthorey, J., et al. (2009). RAG-1 and ATM coordinate monoallelic recombination and nuclear positioning of immunoglobulin loci. *Nature Immunology, 10*, 655.

Hicar, M. D., Liu, Y., Allen, C. E., & Wu, L. C. (2001). Structure of the human zinc finger protein HIVEP3: Molecular cloning, expression, exon-intron structure, and comparison with paralogous genes HIVEP1 and HIVEP2. *Genomics, 71*, 89.

Hieter, P. A., Korsmeyer, S. J., Waldmann, T. A., & Leder, P. (1981). Human immunoglobulin kappa light-chain genes are deleted or rearranged in lambda-producing B cells. *Nature, 290*, 368.

Hobeika, E., Thiemann, S., Storch, B., Jumaa, H., Nielsen, P. J., Pelanda, R., et al. (2006). Testing gene function early in the B cell lineage in mb1-cre mice. *Proceedings of the National Academy of Sciences of the United States of America, 103*, 13789.

Hodawadekar, S., Park, K., Farrar, M. A., & Atchison, M. L. (2012). A developmentally controlled competitive STAT5-PU.1 DNA binding mechanism regulates activity of the Ig kappa E3′ enhancer. *Journal of Immunology, 188*, 2276.

Hoflinger, S., Kesavan, K., Fuxa, M., Hutter, C., Heavey, B., Radtke, F., et al. (2004). Analysis of Notch1 function by in vitro T cell differentiation of Pax5 mutant lymphoid progenitors. *Journal of Immunology, 173*, 3935.

Holmes, M. L., Carotta, S., Corcoran, L. M., & Nutt, S. L. (2006). Repression of Flt3 by Pax5 is crucial for B-cell lineage commitment. *Genes & Development, 20*, 933.

Holwerda, S. J., van de Werken, H. J., Ribeiro de Almeida, C., Bergen, I. M., de Bruijn, M. J., Verstegen, M. J., et al. (2013). Allelic exclusion of the immunoglobulin heavy chain locus is independent of its nuclear localization in mature B cells. *Nucleic Acids Research, 41*, 6905.

Hosaka, T., Biggs, W. H., 3rd, Tieu, D., Boyer, A. D., Varki, N. M., Cavenee, W. K., et al. (2004). Disruption of forkhead transcription factor (FOXO) family members in mice reveals their functional diversification. *Proceedings of the National Academy of Sciences of the United States of America, 101*, 2975.

Houston, I. B., Kamath, M. B., Schweitzer, B. L., Chlon, T. M., & DeKoter, R. P. (2007). Reduction in PU.1 activity results in a block to B-cell development, abnormal myeloid proliferation, and neonatal lethality. *Experimental Hematology, 35*, 1056.

Hozumi, N., & Tonegawa, S. (1976). Evidence for somatic rearrangement of immunoglobulin genes coding for variable and constant regions. *Proceedings of the National Academy of Sciences of the United States of America, 73*, 3628.

Hu, C. J., Rao, S., Ramirez-Bergeron, D. L., Garrett-Sinha, L. A., Gerondakis, S., Clark, M. R., et al. (2001). PU.1/Spi-B regulation of c-rel is essential for mature B cell survival. *Immunity, 15*, 545.

Hu, H., Wang, B., Borde, M., Nardone, J., Maika, S., Allred, L., et al. (2006). Foxp1 is an essential transcriptional regulator of B cell development. *Nature Immunology*, 7, 819.

Huang, H., Kearney, J. F., Grusby, M. J., Benoist, C., & Mathis, D. (2006). Induction of tolerance in arthritogenic B cells with receptors of differing affinity for self-antigen. *Proceedings of the National Academy of Sciences of the United States of America*, *103*, 3734.

Inlay, M., Alt, F. W., Baltimore, D., & Xu, Y. (2002). Essential roles of the kappa light chain intronic enhancer and 3′ enhancer in kappa rearrangement and demethylation. *Nature Immunology*, *3*, 463.

Inlay, M. A., Gao, H. H., Odegard, V. H., Lin, T., Schatz, D. G., & Xu, Y. (2006). Roles of the Ig kappa light chain intronic and 3′ enhancers in Igk somatic hypermutation. *Journal of Immunology*, *177*, 1146.

Inlay, M. A., Lin, T., Gao, H. H., & Xu, Y. (2006). Critical roles of the immunoglobulin intronic enhancers in maintaining the sequential rearrangement of IgH and Igk loci. *The Journal of Experimental Medicine*, *203*, 1721.

Inlay, M. A., Tian, H., Lin, T., & Xu, Y. (2004). Important roles for E protein binding sites within the immunoglobulin kappa chain intronic enhancer in activating Vkappa Jkappa rearrangement. *The Journal of Experimental Medicine*, *200*, 1205.

Iwasaki, H., Somoza, C., Shigematsu, H., Duprez, E. A., Iwasaki-Arai, J., Mizuno, S., et al. (2005). Distinctive and indispensable roles of PU.1 in maintenance of hematopoietic stem cells and their differentiation. *Blood*, *106*, 1590.

Jhunjhunwala, S., van Zelm, M. C., Peak, M. M., & Murre, C. (2009). Chromatin architecture and the generation of antigen receptor diversity. *Cell*, *138*, 435.

Ji, Y., Resch, W., Corbett, E., Yamane, A., Casellas, R., & Schatz, D. G. (2010). The in vivo pattern of binding of RAG1 and RAG2 to antigen receptor loci. *Cell*, *141*, 419.

John, L. B., & Ward, A. C. (2011). The Ikaros gene family: Transcriptional regulators of hematopoiesis and immunity. *Molecular Immunology*, *48*, 1272.

Johnson, K., Hashimshony, T., Sawai, C. M., Pongubala, J. M., Skok, J. A., Aifantis, I., et al. (2008). Regulation of immunoglobulin light-chain recombination by the transcription factor IRF-4 and the attenuation of interleukin-7 signaling. *Immunity*, *28*, 335.

Johnson, K., Pflugh, D. L., Yu, D., Hesslein, D. G., Lin, K. I., Bothwell, A. L., et al. (2004). B cell-specific loss of histone 3 lysine 9 methylation in the V(H) locus depends on Pax5. *Nature Immunology*, *5*, 853.

Joshi, I., Yoshida, T., Jena, N., Qi, X., Zhang, J., Van Etten, R. A., et al. (2014). Loss of Ikaros DNA-binding function confers integrin-dependent survival on pre-B cells and progression to acute lymphoblastic leukemia. *Nature Immunology*, *15*, 294.

Jumaa, H., Mitterer, M., Reth, M., & Nielsen, P. J. (2001). The absence of SLP65 and Btk blocks B cell development at the preB cell receptor-positive stage. *European Journal of Immunology*, *31*, 2164.

Jumaa, H., Wollscheid, B., Mitterer, M., Wienands, J., Reth, M., & Nielsen, P. J. (1999). Abnormal development and function of B lymphocytes in mice deficient for the signaling adaptor protein SLP-65. *Immunity*, *11*, 547.

Jung, D., & Alt, F. W. (2004). Unraveling V(D)J recombination. Insights into gene regulation. *Cell*, *116*, 299.

Kalled, S. L., & Brodeur, P. H. (1991). Utilization of V kappa families and V kappa exons. Implications for the available B cell repertoire. *Journal of Immunology*, *147*, 3194.

Karnowski, A., Cao, C., Matthias, G., Carotta, S., Corcoran, L. M., Martensson, I. L., et al. (2008). Silencing and nuclear repositioning of the lambda5 gene locus at the pre-B cell stage requires Aiolos and OBF-1. *PLoS One*, *3*, e3568.

Kee, B. L., & Murre, C. (1998). Induction of early B cell factor (EBF) and multiple B lineage genes by the basic helix-loop-helix transcription factor E12. *The Journal of Experimental Medicine*, *188*, 699.

Kee, B. L., Quong, M. W., & Murre, C. (2000). E2A proteins: Essential regulators at multiple stages of B-cell development. *Immunological Reviews*, *175*, 138.

Keenan, R. A., De Riva, A., Corleis, B., Hepburn, L., Licence, S., Winkler, T. H., et al. (2008). Censoring of autoreactive B cell development by the pre-B cell receptor. *Science, 321*, 696.

Kersseboom, R., Ta, V. B., Zijlstra, A. J., Middendorp, S., Jumaa, H., van Loo, P. F., et al. (2006). Bruton's tyrosine kinase and SLP-65 regulate pre-B cell differentiation and the induction of Ig light chain gene rearrangement. *Journal of Immunology, 176*, 4543.

Kikuchi, K., Lai, A. Y., Hsu, C. L., & Kondo, M. (2005). IL-7 receptor signaling is necessary for stage transition in adult B cell development through up-regulation of EBF. *The Journal of Experimental Medicine, 201*, 1197.

Kim, T. H., Abdullaev, Z. K., Smith, A. D., Ching, K. A., Loukinov, D. I., Green, R. D., et al. (2007). Analysis of the vertebrate insulator protein CTCF-binding sites in the human genome. *Cell, 128*, 1231.

Kim, U., Qin, X. F., Gong, S., Stevens, S., Luo, Y., Nussenzweig, M., et al. (1996). The B-cell-specific transcription coactivator OCA-B/OBF-1/Bob-1 is essential for normal production of immunoglobulin isotypes. *Nature, 383*, 542.

Kim, J., Sif, S., Jones, B., Jackson, A., Koipally, J., Heller, E., et al. (1999). Ikaros DNA-binding proteins direct formation of chromatin remodeling complexes in lymphocytes. *Immunity, 10*, 345.

Kirstetter, P., Thomas, M., Dierich, A., Kastner, P., & Chan, S. (2002). Ikaros is critical for B cell differentiation and function. *European Journal of Immunology, 32*, 720.

Klein, S., Sablitzky, F., & Radbruch, A. (1984). Deletion of the IgH enhancer does not reduce immunoglobulin heavy chain production of a hybridoma IgD class switch variant. *The EMBO Journal, 3*, 2473.

Kohler, F., Hug, E., Eschbach, C., Meixlsperger, S., Hobeika, E., Kofer, J., et al. (2008). Autoreactive B cell receptors mimic autonomous pre-B cell receptor signaling and induce proliferation of early B cells. *Immunity, 29*, 912.

Kosak, S. T., Skok, J. A., Medina, K. L., Riblet, R., Le Beau, M. M., Fisher, A. G., et al. (2002). Subnuclear compartmentalization of immunoglobulin loci during lymphocyte development. *Science, 296*, 158.

Kumar, P. P., Bischof, O., Purbey, P. K., Notani, D., Urlaub, H., Dejean, A., et al. (2007). Functional interaction between PML and SATB1 regulates chromatin-loop architecture and transcription of the MHC class I locus. *Nature Cell Biology, 9*, 45.

Kung, J. T., Kesner, B., An, J. Y., Ahn, J. Y., Cifuentes-Rojas, C., Colognori, D., et al. (2015). Locus-specific targeting to the X chromosome revealed by the RNA interactome of CTCF. *Molecular Cell, 57*, 361.

Kurukuti, S., Tiwari, V. K., Tavoosidana, G., Pugacheva, E., Murrell, A., Zhao, Z., et al. (2006). CTCF binding at the H19 imprinting control region mediates maternally inherited higher-order chromatin conformation to restrict enhancer access to Igf2. *Proceedings of the National Academy of Sciences of the United States of America, 103*, 10684.

Kwon, K., Hutter, C., Sun, Q., Bilic, I., Cobaleda, C., Malin, S., et al. (2008). Instructive role of the transcription factor E2A in early B lymphopoiesis and germinal center B cell development. *Immunity, 28*, 751.

Laemmli, U. K., Kas, E., Poljak, L., & Adachi, Y. (1992). Scaffold-associated regions: cis-Acting determinants of chromatin structural loops and functional domains. *Current Opinion in Genetics & Development, 2*, 275.

Lai, F., Orom, U. A., Cesaroni, M., Beringer, M., Taatjes, D. J., Blobel, G. A., et al. (2013). Activating RNAs associate with Mediator to enhance chromatin architecture and transcription. *Nature, 494*, 497.

Lang, J., Arnold, B., Hammerling, G., Harris, A. W., Korsmeyer, S., Russell, D., et al. (1997). Enforced Bcl-2 expression inhibits antigen-mediated clonal elimination of peripheral B cells in an antigen dose-dependent manner and promotes receptor editing in autoreactive, immature B cells. *The Journal of Experimental Medicine, 186*, 1513.

Lazorchak, A. S., Schlissel, M. S., & Zhuang, Y. (2006). E2A and IRF-4/Pip promote chromatin modification and transcription of the immunoglobulin kappa locus in pre-B cells. *Molecular and Cellular Biology, 26*, 810.

Lefranc, M. P., Giudicelli, V., Kaas, Q., Duprat, E., Jabado-Michaloud, J., Scaviner, D., et al. (2005). IMGT, the international ImMunoGeneTics information system. *Nucleic Acids Research, 33*, D593.

Lenhard, B., Sandelin, A., & Carninci, P. (2012). Metazoan promoters: Emerging characteristics and insights into transcriptional regulation. *Nature Reviews. Genetics, 13*, 233.

Lennon, G. G., & Perry, R. P. (1985). C mu-containing transcripts initiate heterogeneously within the IgH enhancer region and contain a novel 5′-nontranslatable exon. *Nature, 318*, 475.

Levin-Klein, R., Kirillov, A., Rosenbluh, C., Cedar, H., & Bergman, Y. (2014). A novel pax5-binding regulatory element in the igkappa locus. *Frontiers in Immunology, 5*, 240.

Li, Z., Dordai, D. I., Lee, J., & Desiderio, S. (1996). A conserved degradation signal regulates RAG-2 accumulation during cell division and links V(D)J recombination to the cell cycle. *Immunity, 5*, 575.

Li, W., Notani, D., Ma, Q., Tanasa, B., Nunez, E., Chen, A. Y., et al. (2013). Functional roles of enhancer RNAs for oestrogen-dependent transcriptional activation. *Nature, 498*, 516.

Lichtenstein, M., Keini, G., Cedar, H., & Bergman, Y. (1994). B cell-specific demethylation: A novel role for the intronic kappa chain enhancer sequence. *Cell, 76*, 913.

Lin, Y. C., Benner, C., Mansson, R., Heinz, S., Miyazaki, K., Miyazaki, M., et al. (2012). Global changes in the nuclear positioning of genes and intra- and interdomain genomic interactions that orchestrate B cell fate. *Nature Immunology, 13*, 1196.

Lin, H., & Grosschedl, R. (1995). Failure of B-cell differentiation in mice lacking the transcription factor EBF. *Nature, 376*, 263.

Lin, S. G., Guo, C., Su, A., Zhang, Y., & Alt, F. W. (2015). CTCF-binding elements 1 and 2 in the Igh intergenic control region cooperatively regulate V(D)J recombination. *Proceedings of the National Academy of Sciences of the United States of America, 112*, 1815.

Lin, Y. C., Jhunjhunwala, S., Benner, C., Heinz, S., Welinder, E., Mansson, R., et al. (2010). A global network of transcription factors, involving E2A, EBF1 and Foxo1, that orchestrates B cell fate. *Nature Immunology, 11*, 635.

Liu, Z. M., George-Raizen, J. B., Li, S., Meyers, K. C., Chang, M. Y., & Garrard, W. T. (2002). Chromatin structural analyses of the mouse Igkappa gene locus reveal new hypersensitive sites specifying a transcriptional silencer and enhancer. *The Journal of Biological Chemistry, 277*, 32640.

Liu, H., Schmidt-Supprian, M., Shi, Y., Hobeika, E., Barteneva, N., Jumaa, H., et al. (2007). Yin Yang 1 is a critical regulator of B-cell development. *Genes & Development, 21*, 1179.

Liu, S., Velez, M. G., Humann, J., Rowland, S., Conrad, F. J., Halverson, R., et al. (2005). Receptor editing can lead to allelic inclusion and development of B cells that retain antibodies reacting with high avidity autoantigens. *Journal of Immunology, 175*, 5067.

Liu, Z., Widlak, P., Zou, Y., Xiao, F., Oh, M., Li, S., et al. (2006). A recombination silencer that specifies heterochromatin positioning and ikaros association in the immunoglobulin kappa locus. *Immunity, 24*, 405.

Lu, R. (2008). Interferon regulatory factor 4 and 8 in B-cell development. *Trends in Immunology, 29*, 487.

Lu, R., Medina, K. L., Lancki, D. W., & Singh, H. (2003). IRF-4,8 orchestrate the pre-B-to-B transition in lymphocyte development. *Genes & Development, 17*, 1703.

Lucas, J. S., Bossen, C., & Murre, C. (2011). Transcription and recombination factories: Common features? *Current Opinion in Cell Biology, 23*, 318.

Lukin, K., Fields, S., Guerrettaz, L., Straign, D., Rodriguez, V., Zandi, S., et al. (2011). A dose-dependent role for EBF1 in repressing non-B-cell-specific genes. *European Journal of Immunology, 41*, 1787.

Luo, Y., Fujii, H., Gerster, T., & Roeder, R. G. (1992). A novel B cell-derived coactivator potentiates the activation of immunoglobulin promoters by octamer-binding transcription factors. *Cell, 71*, 231.

Ma, S., Pathak, S., Mandal, M., Trinh, L., Clark, M. R., & Lu, R. (2010). Ikaros and Aiolos inhibit pre-B-cell proliferation by directly suppressing c-Myc expression. *Molecular and Cellular Biology, 30*, 4149.

Ma, S., Pathak, S., Trinh, L., & Lu, R. (2008). Interferon regulatory factors 4 and 8 induce the expression of Ikaros and Aiolos to down-regulate pre-B-cell receptor and promote cell-cycle withdrawal in pre-B-cell development. *Blood, 111*, 1396.

Ma, S., Turetsky, A., Trinh, L., & Lu, R. (2006). IFN regulatory factor 4 and 8 promote Ig light chain kappa locus activation in pre-B cell development. *Journal of Immunology, 177*, 7898.

Maier, H., Ostraat, R., Gao, H., Fields, S., Shinton, S. A., Medina, K. L., et al. (2004). Early B cell factor cooperates with Runx1 and mediates epigenetic changes associated with mb-1 transcription. *Nature Immunology, 5*, 1069.

Maitra, S., & Atchison, M. (2000). BSAP can repress enhancer activity by targeting PU.1 function. *Molecular and Cellular Biology, 20*, 1911.

Majumder, P., Gomez, J. A., Chadwick, B. P., & Boss, J. M. (2008). The insulator factor CTCF controls MHC class II gene expression and is required for the formation of long-distance chromatin interactions. *The Journal of Experimental Medicine, 205*, 785.

Majumder, K., Koues, O. I., Chan, E. A., Kyle, K. E., Horowitz, J. E., Yang-Iott, K., et al. (2015). Lineage-specific compaction of Tcrb requires a chromatin barrier to protect the function of a long-range tethering element. *The Journal of Experimental Medicine, 212*, 107.

Malin, S., McManus, S., Cobaleda, C., Novatchkova, M., Delogu, A., Bouillet, P., et al. (2010). Role of STAT5 in controlling cell survival and immunoglobulin gene recombination during pro-B cell development. *Nature Immunology, 11*, 171.

Mandal, M., Powers, S. E., Maienschein-Cline, M., Bartom, E. T., Hamel, K. M., Kee, B. L., et al. (2011). Epigenetic repression of the Igk locus by STAT5-mediated recruitment of the histone methyltransferase Ezh2. *Nature Immunology, 12*, 1212.

Mandal, M., Powers, S. E., Ochiai, K., Georgopoulos, K., Kee, B. L., Singh, H., et al. (2009). Ras orchestrates exit from the cell cycle and light-chain recombination during early B cell development. *Nature Immunology, 10*, 1110.

Mansson, R., Welinder, E., Ahsberg, J., Lin, Y. C., Benner, C., Glass, C. K., et al. (2012). Positive intergenic feedback circuitry, involving EBF1 and FOXO1, orchestrates B-cell fate. *Proceedings of the National Academy of Sciences of the United States of America, 109*, 21028.

Marshall, A. J., Fleming, H. E., Wu, G. E., & Paige, C. J. (1998). Modulation of the IL-7 dose-response threshold during pro-B cell differentiation is dependent on pre-B cell receptor expression. *Journal of Immunology, 161*, 6038.

Martin, D. J., & van Ness, B. G. (1990). Initiation and processing of two kappa immunoglobulin germ line transcripts in mouse B cells. *Molecular and Cellular Biology, 10*, 1950.

Mason, J. O., Williams, G. T., & Neuberger, M. S. (1985). Transcription cell type specificity is conferred by an immunoglobulin VH gene promoter that includes a functional consensus sequence. *Cell, 41*, 479.

Massari, M. E., Grant, P. A., Pray-Grant, M. G., Berger, S. L., Workman, J. L., & Murre, C. (1999). A conserved motif present in a class of helix-loop-helix proteins activates transcription by direct recruitment of the SAGA complex. *Molecular Cell, 4*, 63.

Maston, G. A., Evans, S. K., & Green, M. R. (2006). Transcriptional regulatory elements in the human genome. *Annual Review of Genomics and Human Genetics, 7*, 29.

McDevit, D. C., Perkins, L., Atchison, M. L., & Nikolajczyk, B. S. (2005). The Ig kappa 3′ enhancer is activated by gradients of chromatin accessibility and protein association. *Journal of Immunology, 174*, 2834.

Medina, K. L., Pongubala, J. M., Reddy, K. L., Lancki, D. W., Dekoter, R., Kieslinger, M., et al. (2004). Assembling a gene regulatory network for specification of the B cell fate. *Developmental Cell*, *7*, 607.

Medvedovic, J., Ebert, A., Tagoh, H., Tamir, I. M., Schwickert, T. A., Novatchkova, M., et al. (2013). Flexible long-range loops in the VH gene region of the Igh locus facilitate the generation of a diverse antibody repertoire. *Immunity*, *39*, 229.

Melamed, D., Benschop, R. J., Cambier, J. C., & Nemazee, D. (1998). Developmental regulation of B lymphocyte immune tolerance compartmentalizes clonal selection from receptor selection. *Cell*, *92*, 173.

Melchers, F. (2005). The pre-B-cell receptor: Selector of fitting immunoglobulin heavy chains for the B-cell repertoire. *Nature Reviews. Immunology*, *5*, 578.

Melchers, F., ten Boekel, E., Seidl, T., Kong, X. C., Yamagami, T., Onishi, K., et al. (2000). Repertoire selection by pre-B-cell receptors and B-cell receptors, and genetic control of B-cell development from immature to mature B cells. *Immunological Reviews*, *175*, 33.

Mercer, T. R., Dinger, M. E., & Mattick, J. S. (2009). Long non-coding RNAs: Insights into functions. *Nature Reviews. Genetics*, *10*, 155.

Meyer, K. B., & Neuberger, M. S. (1989). The immunoglobulin kappa locus contains a second, stronger B-cell-specific enhancer which is located downstream of the constant region. *The EMBO Journal*, *8*, 1959.

Middendorp, S., Dingjan, G. M., & Hendriks, R. W. (2002). Impaired precursor B cell differentiation in Bruton's tyrosine kinase-deficient mice. *Journal of Immunology*, *168*, 2695.

Morgan, B., Sun, L., Avitahl, N., Andrikopoulos, K., Ikeda, T., Gonzales, E., et al. (1997). Aiolos, a lymphoid restricted transcription factor that interacts with Ikaros to regulate lymphocyte differentiation. *The EMBO Journal*, *16*, 2004.

Mostoslavsky, R., Alt, F. W., & Rajewsky, K. (2004). The lingering enigma of the allelic exclusion mechanism. *Cell*, *118*, 539.

Mostoslavsky, R., Singh, N., Kirillov, A., Pelanda, R., Cedar, H., Chess, A., et al. (1998). Kappa chain monoallelic demethylation and the establishment of allelic exclusion. *Genes & Development*, *12*, 1801.

Mostoslavsky, R., Singh, N., Tenzen, T., Goldmit, M., Gabay, C., Elizur, S., et al. (2001). Asynchronous replication and allelic exclusion in the immune system. *Nature*, *414*, 221.

Muljo, S. A., & Schlissel, M. S. (2003). A small molecule Abl kinase inhibitor induces differentiation of Abelson virus-transformed pre-B cell lines. *Nature Immunology*, *4*, 31.

Murre, C., McCaw, P. S., & Baltimore, D. (1989). A new DNA binding and dimerization motif in immunoglobulin enhancer binding, daughterless, MyoD, and myc proteins. *Cell*, *56*, 777.

Murrell, A., Heeson, S., & Reik, W. (2004). Interaction between differentially methylated regions partitions the imprinted genes Igf2 and H19 into parent-specific chromatin loops. *Nature Genetics*, *36*, 889.

Nadel, B., Cazenave, P. A., & Sanchez, P. (1990). Murine lambda gene rearrangements: The stochastic model prevails over the ordered model. *The EMBO Journal*, *9*, 435.

Nagano, T., Lubling, Y., Stevens, T. J., Schoenfelder, S., Yaffe, E., Dean, W., et al. (2013). Single-cell Hi-C reveals cell-to-cell variability in chromosome structure. *Nature*, *502*, 59.

Nagaoka, H., Gonzalez-Aseguinolaza, G., Tsuji, M., & Nussenzweig, M. C. (2000). Immunization and infection change the number of recombination activating gene (RAG)-expressing B cells in the periphery by altering immature lymphocyte production. *The Journal of Experimental Medicine*, *191*, 2113.

Nagulapalli, S., & Atchison, M. L. (1998). Transcription factor Pip can enhance DNA binding by E47, leading to transcriptional synergy involving multiple protein domains. *Molecular and Cellular Biology*, *18*, 4639.

Nakayama, J., Yamamoto, M., Hayashi, K., Satoh, H., Bundo, K., Kubo, M., et al. (2009). BLNK suppresses pre-B-cell leukemogenesis through inhibition of JAK3. *Blood, 113*, 1483.

Nechanitzky, R., Akbas, D., Scherer, S., Gyory, I., Hoyler, T., Ramamoorthy, S., et al. (2013). Transcription factor EBF1 is essential for the maintenance of B cell identity and prevention of alternative fates in committed cells. *Nature Immunology, 14*, 867.

Nelson, K. J., Haimovich, J., & Perry, R. P. (1983). Characterization of productive and sterile transcripts from the immunoglobulin heavy-chain locus: Processing of micron and muS mRNA. *Molecular and Cellular Biology, 3*, 1317.

Nemazee, D. (2006). Receptor editing in lymphocyte development and central tolerance. *Nature Reviews. Immunology, 6*, 728.

Nemazee, D. A., & Burki, K. (1989). Clonal deletion of B lymphocytes in a transgenic mouse bearing anti-MHC class I antibody genes. *Nature, 337*, 562.

Nielsen, P. J., Georgiev, O., Lorenz, B., & Schaffner, W. (1996). B lymphocytes are impaired in mice lacking the transcriptional co-activator Bob1/OCA-B/OBF1. *European Journal of Immunology, 26*, 3214.

Novobrantseva, T. I., Martin, V. M., Pelanda, R., Muller, W., Rajewsky, K., & Ehlich, A. (1999). Rearrangement and expression of immunoglobulin light chain genes can precede heavy chain expression during normal B cell development in mice. *The Journal of Experimental Medicine, 189*, 75.

Nutt, S. L., Heavey, B., Rolink, A. G., & Busslinger, M. (1999). Commitment to the B-lymphoid lineage depends on the transcription factor Pax5. *Nature, 401*, 556.

Nutt, S. L., & Kee, B. L. (2007). The transcriptional regulation of B cell lineage commitment. *Immunity, 26*, 715.

Nutt, S. L., Morrison, A. M., Dorfler, P., Rolink, A., & Busslinger, M. (1998). Identification of BSAP (Pax-5) target genes in early B-cell development by loss- and gain-of-function experiments. *The EMBO Journal, 17*, 2319.

Nutt, S. L., Urbanek, P., Rolink, A., & Busslinger, M. (1997). Essential functions of Pax5 (BSAP) in pro-B cell development: Difference between fetal and adult B lymphopoiesis and reduced V-to-DJ recombination at the IgH locus. *Genes & Development, 11*, 476.

Ochiai, K., Maienschein-Cline, M., Mandal, M., Triggs, J. R., Bertolino, E., Sciammas, R., et al. (2012). A self-reinforcing regulatory network triggered by limiting IL-7 activates pre-BCR signaling and differentiation. *Nature Immunology, 13*, 300.

Odegard, V. H., & Schatz, D. G. (2006). Targeting of somatic hypermutation. *Nature Reviews Immunology, 6*, 573.

Oettinger, M. A., Schatz, D. G., Gorka, C., & Baltimore, D. (1990). RAG-1 and RAG-2, adjacent genes that synergistically activate V(D)J recombination. *Science, 248*, 1517.

Ohnishi, K., & Melchers, F. (2003). The nonimmunoglobulin portion of lambda5 mediates cell-autonomous pre-B cell receptor signaling. *Nature Immunology, 4*, 849.

Ong, C. T., & Corces, V. G. (2011). Enhancer function: New insights into the regulation of tissue-specific gene expression. *Nature Reviews. Genetics, 12*, 283.

Ong, C. T., & Corces, V. G. (2014). CTCF: An architectural protein bridging genome topology and function. *Nature Reviews. Genetics, 15*, 234.

O'Riordan, M., & Grosschedl, R. (1999). Coordinate regulation of B cell differentiation by the transcription factors EBF and E2A. *Immunity, 11*, 21.

Palstra, R. J., Tolhuis, B., Splinter, E., Nijmeijer, R., Grosveld, F., & de Laat, W. (2003). The beta-globin nuclear compartment in development and erythroid differentiation. *Nature Genetics, 35*, 190.

Pan, X., Papasani, M., Hao, Y., Calamito, M., Wei, F., Quinn Iii, W. J., et al. (2013). YY1 controls Igkappa repertoire and B-cell development, and localizes with condensin on the Igkappa locus. *The EMBO Journal, 32*, 1168.

Parelho, V., Hadjur, S., Spivakov, M., Leleu, M., Sauer, S., Gregson, H. C., et al. (2008). Cohesins functionally associate with CTCF on mammalian chromosome arms. *Cell, 132*, 422.

Park, K., & Atchison, M. L. (1991). Isolation of a candidate repressor/activator, NF-E1 (YY-1, delta), that binds to the immunoglobulin kappa 3′ enhancer and the immunoglobulin heavy-chain mu E1 site. *Proceedings of the National Academy of Sciences of the United States of America, 88*, 9804.

Park, S. K., Xiang, Y., Feng, X., & Garrard, W. T. (2014). Pronounced cohabitation of active immunoglobulin genes from three different chromosomes in transcription factories during maximal antibody synthesis. *Genes & Development, 28*, 1159.

Peschon, J. J., Morrissey, P. J., Grabstein, K. H., Ramsdell, F. J., Maraskovsky, E., Gliniak, B. C., et al. (1994). Early lymphocyte expansion is severely impaired in interleukin 7 receptor-deficient mice. *The Journal of Experimental Medicine, 180*, 1955.

Picard, D., & Schaffner, W. (1984). A lymphocyte-specific enhancer in the mouse immunoglobulin kappa gene. *Nature, 307*, 80.

Pierani, A., Heguy, A., Fujii, H., & Roeder, R. G. (1990). Activation of octamer-containing promoters by either octamer-binding transcription factor 1 (OTF-1) or OTF-2 and requirement of an additional B-cell-specific component for optimal transcription of immunoglobulin promoters. *Molecular and Cellular Biology, 10*, 6204.

Polli, M., Dakic, A., Light, A., Wu, L., Tarlinton, D. M., & Nutt, S. L. (2005). The development of functional B lymphocytes in conditional PU.1 knock-out mice. *Blood, 106*, 2083.

Pongubala, J. M., Nagulapalli, S., Klemsz, M. J., McKercher, S. R., Maki, R. A., & Atchison, M. L. (1992). PU.1 recruits a second nuclear factor to a site important for immunoglobulin kappa 3′ enhancer activity. *Molecular and Cellular Biology, 12*, 368.

Pongubala, J. M., Northrup, D. L., Lancki, D. W., Medina, K. L., Treiber, T., Bertolino, E., et al. (2008). Transcription factor EBF restricts alternative lineage options and promotes B cell fate commitment independently of Pax5. *Nature Immunology, 9*, 203.

Predeus, A. V., Gopalakrishnan, S., Huang, Y., Tang, J., Feeney, A. J., Oltz, E. M., et al. (2014). Targeted chromatin profiling reveals novel enhancers in Ig H and Ig L chain Loci. *Journal of Immunology, 192*, 1064.

Pridans, C., Holmes, M. L., Polli, M., Wettenhall, J. M., Dakic, A., Corcoran, L. M., et al. (2008). Identification of Pax5 target genes in early B cell differentiation. *Journal of Immunology, 180*, 1719.

Queen, C., & Baltimore, D. (1983). Immunoglobulin gene transcription is activated by downstream sequence elements. *Cell, 33*, 741.

Queen, C., & Stafford, J. (1984). Fine mapping of an immunoglobulin gene activator. *Molecular and Cellular Biology, 4*, 1042.

Quong, M. W., Martensson, A., Langerak, A. W., Rivera, R. R., Nemazee, D., & Murre, C. (2004). Receptor editing and marginal zone B cell development are regulated by the helix-loop-helix protein, E2A. *The Journal of Experimental Medicine, 199*, 1101.

Radic, M. Z., & Zouali, M. (1996). Receptor editing, immune diversification, and self-tolerance. *Immunity, 5*, 505.

Rao, S. S., Huntley, M. H., Durand, N. C., Stamenova, E. K., Bochkov, I. D., Robinson, J. T., et al. (2014). A 3D map of the human genome at kilobase resolution reveals principles of chromatin looping. *Cell, 159*, 1665.

Ray, D., Bosselut, R., Ghysdael, J., Mattei, M. G., Tavitian, A., & Moreau-Gachelin, F. (1992). Characterization of Spi-B, a transcription factor related to the putative oncoprotein Spi-1/PU.1. *Molecular and Cellular Biology, 12*, 4297.

Remenyi, A., Tomilin, A., Pohl, E., Lins, K., Philippsen, A., Reinbold, R., et al. (2001). Differential dimer activities of the transcription factor Oct-1 by DNA-induced interface swapping. *Molecular Cell, 8*, 569.

Retter, M. W., & Nemazee, D. (1998). Receptor editing occurs frequently during normal B cell development. *The Journal of Experimental Medicine, 188*, 1231.

Revilla, I. D. R., Bilic, I., Vilagos, B., Tagoh, H., Ebert, A., Tamir, I. M., et al. (2012). The B-cell identity factor Pax5 regulates distinct transcriptional programmes in early and late B lymphopoiesis. *The EMBO Journal, 31*, 3130.

Reynaud, D., Demarco, I. A., Reddy, K. L., Schjerven, H., Bertolino, E., Chen, Z., et al. (2008). Regulation of B cell fate commitment and immunoglobulin heavy-chain gene rearrangements by Ikaros. *Nature Immunology, 9*, 927.

Ribeiro de Almeida, C., Stadhouders, R., de Bruijn, M. J., Bergen, I. M., Thongjuea, S., Lenhard, B., et al. (2011). The DNA-binding protein CTCF limits proximal Vkappa recombination and restricts kappa enhancer interactions to the immunoglobulin kappa light chain locus. *Immunity, 35*, 501.

Ribeiro de Almeida, C., Stadhouders, R., Thongjuea, S., Soler, E., & Hendriks, R. W. (2012). DNA-binding factor CTCF and long-range gene interactions in V(D)J recombination and oncogene activation. *Blood, 119*, 6209.

Roessler, S., Gyory, I., Imhof, S., Spivakov, M., Williams, R. R., Busslinger, M., et al. (2007). Distinct promoters mediate the regulation of Ebf1 gene expression by interleukin-7 and Pax5. *Molecular and Cellular Biology, 27*, 579.

Roldan, E., Fuxa, M., Chong, W., Martinez, D., Novatchkova, M., Busslinger, M., et al. (2005). Locus 'decontraction' and centromeric recruitment contribute to allelic exclusion of the immunoglobulin heavy-chain gene. *Nature Immunology, 6*, 31.

Rolink, A., Karasuyama, H., Grawunder, U., Haasner, D., Kudo, A., & Melchers, F. (1993). B cell development in mice with a defective lambda 5 gene. *European Journal of Immunology, 23*, 1284.

Rolink, A., Kudo, A., Karasuyama, H., Kikuchi, Y., & Melchers, F. (1991). Long-term proliferating early pre B cell lines and clones with the potential to develop to surface Ig-positive, mitogen reactive B cells in vitro and in vivo. *The EMBO Journal, 10*, 327.

Rolink, A. G., Nutt, S. L., Melchers, F., & Busslinger, M. (1999). Long-term in vivo reconstitution of T-cell development by Pax5-deficient B-cell progenitors. *Nature, 401*, 603.

Romanow, W. J., Langerak, A. W., Goebel, P., Wolvers-Tettero, I. L., van Dongen, J. J., Feeney, A. J., et al. (2000). E2A and EBF act in synergy with the V(D)J recombinase to generate a diverse immunoglobulin repertoire in nonlymphoid cells. *Molecular Cell, 5*, 343.

Roque, M. C., Smith, P. A., & Blasquez, V. C. (1996). A developmentally modulated chromatin structure at the mouse immunoglobulin kappa 3′ enhancer. *Molecular and Cellular Biology, 16*, 3138.

Rosenbauer, F., Owens, B. M., Yu, L., Tumang, J. R., Steidl, U., Kutok, J. L., et al. (2006). Lymphoid cell growth and transformation are suppressed by a key regulatory element of the gene encoding PU.1. *Nature Genetics, 38*, 27.

Rubio, E. D., Reiss, D. J., Welcsh, P. L., Disteche, C. M., Filippova, G. N., Baliga, N. S., et al. (2008). CTCF physically links cohesin to chromatin. *Proceedings of the National Academy of Sciences of the United States of America, 105*, 8309.

Rumfelt, L. L., Zhou, Y., Rowley, B. M., Shinton, S. A., & Hardy, R. R. (2006). Lineage specification and plasticity in CD19- early B cell precursors. *The Journal of Experimental Medicine, 203*, 675.

Sakamoto, S., Wakae, K., Anzai, Y., Murai, K., Tamaki, N., Miyazaki, M., et al. (2012). E2A and CBP/p300 act in synergy to promote chromatin accessibility of the immunoglobulin kappa locus. *Journal of Immunology, 188*, 5547.

Satijn, D. P., Hamer, K. M., den Blaauwen, J., & Otte, A. P. (2001). The polycomb group protein EED interacts with YY1, and both proteins induce neural tissue in Xenopus embryos. *Molecular and Cellular Biology, 21*, 1360.

Sato, H., Saito-Ohara, F., Inazawa, J., & Kudo, A. (2004). Pax-5 is essential for kappa sterile transcription during Ig kappa chain gene rearrangement. *Journal of Immunology*, *172*, 4858.

Sato, H., Wang, D., & Kudo, A. (2001). Dissociation of Pax-5 from KI and KII sites during kappa-chain gene rearrangement correlates with its association with the underphosphorylated form of retinoblastoma. *Journal of Immunology*, *166*, 6704.

Schable, K. F., & Zachau, H. G. (1993). The variable genes of the human immunoglobulin kappa locus. *Biological Chemistry Hoppe-Seyler*, *374*, 1001.

Schaniel, C., Bruno, L., Melchers, F., & Rolink, A. G. (2002). Multiple hematopoietic cell lineages develop in vivo from transplanted Pax5-deficient pre-B I-cell clones. *Blood*, *99*, 472.

Schaniel, C., Gottar, M., Roosnek, E., Melchers, F., & Rolink, A. G. (2002). Extensive in vivo self-renewal, long-term reconstitution capacity, and hematopoietic multipotency of Pax5-deficient precursor B-cell clones. *Blood*, *99*, 2760.

Schatz, D. G., & Swanson, P. C. (2011). V(D)J recombination: Mechanisms of initiation. *Annual Review of Genetics*, *45*, 167.

Schebesta, A., McManus, S., Salvagiotto, G., Delogu, A., Busslinger, G. A., & Busslinger, M. (2007). Transcription factor Pax5 activates the chromatin of key genes involved in B cell signaling, adhesion, migration, and immune function. *Immunity*, *27*, 49.

Schebesta, M., Pfeffer, P. L., & Busslinger, M. (2002). Control of pre-BCR signaling by Pax5-dependent activation of the BLNK gene. *Immunity*, *17*, 473.

Schlissel, M. S. (2004). Regulation of activation and recombination of the murine Igkappa locus. *Immunological Reviews*, *200*, 215.

Schlissel, M. S., & Baltimore, D. (1989). Activation of immunoglobulin kappa gene rearrangement correlates with induction of germline kappa gene transcription. *Cell*, *58*, 1001.

Schubart, K., Massa, S., Schubart, D., Corcoran, L. M., Rolink, A. G., & Matthias, P. (2001). B cell development and immunoglobulin gene transcription in the absence of Oct-2 and OBF-1. *Nature Immunology*, *2*, 69.

Schubart, D. B., Rolink, A., Kosco-Vilbois, M. H., Botteri, F., & Matthias, P. (1996). B-cell-specific coactivator OBF-1/OCA-B/Bob1 required for immune response and germinal centre formation. *Nature*, *383*, 538.

Schubart, D. B., Rolink, A., Schubart, K., & Matthias, P. (2000). Cutting edge: lack of peripheral B cells and severe agammaglobulinemia in mice simultaneously lacking Bruton's tyrosine kinase and the B cell-specific transcriptional coactivator OBF-1. *Journal of Immunology*, *164*, 18.

Schulz, D., Vassen, L., Chow, K. T., McWhirter, S. M., Amin, R. H., Moroy, T., et al. (2012). Gfi1b negatively regulates Rag expression directly and via the repression of FoxO1. *The Journal of Experimental Medicine*, *209*, 187.

Schwarzenbach, H., Newell, J. W., & Matthias, P. (1995). Involvement of the Ets family factor PU.1 in the activation of immunoglobulin promoters. *The Journal of Biological Chemistry*, *270*, 898.

Schweitzer, B. L., & DeKoter, R. P. (2004). Analysis of gene expression and Ig transcription in PU.1/Spi-B-deficient progenitor B cell lines. *Journal of Immunology*, *172*, 144.

Schwickert, T. A., Tagoh, H., Gultekin, S., Dakic, A., Axelsson, E., Minnich, M., et al. (2014). Stage-specific control of early B cell development by the transcription factor Ikaros. *Nature Immunology*, *15*, 283.

Scott, E. W., Fisher, R. C., Olson, M. C., Kehrli, E. W., Simon, M. C., & Singh, H. (1997). PU.1 functions in a cell-autonomous manner to control the differentiation of multipotential lymphoid-myeloid progenitors. *Immunity*, *6*, 437.

Scott, E. W., Simon, M. C., Anastasi, J., & Singh, H. (1994). Requirement of transcription factor PU.1 in the development of multiple hematopoietic lineages. *Science*, *265*, 1573.

Seet, C. S., Brumbaugh, R. L., & Kee, B. L. (2004). Early B cell factor promotes B lymphopoiesis with reduced interleukin 7 responsiveness in the absence of E2A. *The Journal of Experimental Medicine, 199*, 1689.

Sekimata, M., Perez-Melgosa, M., Miller, S. A., Weinmann, A. S., Sabo, P. J., Sandstrom, R., et al. (2009). CCCTC-binding factor and the transcription factor T-bet orchestrate T helper 1 cell-specific structure and function at the interferon-gamma locus. *Immunity, 31*, 551.

Sellars, M., Kastner, P., & Chan, S. (2011). Ikaros in B cell development and function. *World Journal of Biological Chemistry, 2*, 132.

Sen, R. (2004). NF-kappaB and the immunoglobulin kappa gene enhancer. *The Journal of Experimental Medicine, 200*, 1099.

Sen, R., & Baltimore, D. (1986a). Inducibility of kappa immunoglobulin enhancer-binding protein Nf-kappa B by a posttranslational mechanism. *Cell, 47*, 921.

Sen, R., & Baltimore, D. (1986b). Multiple nuclear factors interact with the immunoglobulin enhancer sequences. *Cell, 46*, 705.

Serwe, M., & Sablitzky, F. (1993). V(D)J recombination in B cells is impaired but not blocked by targeted deletion of the immunoglobulin heavy chain intron enhancer. *The EMBO Journal, 12*, 2321.

Shaffer, A. L., Peng, A., & Schlissel, M. S. (1997). In vivo occupancy of the kappa light chain enhancers in primary pro- and pre-B cells: A model for kappa locus activation. *Immunity, 6*, 131.

Shaw, A. C., Swat, W., Davidson, L., & Alt, F. W. (1999). Induction of Ig light chain gene rearrangement in heavy chain-deficient B cells by activated Ras. *Proceedings of the National Academy of Sciences of the United States of America, 96*, 2239.

Shimizu, T., Mundt, C., Licence, S., Melchers, F., & Martensson, I. L. (2002). VpreB1/VpreB2/lambda 5 triple-deficient mice show impaired B cell development but functional allelic exclusion of the IgH locus. *Journal of Immunology, 168*, 6286.

Sigvardsson, M. (2000). Overlapping expression of early B-cell factor and basic helix-loop-helix proteins as a mechanism to dictate B-lineage-specific activity of the lambda5 promoter. *Molecular and Cellular Biology, 20*, 3640.

Sigvardsson, M., Akerblad, P., & Leanderson, T. (1996). Early B cell factor interacts with a subset of kappa promoters. *Journal of Immunology, 156*, 3788.

Sigvardsson, M., Bemark, M., & Leanderson, T. (1995). Stimulation of kappa transcription by a decamer-dependent, synergistic mechanism. *European Journal of Immunology, 25*, 298.

Sigvardsson, M., O'Riordan, M., & Grosschedl, R. (1997). EBF and E47 collaborate to induce expression of the endogenous immunoglobulin surrogate light chain genes. *Immunity, 7*, 25.

Siminovitch, K. A., Bakhshi, A., Goldman, P., & Korsmeyer, S. J. (1985). A uniform deleting element mediates the loss of kappa genes in human B cells. *Nature, 316*, 260.

Singh, N., Bergman, Y., Cedar, H., & Chess, A. (2003). Biallelic germline transcription at the kappa immunoglobulin locus. *The Journal of Experimental Medicine, 197*, 743.

Sitnikova, T., & Nei, M. (1998). Evolution of immunoglobulin kappa chain variable region genes in vertebrates. *Molecular Biology and Evolution, 15*, 50.

Skok, J. A., Brown, K. E., Azuara, V., Caparros, M. L., Baxter, J., Takacs, K., et al. (2001). Nonequivalent nuclear location of immunoglobulin alleles in B lymphocytes. *Nature Immunology, 2*, 848.

Smith, E. M., Gisler, R., & Sigvardsson, M. (2002). Cloning and characterization of a promoter flanking the early B cell factor (EBF) gene indicates roles for E-proteins and autoregulation in the control of EBF expression. *Journal of Immunology, 169*, 261.

Souabni, A., Cobaleda, C., Schebesta, M., & Busslinger, M. (2002). Pax5 promotes B lymphopoiesis and blocks T cell development by repressing Notch1. *Immunity, 17*, 781.

Splinter, E., Heath, H., Kooren, J., Palstra, R. J., Klous, P., Grosveld, F., et al. (2006). CTCF mediates long-range chromatin looping and local histone modification in the beta-globin locus. *Genes & Development, 20*, 2349.

Stadhouders, R., de Bruijn, M. J., Rother, M. B., Yuvaraj, S., Ribeiro de Almeida, C., Kolovos, P., et al. (2014). Pre-B cell receptor signaling induces immunoglobulin kappa locus accessibility by functional redistribution of enhancer-mediated chromatin interactions. *PLoS Biology, 12*, e1001791.

Stanhope-Baker, P., Hudson, K. M., Shaffer, A. L., Constantinescu, A., & Schlissel, M. S. (1996). Cell type-specific chromatin structure determines the targeting of V(D)J recombinase activity in vitro. *Cell, 85*, 887.

Staudt, L. M., & Lenardo, M. J. (1991). Immunoglobulin gene transcription. *Annual Review of Immunology, 9*, 373.

Stevens, S., Ong, J., Kim, U., Eckhardt, L. A., & Roeder, R. G. (2000). Role of OCA-B in 3′-IgH enhancer function. *Journal of Immunology, 164*, 5306.

Strubin, M., Newell, J. W., & Matthias, P. (1995). OBF-1, a novel B cell-specific coactivator that stimulates immunoglobulin promoter activity through association with octamer-binding proteins. *Cell, 80*, 497.

Su, I. H., Basavaraj, A., Krutchinsky, A. N., Hobert, O., Ullrich, A., Chait, B. T., et al. (2003). Ezh2 controls B cell development through histone H3 methylation and Igh rearrangement. *Nature Immunology, 4*, 124.

Takeda, S., Zou, Y. R., Bluethmann, H., Kitamura, D., Muller, U., & Rajewsky, K. (1993). Deletion of the immunoglobulin kappa chain intron enhancer abolishes kappa chain gene rearrangement in cis but not lambda chain gene rearrangement in trans. *The EMBO Journal, 12*, 2329.

Taniguchi, T., Ogasawara, K., Takaoka, A., & Tanaka, N. (2001). IRF family of transcription factors as regulators of host defense. *Annual Review of Immunology, 19*, 623.

Taylor, B., Cobb, B. S., Bruno, L., Webster, Z., Fisher, A. G., & Merkenschlager, M. (2009). A reappraisal of evidence for probabilistic models of allelic exclusion. *Proceedings of the National Academy of Sciences of the United States of America, 106*, 516.

ten Boekel, E., Yamagami, T., Andersson, J., Rolink, A. G., & Melchers, F. (1999). The formation and selection of cells expressing preB cell receptors and B cell receptors. *Current Topics in Microbiology and Immunology, 246*, 3.

Thevenin, C., Nutt, S. L., & Busslinger, M. (1998). Early function of Pax5 (BSAP) before the pre-B cell receptor stage of B lymphopoiesis. *The Journal of Experimental Medicine, 188*, 735.

Thompson, E. C., Cobb, B. S., Sabbattini, P., Meixlsperger, S., Parelho, V., Liberg, D., et al. (2007). Ikaros DNA-binding proteins as integral components of B cell developmental-stage-specific regulatory circuits. *Immunity, 26*, 335.

Thompson, A., Timmers, E., Kenter, M. J., Kraakman, M. E., Hendriks, R. W., & Schuurman, R. K. (1992). Immunoglobulin kappa light chain germ-line transcripts in human precursor B lymphocytes. *European Journal of Immunology, 22*, 3167.

Tian, J., Okabe, T., Miyazaki, T., Takeshita, S., & Kudo, A. (1997). Pax-5 is identical to EBB-1/KLP and binds to the VpreB and lambda5 promoters as well as the KI and KII sites upstream of the Jkappa genes. *European Journal of Immunology, 27*, 750.

Tiegs, S. L., Russell, D. M., & Nemazee, D. (1993). Receptor editing in self-reactive bone marrow B cells. *The Journal of Experimental Medicine, 177*, 1009.

Timblin, G. A., & Schlissel, M. S. (2013). Ebf1 and c-Myb repress rag transcription downstream of Stat5 during early B cell development. *Journal of Immunology, 191*, 4676.

Tiwari, V. K., McGarvey, K. M., Licchesi, J. D., Ohm, J. E., Herman, J. G., Schubeler, D., et al. (2008). PcG proteins, DNA methylation, and gene repression by chromatin looping. *PLoS Biology, 6*, 2911.

Tokoyoda, K., Egawa, T., Sugiyama, T., Choi, B. I., & Nagasawa, T. (2004). Cellular niches controlling B lymphocyte behavior within bone marrow during development. *Immunity, 20*, 707.

Tolhuis, B., Palstra, R. J., Splinter, E., Grosveld, F., & de Laat, W. (2002). Looping and interaction between hypersensitive sites in the active beta-globin locus. *Molecular Cell, 10*, 1453.

Tomilin, A., Remenyi, A., Lins, K., Bak, H., Leidel, S., Vriend, G., et al. (2000). Synergism with the coactivator OBF-1 (OCA-B, BOB-1) is mediated by a specific POU dimer configuration. *Cell, 103*, 853.

Tonegawa, S. (1983). Somatic generation of antibody diversity. *Nature, 302*, 575.

Treiber, T., Mandel, E. M., Pott, S., Gyory, I., Firner, S., Liu, E. T., et al. (2010). Early B cell factor 1 regulates B cell gene networks by activation, repression, and transcription- independent poising of chromatin. *Immunity, 32*, 714.

Ubelhart, R., Bach, M. P., Eschbach, C., Wossning, T., Reth, M., & Jumaa, H. (2010). N-linked glycosylation selectively regulates autonomous precursor BCR function. *Nature Immunology, 11*, 759.

Urbanek, P., Wang, Z. Q., Fetka, I., Wagner, E. F., & Busslinger, M. (1994). Complete block of early B cell differentiation and altered patterning of the posterior midbrain in mice lacking Pax5/BSAP. *Cell, 79*, 901.

van der Stoep, N., Gorman, J. R., & Alt, F. W. (1998). Reevaluation of 3′Ekappa function in stage- and lineage-specific rearrangement and somatic hypermutation. *Immunity, 8*, 743.

van Loo, P. F., Dingjan, G. M., Maas, A., & Hendriks, R. W. (2007). Surrogate-light-chain silencing is not critical for the limitation of pre-B cell expansion but is for the termination of constitutive signaling. *Immunity, 27*, 468.

Van Ness, B. G., Weigert, M., Coleclough, C., Mather, E. L., Kelley, D. E., & Perry, R. P. (1981). Transcription of the unrearranged mouse C kappa locus: Sequence of the initiation region and comparison of activity with a rearranged V kappa-C kappa gene. *Cell, 27*, 593.

van Zelm, M. C., Szczepanski, T., van der Burg, M., & van Dongen, J. J. (2007). Replication history of B lymphocytes reveals homeostatic proliferation and extensive antigen-induced B cell expansion. *The Journal of Experimental Medicine, 204*, 645.

Vela, J. L., Ait-Azzouzene, D., Duong, B. H., Ota, T., & Nemazee, D. (2008). Rearrangement of mouse immunoglobulin kappa deleting element recombining sequence promotes immune tolerance and lambda B cell production. *Immunity, 28*, 161.

Venkitaraman, A. R., & Cowling, R. J. (1994). Interleukin-7 induces the association of phosphatidylinositol 3-kinase with the alpha chain of the interleukin-7 receptor. *European Journal of Immunology, 24*, 2168.

Verma-Gaur, J., Torkamani, A., Schaffer, L., Head, S. R., Schork, N. J., & Feeney, A. J. (2012). Noncoding transcription within the Igh distal V(H) region at PAIR elements affects the 3D structure of the Igh locus in pro-B cells. *Proceedings of the National Academy of Sciences of the United States of America, 109*, 17004.

Vettermann, C., Timblin, G. A., Lim, V., Lai, E. C., & Schlissel, M. S. (2015). The proximal J kappa germline-transcript promoter facilitates receptor editing through control of ordered recombination. *PLoS One, 10*, e0113824.

Wabl, M. R., & Burrows, P. D. (1984). Expression of immunoglobulin heavy chain at a high level in the absence of a proposed immunoglobulin enhancer element in cis. *Proceedings of the National Academy of Sciences of the United States of America, 81*, 2452.

Wang, J. H., Avitahl, N., Cariappa, A., Friedrich, C., Ikeda, T., Renold, A., et al. (1998). Aiolos regulates B cell activation and maturation to effector state. *Immunity, 9*, 543.

Wang, V. E., Tantin, D., Chen, J., & Sharp, P. A. (2004). B cell development and immunoglobulin transcription in Oct-1-deficient mice. *Proceedings of the National Academy of Sciences of the United States of America, 101*, 2005.

Wardemann, H., Yurasov, S., Schaefer, A., Young, J. W., Meffre, E., & Nussenzweig, M. C. (2003). Predominant autoantibody production by early human B cell precursors. *Science*, *301*, 1374.

Weaver, D., & Baltimore, D. (1987). B lymphocyte-specific protein binding near an immunoglobulin kappa-chain gene J segment. *Proceedings of the National Academy of Sciences of the United States of America*, *84*, 1516.

Weigert, M., Gatmaitan, L., Loh, E., Schilling, J., & Hood, L. (1978). Rearrangement of genetic information may produce immunoglobulin diversity. *Nature*, *276*, 785.

Weill, J. C., & Reynaud, C. A. (2008). DNA polymerases in adaptive immunity. *Nature Reviews. Immunology*, *8*, 302.

Welinder, E., Mansson, R., Mercer, E. M., Bryder, D., Sigvardsson, M., & Murre, C. (2011). The transcription factors E2A and HEB act in concert to induce the expression of FOXO1 in the common lymphoid progenitor. *Proceedings of the National Academy of Sciences of the United States of America*, *108*, 17402.

Wendt, K. S., Yoshida, K., Itoh, T., Bando, M., Koch, B., Schirghuber, E., et al. (2008). Cohesin mediates transcriptional insulation by CCCTC-binding factor. *Nature*, *451*, 796.

Whitehurst, C., Henney, H. R., Max, E. E., Schroeder, H. W., Jr., Stuber, F., Siminovitch, K. A., et al. (1992). Nucleotide sequence of the intron of the germline human kappa immunoglobulin gene connecting the J and C regions reveals a matrix association region (MAR) next to the enhancer. *Nucleic Acids Research*, *20*, 4929.

Whyte, W. A., Orlando, D. A., Hnisz, D., Abraham, B. J., Lin, C. Y., Kagey, M. H., et al. (2013). Master transcription factors and mediator establish super-enhancers at key cell identity genes. *Cell*, *153*, 307.

Wilkinson, F. H., Park, K., & Atchison, M. L. (2006). Polycomb recruitment to DNA in vivo by the YY1 REPO domain. *Proceedings of the National Academy of Sciences of the United States of America*, *103*, 19296.

Witsch, E. J., & Bettelheim, E. (2008). Allelic and isotypic light chain inclusion in peripheral B cells from anti-DNA antibody transgenic C57BL/6 and BALB/c mice. *Journal of Immunology*, *180*, 3708.

Wood, D. L., & Coleclough, C. (1984). Different joining region J elements of the murine kappa immunoglobulin light chain locus are used at markedly different frequencies. *Proceedings of the National Academy of Sciences of the United States of America*, *81*, 4756.

Xiang, Y., & Garrard, W. T. (2008). The Downstream Transcriptional Enhancer, Ed, positively regulates mouse Ig kappa gene expression and somatic hypermutation. *Journal of Immunology*, *180*, 6725.

Xiang, Y., Park, S. K., & Garrard, W. T. (2013). Vkappa gene repertoire and locus contraction are specified by critical DNase I hypersensitive sites within the Vkappa-Jkappa intervening region. *Journal of Immunology*, *190*, 1819.

Xiang, Y., Park, S. K., & Garrard, W. T. (2014). A major deletion in the Vkappa-Jkappa intervening region results in hyperelevated transcription of proximal Vkappa genes and a severely restricted repertoire. *Journal of Immunology*, *193*, 3746.

Xiang, Y., Zhou, X., Hewitt, S. L., Skok, J. A., & Garrard, W. T. (2011). A multifunctional element in the mouse Igkappa locus that specifies repertoire and Ig loci subnuclear location. *Journal of Immunology*, *186*, 5356.

Xu, Y., Davidson, L., Alt, F. W., & Baltimore, D. (1996). Deletion of the Ig kappa light chain intronic enhancer/matrix attachment region impairs but does not abolish V kappa J kappa rearrangement. *Immunity*, *4*, 377.

Yamagami, T., ten Boekel, E., Andersson, J., Rolink, A., & Melchers, F. (1999). Frequencies of multiple IgL chain gene rearrangements in single normal or kappaL chain-deficient B lineage cells. *Immunity*, *11*, 317.

Yancopoulos, G. D., & Alt, F. W. (1985). Developmentally controlled and tissue-specific expression of unrearranged VH gene segments. *Cell*, *40*, 271.

Yancopoulos, G. D., Blackwell, T. K., Suh, H., Hood, L., & Alt, F. W. (1986). Introduced T cell receptor variable region gene segments recombine in pre-B cells: Evidence that B and T cells use a common recombinase. *Cell*, *44*, 251.

Yao, Z., Cui, Y., Watford, W. T., Bream, J. H., Yamaoka, K., Hissong, B. D., et al. (2006). Stat5a/b are essential for normal lymphoid development and differentiation. *Proceedings of the National Academy of Sciences of the United States of America*, *103*, 1000.

Ye, M., Ermakova, O., & Graf, T. (2005). PU.1 is not strictly required for B cell development and its absence induces a B-2 to B-1 cell switch. *The Journal of Experimental Medicine*, *202*, 1411.

Yi, M., Wu, P., Trevorrow, K. W., Claflin, L., & Garrard, W. T. (1999). Evidence that the Igkappa gene MAR regulates the probability of premature V-J joining and somatic hypermutation. *Journal of Immunology*, *162*, 6029.

Zandi, S., Mansson, R., Tsapogas, P., Zetterblad, J., Bryder, D., & Sigvardsson, M. (2008). EBF1 is essential for B-lineage priming and establishment of a transcription factor network in common lymphoid progenitors. *Journal of Immunology*, *181*, 3364.

Zhang, Z., Espinoza, C. R., Yu, Z., Stephan, R., He, T., Williams, G. S., et al. (2006). Transcription factor Pax5 (BSAP) transactivates the RAG-mediated V(H)-to-DJ (H) rearrangement of immunoglobulin genes. *Nature Immunology*, 7, 616.

Zhang, J., Jackson, A. F., Naito, T., Dose, M., Seavitt, J., Liu, F., et al. (2012). Harnessing of the nucleosome-remodeling-deacetylase complex controls lymphocyte development and prevents leukemogenesis. *Nature Immunology*, *13*, 86.

Zhang, X., Tang, N., Hadden, T. J., & Rishi, A. K. (2011). Akt, FoxO and regulation of apoptosis. *Biochimica et Biophysica Acta*, *1813*, 1978.

Zhao, F., McCarrick-Walmsley, R., Akerblad, P., Sigvardsson, M., & Kadesch, T. (2003). Inhibition of p300/CBP by early B-cell factor. *Molecular and Cellular Biology*, *23*, 3837.

Zhou, X., Xiang, Y., Ding, X., & Garrard, W. T. (2012). A new hypersensitive site, HS10, and the enhancers, E3′ and Ed, differentially regulate Igkappa gene expression. *Journal of Immunology*, *188*, 2722.

Zhou, X., Xiang, Y., Ding, X., & Garrard, W. T. (2013). Loss of an Igkappa gene enhancer in mature B cells results in rapid gene silencing and partial reversible dedifferentiation. *Molecular and Cellular Biology*, *33*, 2091.

Zhou, X., Xiang, Y., & Garrard, W. T. (2010). The Igkappa gene enhancers, E3′ and Ed, are essential for triggering transcription. *Journal of Immunology*, *185*, 7544.

Zhuang, Y., Soriano, P., & Weintraub, H. (1994). The helix-loop-helix gene E2A is required for B cell formation. *Cell*, *79*, 875.

Zou, Y. R., Takeda, S., & Rajewsky, K. (1993). Gene targeting in the Ig kappa locus: Efficient generation of lambda chain-expressing B cells, independent of gene rearrangements in Ig kappa. *The EMBO Journal*, *12*, 811.

CHAPTER SIX

Regulation of *Tcrb* Gene Assembly by Genetic, Epigenetic, and Topological Mechanisms

Kinjal Majumder*, Craig H. Bassing[†,‡], Eugene M. Oltz*[,1]
*Department of Pathology and Immunology, Washington University School of Medicine, Saint Louis, Missouri, USA
[†]Division of Cancer Pathobiology, Department of Pathology and Laboratory Medicine, Center for Childhood Cancer Research, Children's Hospital of Philadelphia, Philadelphia, Pennsylvania, USA
[‡]Abramson Family Cancer Research Institute, Cell and Molecular Biology Graduate Program, Department of Pathology and Laboratory Medicine, Perelman School of Medicine at the University of Pennsylvania, Philadelphia, Pennsylvania, USA
[1]Corresponding author: e-mail address: eoltz@wustl.edu

Contents

Abstract

The adaptive immune system endows mammals with an ability to recognize nearly any foreign invader through antigen receptors that are expressed on the surface of all lymphocytes. This defense network is generated by V(D)J recombination, a set of sequentially controlled DNA cleavage and repair events that assemble antigen receptor genes from physically separated variable (V), joining (J), and sometimes diversity (D) gene segments. The recombination process itself must be stringently regulated to minimize oncogenic translocations involving chromosomes that harbor immunoglobulin and T cell receptor loci. Indeed, V(D)J recombination is controlled at several levels, including

Advances in Immunology, Volume 128
ISSN 0065-2776
http://dx.doi.org/10.1016/bs.ai.2015.07.001
© 2015 Elsevier Inc.
All rights reserved.

tissue-, developmental stage-, allele-, and gene segment-specificity. These levels of control are imposed by a collection of architectural and regulatory elements that are distributed throughout each antigen receptor locus. Together, the genetic elements regulate developmental changes in chromatin, transcription, and locus topology that promote or disfavor long-range recombination. This chapter focuses on the cross talk between these mechanisms at the T cell receptor beta (*Tcrb*) locus, and how they sculpt a diverse TCRβ repertoire while maintaining monospecificity of this antigen receptor on each mature T lymphocyte. We also discuss how insights obtained from studies of *Tcrb* are more generally relevant to our understanding of gene regulation strategies employed by mammals.

1. INTRODUCTION

The generation of antigen receptor (AgR) diversity by V(D)J recombination is a cornerstone of adaptive immunity. Through this process, each developing lymphocyte fuses a single variable (V), diversity (D), and joining (J) gene segment from large arrays to produce an exon encoding the antigen recognition portion (variable region) for each chain of B and T cell receptors (BCR and TCR, respectively) (Alt, Zhang, Meng, Guo, & Schwer, 2013; Bossen, Mansson, & Murre, 2012; Brack, Hirama, Lenhard-Schuller, & Tonegawa, 1978; Choi & Feeney, 2014; Jung, Giallourakis, Mostoslavsky, & Alt, 2006; Seitan, Krangel, & Merkenschlager, 2012; Weigert, Gatmaitan, Loh, Schilling, & Hood, 1978). V(D)J recombination is tightly regulated to strike a balance between sculpting a diverse AgR repertoire, which is required to set up a frontline defense against nearly any pathogen, and producing a subset of specific receptors that provide the signature antigen-binding properties of specialized lymphocyte subsets. At the molecular level, V(D)J recombination relies on a stepwise DNA break and repair process that is mediated by the RAG1/2 (RAG) endonuclease complex (Oettinger, Schatz, Gorka, & Baltimore, 1990; Schatz, Oettinger, & Baltimore, 1989). Stepwise assembly of each AgR locus, in turn, guides precursor B/T cell development and ensures that a single BCR or TCR is expressed by each lymphocyte. This chapter focuses on assembly of the *Tcrb* locus, which transpires in $CD4^-CD8^-$ thymocytes (double negative, DN), culminating in expression of the beta TCR chain on mature α/β T cells. Emphasis will be placed on advances in our understanding of the genetic elements that control tissue-, stage-, and allele-specific recruitment of RAG to regions within *Tcrb*, while sparing rearrangement of other AgR genes in DN thymocytes. We also discuss how these genetic elements supervise epigenetic and topological mechanisms that facilitate long-range Vβ to DβJβ recombination, an important checkpoint for maintaining TCR monospecificity on T lymphocytes.

2. RAG-MEDIATED RECOMBINATION OF AgR GENE SEGMENTS

The variable region exons of *Ig* and *Tcr* genes must be pieced together by somatic recombination of V, J, and sometimes D segments in developing B and T cells. There are two critical components of the lymphocyte-specific V(D)J recombinase, RAG1 and RAG2 (heretofore simply called RAG), which form a recognition/endonuclease complex that creates DNA double-strand breaks (DSBs) adjacent to AgR gene segments. Initially, RAG must gain access to its cognate DNA targets, known as recombination signal sequences (RSSs), which flank all V, D, and J gene segments and are composed of conserved heptamer and nonamer motifs (consensus sequences CACAGTG and ACAAAAACC, respectively) (Schatz & Ji, 2011). For each RSS, the heptamer and nonamer are separated by a less conserved spacer sequence that is 12 or 23 base pairs in length (12-RSS or 23-RSS, respectively). Synapsis of two gene segments and their coordinate cleavage by RAG only occurs between segments flanked by a 12-RSS and a 23-RSS (12/23 rule), generating DSBs precisely at the two RSS/coding region borders.

The cleaved gene segments, which remain on two pieces of the targeted chromosome, are rapidly fused together by DNA repair proteins belonging to the nonhomologous end joining machinery, forming a coding join. Depending on the orientation of targeted gene segments, intervening DNA is either excised as an extrachromosomal signal join or it is inverted and retained within the chromosome (Gellert, 2002; Schatz & Ji, 2011). Importantly, before gene segments are fused back together, their ends are processed randomly by endonucleases and by addition of nucleotides. Random processing of the V(D)J junctions can disrupt the reading frame of variable region exons, resulting in nonfunctional coding joins for 2/3 rearranged alleles. Although seemingly wasteful, diversity arising from the processed coding joins is critical because this portion of a variable exon encodes for the complementarity determining region 3 (CDR3) of AgRs, which normally forms the most intimate contacts with cognate antigens.

3. *Tcrb* GENE ASSEMBLY DURING THYMOCYTE DEVELOPMENT

Thymocytes destined to become α/β T cells first assemble *Tcrb* in DN thymocytes from its composite Vβ, Dβ, and Jβ gene segments. As shown in Fig. 1, mouse (m*Tcrb*, chr6, 650 kb) and human *Tcrb* (h*Tcrb*, chr7, 575 kb) exhibit similar linear arrangements, suggesting that common mechanisms may regulate their assembly and expression (Glusman et al., 2001; Lane, Duroux, & Lefranc, 2010). The 3′ end of both loci harbor two Dβ–Jβ clusters (see also Fig. 2), spanning 25 kb, which each also possess Dβ promoter elements that drive transcription of Dβ–Jβ gene segments in thymocytes (Doty, Xia, Nguyen, Hathaway, & Willerford, 1999; Sikes, Suarez, & Oltz, 1999; Whitehurst, Chattopadhyay, & Chen, 1999). A single enhancer element (Eβ) and a Vβ gene segment (*Trbv31*) lie downstream of the Dβ–Jβ clusters in both loci (Figs. 1 and 2). The *Trbv31* orientation is opposite that of the Dβ–Jβ clusters, requiring *Trbv31* to rearrange by inversion with Dβ–Jβ segments. The main Vβ cluster, whose Vβ segments all rearrange by a

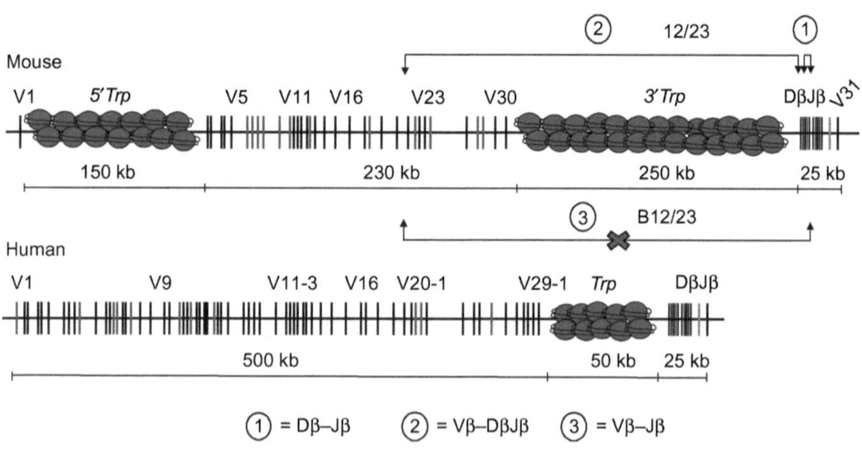

Figure 1 Comparison of the mouse and human T cell receptor beta loci. The linear arrangement of V, D, and J gene segments on mouse (upper) and human (lower) *Tcrb* loci. Functional Vβ gene segments are depicted as black bars, nonfunctional, pseudogene segments as red bars, and the enhancer, Eβ, as a blue bar. Large regions of repressive chromatin spanning the silent trypsinogen gene clusters (5′ and 3′ *Trp*) are represented by red nucleosomes. The steps of ordered *Tcrb* assembly are shown as 1 (Dβ to Jβ) and 2 (Vβ to DβJβ), and the red "X" signifies a block in direct Vβ to Jβ rearrangement in accordance with "Beyond 12/23" restrictions (B12/23). (See the color plate.)

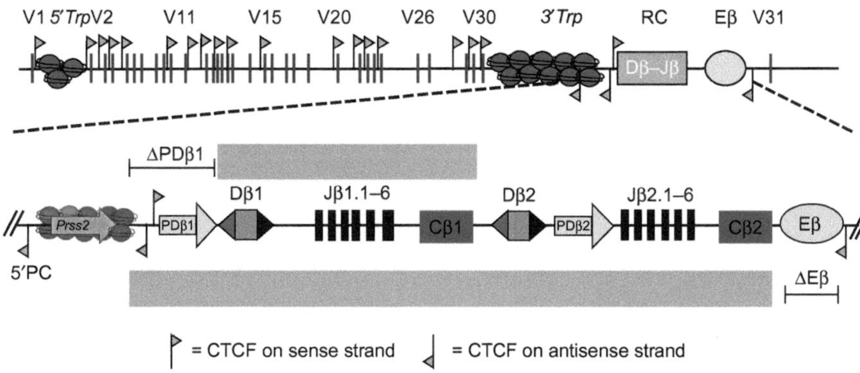

Figure 2 Schematic of *Tcrb* regulatory and architectural elements. (Top) Overview showing relative locations of the DβJβEβ recombination center (RC) and major Vβ cluster. Highlighted are the approximate positions of CTCF sites in *Tcrb*, which are depicted as blue flags indicating their orientations (on sense or antisense strands). (Bottom) Zoomed-in view of the *Tcrb*-RC and its neighboring region. Deletions of two regulatory elements are shown (ΔPDβ1 and ΔEβ), and their impact on silencing of gene segments within the RC is indicated by gray bars. (See the color plate.)

deletional mechanism, resides 250–500 kb upstream of the Dβ–Jβ clusters and is composed of the *Trbv1–Trbv30* gene segments in h*Tcrb* or the *Trbv2–Trbv30* gene segments in m*Tcrb* (Fig. 1). In m*Tcrb*, the *Trbv1* gene segment resides an additional 150 kb upstream of the main Vβ cluster. The region separating the main Vβ cluster from the Dβ–Jβ clusters in both m*Tcrb* and h*Tcrb* encompasses several trypsinogen (*Trp*) genes. Likewise, the region separating *Trbv1* from the main Vβ cluster in m*Tcrb* contains multiple *Trp* genes.

From a regulatory perspective, the chromatin and transcriptional landscape of *Tcrb* alternates from active to inactive compartments over its 700 kb span. The Dβ–Jβ cluster, which resides in open chromatin and is highly active transcriptionally in DN thymocytes, is separated from open chromatin that spans the modestly expressed Vβ gene segments by a series of transcriptionally repressed *Trp* genes, which reside in closed chromatin. Interruption of active Vβ and Dβ–Jβ segments by large stretches of inactive chromatin may compartmentalize *Tcrb* to ensure the stepwise assembly of VβDβJβ coding joins. In this regard, DN thymocytes initiate recombination by targeting RAG proteins to *Tcrb*, *Tcrg*, and *Tcrd*. At *Tcrb*, RAG deposition occurs primarily within the two Dβ–Jβ clusters, which, for this reason, has been coined its recombination center (RC) (Ji, Resch et al., 2010). As DN thymocytes differentiate from the $CD44^{+}CD25^{-}$ (DN1) to the

$CD44^{+}CD25^{+}$ (DN2) stage, the RC undergoes efficient rearrangement, generating DβJβ joins on both alleles. Progression to the $CD44^{-}CD25^{+}$ (DN3) stage is accompanied by the less efficient step of long-range Vβ-to-DβJβ rearrangement, which is thought to occur one allele at a time (Brady, Steinel, & Bassing, 2010).

Although stepwise D-to-J, then V-to-DJ, also is a hallmark of *Igh* assembly, additional mechanisms are in place to regulate this ordered process at *Tcrb*. Namely, unlike *Igh*, direct joining of Vβ (23-RSS) to Jβ segments (12-RSS) is allowed by the 12/23 rule, but it rarely occurs, highlighting a form of control dubbed "Beyond 12/23" (B12/23; Fig. 1) (Bassing et al., 2000; Sleckman et al., 2000). Three nonmutually exclusive mechanisms likely contribute to B12/23 restriction at the levels of substrate recognition and synapsis. First, the RSSs themselves favor Dβ–Jβ over direct Vβ–Jβ synapsis. *In vitro*, a set of tested Vβ 23-RSSs rearrange more efficiently with 12-RSSs on the 5′ side of Dβ gene segments when the latter are put into direct competition with 12-RSSs derived from Jβ flanks (Banerjee & Schatz, 2014; Jung et al., 2003; Tillman, Wooley, Hughes, Khor, & Sleckman, 2004). Second, the 23-RSS on the 3′ side of Dβ1 possesses a binding site for the c-Fos transcription factor, which in turn associates with the RAG complex (Wang et al., 2008). Recruitment of RAG to the 23-RSS likely focuses its endonuclease activity to promote Dβ–Jβ recombination at the expense of direct Vβ joining to the 5′RSS of Dβ1 (Wang et al., 2008). Third, the location of Dβ1 and Jβ1.1 segments relative to other elements within Dβ–Jβ clusters, most likely Dβ promoters, is required for preventing Vβ-to-Dβ recombination before Dβ-to-Jβ recombination (Yang-Iott et al., 2010).

Despite the exquisite mechanisms in place to control stepwise *Tcrb* assembly, neither 12/23 nor B12/23 restrictions can explain the recombination of *Tcrb* only in developing thymocytes versus B cell precursors, which both express RAG. Nor can these mechanisms explain restrictions that govern the generation of a functional *Tcrb* gene on only one allele in most cells. In the latter regard, DN3 thymocytes that have assembled a functional VDJβ exon express TCRβ protein in association with pTα, a surrogate for TCRα protein, and CD3 signaling components, which are collectively displayed on the cell surface as a pre-TCR (Aifantis, Buer, von Boehmer, & Azogui, 1997; Von Boehmer, 2005). In DN3 thymocytes, pre-TCR signaling induces robust proliferation and their subsequent differentiation into $CD44^{-}CD25^{-}$ (DN4) cells, which, in turn, differentiate into the $CD4^{+}CD8^{+}$ double positive (DP) thymocyte subset, a process called β-selection (von Boehmer & Melchers, 2010). Despite ample RAG expression, Vβ-to-DJβ rearrangement ceases in DP thymocytes (Brady et al., 2010; Jackson & Krangel, 2006), fixing TCRβ specificity and ensuring monospecificity, a control mechanism frequently referred to as allelic exclusion.

In DP cells, RAG activity is instead directed to *Tcra*, where it targets Vα–Jα gene segments for recombination (Krangel, 2009). Assembly of a productive *Tcra* allele permits DP thymocytes to express surface αβ TCR. Subsequent thymic selection of αβ TCR specificity by MHC and self-antigen drives further maturation into either the $CD4^{+}$ or $CD8^{+}$ T cell lineage, with each functional lymphocyte expressing a unique TCR capable of recognizing foreign peptide antigens (Brady & Bassing, 2011). Although the stepwise, tissue-specific assembly of *Tcrb*/*Tcra*, as well as the phenomenon of *Tcrb* allelic exclusion, each have been appreciated for decades, the underlying mechanisms that control these processes have been difficult to dissect and their study still yields surprising insights into general strategies of mammalian gene regulation.

4. ACCESSIBILITY CONTROL OF *Tcrb* ASSEMBLY

The first clues in deciphering mechanisms that regulate RAG recruitment to AgR loci during lymphocyte development derived from studies showing that unrearranged, germline gene segments often give rise to sterile transcripts in the precise cell types that ultimately target them for recombination (Van Ness et al., 1981; Yancopoulos & Alt, 1985). These observations led Alt and colleagues to hypothesize that recombination of *Ig*, but not *Tcr* gene segments, in precursor B cells reflects their accessibility to V(D)J recombinase. Indeed, this hypothesis is supported by other metrics of chromatin accessibility, especially sensitivity to nucleases, which track nicely with the potential for a given cluster of *Ig* or *Tcr* gene segments to rearrange in the appropriate cell type and at the appropriate stages of development (Chattopadhyay, Whitehurst, Schwenk, & Chen, 1998; Cobb, Oestreich, Osipovich, & Oltz, 2006; Larsen & Weintraub, 1982; Stalder et al., 1980; Stanhope-Baker, Hudson, Shaffer, Constantinescu, & Schlissel, 1996).

The link between germline transcription and recombination of specific gene segments spurred a number of studies to examine the role of *cis*-acting transcriptional elements in regulating RSS accessibility to the RAG complex. For *Tcrb*, the most powerful *cis*-elements control transcription of the RC Fig. 2. Specifically, the powerful enhancer Eβ activates two promoters that independently drive transcription of the Dβ1–Jβ1 and Dβ2–Jβ2 clusters

(McDougall, Peterson, & Calame, 1988; Osipovich & Oltz, 2010). Promoters also reside upstream of the coding portion for each Vβ gene segment. The Vβ promoters drive only modest levels of transcription in DN thymocytes when the gene segments are in their germline configuration, but drive high expression levels when a specific Vβ element is brought into the proximity of Eβ following its rearrangement. Despite extensive chromatin profiling, no enhancer has been discovered in the Vβ cluster (Gopalakrishnan et al., 2013), suggesting that low-level activation of germline Vβ promoters is enhancer independent (Mathieu, Hempel, Spicuglia, Verthuy, & Ferrier, 2000). As we now describe, this collection of *cis*-elements contributes distinct aspects to the regulatory mechanisms that govern *Tcrb* gene assembly in developing thymocytes.

4.1 Enhancer-Mediated Assembly of *Tcrb* Genes

Enhancer elements serve as the primary switches to control gene expression via their communication with target promoters (Ong & Corces, 2011). Indeed, Eβ is activated at the earliest stages of T-lineage commitment in the thymus (McDougall et al., 1988). Numerous lines of evidence underscore the importance of Eβ in regulating not only *Tcrb* transcription, but also its assembly by V(D)J recombinase. The first studies showed that a transgenic *Tcrb* minilocus, composed of a single Vβ, Dβ, and two Jβ segments, fails to recombine in thymocytes unless an active enhancer, such as Eβ, is included in the substrate (Ferrier et al., 1990). More importantly, when Eβ is deleted from the endogenous locus (ΔEβ mice; Fig. 2, bottom), all *Tcrb* gene segments remain frozen in their germline configuration, blocking αβ T cell development at the DN2/3 thymocyte stage (Bories, Demengeot, Davidson, & Alt, 1996; Bouvier et al., 1996). Consistent with the accessibility model of control, deletion of Eβ also abolishes germline DβJβ transcription and renders the RC inaccessible to nucleases, including RAG (Bories et al., 1996; Bouvier et al., 1996; Capone et al., 1993; Ji, Little et al., 2010). Surprisingly, however, Eβ deletion only modestly impacts the expression of germline Vβ transcripts (Mathieu et al., 2000). The latter finding suggested that a complete lack of *Tcrb* recombination in ΔEβ mice is primarily due to the impact of the enhancer deletion on the RC, while adding further support to the notion that the Vβ cluster is regulated in an Eβ-independent manner (Majumder, Koues et al., 2015; Mathieu et al., 2000).

The Eβ enhancer is activated initially upon the commitment of multipotent lymphocyte progenitors to the T cell lineage, a process that likely relates to accompanying changes in transcription factor profiles, including those of cognate Eβ transcription factors KLF, E2A, ETS1, RUNX1, and GATA3 (Spicuglia et al., 2002). However, mutagenesis of the endogenous enhancer has been used to show a direct role for only one of these transcription factors in Eβ activation. Site-directed mutagenesis of the two RUNX1-binding sites in Eβ phenocopies deletion of the entire enhancer, crippling germline transcription and rearrangement of the DβJβ clusters (Majumder, Koues et al., 2015). In Eβ, these two RUNX1-binding sites are composite motifs that permit cooperative binding of RUNX1 and ETS1 (Bonnet et al., 2009). Coupled with recent expression data for transcription factors during T-lineage commitment (Weber et al., 2011), these genetic data suggest the following stepwise model for activation of the *Tcrb*-RC by Eβ.

Multipotent progenitor cells in the thymus constitutively express RUNX1, which cannot bind stably to the composite RUNX/ETS sites in Eβ. Notch signaling in these progenitors specifies T-lineage commitment, at least in part, through induction of TCF7, which triggers the expression of ETS1 and GATA3 (Weber et al., 2011). In turn, ETS1 cooperates with RUNX1 to bind the composite Eβ motifs, activating enhancer function, which is further augmented by GATA3. Cooperative RUNX1–ETS1 binding has been shown to recruit histone acetyltransferases, including p300, at other genetic loci (Hollenhorst et al., 2009). Thus, Eβ activation in DN thymocytes likely generates acetylated, accessible chromatin throughout the RC. Based on chromatin profiling studies, the 25 kb RC region is, in fact, classified as a "superenhancer" in DN thymocytes, due to its extremely high load and wide berth of acetylated histones, and deletion of Eβ destroys this hyperactivated chromatin landscape (Gopalakrishnan et al., 2013; Pott & Lieb, 2015). As the final step of RC activation, transcription factors bound to newly accessible promoters neighboring the Dβ1 and Dβ2 gene segments form contacts with transcription factors on Eβ, generating a platform for transcriptional initiation (Koch et al., 2011; Oestreich et al., 2006; Zacarías-Cabeza et al., 2015).

4.2 Promoter-Directed Assembly of *Tcrb* Genes

Although Eβ serves as an accessibility control element for the entire RC, germline promoters associated with Dβ1 and Dβ2 (PDβ1 and PDβ2) have a more localized function in the regulation of Dβ–Jβ recombination. Deletion of PDβ1 (ΔPDβ1; Fig. 2, bottom) abrogates germline transcription, active chromatin marks, and rearrangement of the neighboring 2 kb-wide Dβ1–Jβ1.1–1.7 cluster (Whitehurst, Schlissel, & Chen, 2000). In contrast, transcription and rearrangement of the Dβ2–Jβ2.1–2.7 cluster, located only 10 kb downstream, is unaffected on ΔPDβ1 alleles, due to the sustained activity of germline promoters surrounding the Dβ2 gene segment (McMillan & Sikes, 2008, 2009).

At the level of chromatin, PDβ1 deletion has only a highly localized effect on histone acetylation, perhaps limited to one or several nucleosomes, further supporting that Eβ is the primary mediator of accessible chromatin within the RC (Oestreich et al., 2006). The highly localized function of PDβ1 is also observed when recombination of model substrates is used as an independent readout. Repositioning of PDβ1 in substrates to a location that is only 400 bp downstream of Dβ1 dramatically impairs Dβ–Jβ rearrangement, despite the retention of germline transcription through neighboring Jβ gene segments (Sikes, Meade, Tripathi, Krangel, & Oltz, 2002). In contrast, inversion of PDβ1 in its natural location upstream of Dβ1 drives efficient Dβ1 to Jβ1 rearrangement, despite the inability of the inverted promoter to support transcription through downstream Dβ–Jβ targets (Sikes et al., 2002). Thus, germline promoters by themselves can function as highly localized accessibility control elements for RAG recruitment to RSSs. However, these studies do not preclude a role for transcriptional readthrough of target gene segments in promoting efficient recombination. In this context, at *Tcra*, transcriptional readthrough from germline Jα promoters is required for efficient rearrangement of more distant Jα gene segments (Abarrategui & Krangel, 2006). Taken together, these observations indicate that promoters play a localized role in augmenting chromatin accessibility of their nearest downstream gene segments, but promoter-directed transcription propagates accessibility to enhance recombination of gene segments located further downstream.

The mechanistic underpinnings by which germline promoters direct recombination likely relate to their epigenetic landscapes; specifically, their decoration in DN thymocytes by nucleosomes containing H3 histones trimethylated at lysine 4 (the H3K4me3 modification). This histone modification is highly enriched near active promoters and is bound tightly by the plant homeodomain of RAG2 (Liu, Subrahmanyam, Chakraborty, Sen, & Desiderio, 2007; Matthews et al., 2007). Indeed, RAG2–H3K4me3 interactions not only stabilize RAG recruitment to transcriptionally active regions genome wide (Ji, Little, et al., 2010) but also stimulate its endonuclease function (Lu, Ward, Bettridge, Liu, & Desiderio, 2015; Shimazaki, Tsai, & Lieber, 2009). Mechanistic insights into promoter-directed control of V(D)J recombination have also emerged from reductionist studies, in which chromatin modifiers were recruited to recombination substrates. For example, recruitment of a histone methyl transferase, called G9a, to the germline promoter of an active substrate cripples its recombination by generating an epigenetic landscape similar to that found on *Tcrb* in cell types where the RC is refractory to RAG-mediated rearrangement (Osipovich et al., 2004). Conversely, the localized effects of a germline promoter on RSS accessibility can be recapitulated by tethering the SWI/SNF chromatin remodeling complex to a promoterless substrate, which mimics SWI/SNF recruitment to PDβ1 upon activation of the *Tcrb*-RC in DN thymocytes (Osipovich et al., 2007). Thus, SWI/SNF recruitment and opening of Dβ-associated chromatin seems to be the major function of PDβ1, at least in its role as a local accessibility control element for the Dβ1–RSSs. SWI/SNF-mediated chromatin opening also provides access to the TATA box of PDβ1, activating germline transcription. Nucleosome mapping studies indicate that PDβ1-mediated transcription is essential for reducing chromatin density and disrupting nucleosome phasing in the Dβ1–Jβ cluster, presumably augmenting access of its RSSs to the RAG complex (Kondilis-Mangum et al., 2010).

Together, these studies provided a common framework for RC accessibility at all AgR loci, in which: (i) initial activation of powerful enhancers opens chromatin over the entire RC region, (ii) induction of germline D or J promoter(s) generates highly localized chromatin access, and (iii) promoter-driven transcription propagates the H3K4me3 mark and general chromatin accessibility throughout the RC, setting up an optimal environment for RAG recruitment, as well as RSS cleavage following the capture of an accessible V segment.

4.3 Regulation of *Tcrb* Chromatin Accessibility by Insulators

The mammalian genome is cordoned off into chromatin domains that help maintain gene expression patterns; in essence, minimizing cross talk between regulatory elements associated with regions of active versus inactive genes (West, Gaszner, & Felsenfeld, 2002). The borders of chromatin domains are enforced by boundary elements (BEs) in DNA, which prevent the spread of active chromatin into repressed domains, or insulators, which preclude interactions between enhancers and promoters (Capelson & Corces, 2004). However, functional analyses have demonstrated that many BEs have insulator function and vice versa (Emery, Yannaki, Tubb, & Stamatoyannopoulos, 2000). Likewise, the mechanisms underlying these two functions are varied and remain somewhat controversial (Yang & Corces, 2012). One common feature of these *cis*-elements is their binding to the architectural protein, CTCF, which is observed at many, but not all BEs and insulators (Bell, West, & Felsenfeld, 1999; Cuddapah et al., 2009). Indeed, the prototypical BE/insulator from the chicken β-globin locus, termed HS4, functions in a CTCF-independent manner. The boundary function of HS4 relies on two transcription factors, USF and VEZF1 (Chung, Bell, & Felsenfeld, 1997; Chung, Whiteley, & Felsenfeld, 1993; Dickson et al., 2010; Huang, Li, Yusufzai, Qiu, & Felsenfeld, 2007).

For AgR loci, including *Tcrb*, insulators may segregate the powerful regulatory domain of the RC from upstream clusters of Vβ segments, which may be a critical mechanism for ensuring stepwise RC, then Vβ to RC rearrangement. Moreover, the transcriptional and recombinational activity of several RCs is not completely restricted to a single lineage, whereas V clusters adhere to lineage-specific restriction of these processes (Born, White, Kappler, & Marrack, 1988; Froňková et al., 2005). Thus, protecting the V region from powerful regulatory elements in their RCs may be an important component of maintaining tissue-specific assembly of a complete *Ig* or *Tcr* gene. For example, DN thymocytes activate Eμ, germline RC transcription, and DH-to-JH recombination (Born et al., 1988). However, the VH cluster is transcriptionally and recombinationally inert in these cells. Functional segregation of the VH cluster and the RC relies on a chromatin BE/insulator, called IGCR, which has two potent CTCF-binding sites. Deletion of IGCR permits VH to DHJH recombination in developing T cells, as well as allowing premature VH to DH recombination on alleles before DH-to-JH joining (Guo, Yoon et al., 2011; Lin, Guo, Su, Zhang, & Alt, 2015). These studies indicate that, at least for *Igh*, insulators or BEs are important *cis*-elements for ensuring stepwise and tissue-specific assembly of AgR genes.

The effect of insulators on sculpting TCR repertoires was first tested using gain-of-function studies, in which the powerful CTCF-bound insulator H19-ICR (from the *Igf2/H19* locus) was inserted between the endogenous Dβ1–Jβ1 and Dβ2–Jβ2 clusters. Consistent with its enhancer-blocking function, H19-ICR insertion largely phenocopies the ΔPDβ1 promoter deletion, dramatically inhibiting transcription, chromatin accessibility, and recombination of the Dβ1–Jβ1 cluster (Shrimali et al., 2012). Assembled *Tcrb* genes in mature T lymphocytes from the mutant mice are overwhelmingly skewed toward Dβ2–Jβ2 usage. Importantly, the insulator insertion also interfered with communication between the RC and the upstream Vβ cluster; most of the completed *Tcrb* genes used the lone *Trbv31* gene segment located downstream of the enhancer. However, the precise modes of long-range communication disrupted by the insulator, including enhancer-Vβ promoter cross talk and topological control, remain to be established.

The natural BEs and insulators for *Tcrb* are also beginning to emerge from recent chromatin analyses. The long stretch of inactive *Trp* genes upstream of the Dβ1–Jβ cluster (Fig. 1) may function as one barrier that isolates hyperactive RC chromatin from the Vβ region. In addition, two more well-defined insulator elements exist upstream of the RC. The first is an LTR/LINE element positioned 5 kb upstream of Dβ1 (Carabana, Watanabe, Hao, & Krangel, 2011). Chromatin profiling suggests that the LTR/LINE blocks the spread of inactive chromatin into the RC (Majumder, Koues et al., 2015). However, the most potent boundary in this region is upstream of PDβ1, which contains two modest CTCF peaks in ChIP-seq profiles (Shih et al., 2012). Deletion of this 5′RC region leads to an Eβ-dependent spread of active RC chromatin upstream, into the *Trp* cluster, which induces expression of the most Dβ–Jβ proximal *Trp* gene, called *Prss2* (Majumder, Koues et al., 2015; Fig. 3). The impact of this BE/insulator region on *Tcrb* gene assembly is described in more detail below.

The compartmentalization of *Tcrb* chromatin domains, and its role in the enforcement of ordered Dβ-to-Jβ then Vβ-to-DβJβ rearrangement, remains to be determined. One clue to answering this question may be taken from a knockout mouse strain called $Tcrb^{LD/LD}$, in which a large swath of *Tcrb* was deleted, spanning from the *Trbv5* gene segment to the Dβ2–Jβ2 cluster (Senoo et al., 2003). Thymocytes in the mutant animals inappropriately rearrange the remaining *Trbv4* segment to Dβ2 prior to Dβ2-to-Jβ recombination, disrupting stepwise assembly. Although interpretations are complicated by the sheer extent of this deletion, the main finding is consistent with models in which 5′RC boundary regions play a regulatory role in maintaining ordered assembly of *Tcrb* genes (Senoo et al., 2003).

Closer to the RC, the *Trbv31* gene segment is not separated from the powerful RC regulatory domain by large stretches of inactive *Trp* genes. However, *Trbv31* adheres to stepwise assembly, rearranging by inversion only after DβJβ rearrangements form on a given allele (Malissen et al., 1986;

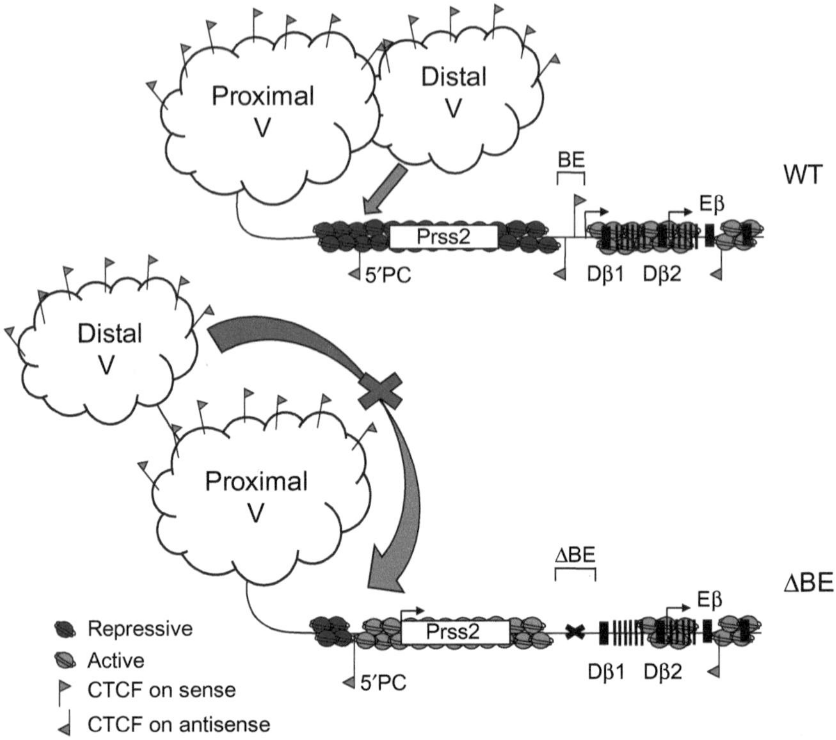

Figure 3 Regulation of *Tcrb* architecture by boundary-tethering elements. Impact of deleting the 5′RC boundary element (ΔPDβ1 mutation) on long-range interactions between the Vβ and DβJβ clusters. Regions of transcriptionally active or inactive chromatin are presented as green or red nucleosomes, respectively. Deletion of the boundary element impairs long-range tethering by the 5′PC CTCF site of gene segments in the distal Vβ topological domain (*Trbv1-14*, Distal V cloud), while RC interactions with the proximal Vβ segments remain unaffected (*Trbv16-30*, Proximal V cloud). (See the color plate.)

Wu et al., 2003). One possible explanation for staving off premature *Trbv31* recombination is that an insulator/BE may exist in the 3 kb stretch between Eβ and *Trbv31*. Indeed, robust CTCF binding is observed at a site in this region. However, because there is no break in the spread of active chromatin from the RC to *Trbv31* in DN thymocytes (Gopalakrishnan et al., 2013; Zacarías-Cabeza et al., 2015), it is unlikely that the CTCF site functions as a chromatin boundary. Instead, this element may be a classical insulator as judged by experiments in which it was deleted from transgenic reporter substrates. The deletion permitted communication between Eμ and the *Trbv31* segment that was incorporated as part of an artificial minilocus

(Schlimgen, Reddy, Singh, & Krangel, 2008). Future work in this area is needed to further elucidate the role of these exciting chromatin architectural elements in enforcing ordered, tissue-specific assembly of *Tcrb* during thymocyte development.

5. GENERAL ASPECTS OF GENOME TOPOLOGY

Long-range recombination of V-to-(D)J segments requires additional changes in the 3D structure of AgR loci, bringing V clusters into spatial proximity with RCs that can be located up to 3.2 Mb away. Such advanced topological control likely takes advantage of general strategies used to fold the genome so that it is compact enough to fit within the cellular nucleus. The recent advent of chromosome conformation capture (3C) technology, which uses cross-linking frequency as a readout for spatial proximity, has enabled rapid progress and an unprecedented level of resolution in probing how genomes fold (Dekker, Marti-Renom, & Mirny, 2013; Dekker, Rippe, Dekker, & Kleckner, 2002). One major insight from these studies is that, in interphase nuclei, chromosomes are structurally compartmentalized into topologically associated domains (TADs), which average 1–2 Mb in size and are largely conserved between cell types and species (Naumova et al., 2013). TADs are generally enriched for either active or inactive genes and these two TAD "flavors" tend to form alternating units along mammalian chromosomes (Dixon et al., 2012). Loci within a given TAD interact with one another, forming discrete regulatory structures, but loci in neighboring TADs are spatially segregated, prohibiting significant association (Phillips-Cremins et al., 2013). Notably, *Tcrb* and several other AgR loci are contained within a single TAD, which presumably promotes long-range contacts between regulatory elements and gene segments in the appropriate cell types.

Many aspects of locus conformation and genome-wide interactions, including the formation of TADs, are determined by the transcription factor CTCF, which generates DNA loops by forming homotypic interactions between distal DNA-binding sites. Efficient CTCF homodimerization between two structural elements appears to require that binding motifs in the two participating elements are oriented convergently (Rao et al., 2014). Once formed, transient CTCF–CTCF interactions are stabilized by association with the collar-like cohesin complex, which locks the base of a new DNA loop into place (Merkenschlager & Odom, 2013). In *Tcrb*, all of the CTCF sites distributed throughout the main Vβ cluster are in the same orientation, whereas the RC-flanking CTCF sites are in the opposite orientation (Fig. 2, top), perhaps facilitating Vβ–RC loop formation.

A second major type of structural building block that sculpts chromosome conformations is the regulatory loop, which is generated by long-range contacts between promoters and enhancers. Regulatory loops are normally cohesin-dependent, but CTCF-independent, instead relying on recruitment of cohesin by the large transcriptional mediator complex that bridges enhancers and promoters (Kagey et al., 2010). Promoter–enhancer communication via stable formation of a regulatory loop augments recruitment of the transcriptional machinery and is therefore a critical architectural component for licensing gene expression (Gorkin, Leung, & Ren, 2014). A specific role for cohesin in forming AgR regulatory loops has emerged from studies of *Tcra*. A thymocyte-specific deletion of the gene encoding RAD21, an essential cohesin subunit, disrupted association between the enhancer (Eα) and the RC promoter (TEA), impairing transcription and *Tcra* gene assembly (Seitan et al., 2011). Within *Tcrb*, the RC forms two major regulatory loops resulting from contacts between factors bound to Eβ and the two Dβ germline promoters, PDβ1 and PDβ2 (Majumder, Koues et al., 2015; Oestreich et al., 2006). Deletion of either PDβ1 or the enhancer is sufficient to dissociate the relevant RC-regulatory loops (Oestreich et al., 2006). These *Tcrb* regulatory loops play an important role not only in initial Dβ to Jβ recombination but also may function in topological mechanisms that promote efficient long-range rearrangements between Vβ and DβJβ segments.

6. SPATIAL MECHANISMS CONTROLLING LONG-RANGE *Tcrb* RECOMBINATION

The first indication that spatial mechanisms regulate long-range Vβ to DβJβ recombination came from 3D-FISH assays (Skok et al., 2007), which revealed that distances between the 5′ and 3′ termini of *Tcrb* differ when comparing DN thymocytes (closer) with other cell lineages (more distant). The DN thymocyte-specific change in *Tcrb* topology, coined locus contraction, is thought to promote Vβ to DβJβ recombination by bringing the two clusters of gene segments into spatial proximity (Shih & Krangel, 2013). Indeed, 3C experiments confirmed that *Tcrb* contraction correlates with more robust interactions between the RC and gene segments throughout the main Vβ cluster (Gopalakrishnan et al., 2013; Majumder, Koues et al., 2015; Majumder, Rupp et al., 2015; Skok et al., 2007). The link between general locus topology and long-range recombination was further supported by the observation that *Tcrb* termini resegregate upon differentiation to the DP thymocyte stage (locus decontraction) (Majumder, Koues et al., 2015; Majumder, Rupp et al., 2015; Skok et al., 2007), where Vβ-to-DβJβ rearrangement is inhibited. However, as outlined below, locus decontraction does not involve complete resegregation of most Vβ segments from the *Tcrb*-RC.

A prime candidate for mechanisms controlling long-range interactions within *Tcrb* was the deposition of RAG, which should promote synapsis of DβJβ rearrangements with the large array of Vβ-RSSs. However, *Tcrb* contraction and Vβ–RC cross-linking are virtually identical in DN thymocytes from RAG-deficient mice versus those expressing RAG2 and a nuclease incompetent version of RAG1, effectively ruling out the possibility that RAG deposition is a major determinant of *Tcrb* topology (Majumder, Koues et al., 2015). A second set of candidates for *Tcrb* topological determinants is the architectural factors CTCF and cohesin. However, a direct test of this possibility has been hampered by technical obstacles, including the current lack of a Cre-driver that would completely ablate CTCF or RAD21 conditional alleles in DN thymocytes and complications that might arise from pleitropic effects on *Tcrb* topology following depletion of these general architectural factors. Accordingly, future studies will need to focus on direct deletion of CTCF sites and other *cis*-acting elements predicted to regulate chromatin conformation.

6.1 *Cis*-Elements That Enforce *Tcrb* Topology

AgR loci adopt specific, more compact conformations in cell types that target them for RAG-mediated assembly. Likewise, these loci express germline transcripts derived from RCs and upstream V clusters in the same cell types. These correlations gave rise to mechanistic models in which transcriptional regulatory elements, or the hyperactive RC chromatin landscape, might drive compaction of AgR loci (Lin et al., 2012; Shih & Krangel, 2013). Consistent with this notion, genome-wide conformational analyses have indicated that regions with similar transcriptional activity in a TAD (e.g., the *Tcrb*-RC and Vβ cluster in DN thymocytes) tend to interact with each other at higher frequencies (Ong & Corces, 2014). However, the impact of regulatory elements on modulating AgR topologies is conflicting. For example, one study has shown that deletion of the *Igh* enhancer, Eμ, perturbs locus contraction in pro-B cells (Guo, Gerasimova et al., 2011), while a second report reached the opposite conclusion (Medvedovic et al., 2013). Similar to the latter study, *Tcra* conformation in DP thymocytes is unaffected by deletion of the powerful Eα enhancer (Shih et al., 2012). One potential source of these conflicts is that most AgR loci, including *Igh* and *Tcra*, harbor more than one enhancer, which may partially compensate for the activities of a deleted element.

To understand the dynamic interplay between regulatory elements, chromosome conformation, and transcription, we have focused on *Tcrb*, a complex, yet tractable locus that undergoes discrete changes in topology during thymocyte development. The tractable features of *Tcrb* include a single enhancer that promotes all recombination steps and a limited number of Vβ segments that recombine. Our more recent studies of *Tcrb* mutant alleles have provided key clues into mechanisms of topological control that may be more broadly applicable for understanding the interplay between genome architecture and gene regulation. In brief, we found that long-range interactions between the Vβ and DβJβ clusters are independent of Eβ function and, therefore, transcriptional activation of the RC (Majumder, Koues et al., 2015). Consistent with this observation, CTCF and RAD21 deposition throughout *Tcrb* is unaffected on alleles lacking Eβ (Majumder, Koues et al., 2015). An implication of these findings is that tissue-specific conformational changes in *Tcrb*, including contraction, must depend on other factors or elements that are independent of RC function in DN thymocytes. One possibility, which remains to be tested, is that transcriptional activation of the Vβ cluster drives these gross topological changes. In turn, cluster-wide Vβ activation may be mediated by the induction of specific transcription factors, such as E2A, which targets many Vβ promoters and has been implicated in the control of Vβ rearrangement (Agata et al., 2007).

Conformational analyses of additional *Tcrb* alleles harboring mutations of RC *cis*-elements yielded several surprises. Specifically, we examined two forms of PDβ1 deletion, a minimal promoter knockout and alleles lacking the promoter and 2.5 kb of additional upstream flanking sequences, called ΔPDβ1 (Fig. 2, bottom) (Whitehurst et al., 1999, 2000). The minimal promoter knockout dramatically impairs germline Dβ1, but not Dβ2, transcription. Consistent with a lack of conformational impact in Eβ deletions, the minimal PDβ1 knockout also had no effect on long-range *Tcrb* looping (Majumder, Koues et al., 2015). However, analyses of the larger ΔPDβ1 deletion revealed a significant loss of interactions between the RC and the most distal half of the Vβ gene cluster in DN thymocytes, while its interactions with the proximal half of the Vβ cluster remained unaffected (Majumder, Koues et al., 2015). These and other data indicated that the upstream Vβ cluster folds into at least two topological domains, a distal (*Trbv1-14*) and a proximal (*Trbv15-30*), with only the distal domain depending on the region deleted from ΔPDβ1 alleles for its interaction with the RC (Majumder, Koues et al., 2015; Fig. 3).

Additional chromatin and conformational profiling of *Tcrb* alleles in DN thymocytes led to deeper mechanistic insights about the function of the 5′RC region deleted from ΔPDβ1 mutants. In wild-type loci, gene segments throughout the distal Vβ domain make robust contacts with a CTCF peak located ~20 kb upstream of the ΔPDβ1 deletion, which is situated in a broad region of silent chromatin and is positioned 5′ to the inert *Prss2* gene (Fig. 2; 5′PC). Contacts between the distal *Trbv* domain and 5′PC are lost in the ΔPDβ1 mutant, despite complete retention of CTCF and cohesin binding at the putative tethering site (Majumder, Koues et al., 2015). Importantly, the ΔPDβ1 deletion disrupted a chromatin boundary, permitting the spread of highly active RC chromatin upstream, inducing *Prss2* expression (Majumder, Koues et al., 2015). In ΔPDβ1 thymocytes, the spread of active chromatin halted precisely at 5′PC, the normal focal point for RC-distal *Trbv* interactions (Fig. 3).

These discoveries suggest that the long-range tethering function of 5′PC is disarmed when it is forced to become a chromatin BE. The separation of function between chromatin boundaries and long-range tethers upstream of RCs may be a common feature of AgR loci and could, perhaps, coordinate changes in topology within other TADs. For example, CTCF-bound elements located upstream of *Igh*- and *Igk*-RCs, IGCR and Cer-Sis, respectively, form chromatin barriers separating the hyperactive RC from the proximal cluster of V gene segments (Guo, Yoon, et al., 2011; Xiang, Park, & Garrard, 2013). Each of these elements have at least two CTCF sites, one of which may serve primarily as a chromatin boundary, protecting the tethering function of the other, a possibility that can be explored now that individual CTCF sites knockouts are available (Lin et al., 2015; Xiang et al., 2013). Notwithstanding, it has become clear that topology, in addition to chromatin accessibility, are important components of the mechanisms that regulate tissue-, stage-, and allele-specific aspects of AgR gene assembly and likely influence the usage of gene segments in shaping primary *Ig* and *Tcr* repertoires.

7. SCULPTING THE PRESELECTION *Tcrb* REPERTOIRE

The relative use of gene segments within AgR loci has an obvious impact on preselection repertoires expressed by emerging B and T lymphocytes. Thus, factors that determine the relative efficiencies of V rearrangement must be balanced not only to maximize AgR diversification but also to favor gene segments that endow specificities for antigens derived from common pathogens or gene segments used in highly specialized lymphocyte subsets (Godfrey, Hammond, Poulton, Smyth, & Baxter, 2000). These factors likely are most important for V rearrangements, which encode for signature CDR1 and CDR2 regions in each AgR chain. Thus, insights into the topological and epigenetic mechanisms that regulate long-range recombination efficiencies have a considerable impact on our understanding of how mammal generate their primary BCR and TCR repertoires, which will then be funneled through positive and negative selection checkpoints.

We recently used a series of quantitative, DNA-based assays to define the preselection usage of Vβ gene segments in mouse DN thymocytes (Gopalakrishnan et al., 2013). In brief, only two of the 25 functional Vβ segments were overrepresented in the preselection *Tcrb* repertoire. These two Vβ gene segments, *Trbv13-2* and *Trbv19*, are enriched in iNKT and MAIT cells, respectively. Both of these innate lymphocyte subsets respond rapidly to common microbial antigens and produce a robust cytokine response (Godfrey et al., 2000; Le Bourhis et al., 2011). These findings suggest that some Vβ segments have evolved features that potentiate their long-range recombination, ensuring sufficient production of key first-responder cells.

In this regard, several parameters were predicted to have the most significant effects on long-range recombination and, therefore, relative Vβ usage. For *Tcrb*, these parameters for each Vβ gene segment include its relative spatial proximity to the RC, its chromatin landscape, and the quality of its RSS with respect to RAG-mediated cleavage. Using a combination of chromatin, transcriptional, and conformational profiling, we defined the contributions of these major parameters to forming the preselection *Trbv* repertoire (Gopalakrishnan et al., 2013). The data coalesced into a two-tiered model for determinants of long-range *Trbv* recombination. In the first tier, RSS qualities, in combination with chromatin accessibility, distinguished active from inert Vβ gene segments (Fig. 4, right). In the second tier, relative usage of active Vβ segments is determined by their particular epigenetic landscape,

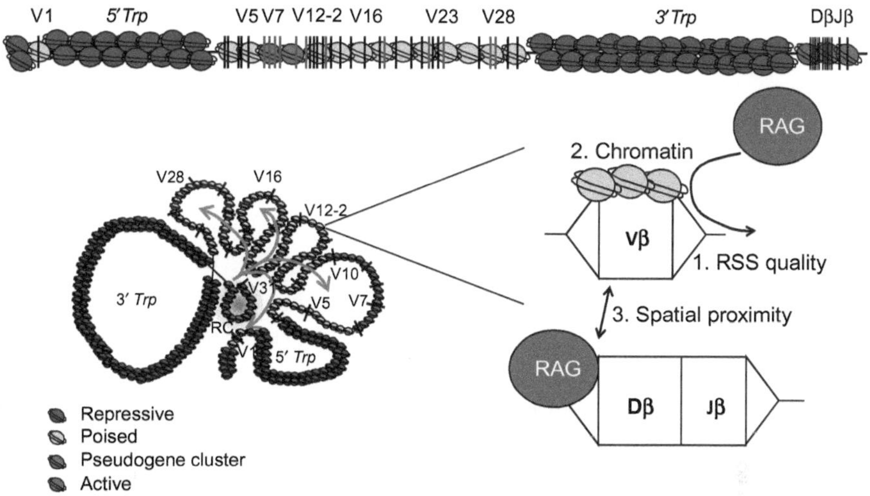

Figure 4 Model for molecular mechanisms that sculpt the preselection *Tcrb* repertoire. (Top) Linear distribution of elements along *Tcrb*, with its epigenetic landscape represented by nucleosomes as indicated (legend). In the lower left panel, hyperactive chromatin in the RC is shown as green. Locus contraction and compaction of the two Vβ architectural domains largely normalizes topological access, allowing the RC to "sample" the entire collection of Vβ segments (orange arrows), which are associated with modestly active chromatin (yellow). The right panel summarizes the relative contributions of variables that contribute to relative Vβ usage in generating the preselection TCRβ repertoire (refer to text for details). (See the color plate.)

with features of active regulatory elements dominating (H3K4me3, H3K4me2, transcription, and the acetyltransferase P300).

This unifying model of repertoire formation predicted that promoters and transcription play a dominant role in Vβ usage. In keeping with this predictive model, deletion of one Vβ regulatory element, the promoter associated with *Trbv14*, significantly impairs its recombination in DN thymocytes (Ryu et al., 2004). Moreover, the large deletion in *Tcrb*$^{LD/LD}$ mice, which places *Trbv4* much closer to Eβ, leads to its increased transcription, accessibility, and rearrangement relative to the other remaining Vβ gene segments (Senoo et al., 2003). Promoter swaps in *Tcra* provide further support for this model, showing that recombination of an underutilized TRAV12 segment is enhanced by replacing its promoter with one from the highly utilized TRAV15 family (Naik, Hawwari, & Krangel, 2015). However, replacement of the *Trbv31*-RSS with the exceedingly efficient 3′Dβ1-RSS promotes 10-fold greater representation of *Trbv31* in the primary Vβ repertoire, indicating that RSS qualities also can contribute significantly to shaping *Tcrb* repertoires (Wu et al., 2003).

An unexpected aspect of the predictive model is a lack of correlation between Vβ–RC spatial proximity, as measured by 3C cross-linking, and rearrangement efficiencies. This finding suggested to us that contributions of spatial proximity to long-range Vβ recombination are likely binary rather than digital, mainly requiring *Trbv* segments to cross a threshold of "closeness" to the RC in order for them to be "sampled" for rearrangement. Presumably, the closeness threshold is crossed when *Tcrb* contracts in DN thymocytes, allowing the folded Vβ clusters to interact more efficiently with the RC (Fig. 4). In support of this model, rearrangement of gene segments within the distal Vβ domain is attenuated in ΔPDβ1 thymocytes, even though their transcriptional and epigenetic landscapes are largely unaltered (Majumder, Koues et al., 2015). Thus, both proximity thresholds and chromatin-based mechanisms contribute to formation of the preselection *Tcrb* repertoire in DN thymocytes by modulating the recombination efficiencies of specific Vβ gene segments. Similarly, these features are likely major determinants in processes that preclude ongoing Vβ to DβJβ rearrangements in RAG-expressing DP thymocytes, which must enforce *Tcrb* allelic exclusion to ensure the monospecificity of AgRs displayed by emerging T cells.

8. DEVELOPMENTAL CHANGES IN *Tcrb* CHROMATIN AND TOPOLOGY

The developmental transition between DN and DP thymocytes is accompanied by the concomitant repression of Vβ-to-DβJβ rearrangement at *Tcrb* and activation of Vα-to-Jα recombination at *Tcra* (Krangel, 2009). The suppression of long-range Vβ-to-DβJβ rearrangement in DP thymocytes, often referred to as allelic exclusion, is linked mechanistically to negative feedback signaling from the pre-TCR (Von Boehmer, 2005). Pre-TCR signaling impacts *Tcrb* assembly during β-selection through two mechanisms that attenuate RAG expression: (i) pre-TCR signaling represses *Rag1* and *Rag2* transcription (Wilson, Held, & MacDonald, 1994), likely preventing new RAG breaks in rapidly dividing thymocytes (Wayne et al., 1994), and (ii) pre-TCR activation leads to phosphorylation and degradation of RAG2 when DN3 cells progress into S-phase (Jiang et al., 2005).

In this regard, pre-TCR signaling drives S-phase entry of DN3 thymocytes via transcriptional induction of Cyclin D3 (Sicinska et al., 2003), which, in turn, binds to and activates Cyclin-dependent kinase 4 and 6. Indeed, Cyclin D3 helps enforce *Tcrb* allelic exclusion (Steinel, Fisher, Yang-Iott, & Bassing, 2014), perhaps by driving cells into S-phase, and thereby inactivating RAG2. However, recent studies suggest a more direct role for Cyclin D3 in the transcriptional repression of *Ig* variable gene segments through its direct binding in cycling B cell precursors (Powers et al., 2012). This exciting finding may be equally applicable to *Tcrb*, where Cyclin D3 may bind to gene segments in the Vβ cluster, impairing their recombination during the proliferative phases of β-selection. Notwithstanding, after transition of DN cells to the DP stage, Cyclin D3 is downregulated, DP thymocytes return to their noncycling state, and RAG proteins are restored to DN levels, but V(D)J recombinase can no longer target Vβ gene segments (Senoo & Shinkai, 1998).

The mechanisms that render *Tcrb* alleles refractory to long-range Vβ-to-DβJβ rearrangement in DP cells have been studied for decades, but a complete picture is still lacking. Early studies demonstrated that DP differentiation is accompanied by attenuation of germline transcription and histone acetylation throughout the Vβ cluster (Chattopadhyay et al., 1998; Tripathi, Jackson, & Krangel, 2002). Although these findings suggest that transcriptional repression of Vβ segments is a key aspect of allelic exclusion, a seminal experiment by Krangel and colleagues indicated that this is not the entire story. Specifically, they generated mice in which the *Tcra* enhancer, Eα, is inserted near the *Trbv14* gene segment, which maintains its transcriptionally active, accessible status in DP thymocytes (Jackson, Kondilis, Khor, Sleckman, & Krangel, 2005). Despite having a favorable chromatin environment for recombination, allelic exclusion of *Trbv14* is largely maintained in DP cells from the mutant mice, indicating a level of control beyond simple accessibility to RAG.

The first clues to this additional level of control came from 3D-FISH and 3C strategies, which suggested that topological mechanisms suppress *Tcrb* assembly in DP thymocytes. Specifically, these early studies showed that *Tcrb* decontracts and that Vβ–RC associations are reduced when comparing DP thymocytes to cultured DN cells (Skok et al., 2007). We have recently reexamined conformational changes in *Tcrb* during the DN-to-DP transition using primary thymocytes and more quantitative 3C approaches. Contrary to expectations, we found that spatial proximity of the RC to most Vβ gene segments is unaltered when comparing DN and DP thymocytes (Majumder, Rupp et al., 2015). Only the distal domain of the Vβ cluster (*Trbv1-11*) segregates spatially from the RC in DP cells, and this separation does not require a massive proliferative burst during β-selection. We drew a similar conclusion from 3D-FISH assays, in which probes for the 5′ terminus of *Tcrb* and the RC exhibit an increased separation in DP thymocytes (decontraction). In contrast, probes for the middle portion of the Vβ cluster (the proximal *Trbv* domain) and the RC remain closely associated when comparing DN and DP thymocytes, but were separated in fibroblasts (Kinjal Majumder, Craig H. Bassing and Eugene M. Oltz, unpublished data). Our new data underscore emerging concerns about conclusions that can be drawn from 3D-FISH or 3C data alone, which frequently do not provide a complete, or even correct, picture of spatial interactions within a complex locus (Williamson et al., 2014). However, current evidence collectively suggests that transcriptional repression in DP thymocytes is a primary determinant for enforcing allelic exclusion of most proximal Vβ segments, since they remain in contact with the RC. By comparison, the most distal Vβ segments are likely excluded by a combination of transcriptional silencing and spatial segregation following β-selection.

Many of these studies were performed on DP thymocytes that harbor two germline *Tcrb* alleles (i.e., RAG-deficient mice with *Tcrb* transgenes or injected with anti-CD3). However, suppression of Vβ recombination in DP cells also prevents secondary joining on an assembled *Tcrb* allele, in which remaining upstream Vβ segments could rearrange with the Dβ2Jβ cluster, deleting a functional or nonfunctional VβDβ1Jβ coding exon. In the absence of such a regulatory mechanism, secondary Vβ rearrangement events could have several seemingly deleterious consequences including (i) restriction of the *Tcrb* repertoire by favoring more distal Vβ segments, (ii) replacements of innocuous with autoreact α/β TCRs, and (iii) creating a higher potential for oncogenic *Tcrb* translocations (Brady et al., 2010).

A large body of evidence now indicates that similar chromatin, transcriptional, and topological mechanisms are at work in DP thymocytes for repressing secondary rearrangements between remaining Vβ gene segments and Dβ2Jβ joins. Specifically, in nuclear transfer mice, which harbor functionally rearranged *Tcrb* loci in their germline (Vβ1NT mice, *Trbv5*–Dβ1–Jβ1.4 rearrangement), the five remaining upstream Vβ segments are transcriptionally and epigenetically active in DN cells, but all except *Trbv4*, located immediately upstream of the functional VDJβ exon, are repressed in DP thymocytes (Brady & Bassing, 2011; Majumder, Rupp et al., 2015). Current evidence also suggests that Vβ-associated CTCF sites remaining on rearranged *Tcrb* alleles may function as BE/insulators to block the regulatory influence of Eβ on upstream Vβ promoters, allowing for their transcriptional repression in DP thymocytes (Brady & Bassing, 2011; Senoo et al., 2003). Likewise, the remaining Vβ gene segments separate spatially from the Dβ2Jβ cluster on Vβ1NT alleles upon progression from the DN to DP stage (Majumder, Rupp et al., 2015).

These findings indicate that all known mechanisms for allelic exclusion are DP intrinsic, since they are deployed at both germline and rearranged *Tcrb* alleles, the latter of which are architecturally very distinct from the former. A possible underlying source for coordinating these suppressive changes, again, may be the E47 isoform of the E2A transcription factor, whose activity is significantly diminished in DP versus DN thymocytes (Agata et al., 2007). Indeed, E47 expression in DP thymocytes overrides feedback inhibition of Vβ rearrangement, increasing the accessibility of *Trbv* gene segments (Agata et al., 2007). However, the topological consequences of enforced E2A expression remain unknown.

Allelic exclusion of *Tcrb* may also be regulated at an even higher level of topology in DP thymocytes—nuclear localization. Several studies have shown that, upon transitioning to the DP stage, unrearranged *Tcrb* loci tend to reposition themselves to the nuclear periphery (Skok et al., 2007), which is a less permissive environment for transcription and for maintaining active chromatin domains (Reddy, Zullo, Bertolino, & Singh, 2008). A similar mechanism of allelic exclusion has been proposed for *Igh* locus (Kosak et al., 2002; Roldán et al., 2005). In addition to transcriptional repression, Krangel and colleagues recently showed that *Tcrb* loci positioned at perinuclear locations are refractory to recombination and the nuclear periphery is depleted for RAG proteins, which alone would suppress recombination of relocalized *Tcrb* alleles (Chan et al., 2013).

More recent evidence suggests that DNA damage responses emanating from *Tcrb* breaks may play a role in suppressing biallelic rearrangement of *Tcrb* in DN thymocytes (allelic inclusion) (Steinel et al., 2014). Specifically, RAG cleavage events during Vβ-to-DβJβ recombination may signal via the damage-dependent ATM protein kinase, providing a transient feedback inhibition of additional Vβ rearrangement on either *Tcrb* allele. Indeed, *Tcrb* allelic inclusion is elevated significantly in ATM-deficient mice (Steinel et al., 2014). This proposed mechanism, called lateral inhibition of V(D)J recombination (Brady et al., 2010), was documented further in studies of *Igk*, which also exhibits higher levels of allelic inclusion in B lineage cells from ATM-deficient mice (Steinel et al., 2013). At a more mechanistic level, RAG-mediated DSBs in *Igk* were shown to signal transiently through

ATM to suppress RAG expression and additional cleavage of the *Igk* locus (Steinel et al., 2013).

While lateral inhibition is associated with downregulation of RAG expression and, therefore, recombinase activity, ATM activation may also lead to the phosphorylation of numerous proteins that contribute to the enforcement of allelic exclusion, including transcription factors, chromatin modifiers, and the RAG recombinase itself (Brady et al., 2010; Hewitt, Chaumeil, & Skok, 2010; Steinel et al., 2013). However, this exciting model for explaining monoallelic V-to-(D)J rearrangement at AgR loci, and its molecular underpinnings, remains to be tested fully. In this regard, inactivation of ATM also leads to impaired DNA repair during V(D)J recombination (Bredemeyer et al., 2006). Hence, increased allelic inclusion in ATM-deficient mice could simply arise from inefficient repair of the first RAG-cleaved allele, leading to V(D)J recombination on the second allele before feedback inhibition signals can be activated (Brady et al., 2010; Steinel et al., 2013). Importantly, ATM-null lymphocytes harbor elevated levels of interlocus recombination events and oncogenic AgR translocations, contributing to the predisposition of ATM-deficient animal for developing lymphoid cancers (Bowen, Wangsa, Ried, Livak, & Hodes, 2013; Callén et al., 2009; Ehrlich, Yang-Iott, & Bassing, 2014; Liyanage et al., 2000; Mahowald et al., 2009; Zha et al., 2010). Thus, in addition to controlling allelic exclusion at AgR loci, lateral inhibition mechanisms of damage control are almost certainly important for preserving genomic stability and suppressing cancer in many cells.

9. FUTURE DIRECTIONS AND CONCLUSIONS

Over the years, we have learned a great deal about general mechanisms of gene regulation through studies of AgR assembly during lymphocyte development. The *Tcrb* locus is especially well suited for such studies because it has an intermediate size and complexity compared with other AgR loci and has a well-defined regulatory architecture. These characteristics make it equally suitable for low- and high-throughput analyses of transcription, chromatin, and conformation, which will allow the field to refine how each of these features contribute to TCRβ repertoires. Recent studies have coalesced into a model that highlights the importance of local chromatin environments and topological changes in *Tcrb* diversification by long-range recombination. In future years, we expect that the field will build on these insights, answering many of the exciting questions that remain. A few of the outstanding questions to be addressed are as follows:

1. What initiates thymocyte-specific folding of *Tcrb* given that RC activation is dispensable? One important avenue to pursue would be the possibility that transcriptional activation of the Vβ cluster, perhaps by E2A, induces deposition of CTCF/cohesin at key sites throughout the region.
2. What are the attachment site(s) for the proximal Vβ domain near the RC, given that perturbation of 5′PC's tethering function only disrupts RC association with the distal Vβ domain?
3. Are there critical anchoring points for the RC within each of the Vβ domains and, similarly, do certain CTCF sites within the Vβ cluster nucleate its folding into separately regulated topological entities (distal and proximal)?
4. On a more global scale, is CTCF-mediated tethering really incompatible with chromatin boundary function, at least at some locations in the genome?
5. What are the DP-intrinsic mechanisms that release Vβ gene segments from the RC interactome, regardless of *Tcrb* configuration?
6. Do RAG cleavage events on one AgR allele signal via ATM to alter chromatin accessibility and/or topology of the other allele?

Given the rapid progress in chromatin profiling, CRISPR-based genome engineering, and high-throughput conformational analyses, which can simultaneously probe from multiple viewpoints, answers to these questions are now within reach.

ACKNOWLEDGMENTS

We are grateful to many of our colleagues for ongoing, valuable discussions and apologize that we could not cite all of your work. The authors were supported by the following grants from the National Institutes of Health: AI118852 and AI115734 to E.M.O.; AI112621 to C.H.B.

REFERENCES

Abarrategui, I., & Krangel, M. S. (2006). Regulation of T cell receptor-α gene recombination by transcription. *Nature Immunology*, *7*, 1109–1115.

Agata, Y., Tamaki, N., Sakamoto, S., Ikawa, T., Masuda, K., Kawamoto, H., et al. (2007). Regulation of T cell receptor β gene rearrangements and allelic exclusion by the helix-loop-helix protein, E47. *Immunity*, *27*, 871–884.

Aifantis, I., Buer, J., von Boehmer, H., & Azogui, O. (1997). Essential role of the pre-T cell receptor in allelic exclusion of the T cell receptor beta locus. *Immunity*, *7*, 601–607.

Alt, F. W., Zhang, Y., Meng, F. L., Guo, C., & Schwer, B. (2013). Mechanisms of programmed DNA lesions and genomic instability in the immune system. *Cell*, *152*, 417–429.

Banerjee, J. K., & Schatz, D. G. (2014). Synapsis alters RAG-mediated nicking at Tcrb recombination signal sequences: Implications for the "beyond 12/23" rule. *Molecular and Cellular Biology*, *34*, 2566–2580.

Bassing, C. H., Alt, F. W., Hughes, M. M., D'Auteuil, M., Wehrly, T. D., Woodman, B. B., et al. (2000). Recombination signal sequences restrict chromosomal V (D) J recombination beyond the 12/23 rule. *Nature*, *405*, 583–586.

Bell, A. C., West, A. G., & Felsenfeld, G. (1999). The protein CTCF is required for the enhancer blocking activity of vertebrate insulators. *Cell*, *98*, 387–396.

Bonnet, M., Huang, F., Benoukraf, T., Cabaud, O., Verthuy, C., Boucher, A., et al. (2009). Duality of enhancer functioning mode revealed in a reduced TCRβ gene enhancer knockin mouse model. *The Journal of Immunology*, *183*, 7939–7948.

Bories, J. C., Demengeot, J., Davidson, L., & Alt, F. W. (1996). Gene-targeted deletion and replacement mutations of the T-cell receptor beta-chain enhancer: The role of enhancer elements in controlling V (D) J recombination accessibility. *Proceedings of the National Academy of Sciences*, *93*, 7871–7876.

Born, W., White, J., Kappler, J., & Marrack, P. (1988). Rearrangement of IgH genes in normal thymocyte development. *The Journal of Immunology*, *140*, 3228–3232.

Bossen, C., Mansson, R., & Murre, C. (2012). Chromatin topology and the regulation of antigen receptor assembly. *Annual Review of Immunology*, *30*, 337–356.

Bouvier, G., Watrin, F., Naspetti, M., Verthuy, C., Naquet, P., & Ferrier, P. (1996). Deletion of the mouse T-cell receptor beta gene enhancer blocks alphabeta T-cell development. *Proceedings of the National Academy of Sciences*, *93*, 7877–7881.

Bowen, S., Wangsa, D., Ried, T., Livak, F., & Hodes, R. J. (2013). Concurrent V (D) J recombination and DNA end instability increase interchromosomal trans-rearrangements in ATM-deficient thymocytes. *Nucleic Acids Research*, *41*, 4535–4548.

Brack, C., Hirama, M., Lenhard-Schuller, R., & Tonegawa, S. (1978). A complete immunoglobulin gene is created by somatic recombination. *Cell*, *15*, 1–14.

Brady, B. L., & Bassing, C. H. (2011). Differential regulation of proximal and distal Vβ segments upstream of a functional VDJβ1 rearrangement upon β-selection. *The Journal of Immunology*, *187*, 3277–3285.

Brady, B. L., Steinel, N. C., & Bassing, C. H. (2010). Antigen receptor allelic exclusion: An update and reappraisal. *The Journal of Immunology*, *185*, 3801–3808.

Bredemeyer, A. L., Sharma, G. G., Huang, C. Y., Helmink, B. A., Walker, L. M., Khor, K. C., et al. (2006). ATM stabilizes DNA double-strand-break complexes during V (D) J recombination. *Nature*, *442*, 466–470.

Callén, E., Bunting, S., Huang, C. Y., Difilippantonio, M. J., Wong, N., Khor, B., et al. (2009). Chimeric IgH-TCRα/δ translocations in T lymphocytes mediated by RAG. *Cell Cycle*, *8*, 2408–2412.

Capelson, M., & Corces, V. G. (2004). Boundary elements and nuclear organization. *Biology of the Cell*, *96*, 617–629.

Capone, M., Watrin, F., Fernex, C., Horvat, B., Krippl, B., Wu, L., et al. (1993). TCR beta and TCR alpha gene enhancers confer tissue-and stage-specificity on V (D) J recombination events. *The EMBO Journal*, *12*, 4335.

Carabana, J., Watanabe, A., Hao, B., & Krangel, M. S. (2011). A barrier-type insulator forms a boundary between active and inactive chromatin at the murine TCRβ locus. *The Journal of Immunology*, *186*, 3556–3562.

Chan, E. A., Teng, G., Corbett, E., Choudhury, K. R., Bassing, C. H., Schatz, D. G., et al. (2013). Peripheral subnuclear positioning suppresses Tcrb recombination and segregates Tcrb alleles from RAG2. *Proceedings of the National Academy of Sciences*, *110*, E4628–E4637.

Chattopadhyay, S., Whitehurst, C. E., Schwenk, F., & Chen, J. (1998). Biochemical and functional analyses of chromatin changes at the TCR-β gene locus during $CD4^-CD8^-$ to $CD4^+$ $CD8^+$ thymocyte differentiation. *The Journal of Immunology*, *160*, 1256–1267.

Choi, N. M., & Feeney, A. J. (2014). CTCF and ncRNA regulate the three-dimensional structure of antigen receptor loci to facilitate V (D) J recombination. *Frontiers in Immunology*, *5*.

Chung, J. H., Bell, A. C., & Felsenfeld, G. (1997). Characterization of the chicken beta-globin insulator. *Proceedings of the National Academy of Sciences, 94*, 575–580.

Chung, J. H., Whiteley, M., & Felsenfeld, G. (1993). A 5′ element of the chicken beta-globin domain serves as an insulator in human erythroid cells and protects against position effect in *Drosophila*. *Cell, 74*, 505–514.

Cobb, R. M., Oestreich, K. J., Osipovich, O. A., & Oltz, E. M. (2006). Accessibility control of V (D) J recombination. *Advances in Immunology, 91*, 45–109.

Cuddapah, S., Jothi, R., Schones, D. E., Roh, T. Y., Cui, K., & Zhao, K. (2009). Global analysis of the insulator binding protein CTCF in chromatin barrier regions reveals demarcation of active and repressive domains. *Genome Research, 19*, 24–32.

Dekker, J., Marti-Renom, M. A., & Mirny, L. A. (2013). Exploring the three-dimensional organization of genomes: Interpreting chromatin interaction data. *Nature Reviews. Genetics, 14*, 390–403.

Dekker, J., Rippe, K., Dekker, M., & Kleckner, N. (2002). Capturing chromosome conformation. *Science, 295*, 1306–1311.

Dickson, J., Gowher, H., Strogantsev, R., Gaszner, M., Hair, A., Felsenfeld, G., et al. (2010). VEZF1 elements mediate protection from DNA methylation. *PLoS Genetics, 6*, e1000804.

Dixon, J. R., Selvaraj, S., Yue, F., Kim, A., Li, Y., Shen, Y., et al. (2012). Topological domains in mammalian genomes identified by analysis of chromatin interactions. *Nature, 485*, 376–380.

Doty, R. T., Xia, D., Nguyen, S. P., Hathaway, T. R., & Willerford, D. M. (1999). Promoter element for transcription of unrearranged T-cell receptor β-chain gene in pro-T cells. *Blood, 93*, 3017–3025.

Ehrlich, L. A., Yang-Iott, K., & Bassing, C. H. (2014). Tcrδ translocations that delete the Bcl11b haploinsufficient tumor suppressor gene promote atm-deficient T cell acute lymphoblastic leukemia. *Cell Cycle, 13*, 3076–3082.

Emery, D. W., Yannaki, E., Tubb, J., & Stamatoyannopoulos, G. (2000). A chromatin insulator protects retrovirus vectors from chromosomal position effects. *Proceedings of the National Academy of Sciences, 97*, 9150–9155.

Ferrier, P., Krippl, B., Blackwell, T. K., Furley, A. J., Suh, H., Winoto, A., et al. (1990). Separate elements control DJ and VDJ rearrangement in a transgenic recombination substrate. *The EMBO Journal, 9*, 117–125.

Froňková, E., Krejčí, O., Kalina, T., Horváth, O., Trka, J., & Hrušák, O. (2005). Lymphoid differentiation pathways can be traced by TCR δ rearrangements. *The Journal of Immunology, 175*, 2495–2500.

Gellert, M. (2002). V (D) J recombination: RAG proteins, repair factors, and regulation. *Annual Review of Biochemistry, 71*, 101–132.

Glusman, G., Rowen, L., Lee, I., Boysen, C., Roach, J. C., Smit, A. F., et al. (2001). Comparative genomics of the human and mouse T cell receptor loci. *Immunity, 15*, 337–349.

Godfrey, D. I., Hammond, K. J., Poulton, L. D., Smyth, M. J., & Baxter, A. G. (2000). NKT cells: Facts, functions and fallacies. *Immunology Today, 21*, 573–583.

Gopalakrishnan, S., Majumder, K., Predeus, A., Huang, Y., Koues, O. I., Verma-Gaur, J., et al. (2013). Unifying model for molecular determinants of the preselection Vβ repertoire. *Proceedings of the National Academy of Sciences, 110*, E3206–E3215.

Gorkin, David U., Leung, Danny, & Ren, Bing. (2014). The 3D genome in transcriptional regulation and pluripotency. *Cell Stem Cell, 14*, 762–775.

Guo, C., Gerasimova, T., Hao, H., Ivanova, I., Chakraborty, T., Selimyan, R., et al. (2011). Two forms of loops generate the chromatin conformation of the immunoglobulin heavy-chain gene locus. *Cell, 147*, 332–343.

Guo, C., Yoon, H. S., Franklin, A., Jain, S., Ebert, A., Cheng, H. L., et al. (2011). CTCF-binding elements mediate control of V (D) J recombination. *Nature, 477*, 424–430.

Hewitt, S. L., Chaumeil, J., & Skok, J. A. (2010). Chromosome dynamics and the regulation of V (D) J recombination. *Immunological reviews, 237*, 43–54.

Hollenhorst, P. C., Chandler, K. J., Poulsen, R. L., Johnson, W. E., Speck, N. A., & Graves, B. J. (2009). DNA specificity determinants associate with distinct transcription factor functions. *PLoS Genetics*, *5*, e1000778.

Huang, S., Li, X., Yusufzai, T. M., Qiu, Y., & Felsenfeld, G. (2007). USF1 recruits histone modification complexes and is critical for maintenance of a chromatin barrier. *Molecular and Cellular Biology*, *27*, 7991–8002.

Jackson, A., Kondilis, H. D., Khor, B., Sleckman, B. P., & Krangel, M. S. (2005). Regulation of T cell receptor β allelic exclusion at a level beyond accessibility. *Nature Immunology*, *6*, 189–197.

Jackson, A. M., & Krangel, M. S. (2006). Turning T-cell receptor β recombination on and off: More questions than answers. *Immunological Reviews*, *209*, 129–141.

Ji, Y., Little, A. J., Banerjee, J. K., Hao, B., Oltz, E. M., Krangel, M. S., et al. (2010). Promoters, enhancers, and transcription target RAG1 binding during V (D) J recombination. *The Journal of Experimental Medicine*, *207*, 2809–2816.

Ji, Y., Resch, W., Corbett, E., Yamane, A., Casellas, R., & Schatz, D. G. (2010). The in vivo pattern of binding of RAG1 and RAG2 to antigen receptor loci. *Cell*, *141*, 419–431.

Jiang, H., Chang, F. C., Ross, A. E., Lee, J., Nakayama, K., Nakayama, K., et al. (2005). Ubiquitylation of RAG-2 by Skp2-SCF links destruction of the V (D) J recombinase to the cell cycle. *Molecular Cell*, *18*, 699–709.

Jung, D., Bassing, C. H., Fugmann, S. D., Cheng, H. L., Schatz, D. G., & Alt, F. W. (2003). Extrachromosomal recombination substrates recapitulate beyond 12/23 restricted V (D) J recombination in nonlymphoid cells. *Immunity*, *18*, 65–74.

Jung, D., Giallourakis, C., Mostoslavsky, R., & Alt, F. W. (2006). Mechanism and control of V (D) J recombination at the immunoglobulin heavy chain locus. *Annual Review of Immunology*, *24*, 541–570.

Kagey, M. H., Newman, J. J., Bilodeau, S., Zhan, Y., Orlando, D. A., van Berkum, N. L., et al. (2010). Mediator and cohesin connect gene expression and chromatin architecture. *Nature*, *467*, 430–435.

Koch, F., Fenouil, R., Gut, M., Cauchy, P., Albert, T. K., Zacarias-Cabeza, J., et al. (2011). Transcription initiation platforms and GTF recruitment at tissue-specific enhancers and promoters. *Nature Structural and Molecular Biology*, *18*, 956–963.

Kondilis-Mangum, H. D., Cobb, R. M., Osipovich, O., Srivatsan, S., Oltz, E. M., & Krangel, M. S. (2010). Transcription-dependent mobilization of nucleosomes at accessible TCR gene segments in vivo. *The Journal of Immunology*, *184*, 6970–6977.

Kosak, S. T., Skok, J. A., Medina, K. L., Riblet, R., Le Beau, M. M., Fisher, A. G., et al. (2002). Subnuclear compartmentalization of immunoglobulin loci during lymphocyte development. *Science*, *296*, 158–162.

Krangel, M. S. (2009). Mechanics of T cell receptor gene rearrangement. *Current Opinion in Immunology*, *21*, 133–139.

Lane, J., Duroux, P., & Lefranc, M. P. (2010). From IMGT-ONTOLOGY to IMGT/LIGMotif: The IMGT® standardized approach for immunoglobulin and T cell receptor gene identification and description in large genomic sequences. *BMC Bioinformatics*, *11*, 223.

Larsen, A., & Weintraub, H. (1982). An altered DNA conformation detected by S1 nuclease occurs at specific regions in active chick globin chromatin. *Cell*, *29*, 609–622.

Le Bourhis, L., Guerri, L., Dusseaux, M., Martin, E., Soudais, C., & Lantz, O. (2011). Mucosal-associated invariant T cells: Unconventional development and function. *Trends in Immunology*, *32*, 212–218.

Lin, Y. C., Benner, C., Mansson, R., Heinz, S., Miyazaki, K., Miyazaki, M., et al. (2012). Global changes in the nuclear positioning of genes and intra-and interdomain genomic interactions that orchestrate B cell fate. *Nature Immunology*, *13*, 1196–1204.

Lin, S. G., Guo, C., Su, A., Zhang, Y., & Alt, F. W. (2015). CTCF-binding elements 1 and 2 in the Igh intergenic control region cooperatively regulate V (D) J recombination. *Proceedings of the National Academy of Sciences*, *112*, 1815–1820.

Liu, Y., Subrahmanyam, R., Chakraborty, T., Sen, R., & Desiderio, S. (2007). A plant homeodomain in RAG-2 that binds Hypermethylated lysine 4 of histone H3 is necessary for efficient antigen-receptor-gene rearrangement. *Immunity*, *27*, 561–571.

Liyanage, M., Weaver, Z., Barlow, C., Coleman, A., Pankratz, D. G., Anderson, S., et al. (2000). Abnormal rearrangement within the α/δ T-cell receptor locus in lymphomas from Atm-deficient mice. *Blood*, *96*, 1940–1946.

Lu, C., Ward, A., Bettridge, J., Liu, Y., & Desiderio, S. (2015). An autoregulatory mechanism imposes allosteric control on the V(D)J recombinase by histone H3 methylation. *Cell Reports*, *10*, 29–38.

Mahowald, G. K., Baron, J. M., Mahowald, M. A., Kulkarni, S., Bredemeyer, A. L., Bassing, C. H., et al. (2009). Aberrantly resolved RAG-mediated DNA breaks in Atm-deficient lymphocytes target chromosomal breakpoints in cis. *Proceedings of the National Academy of Sciences*, *106*, 18339–18344.

Majumder, K., Koues, O. I., Chan, E. A., Kyle, K. E., Horowitz, J. E., Yang-Iott, K., et al. (2015). Lineage-specific compaction of Tcrb requires a chromatin barrier to protect the function of a long-range tethering element. *The Journal of Experimental Medicine*, *212*, 107–120.

Majumder, K., Rupp, L. J., Yang-Iott, K. S., Koues, O. I., Kyle, K. E., Bassing, C. H., et al. (2015). Domain-specific and stage-intrinsic changes in Tcrb conformation during thymocyte development. *The Journal of Immunology*, *195*, 1262–1272.

Malissen, M., McCoy, C., Blanc, D., Trucy, J., Devaux, C., Schmitt-Verhulst, A. M., et al. (1986). Direct evidence for chromosomal inversion during T-cell receptor β-gene rearrangements. *Nature*, *319*, 28–33.

Mathieu, N., Hempel, W. M., Spicuglia, S., Verthuy, C., & Ferrier, P. (2000). Chromatin remodeling by the T cell receptor (Tcr)-β gene enhancer during early T cell development implications for the control of Tcr-β locus recombination. *The Journal of Experimental Medicine*, *192*, 625–636.

Matthews, A. G., Kuo, A. J., Ramón-Maiques, S., Han, S., Champagne, K. S., Ivanov, D., et al. (2007). RAG2 PHD finger couples histone H3 lysine 4 trimethylation with V (D) J recombination. *Nature*, *450*, 1106–1110.

McDougall, S., Peterson, C. L., & Calame, K. (1988). A transcriptional enhancer 3′of C beta 2 in the T cell receptor beta locus. *Science*, *241*, 205–208.

McMillan, R. E., & Sikes, M. L. (2008). Differential activation of dual promoters alters Dβ2 germline transcription during thymocyte development. *The Journal of Immunology*, *180*, 3218–3228.

McMillan, R. E., & Sikes, M. L. (2009). Promoter activity 5′ of Dβ2 is coordinated by E47, Runx1 and GATA-3. *Molecular Immunology*, *46*, 3009–3017.

Medvedovic, J., Ebert, A., Tagoh, H., Tamir, I. M., Schwickert, T. A., Novatchkova, M., et al. (2013). Flexible long-range loops in the VH gene region of the Igh locus facilitate the generation of a diverse antibody repertoire. *Immunity*, *39*, 229–244.

Merkenschlager, M., & Odom, D. T. (2013). CTCF and cohesin: Linking gene regulatory elements with their targets. *Cell*, *152*, 1285–1297.

Naik, A. K., Hawwari, A., & Krangel, M. S. (2015). Specification of Vδ and Vα usage by Tcra/Tcrd locus V gene segment promoters. *The Journal of Immunology*, *194*, 790–794.

Naumova, N., Imakaev, M., Fudenberg, G., Zhan, Y., Lajoie, B. R., Mirny, L. A., et al. (2013). Organization of the mitotic chromosome. *Science*, *342*, 948–953.

Oestreich, K. J., Cobb, R. M., Pierce, S., Chen, J., Ferrier, P., & Oltz, E. M. (2006). Regulation of TCRβ gene assembly by a promoter/enhancer holocomplex. *Immunity*, *24*, 381–391.

Oettinger, M. A., Schatz, D. G., Gorka, C., & Baltimore, D. (1990). RAG-1 and RAG-2, adjacent genes that synergistically activate V (D) J recombination. *Science*, *248*, 1517–1523.

Ong, C. T., & Corces, V. G. (2011). Enhancer function: New insights into the regulation of tissue-specific gene expression. *Nature Reviews. Genetics*, *12*, 283–293.

Ong, C. T., & Corces, V. G. (2014). CTCF: an architectural protein bridging genome topology and function. *Nature Reviews. Genetics*, *15*, 234–246.

Osipovich, O., Cobb, R. M., Oestreich, K. J., Pierce, S., Ferrier, P., & Oltz, E. M. (2007). Essential function for SWI-SNF chromatin-remodeling complexes in the promoter-directed assembly of Tcrb genes. *Nature Immunology*, *8*, 809–816.

Osipovich, O., Milley, R., Meade, A., Tachibana, M., Shinkai, Y., Krangel, M. S., et al. (2004). Targeted inhibition of V (D) J recombination by a histone methyltransferase. *Nature Immunology*, *5*, 309–316.

Osipovich, O., & Oltz, E. M. (2010). Regulation of antigen receptor gene assembly by genetic–epigenetic crosstalk. *Seminars in Immunology*, *22*, 313–322.

Phillips-Cremins, J. E., Sauria, M. E., Sanyal, A., Gerasimova, T. I., Lajoie, B. R., Bell, J. S., et al. (2013). Architectural protein subclasses shape 3D organization of genomesduring lineage commitment. *Cell*, *153*, 1281–1295.

Pott, S., & Lieb, J. D. (2015). What are super-enhancers? *Nature Genetics*, *47*, 8–12.

Powers, S. E., Mandal, M., Matsuda, S., Miletic, A. V., Cato, M. H., Tanaka, A., et al. (2012). Subnuclear cyclin D3 compartments and the coordinated regulation of proliferation and immunoglobulin variable gene repression. *The Journal of Experimental Medicine*, *209*, 2199–2213.

Rao, S. S., Huntley, M. H., Durand, N. C., Stamenova, E. K., Bochkov, I. D., Robinson, J. T., et al. (2014). A 3D map of the human genome at kilobase resolution reveals principles of chromatin looping. *Cell*, *159*, 1665–1680.

Reddy, K. L., Zullo, J. M., Bertolino, E., & Singh, H. (2008). Transcriptional repression mediated by repositioning of genes to the nuclear lamina. *Nature*, *452*, 243–247.

Roldán, E., Fuxa, M., Chong, W., Martinez, D., Novatchkova, M., Busslinger, M., et al. (2005). Locus "decontraction" and centromeric recruitment contribute to allelic exclusion of the immunoglobulin heavy-chain gene. *Nature Immunology*, *6*, 31–41.

Ryu, C. J., Haines, B. B., Lee, H. R., Kang, Y. H., Draganov, D. D., Lee, M., et al. (2004). The T-cell receptor beta variable gene promoter is required for efficient V beta rearrangement but not allelic exclusion. *Molecular and Cellular Biology*, *24*, 7015–7023.

Schatz, D. G., & Ji, Y. (2011). Recombination centres and the orchestration of V (D) J recombination. *Nature Reviews. Immunology*, *11*, 251–263.

Schatz, D. G., Oettinger, M. A., & Baltimore, D. (1989). The V (D) J recombination activating gene, Rag-1. *Cell*, *59*, 1035–1048.

Schlimgen, R. J., Reddy, K. L., Singh, H., & Krangel, M. S. (2008). Initiation of allelic exclusion by stochastic interaction of Tcrb alleles with repressive nuclear compartments. *Nature Immunology*, *9*, 802–809.

Seitan, V. C., Hao, B., Tachibana-Konwalski, K., Lavagnolli, T., Mira-Bontenbal, H., Brown, K. E., et al. (2011). A role for cohesin in T-cell-receptor rearrangement and thymocyte differentiation. *Nature*, *476*, 467–471.

Seitan, V. C., Krangel, M. S., & Merkenschlager, M. (2012). Cohesin, CTCF and lymphocyte antigen receptor locus rearrangement. *Trends in Immunology*, *33*, 153–159.

Senoo, M., & Shinkai, Y. (1998). Regulation of Vbeta germline transcription in RAG-deficient mice by the CD3epsilon-mediated signals: Implication of Vbeta transcriptional regulation in TCR beta allelic exclusion. *International Immunology*, *10*, 553–560.

Senoo, M., Wang, L. L., Suzuki, D., Takeda, N., Shinkai, Y., & Habu, S. (2003). Increase of TCR Vb accessibility within Eb regulatory region influences its recombination frequency but not allelic exclusion. *The Journal of Immunology*, *171*, 829–835.

Shih, H. Y., & Krangel, M. S. (2013). Chromatin architecture, CCCTC-binding factor, and V (D) J recombination: Managing long-distance relationships at antigen receptor loci. *The Journal of Immunology*, *190*, 4915–4921.

Shih, H. Y., Verma-Gaur, J., Torkamani, A., Feeney, A. J., Galjart, N., & Krangel, M. S. (2012). Tcra gene recombination is supported by a Tcra enhancer-and CTCF-dependent chromatin hub. *Proceedings of the National Academy of Sciences*, *109*, E3493–E3502.

Shimazaki, N., Tsai, A. G., & Lieber, M. R. (2009). H3K4me3 stimulates the V(D)J RAG complex for both nicking and hairpinning in trans in addition to tethering in cis: Implications for translocations. *Molecular Cell*, *34*, 535–544.

Shrimali, S., Srivastava, S., Varma, G., Grinberg, A., Pfeifer, K., & Srivastava, M. (2012). An ectopic CTCF-dependent transcriptional insulator influences the choice of Vβ gene segments for VDJ recombination at TCRβ locus. *Nucleic Acids Research*, *40*, 7753–7765.

Sicinska, E., Aifantis, I., Le Cam, L., Swat, W., Borowski, C., Yu, Q., et al. (2003). Requirement for cyclin D3 in lymphocyte development and T cell leukemias. *Cancer Cell*, *4*, 451–461.

Sikes, M. L., Meade, A., Tripathi, R., Krangel, M. S., & Oltz, E. M. (2002). Regulation of V (D) J recombination: A dominant role for promoter positioning in gene segment accessibility. *Proceedings of the National Academy of Sciences*, *99*, 12309–12314.

Sikes, M. L., Suarez, C. C., & Oltz, E. M. (1999). Regulation of V (D) J recombination by transcriptional promoters. *Molecular and Cellular Biology*, *19*, 2773–2781.

Skok, J. A., Gisler, R., Novatchkova, M., Farmer, D., de Laat, W., & Busslinger, M. (2007). Reversible contraction by looping of the Tcra and Tcrb loci in rearranging thymocytes. *Nature Immunology*, *8*, 378–387.

Sleckman, B. P., Bassing, C. H., Hughes, M. M., Okada, A., D'Auteuil, M., Wehrly, T. D., et al. (2000). Mechanisms that direct ordered assembly of T cell receptor β locus V, D, and J gene segments. *Proceedings of the National Academy of Sciences*, *97*, 7975–7980.

Spicuglia, S., Kumar, S., Yeh, J. H., Vachez, E., Chasson, L., Gorbatch, S., et al. (2002). Promoter activation by enhancer-dependent and -independent loading of activator and coactivator complexes. *Molecular Cell*, *10*, 1479–1487.

Stalder, J., Larsen, A., Engel, J. D., Dolan, M., Groudine, M., & Weintraub, H. (1980). Tissue-specific DNA cleavages in the globin chromatin domain introduced by DNAase I. *Cell*, *20*, 451–460.

Stanhope-Baker, P., Hudson, K. M., Shaffer, A. L., Constantinescu, A., & Schlissel, M. S. (1996). Cell type-specific chromatin structure determines the targeting of V(D)J recombinase activity in vitro. *Cell*, *85*, 887–897.

Steinel, N. C., Fisher, M. R., Yang-Iott, K. S., & Bassing, C. H. (2014). The ataxia telangiectasia mutated and cyclin D3 proteins cooperate to help enforce TCRβ and IgH allelic exclusion. *The Journal of Immunology*, *193*, 2881–2890.

Steinel, N. C., Lee, B. S., Tubbs, A. T., Bednarski, J. J., Schulte, E., Yang-Iott, K. S., et al. (2013). The ataxia telangiectasia mutated kinase controls Igκ allelic exclusion by inhibiting secondary Vκ-to-Jκ rearrangements. *The Journal of Experimental Medicine*, *210*, 233–239.

Tillman, R. E., Wooley, A. L., Hughes, M. M., Khor, B., & Sleckman, B. P. (2004). Regulation of T-cell receptor β-chain gene assembly by recombination signals: The beyond 12/23 restriction. *Immunological Reviews*, *200*, 36–43.

Tripathi, R., Jackson, A., & Krangel, M. S. (2002). A change in the structure of Vβ chromatin associated with TCR β allelic exclusion. *The Journal of Immunology*, *168*, 2316–2324.

Van Ness, B. G., Weigert, M., Coleclough, C., Mather, E. L., Kelley, D. E., & Perry, R. P. (1981). Transcription of the unrearranged mouse C κ locus: Sequence of the initiation region and comparison of activity with a rearranged V κ-C κ gene. *Cell*, *27*, 593–602.

Von Boehmer, H. (2005). Unique features of the pre-T-cell receptor α-chain: Not just a surrogate. *Nature Reviews Immunology*, *5*, 571–577.

von Boehmer, H., & Melchers, F. (2010). Checkpoints in lymphocyte development and autoimmune disease. *Nature Immunology*, *11*, 14–20.

Wang, X., Xiao, G., Zhang, Y., Wen, X., Gao, X., Okada, S., et al. (2008). Regulation of Tcrb recombination ordering by c-Fos-dependent RAG deposition. *Nature Immunology*, *9*, 794–801.

Wayne, J., Suh, H., Sokol, K. A., Petrie, H. T., Witmer-Pack, M., Edelhoff, S., et al. (1994). TCR selection and allelic exclusion in RAG transgenic mice that exhibit abnormal T cell localization in lymph nodes and lymphatics. *The Journal of Immunology*, *153*, 5491–5502.

Weber, B. N., Chi, A. W. S., Chavez, A., Yashiro-Ohtani, Y., Yang, Q., Shestova, O., et al. (2011). A critical role for TCF-1 in T-lineage specification and differentiation. *Nature*, *476*, 63–68.

Weigert, M., Gatmaitan, L., Loh, E., Schilling, J., & Hood, L. (1978). Rearrangement of genetic information may produce immunoglobulin diversity. *Nature*, *276*, 785–790.

West, A. G., Gaszner, M., & Felsenfeld, G. (2002). Insulators: Many functions, many mechanisms. *Genes & Development*, *16*, 271–288.

Whitehurst, C. E., Chattopadhyay, S., & Chen, J. (1999). Control of V(D)J recombinational accessibility of the Db1 gene segment at the TCR beta locus by a germline promoter. *Immunity*, *10*, 313–322.

Whitehurst, C. E., Schlissel, M. S., & Chen, J. (2000). Deletion of germline promoter PDβ1 from the TCRβ locus causes hypermethylation that impairs Dβ1 recombination by multiple mechanisms. *Immunity*, *13*, 703–714.

Williamson, I., Berlivet, S., Eskeland, R., Boyle, S., Illingworth, R. S., Paquette, D., et al. (2014). Spatial genome organization: Contrasting views from chromosome conformation capture and fluorescence in situ hybridization. *Genes and Development*, *28*, 2778–2791.

Wilson, A., Held, W., & MacDonald, H. R. (1994). Two waves of recombinase gene expression in developing thymocytes. *The Journal of Experimental Medicine*, *179*, 1355–1360.

Wu, C., Bassing, C. H., Jung, D., Woodman, B. B., Foy, D., & Alt, F. W. (2003). Dramatically increased rearrangement and peripheral representation of Vβ14 driven by the 3′ Dβ1 recombination signal sequence. *Immunity*, *18*, 75–85.

Xiang, Y., Park, S. K., & Garrard, W. T. (2013). Vκ gene repertoire and locus contraction are specified by critical DNase I hypersensitive sites within the Vκ-Jκ intervening region. *The Journal of Immunology*, *190*, 1819–1826.

Yancopoulos, G. D., & Alt, F. W. (1985). Developmentally controlled and tissue-specific expression of unrearranged VH gene segments. *Cell*, *40*, 271–281.

Yang, J., & Corces, V. G. (2012). Insulators, long-range interactions, and genome function. *Current Opinion in Genetics and Development*, *22*, 86–92.

Yang-Iott, K. S., Carpenter, A. C., Rowh, M. A., Steinel, N., Brady, B. L., Hochedlinger, K., et al. (2010). TCRβ feedback signals inhibit the coupling of recombinationally accessible Vβ14 segments with DJβ complexes. *The Journal of Immunology*, *184*, 1369–1378.

Zacarías-Cabeza, J., Belhocine, M., Vanhille, L., Cauchy, P., Koch, F., Pekowska, A., et al. (2015). Transcription-dependent generation of a specialized chromatin structure at the TCRβ locus. *The Journal of Immunology*, *194*, 3432–3443.

Zha, S., Bassing, C. H., Sanda, T., Brush, J. W., Patel, H., Goff, P. H., et al. (2010). ATM-deficient thymic lymphoma is associated with aberrant tcrd rearrangement and gene amplification. *The Journal of Experimental Medicine*, *207*, 1369–1380.

CHAPTER SEVEN

Chromatin Dynamics and the Development of the TCRα and TCRδ Repertoires

Zachary Carico, Michael S. Krangel[1]
Department of Immunology, Duke University Medical Center, Durham, North Carolina, USA
[1]Corresponding author: e-mail address: krang001@mc.duke.edu

Contents

Abstract

The adaptive immune system allows vertebrates to orchestrate highly specific responses to a virtually unlimited milieu of antigens. Effective adaptive immune responses depend on the capacity of T and B lymphocytes to generate diverse repertoires of antigen receptors through the recombination of variable (V), diversity (D), and joining (J) gene segments at antigen receptor loci. V(D)J recombination must be

Advances in Immunology, Volume 128
ISSN 0065-2776
http://dx.doi.org/10.1016/bs.ai.2015.07.005
© 2015 Elsevier Inc.
All rights reserved.

carefully regulated during the early stages of T and B lymphocyte development to ensure the proper development of lymphocyte subsets and to maximize antigen receptor combinatorial diversity. Among all T cell receptor (TCR) and immunoglobulin loci, the TCRα/δ (*Tcra/Tcrd*) locus is unique in its complexity since it undergoes recombination at two distinct stages of T cell development to create distinct TCR proteins that are used by different lineages of T cells. Here, we review the mechanisms that regulate V(D)J recombination at the *Tcra/Tcrd* locus, with a focus on the dynamic chromatin environment and how it instructs the assembly of the *Tcra* and *Tcrd* repertoires. We discuss the dynamics of *Tcra* and *Tcrd* repertoire formation in the context of T cell development, and we consider how the recombination program is directed by localized changes in chromatin structure that regulate the accessibility of *Tcra* and *Tcrd* gene segments to the V(D)J recombinase. We then move beyond local to address spatial relationships in the nucleus, emphasizing the three-dimensional organization of the *Tcra/Tcrd* locus as a critical player in understanding long-distance interactions between chromatin regulatory elements as well as long-distance interactions between recombination substrates.

1. INTRODUCTION

The hallmark of an adaptive immune response is the ability to precisely recognize pathogens using clonally distributed, antigen-specific receptors expressed by T and B lymphocytes. Antigen receptor genes are created from arrays of variable (V), diversity (D), and joining (J) gene segments by a process called V(D)J recombination. This process depends on expression of the lymphocyte-specific recombinase proteins RAG1 and RAG2, which can recognize recombination signal sequences (RSSs) that flank antigen receptor V, D, and J gene segments, and with two RSSs juxtaposed in a synaptic complex, generate double-strand breaks between the RSSs and antigen receptor coding gene segments (Schatz & Swanson, 2011). Complete variable genes are then assembled through the rejoining of coding ends by nonhomologous end joining proteins (Helmink & Sleckman, 2012). V(D)J recombination imparts tremendous combinatorial diversity because any of a large number of V, D, and J gene segments within a locus can be selected for recombination in a given lymphocyte precursor. Additional diversity is imparted at the junctions between V, D, and J gene segments due to heterogeneous processing of coding ends prior to joining.

B cell development is characterized by the stepwise rearrangement of immunoglobulin (Ig) heavy chain (*Igh*) and either Igκ (*Igk*) or Igλ (*Igl*) light chain genes to form antibodies composed of two heavy and two light chains. T cell development is characterized by stepwise rearrangement of T cell receptor (TCR) genes to form heterodimeric TCRs expressed by two distinct lineages of T lymphocytes (Krangel, 2009). αβ T cells express TCRα and TCRβ proteins encoded by rearranged *Tcra* and *Tcrb* genes, respectively. These T cells represent the largest T cell population in most vertebrates and have the ability to recognize and respond to peptide-major histocompatibility complex (MHC) protein complexes on other cells. γδ T cells express TCRγ and TCRδ proteins encoded by rearranged *Tcrg* and *Tcrd* genes, respectively. Although these T cells represent a smaller population in most species, they make unique contributions to host defense (Bonneville, O'Brien, & Born, 2010).

TCR gene rearrangement is tightly regulated during T cell development, and in turn, T cell development is strictly dependent on the assembly and expression of TCR genes and proteins (Krangel, 2009). TCR gene assembly, like Ig gene assembly, is an inherently dangerous process, because it involves the introduction of double-strand breaks into chromosomes. Because of this, V(D)J recombination is strictly controlled to ensure that these lesions, once generated, are appropriately repaired. Additional developmental controls are in place to enforce an appropriate progression of recombination events at individual TCR loci that properly integrates with T cell development. With this in mind, it is notable that the four TCR genes are encoded within three genetic loci. *Tcrb* and *Tcrg* are unlinked; however, *Tcrd* and *Tcra* gene segments, although distinctly regulated and contributing to two different TCRs, are organized within a single, complex genetic locus. This suggests the need for even greater and more complex layers of regulation to enforce appropriate developmental outcomes.

Multiple mechanisms are thought to contribute to ordered, lineage-, and developmental stage-specific regulation of V(D)J recombination at antigen receptor loci. For many years, research focused on the mechanisms by which chromatin structure is modified to allow the RAG proteins selective access to particular RSSs, and in particular, the roles of *cis*-acting regulatory elements such as enhancers and promoters (Bassing, Swat, & Alt, 2002; Krangel, 2003; Cobb, Oestreich, Osipovich, & Oltz, 2006). More recently, it has been appreciated that the topology of antigen receptor loci must represent an additional level of control, in part because chromatin regulation requires an interplay between *cis*-acting regulators that are widely separated in the linear DNA sequence, and in part because V(D)J recombination requires synapsis of RSSs that are similarly distributed over long distances (Bossen, Mansson, & Murre, 2012; Jhunjhunwala, van Zelm, Peak, & Murre, 2009; Schatz & Ji, 2011; Shih & Krangel, 2013).

Here, we provide a comprehensive view of V(D)J recombination at the *Tcra/Tcrd* locus. We first address the organization of the locus and how the *Tcrd* and *Tcra* repertoires are generated within the context of T cell development. We then turn to locus regulation and consider how controls exerted at the level of RSS accessibility and locus topology may integrate to enforce the complex developmental program and the generation of diverse receptor repertoires.

2. THE LOCUS

2.1 Overview

Given that the αβ and γδ fates represent distinct T cell lineages, it is striking that the genes encoding *Tcra* and *Tcrd* are encoded by a single genetic locus on murine chromosome 14. The *Tcra/Tcrd* locus is polymorphic, with five haplotypes identified amongst laboratory mouse strains on the basis of restriction fragment length polymorphisms (Jouvin-Marche et al., 1989; Singer, McEvilly, Balderas, Dixon, & Theofilopoulos, 1988). The complete sequence of the haplotype unique to C57BL/6 reveals it to span 1.8 Mb. Another well-studied haplotype is shared by strains 129, BALB/c, AKR, CBA/J, C3H, and A/J. An almost complete sequence of the strain 129 locus (NCBI sequence NT_039614.1; the last 100 kb of this file represents BALB/c sequence) reveals it to span 1.6 Mb (Fig. 1). The *Tcra/Tcrd* locus is bounded at one end by a cluster of olfactory receptor genes, whereas the antiapoptotic gene *Dad1* marks the other end. The largest portion of the locus, and the largest contributor to structural variation among haplotypes, contains the V gene segments. The final 0.1 Mb of the locus contains Dδ, Jδ, and TCRδ constant (Cδ) gene segments, followed by Jα and Cα gene segments. All gene segments but one (a Vδ) are in the same transcriptional orientation; thus, rearrangement occurs almost exclusively by deletion.

2.2 Nomenclature

Murine *Tcra/Tcrd* locus nomenclature has been and remains a source of confusion, particularly with respect to the naming of V gene segments. The original Vα/Vδ family nomenclature (Vα1, Vα2, Vα3, Vα4, etc.; Vδ1, Vδ2, Vδ4, Vδ5, etc.) was organized and replaced in 1995 by a WHO-proscribed nomenclature that mostly conserved the numbering scheme (AV1, AV2, AV3, etc., for Vαs; DV101, DV102, DV104,

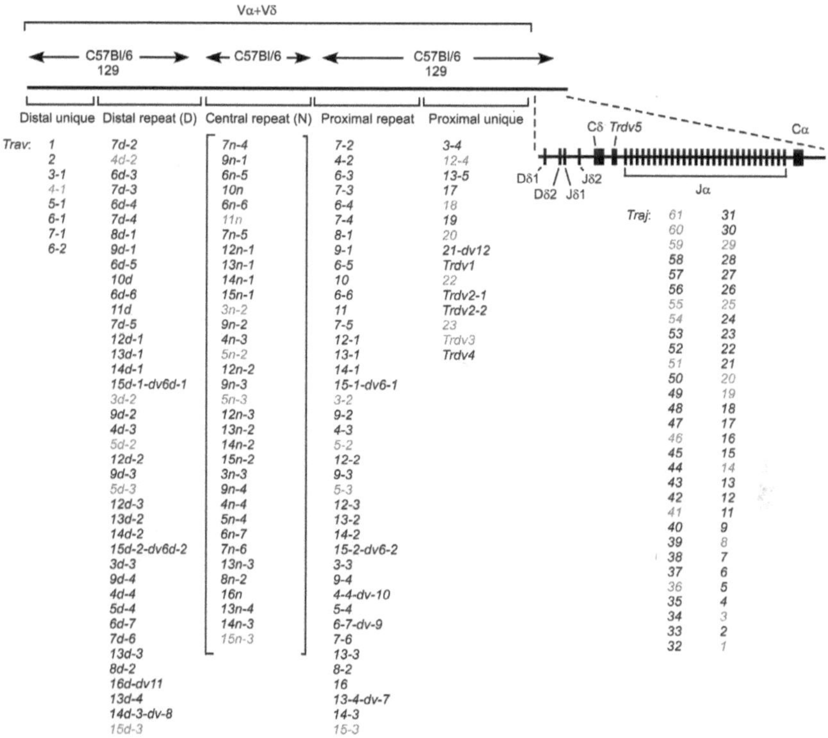

Figure 1 Organization of the murine *Tcra/Tcrd* locus. The structure of the Vα/Vδ region is different in strains 129 and C57/BL6 due to the inclusion of the N repeat in the latter. Gene segments are listed in order of appearance from distal to proximal below each region. V gene segment nomenclature is according to IMGT (www.imgt.org) and NCBI. Functional gene segments are in black lettering. Gene segments that are nonfunctional, based on the presence of a stop codon or failure to rearrange or undergo mRNA splicing, are in gray lettering.

DV105 for proximal Vδs; and ADV7, ADV11, etc., for V segments contributing to both repertoires) (Arden, Clark, Kabelitz, & Mak, 1995). Subsequently, the complete sequence of the locus prompted a renumbering of gene segments in order from distal to proximal based on the first appearance of a family member (*Trav1*, *Trav2*, etc., for Vαs, followed by *Trdv1*, *Trdv2*, *Trdv4*, *Trdv5* for proximal Vδs), with more distal V segments contributing to both repertoires having compound names (*Trav15-dv6*, *Trav16d-dv11*, etc.) (Bosc & Lefranc, 2003; Glusman et al., 2001). Nevertheless, the original Vα nomenclature is still in use, perpetuated at least in part by the names of commercially available antibodies that are commonly used for cell staining. The Arden nomenclature, although not currently in use, is ensconced in the literature of the 1990s. As a result of this history, the original Vα7/Vδ6 family appears in the 1990s literature as ADV7 and is currently referred to as *Trav15-dv6*; the original Vα8 appears in the 1990s literature as AV8 and is currently referred to as *Trav12*; the original Vδ1 appears in the 1990s literature as DV101 and is currently referred to as *Trdv4*, and the original Vδ4 is in the 1990s literature as DV104 and is currently referred to as *Trdv2*. We will use the IMGT/NCBI nomenclature for V gene segments; a conversion chart between the multiple systems is provided (Table 1) and is also maintained at the IMGT web site (www.imgt.org; Lefranc, 2001).

Table 1 *Tcra/Tcrd* V Gene Segment Nomenclature

	Nomenclature		
V Segment Type	**IMGT/NCBI[a]**	**Arden**	**Vα**
Trav + Trav-dv	*Trav1*	AV19	Vα19
	Trav2	AV12	Vα12
	Trav3	AV5	Vα5
	Trav4	AV11	Vα11
	**Trav4-dv10*	ADV11	Vα11
	Trav5	AV13	Vα13
	Trav6	ADV4	Vα4
	**Trav6-7-dv9*	ADV4S8	Vα4
	Trav7	AV1	Vα1
	Trav8	AV18	Vα18
	Trav9	AV3	Vα3
	Trav10	AV15	Vα15
	Trav11	AV14	Vα14
	Trav12	AV8	Vα8
	Trav13	ADV10	Vα10
	**Trav13-4-dv7*	ADV10S7	Vα10/Vδ7
	Trav14	ADV2	Vα2
	**Trav14d-3-dv8*	ADV2S8	Vα2/Vδ8

Table 1 *Tcra/Tcrd* V Gene Segment Nomenclature—cont'd

V Segment Type	Nomenclature		
	IMGT/NCBI	**Arden**	**Vα**
	Trav15-dv6	ADV7	Vα7/Vδ6
	**Trav15d-1-dv6d-1*	ADV7S1	Vα7/Vδ6.3
	**Trav15d-2-dv6d-2*	ADV7S5	Vα7/Vδ6
	**Trav15-1-dv6-1*	ADV7S2	Vα7/Vδ6.1
	**Trav15-2-dv6-2*	ADV7S4	Vα7/Vδ6
	Trav16	AV17	Vα17
	**Trav16d-dv11*	ADV17	Vδ9
	Trav17	AV9	Vα9
	Trav19	AV20	Vα20
	Trav21-dv12	ADV6	Vα6/Vδ3
Trdv	*Trdv1*	DV102	Vδ2
	Trdv2	DV104	Vδ4
	**Trdv2-1*	DV104S1	
	**Trdv2-2*	DV104S2	Vδ4
	Trdv4	DV101	Vδ1
	Trdv5	DV105	Vδ5

[a]V gene segment families are listed. Individual V gene segments within a family in strain 129 mice are also noted and marked (*) when one or more members of the family have a *Trdv* designation.

2.3 *Tcra/Tcrd* Structure

2.3.1 *Vα/Vδ* Array

The strain 129 *Tcra/Tcrd* locus includes 104 V segments distributed across 1.5 Mb of DNA, whereas the C57BL/6 locus includes 138 V segments distributed across 1.7 Mb (Fig. 1). These differences reflect the fact that the 129 locus carries a duplication of a 400-kb portion of the V array (denoted D), whereas the C57BL/6 locus includes an additional, albeit incomplete, copy of this region (denoted N) that sits between the original copy and the duplicated copy (D) (Jouvin-Marche et al., 1989; Singer et al., 1988). Of the 104 V gene segments in strain 129 haplotype, 16 are considered to be nonfunctional pseudogenes; in C57BL/6, 21 of 138 V gene segments are similarly categorized. The majority of V gene segments in either locus is used for *Tcra* recombination; these V gene segments are designated as *Trav*. The *Trav* are further categorized into 21 families defined by the sharing of at least 75% sequence homology (Arden et al., 1995). The largest of these families is comprised of more than a dozen members. In contrast, *Trav1* and *Trav2* at the distal end of the V array and *Trav17*, *Trav19*, and *Trav21* at the proximal end each consist of only a single member. Some V gene segments located in the proximal portion of the V array are thought to be used exclusively in *Tcrd* rearrangements; these are designated *Trdv*. However, a number of *Trav* are used in both the *Tcra* and *Tcrd* repertoires and are designated as such (e.g., *Trav21-dv12*). In total, 16 of the 104 V segments in strain 129 have a *Trdv* or *Trav-dv* designation; 15 of these are functional (Bosc & Lefranc, 2003).

2.3.2 The Business (D, J, C) End of the Locus

Residing downstream of the Vα/Vδ array is a 100-kb region containing two Dδ and two Jδ segments, Cδ, a single, inverted Vδ gene segment (*Trdv5*), a large array of Jα gene segments, and Cα. Jα numbering is from 1 to 61, with Jα61 (or *Traj61*) at the 5′, or Cα-distal end and Jα1 (or *Traj1*) at the 3′, or Cα-proximal end of the array. It should be noted that this numbering scheme was designed to maintain consistency in nomenclature for orthologous J gene segments in the murine and human Jα arrays. In fact, the murine Jα array contains only 60 Jα gene segments; Jα10 does not exist in the mouse. Only 43 of 60 murine Jα gene segments are likely to be functional as judged by their ability to undergo DNA recombination and RNA splicing to Cα and to maintain the correct reading frame. That said, the most 5′ Jα gene segment, Jα61, is classified as a pseudogene but still plays a critical role in the *Tcra* recombination program. As we will discuss later, Jα61 has a functional recombination signal sequence and serves an important role as the dominant site for initial, or primary, Vα to Jα recombination events, even if it does not contribute directly to the expressed *Tcra* repertoire.

The "business end" of the locus also contains two critical *cis*-regulatory elements that play important roles in *Tcrd* and *Tcra* rearrangement and expression. The *Tcrd* enhancer, Eδ, is situated in the Jδ2–Cδ intron; the *Tcra* enhancer, Eα, is positioned several kb downstream of Cα. These will be discussed in greater detail in subsequent sections.

2.4 Comparative Genomics of *Tcra/Tcrd*

The human *Tcra/Tcrd* locus spans 1.1 Mb and is very similar in overall organization to that of the mouse. The major structural difference between the human and mouse loci is the absence of recent Vα/Vδ duplication events in humans, resulting in fewer Vα/Vδ gene segments (55) (Glusman et al., 2001). In addition, the human locus carries three Dδ and four Jδ gene segments.

A genetic locus with linked *Tcra* and *Tcrd* genes exists in all jawed vertebrates, but some diverge from the mammalian *Tcra/Tcrd* locus organization (Flajnik & Kasahara, 2010; Hirano, Das, Guo, & Cooper, 2011; Rast et al., 1997). Chickens carry a *Tcra/Tcrd* locus organized much like that of the mouse and other mammals, with about 30 Vδ, 2 Dδ, and 2 Jδ gene segments embedded between 40 Vα and 25 Jα gene segments (Kubota et al., 1999; Fig. 2). The locus also displays a nested organization in the teleost puffer fish *Tetraodon nigroviridis*, but with an interesting twist (Fischer et al., 2002). The locus is much more compact, extending for only 30 kb and with smaller arrays of Vα/Vδ and Jα gene segments. Most striking, the "business end" of the locus, from Dδ through Cα, is inverted with respect to the Vα/Vδ array, such that initial rearrangement must occur by inversion rather than deletion (Fig. 2). A similarly inverted locus was reported for the Japanese pufferfish *Takifugu rubripes* (Wang et al., 2001). In the amphibian *Xenopus tropicalis*, a 600-kb *Tcra/Tcrd* locus contains *Tcra* and *Tcrd* gene segments that are segregated from each other rather than nested (Fig. 2). The *Tcra* end of the locus is composed of 71 Vα and 75 Jα segments, followed by Cα. Upstream of and inverted with respect to the Vα array are 17 Vδ, 2 Dδ, and 2 Jδ gene segments, followed by

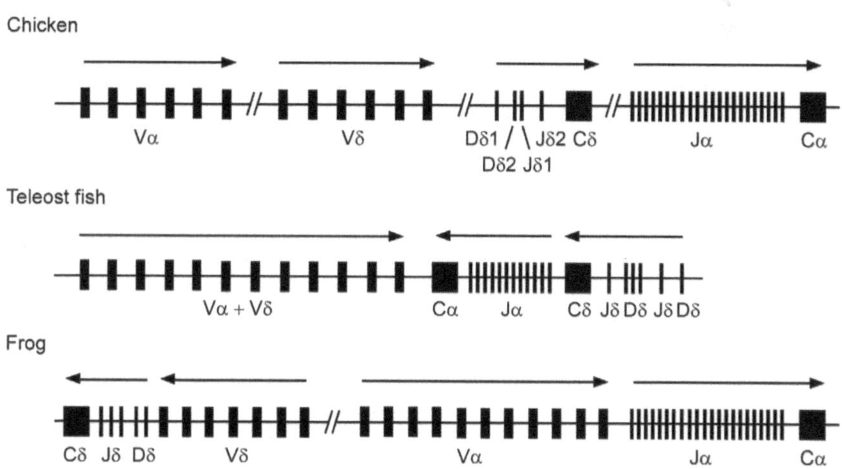

Figure 2 Schematic depictions of the *Tcra/Tcrd* locus in different vertebrates. Arrows denote transcriptional orientation of gene segments. The schematics depict overall locus organization but do not accurately represent gene segment numbers or overall locus size.

Cδ segments (Parra, Ohta, Criscitiello, Flajnik, & Miller, 2010). In sharks, some Vδ gene segments have upstream New Antigen Receptor (NAR) V–D–J cassettes. These contribute to the expression of a double V domain TCR chain, with the amino-terminal V domain encoded by the rearranged NAR VDJ and the second V domain encoded by the downstream Vδ gene segment rearranged to traditionally organized Dδ and Jδ gene segments (Criscitiello, Saltis, & Flajnik, 2006). Marsupials also have a *Tcrd*-like locus termed TCRμ that is on a chromosome distinct from the *Tcra/Tcrd* locus (Parra et al., 2007).

In summary, there is evidence of remarkable evolutionary flexibility in *Tcra/Tcrd* locus organization. Regulatory challenges can be expected to be somewhat distinct for species with embedded versus tandemly organized *Tcra* and *Tcrd* gene arrays, or for species rearranging by deletion versus inversion. It should be noted, however, that a genetically modified murine *Tcra/Tcrd* locus with an inverted Jα–Cα portion supports apparently normal αβ and γδ T cell development in mice (Khor, Wehrly, & Sleckman, 2005). It is widely accepted that the nested organization of *Tcrd* and *Tcra* gene segments allows for a critical lineage commitment step because *Tcrd* gene segments are irreversibly deleted by Vα to Jα recombination. However, the examples of divergently organized *Tcra/Tcrd* loci and a genetically modified murine *Tcra/Tcrd* locus suggest that this supposition is incorrect.

3. REPERTOIRE

3.1 Overview of *Tcra/Tcrd* Rearrangement in Developing T Cells

V(D)J recombination is governed by the so-called 12/23 rule: RSSs are composed of conserved heptamer and nonamer elements separated by a less conserved spacer of either 12 or 23 bp, and recombination events occurs between one gene segment flanked by a 12-bp RSS and another gene segment flanked by a 23-bp RSS (Schatz & Swanson, 2011). *Tcrd* genes, like *Igh* and *Tcrb* genes, are assembled from three different gene segments (V, D, and J) and require two recombination events. Vδ gene segments are flanked by 23 bp RSSs, Jδ gene segments are flanked by 12 bp RSSs, and Dδ gene segments are flanked by 12 bp RSSs on the 5′ side and 23 bp RSSs on the 3′ side. Given this configuration, and with two Dδ segments arranged in tandem, murine *Tcrd* genes are often assembled through three recombination events and can have a V–D–D–J structure. Human *Tcrd* genes, with three Dδ segments arranged in tandem, can be assembled through four recombination events and can have a V–D–D–D–J structure (Loh, Cwirla, Serafini, Phillips, & Lanier, 1988).

Tcrd rearrangement occurs in developing DN thymocytes, initiating at the DN2 stage and continuing into the DN3 stage during an initial period of *Rag1* and *Rag2* gene expression (Capone, Hockett, & Zlotnik, 1998; Livák, Tourigny, Schatz, & Petrie, 1999). An unusual feature of murine *Tcrd* gene assembly is that it is relatively disordered; V–D, D–D, D–J, and D–D–J recombination intermediates can all be detected in early thymocytes (Chien et al., 1987). This is very much unlike *Tcrb* or *Igh* rearrangements, which invariably proceed through D–J intermediates. Notably, human *Tcrd* rearrangements seem to be more ordered than in the mouse, with Dδ2–Dδ3 rearrangement occurring as a first step (Cieslak et al., 2014). Complete *Tcrd* rearrangements occur on both alleles in as many as 70% of murine γδ hybridomas, and a substantial fraction of these carry two in-frame *Tcrd* rearrangements, suggesting that *Tcrd* rearrangement is not subject to allelic exclusion (Sleckman, Khor, Monroe, & Alt, 1998). In contrast, analysis of human γδ cell clones reveals that, depending on the Vδ gene segment used, only 34% or 18% had complete *Tcrd* rearrangements on both alleles. These frequencies are even lower than what would be expected for allelic exclusion based on a feedback inhibition mechanism (Brady, Steinel, & Bassing, 2010; Vettermann & Schlissel, 2010). Nevertheless, among those with two rearrangements, allelic inclusion was fairly common (Couedel, Lippert, Bernardeau, Bonneville, & Davodeau, 2004). Thus, the extent to which allelic exclusion, selection, or other developmental constraints limit human *Tcrd* rearrangements requires further studies.

Tcrd is also extensively rearranged in cells of the αβ T lineage, including DP thymocytes and lymph node T cells (Livak, Petrie, Crisps, & Schatz, 1995; Nakajima, Menetski, Roth, Gellert, & Bosma, 1995). These rearranged *Tcrd* genes are maintained on extrachromosomal circles (sometimes called T cell receptor excision circles, or TRECs) as a result of excision by Vα to Jα rearrangement in DP thymocytes (Nakajima et al., 1995; Okazaki & Sakano, 1988; Takeshita, Toda, & Yamagishi, 1989). Such circles are maintained because, subsequent to Vα to Jα recombination, there is minimal proliferation of DP thymocytes or naive peripheral T lymphocytes; extrachromosomal circles are only diluted out and lost once T cells are activated and undergo extensive cell proliferation. Sequence analysis of excised, Vδ–Dδ–Jδ rearranged *Tcrd* genes in DP thymocytes suggests modest selection against in-frame *Tcrd* sequences, implying that DP thymocytes develop from cells that had failed to express a γδ TCR (Livak et al., 1995). Earlier studies identified an RSS called δRec, located upstream of human Dδ1, that rearranged frequently to Jα61 (then called ΨJα); this rearrangement was hypothesized to preclude Vδ rearrangement and commit thymocytes to undergo Vα to Jα rearrangement (de Villartay, Hockett, Coran, Korsmeyer, & Cohen, 1988; de Villartay, Mossalayi, de Chasseval, Dalloul, & Debré, 1991; Hockett et al., 1988). The high abundance of Vδ–Dδ–Jδ gene rearrangements in murine DP thymocytes (Livak et al., 1995; Nakajima et al., 1995) and total human thymocytes (Krangel, Yssel, Brocklehurst, & Spits, 1990), most of which would delete δRec, and the propensity of proximal Vα gene segments to rearrange to Jα61 (see below), both preclude such a model. Moreover, αβ T cell development does not depend on *Tcrd* deletion, because *Tcra/Tcrd* alleles engineered to undergo Vα to Jα recombination by inversion and retain *Tcrd* on the chromosome support normal αβ T cell development (Khor et al., 2005).

The mechanistic basis for commitment to the γδ and αβ T cell lineages has been extensively reviewed elsewhere (Ciofani & Zúñiga-Pflücker, 2010; Kreslavsky & von Boehmer, 2010; Narayan & Kang, 2010). The prevailing view is of competition in DN3 thymocytes between assembly of in-frame *Tcrb* genes, on the one hand, and in-frame *Tcrg* and *Tcrd* genes, on the other hand, with a modified instructional mechanism of lineage commitment. Thus, assembly of a pre-TCR composed of TCRβ and pre-Tα commits thymocytes to the αβ lineage (defined as differentiation to the DP stage), whereas assembly of a γδ TCR commits thymocytes to the γδ lineage (defined by fixation of the DN phenotype). Nevertheless, lineage commitment is not thought to be strictly tied to TCR identity, but rather to the strength of signals those TCRs generate. Thus, weak signals, which are typical of the pre-TCR, are thought to enforce an αβ T cell fate, whereas strong signals, presumably generated by intrathymic γδ T cell ligands, are thought to enforce a γδ T cell fate (Haks et al., 2005; Hayes, Li, & Love, 2005; Kreslavsky, Garbe, Krueger, & von Boehmer, 2008). Nevertheless, there are some data consistent with γδ T cells arising from pre-committed progenitors, and this could be the case for some subsets of γδ T cells (Narayan & Kang, 2010).

Cells that assemble a pre-TCR and pass the *Tcrb*-selection checkpoint undergo extensive cell proliferation and differentiate through the DN4 and immature single-positive (ISP) stages to the DP stage (Michie & Zúñiga-Pflücker, 2002). *Tcra/Tcrd* locus recombination does not occur during this transition because *Rag1* and *Rag2* gene expression is downregulated and RAG2 protein is degraded in cycling cells (Li, Dordai, Lee, & Desiderio,

1996; Taghon, Yui, Pant, Diamond, & Rothenberg, 2006). As large, proliferating DP thymocytes come out of the cell cycle and transition to become small, resting DP thymocytes, *Rag1* and *Rag2* genes are reexpressed and *Tcra* gene rearrangement commences. *Tcra* rearrangement occurs on both alleles, with no evidence for allelic exclusion (Malissen et al., 1992, 1988). Whereas Vδ to Dδ to Jδ rearrangement eliminates all germline Dδ gene segments and is incompatible with further rearrangement events, Vα to Jα recombination leaves upstream Vα gene segments and downstream Jα gene segments and is therefore compatible with multiple rounds of recombination that would allow "editing" of the initial rearrangement. It has been estimated by computer modeling that *Tcra* alleles can undergo, on average, three to five cycles of Vα to Jα rearrangement (Thuderoz et al., 2010; Warmflash & Dinner, 2006). The process is terminated due to RAG gene downregulation in response to positive selection based on the specificity of the assembled αβ TCR for self-peptide:MHC complexes (Borgulya, Kishi, Uematsu, & von Boehmer, 1992; Brändle, Müller, Rülicke, Hengartner, & Pircher, 1992; Kouskoff, Vonesch, Benoist, & Mathis, 1995; Turka et al., 1991). In DP thymocytes that fail to produce a useful αβ TCR, Vα to Jα rearrangement is limited by cell death (Guo et al., 2002). Given the link between Vα to Jα recombination and the outcome of positive selection, substantial time would be required to allow synthesis and surface expression of an αβ TCR and testing against thymic ligands before the next cycle of Vα to Jα recombination commences. This suggests the existence of a mechanism for pausing recombination after each cycle. Along these lines, ATM activation by RAG-induced double-stranded DNA breaks was shown to temporarily suppress RAG expression and *Igk* and *Igl* rearrangements in pre-B cells (Steinel et al., 2013); it is tempting to speculate that a similar mechanism is employed by DP thymocytes to pace *Tcra* recombination events.

3.2 *Tcrd* Repertoire

The signature feature of γδ T cell development is the programmed appearance of thymocytes that use distinct Vγ and Vδ gene segments and populate distinct sites in the periphery (Bonneville et al., 2010; Xiong & Raulet, 2007). The earliest murine fetal thymocytes develop from distinct fetal progenitor cells (Ikuta et al., 1990; Ramond et al., 2014). These progenitors give rise to γδ subsets that use Vγ3 (nomenclature according to Garman, Doherty, and Raulet (1986); alternatively Vγ5 according to Heilig and Tonegawa (1986) or *Trgv6* according to IMGT) and Vδ1 (*Trdv4*) that populate the epidermis and Vγ4 (alternatively Vγ6 or *Trgv4*) and Vδ1 (*Trdv4*) that populate the female reproductive tract (Havran & Allison, 1988; Itohara et al., 1993). These γδ cells dominate the fetal thymus from days 14–16 and are also notable for the lack of N-nucleotides, selective utilization of Dδ2, and canonical junctional sequences (Itohara et al., 1993; Weber-Arden, Wilbert, Kabelitz, & Arden, 2000). Starting from fetal day 18 and extending into early adulthood, γδ thymocytes rearrange a diverse set of Vδ gene segments including *Trdv* and *Trav-dv*, use both Dδ1 and Dδ2, and display substantial junctional diversity (Elliott, Rock, Patten, Davis, & Chien, 1988; Gallagher, Obeïd, Marche, & Jouvin-Marche, 2001; Weber-Arden et al., 2000). These thymocytes populate a variety of peripheral tissues, with tissue-biased usage of certain Vγ and Vδ combinations (Bonneville et al., 2010; Xiong & Raulet, 2007). In addition, rearrangement of *Trdv4* ceases in neonates and adults (Weber-Arden et al., 2000). Developmentally staged utilization of Vγ and Vδ gene segments is thought to result from programmed rearrangement rather than selection, because the rearrangement profile is unperturbed in thymocytes of mice that cannot express a γδ TCR (Itohara et al., 1993). Similarly, the subset of *Tcra/Tcrd* locus V gene segments that contribute to the *Tcrd* repertoire is unperturbed in thymocytes lacking CD3 expression, and cannot simply be explained by the linear order of V gene segments in the locus (Gallagher et al., 1998).

Although we will not discuss the human *Tcrd* repertoire in detail, there are notable parallels between man and mouse. Specifically, early human fetal thymic γδ T cells can be distinguished from late fetal and postnatal γδ thymocytes due to the lack of junctional N-nucleotides and the selective rearrangement of Vδ2 to Dδ3 (Krangel et al., 1990). The latter is homologous to murine Dδ2, suggesting a conserved mechanism regulating Dδ usage.

3.3 *Tcra* Repertoire

3.3.1 *Usage of Jα Gene Segments*

The defining feature of *Tcra* repertoire is the ability of the locus to undertake multiple cycles of Vα to Jα rearrangement in DP thymocytes. Multiple lines of evidence support the view that Jα segments are used in a highly ordered progression from 5′ to 3′ across the Jα array. *Tcra* rearrangement begins on fetal day 16–18 in the mouse, and these early thymocytes exclusively use 5′ Jα segments (especially Jα61), whereas adults make use of Jα segments dispersed along the entire length of the Jα array (Roth, Holman, & Kranz, 1991; Rytkönen, Hurwitz, Thompson, & Pelkonen, 1996; Thompson,

Pelkonen, & Hurwitz, 1990). *Tcra* rearrangements are similarly biased towards Jα61 in $CD71^+$ early DP thymocytes of adult mice, whereas they are broadly distributed in more mature, $CD71^-$ DP thymocytes (Shih et al., 2012). Rearrangement is also restricted to the most 5′ Jα gene segments in DP thymocytes with defective RAG expression (Yannoutsos et al., 2001) or abbreviated lifespan (Guo et al., 2002). In contrast, Jα usage extends further 3′ upon *in vitro* culture of $CD3^{low}$ DP thymocytes isolated from *bcl-2* transgenic mice (Petrie, Livak, Burtrum, & Mazel, 1995) or *in vivo* extension of DP thymocyte lifespan using a transgene encoding Bcl-X_L (Guo et al., 2002). Consistent with this progression, the analysis of circular excision products resulting from Vα to Jα recombination (Okazaki & Sakano, 1988; Toda et al., 1988) and of mice engineered to carry a rearranged VαJα in the 5′ portion of the Jα array (Buch, Rieux-Laucat, Förster, & Rajewsky, 2002; Wang, Huang, & Kanagawa, 1998) reveal that VαJα rearrangements can be excised by further Vα to Jα rearrangement in the absence of positive selection.

A remarkable feature of Jα usage is that individual T cells tend to have rearranged similarly positioned Jα segments on the two alleles (Davodeau et al., 2001; Huang & Kanagawa, 2001; Mauvieux, Villey, & de Villartay, 2001; Rytkönen, Hurwitz, Tolonen, & Pelkonen, 1994). This suggests a tightly regulated and coordinated program of rearrangements, so as to allow for the three to five rounds of rearrangement estimated to occur during the lifespan of a DP thymocyte (Thuderoz et al., 2010; Warmflash & Dinner, 2006).

3.3.2 Usage of Vα Gene Segments

Studies of the murine Vα repertoire have posed unique challenges due to the repetitive nature of the Vα/Vδ array. As noted previously, the locus contains single member V families at the Jα-proximal and Jα-distal ends of the V array. However, the central portion of the V array has undergone recent duplication or triplication of a 400-kb unit in different strains of mice (Fig. 1). This portion of the locus carries multimembered V families, typically with several closely related members within a 400-kb unit, and each with an even more closely related homologue in the second or third copies of the 400-kb unit. It has proven a challenge to distinguish different family members using any quantitative PCR-based approach or even by deep sequencing with relatively short reads.

The central question that has dominated the field is the extent of Vα/Jα combinatorial diversity. As discussed above, it is well established that there is a temporal progression of recombination events proceeding 5′ to 3′ across the Jα array. If there were a similarly regulated, but reciprocal progression from Jα-proximal to Jα-distal across the Vα array, this would allow thymocytes to efficiently use the Vα array through multiple cycles of Vα to Jα recombination, but would impose substantial constraints on *Tcra* combinatorial diversity. If there were no such regulation and Vα usage was position independent, combinatorial diversity might be greater, but thymocytes that had initially chosen a Jα-distal Vα gene segment would face limitations on their ability to edit their receptors by subsequent rounds of recombination.

A semiquantitative genomic PCR approach using murine Vα family-specific and Jα-specific primers revealed that the unique Jα-proximal Vα segments are biased to rearrange to 5′ Jα gene segments (Jα61–Jα48), the unique Jα-distal Vα segments are biased to rearrange to relatively 3′ Jα gene segments (Jα38–Jα16) and members of a dispersed Vα family tend to rearrange broadly across the Jα array (Pasqual et al., 2002). Application of the same approach to human thymocytes yielded a very similar picture (Fuschiotti et al., 2007), as did analysis of murine thymocyte Jα usage in Vα–Cα cDNAs (Huang & Kanagawa, 2001). Six members of the *Trav14* (then called AV2) family are dispersed across 700-kb of the V array in BALB/c mice; sequence analysis of 268 *Trav14* cDNAs from thymocytes and peripheral lymphocytes indicated that the three most Jα-proximal *Trav14* members rearranged to Jα61–Jα41 81% of the time and to more 3′ Jα segments 19% of the time, whereas the three most Jα-distal *Trav14* members rearranged to Jα61–Jα41 only 39% of the time, and to more 3′ Jα gene segments 60% of the time (Aude-Garcia, Gallagher, Marche, & Jouvin-Marche, 2001). These studies convincingly demonstrate progressive proximal to distal usage of Vα gene segments through multiple cycles of recombination. In different studies, this behavior has been modeled by assuming either a defined and sequential proximal-to-distal opening of discrete regions of the V gene array or a defined proximal-to-distal velocity of availability of individual V gene segments (Thuderoz et al., 2010; Warmflash & Dinner, 2006). However, analysis of panels of human T cell clones showed only weak correlations between the chromosomal positions of Vα and Jα segments that had undergone recombination with each other, and similarly weak correlations between the chromosomal positions of the rearranged V gene segments on the two alleles in individual clones (Davodeau et al., 2001).

More recently, the *Tcra* repertoire was analyzed by high-throughput sequencing of *Tcra* transcripts in C57BL/6 pre-immune CD8$^+$ splenic T cells (Genolet, Stevenson, Farinelli, Østerås, & Luescher, 2012). The data revealed that the three most proximal Vα gene segments are strongly biased to rearrange to the most 5′ Jα segments and that the most distal Vα gene segments rearrange rarely to the 5′ Jα segments but most frequently to Jα segments in the 3′ half of the Jα array. However, among the large, highly repetitive Vα families distributed across the major portion of the locus, there was no significant relationship between Vα chromosomal position and the Jα segments with which they rearranged. Although the authors emphasized the random nature Vα–Jα recombination, in fact, their data do not look too dissimilar to those obtained in most of the earlier studies, given that those studies generally did not attempt to discriminate the individual rearrangement patterns of the dispersed Vα family members. Moreover, some additional cautions should be applied when considering the conclusions of the deep sequencing study. First, this study, as others before it, had difficulty contending with the dispersed but highly similar Vα family members, particularly since the analysis was of C57BL/6, in which the 400-kb unit of the Vα array is triplicated. The short read lengths failed to unambiguously assign some 40% of Vα sequences to a single Vα segment. The authors nevertheless mapped these ambiguous reads by distributing them in proportion to the unambiguous reads for a given V family. Whether this approach skewed the results is difficult to judge, but we note that the authors' emphasis on the random nature of the repertoire depends primarily on the random nature by which the individual members of multimember V families are used. Second, this and other studies that used cDNA to analyze repertoire may introduce biases due to differences in mRNA expression so that the results do not strictly reflect the number of cells bearing particular rearrangements (Robins, 2013). Finally, any expectations of highly coordinated Vα rearrangement in the various repertoire studies must be tempered by the realization that a history of prior *Tcrd* recombination is common in thymocytes that initiate *Tcra* rearrangement. Indeed, an examination of germline retention of *Tcra/Tcrd* sequences in DP thymocytes from *Tcra* enhancer-deleted (E$\alpha^{-/-}$) mice (which fail to rearrange *Tcra*) reveal that a substantial proportion of *Tcra/Tcrd* alleles have undergone *Tcrd* rearrangement to Vδ gene segments upstream of *Trav17* before the onset of *Tcra* rearrangement (Shih et al., 2012). This would scramble the starting point for primary Vα to Jα recombination and would be expected to influence combinatorial diversity through subsequent rounds of recombination in a manner that could obscure overt regulation of the process. In this regard, it would be of interest to analyze the *Tcra* repertoire in mice engineered to prevent *Tcrd* rearrangements.

Considering all of the above, we do not yet have a clear understanding of the rules underpinning the formation of the Vα repertoire. The use of appropriate genetic models, deep sequencing platforms, and bioinformatics workflows designed specifically for analysis of TCR repertoires (Bolotin et al., 2013; Yang et al., 2015) have the potential to address remaining uncertainties.

4. ACCESSIBILITY

4.1 Chromatin Accessibility as a Regulator of Antigen Receptor Recombination

We now turn our attention to the mechanisms that regulate assembly of the *Tcrd* and *Tcra* repertoires in developing thymoctyes. Studies of the developmental regulation of V(D)J recombination have been dominated by the concept of "accessibility" since it was first proposed 30 years ago (Yancopoulos & Alt, 1985). This area has been extensively reviewed over the years (Bassing et al., 2002; Cobb et al., 2006; Krangel, 2003; Schatz & Ji, 2011). Here, we will summarize current understanding of this critical developmental constraint on recombination and we will then discuss in detail its contribution to the developmental program at the *Tcra/Tcrd* locus.

It has long been understood that chromatin should pose a barrier to V(D)J recombination because nucleosomal packaging of DNA and subsequent higher order packing of nucleosomes within chromatin would limit the ability of the RAG recombinase to bind to and cleave at RSSs. The "accessibility hypothesis" originated from studies that correlated the developmental activation of gene segments for V(D)J recombination with transcriptional activation of the unrearranged gene segments, so-called "germline transcription" (Yancopoulos & Alt, 1985). Subsequent studies expanded this correlation to include a set of histone modifications that are associated with transcriptionally active chromatin, hypersensitivity to nuclease digestion, and hypomethylation of DNA (Bassing et al., 2002; Cobb et al., 2006; Krangel, 2003; Schatz & Ji, 2011). The concept of accessibility was solidified by the observation that RAG can cleave at chromosomal RSSs when added to isolated nuclei *in vitro*, but that the specificity of cleavage was dictated by the lineage and developmental stage of the cells from which those nuclei were isolated (Stanhope-Baker, Hudson, Shaffer, Constantinescu, & Schlissel, 1996). This suggested that developmental stage-specific properties of chromatin dictated which RSSs would be available to RAG.

Assembly of RSSs into nucleosomes inhibits RAG cleavage *in vitro* (Golding, Chandler, Ballestar, Wolffe, & Schlissel, 1999; Kwon, Imbalzano, Matthews, & Oettinger, 1999), but this barrier can be overcome both *in vitro* (Du, Ishii, Pazin, & Sen, 2008; Patenge, Elkin, & Oettinger, 2004) and *in vivo* (Osipovich, Subrahmanyam, Pierce, Sen, & Oltz, 2009; Osipovich et al., 2007) by the recruitment of ATP-dependent chromatin remodelers such as SWI/SNF. Recombination also depends directly on the histone modification H3 lysine 4 trimethylation (H3K4me3), since this modification is specifically recognized by the RAG2 PHD domain (Liu, Subrahmanyam, Chakraborty, Sen, & Desiderio, 2007; Matthews et al., 2007) and can stimulate RAG cleavage (Shimazaki, Tsai, & Lieber, 2009).

Notably, the recruitment of chromatin remodelers and the deposition of histone modifications such as H3K4me3 are associated with promoter activation and the initiation of transcription. Many of the gene segments at TCR and Ig loci are tightly linked to promoters, with D and J segment-associated promoters controlled by distant, developmentally regulated enhancers (Bassing et al., 2002; Cobb et al., 2006; Krangel, 2003; Schatz & Ji, 2011). Indeed, gene-targeted elimination of promoters and enhancers has shown them to be critical local and long-distance regulators, respectively, of V(D)J recombination *in vivo* (Bassing et al., 2002; Cobb et al., 2006; Krangel, 2003; Schatz & Ji, 2011). Further, enhancer- and promoter-dependent germline transcription across RSSs has been causally linked to the ability of those RSSs to serve as recombination substrates *in vivo* (Abarrategui & Krangel, 2006, 2007, 2009). Transcriptional elongation by RNA polymerase II deposits histone modifications such as H3ac and H3K4me3 across the 5′ portions of transcription units (Abarrategui & Krangel, 2007, 2009). Transcriptional elongation also disrupts nucleosome organization and may provoke nucleosome removal (Kondilis-Mangum et al., 2010). A recent study indicated that transcription causes transient eviction of one copy each of H2A and H2B from histone octamers, exposing 35–40 bp of nucleosomal DNA to RAG (Bevington & Boyes, 2013). Binding of RAG1 likely depends on the exposure of nucleosomal RSSs (Bevington & Boyes, 2013; Ji, Little, et al., 2010; Ji, Resch, et al., 2010) whereas binding of RAG2 depends critically on nucleosomal H3K4me3 (Ji, Resch, et al., 2010; Liu, Subrahmanyam, et al., 2007; Matthews et al., 2007). RAG1 and RAG2 therefore appear to be corecruited to RSSs associated with strong promoter activity, high H3K4me3, and other signatures of highly accessible chromatin.

High level, focal binding of both RAG1 and RAG2 can be detected *in vivo* at antigen receptor D and J gene segments; these have been referred to as "recombination centers"(Ji, Resch, et al., 2010). Prior work suggested that RAG cleavage *in vitro* proceeds most efficiently with the binding of a RAG1/RAG2 heterotetramer to one RSS, with subsequent capture of a second RSS (Jones & Gellert, 2002). Both of these lines of reasoning inform the current view that RAG is recruited to highly accessible D–J chromatin to form a "recombination center" and that RAG must then capture a second RSS (for example, a V segment RSS) to initiate V(D)J recombination (Schatz & Ji, 2011). We will address the developmental program at the *Tcra/Tcrd* locus in this context.

4.2 Regulation of Gene Segment Accessibility during *Tcrd* Rearrangement

4.2.1 Accessibility of Dδ and Jδ Gene Segments

Developmental regulation of RAG accessibility to Dδ and Jδ gene segments is thought to be directed by the *Tcrd* enhancer (Eδ), which is located in the Jδ3–Cδ intron in humans and the Jδ2–Cδ intron in mice (Gill, Karjalainen, & Zaninetta, 1991; Redondo, Hata, Brocklehurst, & Krangel, 1990; Fig. 3). Using a transgenic minilocus V(D)J recombination reporter substrate, the human Eδ was shown to direct a specific step of transgene rearrangement, VδDδ to Jδ, and it did so in DN thymocyte precursors of both αβ and γδ T cells (Lauzurica & Krangel, 1994a, 1994b). Elimination of Eδ from the mouse germline by gene targeting inhibited γδ T cell development. Moreover, Eδ-deleted alleles displayed substantial reductions in germline transcription and complete VδDδJδ rearrangements, with concomitant increases in Vδ–Dδ, Dδ–Dδ, and Dδ–Jδ partial rearrangement products (Monroe et al., 1999).

Eδ is thought to be activated in DN thymocytes and then inactivated in DP thymocytes (Hernández-Munain, Sleckman, & Krangel, 1999). A combination of *in vitro* and *in vivo* experiments have implicated the binding of RUNX, c-Myb, and GATA-3 transcription factors for enhancer activity (Hernandez-Munain & Krangel, 1994; Hernández-Munain,

Figure 3 Organization of gene segments and *cis*-regulatory elements in the 3′ portion of the murine *Tcra/Tcrd* locus. Arrows denote promoters, large ovals denote enhancers, small ovals denote DNase hypersensitive sites within the *Tcra* locus control region (LCR).

Lauzurica, & Krangel, 1996; Ko et al., 1991; Lauzurica, Zhong, Krangel, & Roberts, 1997; Redondo, Pfohl, & Krangel, 1991; Redondo et al., 1992). Eδ inactivation upon differentiation from the DN to the DP stage is attributable to a loss of transcription factor occupancy (Hernández-Munain et al., 1999). Consistent with this, RUNX binding is substantially reduced at the *Tcrg* enhancer, Eγ1, in association with *Tcrg* silencing in DP thymocytes (Tani-ichi, Satake, & Ikuta, 2011). Furthermore, RUNX1 and RUNX3 are both expressed in DN thymocytes and downregulated in DP thymocytes (Taniuchi et al., 2002).

Eδ promotes accessibility of Dδ and Jδ segments by influencing the activities of promoters associated with these gene segments (Fig. 3). Germline transcripts initiate within murine Dδ2 and upstream of and within murine Jδ1 in DN thymocytes (Carabana, Ortigoza, & Krangel, 2005) and these sites display high levels of histone modifications H3ac, H4ac, H3K4me2, and H3K4me3 (Hao & Krangel, 2011). Although RAG binding has not been examined directly, Dδ2 and Jδ1 would logically constitute the *Tcrd* recombination center. Eδ-deleted alleles display partial reductions in steady state transcript levels and histone modifications at these locations (Hao & Krangel, 2011), results that are compatible with the partially impaired V(D)J recombination phenotype of $E\delta^{-/-}$ mice (Monroe et al., 1999). RUNX1, c-Myb, and USF1 appear to be important mediators of murine Dδ2 promoter activity (Carabana et al., 2005).

RUNX1 may also instigate *Tcrd* rearrangement in a fashion that is independent of any role in transcription. Human *Tcrd* rearrangement differs from mouse in that it is highly ordered, initiating with Dδ2 to Dδ3 rearrangement and proceeding to Dδ2–Dδ3 to Jδ1 rearrangement (Cieslak et al., 2014; Dik et al., 2005). This ordering is dependent on the binding of RUNX1 to the human Dδ2 23 bp RSS. RUNX1 was found to interact with RAG1 and to recruit RAG1 to the Dδ2 23 bp RSS. Mutation of the RUNX1 binding site inhibited Dδ2 to Dδ3 rearrangement and facilitated Dδ3 to Jδ1 rearrangement on recombination substrates *in vitro*. Moreover, at the endogenous locus, RUNX1 was required to enforce a classical Dδ2 to Dδ3 join involving the Dδ2 23 bp RSS (3′ of Dδ2) and the Dδ3 12 bp RSS (5′ of Dδ3) to form a chromosomal Dδ2–Dδ3 coding joint. In the absence of RUNX1, Dδ2 to Dδ3 recombination proceeded through a nonclassical (and nonproductive) pathway that joined the Dδ2 12 bp RSS (5′ of Dδ2) to the Dδ3 23 bp RSS (3′ of Dδ3), thus creating a chromosomal signal joint and excising Dδ2 and Dδ3 from the chromosome (Cieslak et al., 2014). Although the *in vitro* and *in vivo* data differ in detail, both point to a critical role for RUNX1 to ensure proper use of the Dδ2 23 bp RSS. Prior work implicated c-Fos dependent recruitment of RAG proteins to the Dβ 23 bp RSSs as important to enforce ordered *Tcrb* recombination (Wang et al., 2008). These examples suggest that there may be much more to learn about the mechanisms by which RAG is targeted to fine-tune recombination programs *in vivo*.

4.2.2 Accessibility of Vδ Gene Segments

Given that only a subset of *Tcra/Tcrd* locus V gene segments function as Vδ gene segments, it is pertinent to consider what features discriminate Vδ from Vα gene segments. Because several Vδ gene segments (*Trdv1, Trdv2, Trdv4, Trdv5*) are situated in the proximal portion of the V array, one possibility is that linear position along the V array defines Vδ gene segments. This, however, appears at best a very incomplete explanation, because as noted previously, numerous other Vδ gene segments are distributed across the central portion of the V array, interspersed with Vα gene segments. Another potential explanation invokes the phenomenon of "beyond 12/23" regulation, so-named to acknowledge that there may be intrinsic biases to the particular RSS pairs that can undergo rearrangement even among those that satisfy the 12/23 rule (Bassing et al., 2000; Sleckman et al., 2000). At the *Tcra/Tcrd* locus such regulation could dictate that Vδ 23 bp RSSs are intrinsically biased to rearrange to Dδ 12 bp RSSs, whereas Vα 23 bp RSSs are intrinsically biased to rearrange to Jα 12 bp RSSs. Using extrachromosomal recombination substrates in nonlymphoid cells, it was shown that Vδ 23 bp RSSs have a strong preference to rearrange to the Dδ1 as opposed to the Jδ1 12 bp RSS, which would contribute to the appropriate ordering of *Tcrd* gene rearrangement events (Lee et al., 2009). However, Vα 23 bp RSSs showed no bias for Jα 12 bp RSSs versus a Dδ 12 bp RSS. Moreover, endogenous *Tcra/Tcrd* alleles with Dδ1 replaced by Jα38 supported Vδ but not Vα rearrangement to the introduced Jα38 gene segment (Lee et al., 2009).

An alternative possibility is that Vδ gene segments may be distinguished from Vα gene segments by their differential accessibility. In other words, Vδ gene segments may be unique among *Tcra/Tcrd* locus V gene segments in being accessible for RAG cleavage in DN thymocytes when Dδ and Jδ segments are also made accessible for recombination. Indeed, analyzes of germline transcription and chromatin signatures provide at least partial support for this possibility, because several prominent Vδ gene segments (*Trdv2-2, Trdv5, Trav15-dv6*) are among the V gene segments that display relatively high germline transcription and histone modifications (H3ac and H4ac) in DN thymocytes (Hawwari & Krangel, 2005). Nevertheless, several other V gene segments examined also display high levels of these histone modifications, but do not function as Vδ gene segments. To address this question in more definitive fashion, the promoter of a murine Vα/Vδ gene segment (*Trav15-1-dv6-1*) was used to replace the promoter of a pure Vα gene segment (*Trav12-2*) using a gene targeting approach. Introduction of the *Trav15-1-dv6-1* promoter caused increased *Trav12-2* transcription and histone modifications in DN thymocytes, and induced the rearrangement of *Trav12-2* to Dδ and Jδ gene segments as well (Naik, Hawwari, & Krangel, 2015). While it is not certain that this result would apply for all Vδ gene segments, it can be concluded that at least some Vδ gene segments are defined based upon promoter-dependent accessibility profiles in DN thymocytes.

The basis for Vδ promoter activity remains largely unexplored. Eδ appears to play a limited role in this process. In adult DN thymocytes, Eδ-deleted alleles display diminished germline transcription of the nearby *Trdv5*, but no differences in transcription or chromatin signatures at more distant Vδ gene segments (Hao & Krangel, 2011; Huang & Sleckman, 2007). This suggests that Eδ operates over a rather limited range, a result consistent with the observation that Eδ cannot regulate its natural Dδ and Jδ targets when it is moved 100 kb downstream, replacing Eα (Bassing et al., 2003).

As noted previously, Vδ rearrangement is developmentally regulated, with *Trdv4* rearrangement restricted to fetal thymocytes and other Vδ gene segments predominating in adults. This switch appears to be regulated at the level of accessibility. *Trdv4* rearrangement and accessibility in fetal thymocytes (as measured by germline transcription and acetylation of histones H3 and H4) are strongly dependent on Eδ, 55 kb away (Hao & Krangel, 2011). However, despite the presence of an active Eδ, *Trdv4* recombination is suppressed in adult thymocytes. This suppression was associated with reduced H3ac and could be overcome in culture by treatment with the histone deacetylase inhibitor trichostatin A (Hao & Krangel, 2011). This suggests a promoter-specific mechanism of suppression. Notably, E2A-deficient adult thymocytes show similar reactivation of *Trdv4* rearrangement, suggesting that E2A enforces *Trdv4* suppression in adults. Conversely, E2A-deficient adult thymocytes display impaired recombination of adult Vδ gene segments *Trdv2-2* and *Trdv5*. Thus, E2A may function to coordinate the switch in Vδ gene segment usage between fetal and adult thymocytes (Bain, Romanow, Albers, Havran, & Murre, 1999).

4.3 Regulation of Gene Segment Accessibility during *Tcra* Rearrangement

4.3.1 The Tcra *Enhancer (Eα)*

The developmental regulation of accessibility for *Tcra* rearrangement depends critically on the *Tcra* enhancer. Eα-deleted mice display an almost complete suppression of Vα to Jα recombination (Sleckman, Bardon, Ferrini, Davidson, & Alt, 1997). Moreover, inclusion of Eα in transgenic V(D)J recombination reporter substrates instigates developmentally appropriate rearrangement in DP thymocytes (Capone et al., 1993; Lauzurica & Krangel, 1994b). Eα is situated 3′ of Cα and is one component of a series of DNase hypersensitive sites that extend from Cα towards the neighboring *Dad1* gene; together, these elements constitute the *Tcra* locus control region (LCR; Diaz, Cado, & Winoto, 1994; Ho, Yang, Morle, & Leiden, 1989; Winoto & Baltimore, 1989a; Fig. 3). Eα consists of protein binding regions Tα1–Tα4 (Ho et al., 1989; Winoto & Baltimore, 1989a), all of which are needed for developmentally appropriate activation of V(D)J recombination within a transgenic reporter substrate (Balmelle, Zamarreño, Krangel, & Hernández-Munain, 2004), and additional protein binding sites 5′ of Tα1. The enhancer recruits an extensive array of transcription factors: CREB, Ets-1/Fli-1 and E2A to the region 5′ of Tα1; CREB and Sp1 to Tα1; Ets-1, TCF-1/LEF-1 and RUNX1 to Tα2; GATA-3 and E2A to Tα3; and HEB to Tα4 (del Blanco, Roberts, Zamarreño, Balmelle-Devaux, & Hernández-Munain, 2009; Giese, Kingsley, Kirshner, & Grosschedl, 1995; Hernandez-Munain, Roberts, & Krangel, 1998; Mayall, Sheridan, Montminy, & Jones, 1997; Spicuglia et al., 2000; Travis, Amsterdam, Belanger, & Grosschedl, 1991). Enhanceosome assembly is highly cooperative (Giese et al., 1995; Hernandez-Munain et al., 1998; Mayall et al., 1997).

Notably, Eα is extensively occupied by bound transcription factors not only in DP thymocytes, but also in DN3a thymocytes, before the enhancer becomes active (Hernández-Munain et al., 1999; Spicuglia et al., 2000). This raises the question of what triggers enhancer activity in DP thymocytes. Recent work has shown that Eα is responsive to pre-TCR signaling, which stimulates (1) the binding of inducible factors NFAT, AP-1, and Egr-1 to multiple sites in Eα, thus adding to the DN3a complex composed of CREB, Ets-1/Fli-1, GATA-3, TCF-1/LEF-1, and E2A; (2) phosphorylation of CREB and recruitment of coactivator CBP/p300; and (3) increased binding of Ets-1 (del Blanco, García-Mariscal, Wiest, & Hernández-Munain, 2012). The inducible factors are thought to be transiently present in proliferating

DN4 and early DP thymocytes; however, in small resting DP thymocytes, Ets-1 and CBP/p300 binding remains high and GATA-3 and E47 binding are substantially increased (del Blanco et al., 2012). Consistent with Eα induction as a consequence of pre-TCR signaling, germline *Tcra* transcripts are detectable as early as the DN4 stage (Taghon et al., 2006). Nevertheless, substantial *Tcra* rearrangement is normally delayed until the DP stage, due to rapid cell cycling of DN4 thymocytes and cell-cycle-dependent degradation of RAG2. In this regard, RAG2 expression is elevated and *Tcra* rearrangement is extensive in nonproliferating DN4 thymocytes isolated from IL-7-deficient mice (Boudil et al., 2015).

Very recent work indicates that under conditions in which *Tcra* rearrangement is suppressed, Eα-dependent germline transcription is downregulated after the DP stage (del Blanco, Angulo, Krangel, & Hernández-Munain, 2015). This downregulation was attributed to Eα, because transcription and histone modifications directed by an ectopic Eα introduced into the *Tcrb* locus were similarly downregulated. Although enhancer inactivation was associated with reduced binding of E2A and HEB, enforced expression of E2A could not overcome this effect. Thus, the mechanism of enhancer downregulation remains an open question.

As noted previously, Eα represents one of a series of DNase hypersensitive sites that compose the *Tcra* LCR. By definition, an LCR confers copy number-dependent and integration site-independent expression of linked transgenes; the *Tcra* LCR is composed of tissue-specific (e.g., Eα) and nonspecific elements that together stimulate chromatin opening specifically in thymocytes (Diaz et al., 1994; Ortiz, Cado, Chen, Diaz, & Winoto, 1997). The tissue nonspecific elements located between Eα and the neighboring *Dad1* gene have been demonstrated to contribute to LCR activity and to function as insulators *in vitro* (Gomos-Klein, Harrow, Alarcón, & Ortiz, 2007; Magdinier, Yusufzai, & Felsenfeld, 2004; Zhong & Krangel, 1999); *in vivo*, they may protect the ubiquitously expressed, endogenous *Dad1* gene from the influence of Eα (Hawwari & Krangel, 2005). Notably, the expression of an LCR-linked reporter was significantly downregulated in peripheral αβ T cells as compared to thymocytes (Harrow & Ortiz, 2005). Hence, studies of the isolated Eα and the intact LCR both point to a DP stage-specific function for Eα, and imply that the expression of rearranged *Tcra* genes in peripheral T cells must be supported by other elements (del Blanco et al., 2015; Harrow & Ortiz, 2005). Ironically, the *Tcra* LCR is active in peripheral γδ T cells (Harrow & Ortiz, 2005), and Eα is necessary for the normal expression of rearranged *Tcrd* genes in peripheral γδ T cells

(Sleckman et al., 1997), even though it plays no role in *Tcrd* rearrangement in DN thymocytes. Thus, Eα is activated late in γδ T cell development, but appears to be inactivated late in αβ T cell development. It will be important to formally test the latter by conditional deletion of Eα in peripheral αβ T cells. In any event, these data do not support the conclusions of prior studies indicating that Eα activity is suppressed in γδ T cells by linked silencer elements (Winoto & Baltimore, 1989b).

4.3.2 Regulation of Jα Accessibility

The activation of Eα stimulates germline transcription and histone modifications across Jα chromatin in DP thymocytes (McMurry & Krangel, 2000; Sleckman et al., 1997). Germline transcription is directed by promoter elements associated with the 5′ Jα gene segments, primarily T-early-alpha (TEA; De Chasseval & De Villartay, 1993; de Villartay et al., 1987; Shimizu et al., 1993). TEA is situated upstream of Jα61 (Fig. 3). All TEA transcripts include the TEA exon; some terminate immediately downstream of this exon, whereas others extend the entire length of the Jα array and are spliced to Cα (de Villartay et al., 1987; Villey, Quartier, Selz, & de Villartay, 1997). Additional germline transcripts initiate upstream of several of the 5′ Jα gene segments: Jα58, Jα57, Jα56, and Jα49; these transcripts also extend the length of the Jα array and are spliced from Jα to Cα (Hawwari, Bock, & Krangel, 2005; Villey et al., 1997). Elimination of the TEA promoter results in a striking and selective defect in the rearrangement of a discrete set of Jα gene segments that extends from Jα61 through Jα52, with relatively normal usage of more 3′ Jα segments (Abarrategui & Krangel, 2007; Hawwari et al., 2005; Villey, Caillol, Selz, Ferrier, & de Villartay, 1996). TEA promoter deletion also results in the loss of transcripts initiating at Jα58–Jα56 and the loss of histone modifications H3ac, H4ac, H3K4me2, and H3K4me3 across the Jα61–Jα52 interval (Abarrategui & Krangel, 2007; Hawwari et al., 2005; Mauvieux, Villey, & de Villartay, 2003; Villey et al., 1997). Moreover, whereas RAG proteins localize to the most 5′ Jα gene segments to form a recombination center on wild-type alleles, this binding is dramatically suppressed on TEA promoter-knockout alleles (Ji, Little, et al., 2010; Ji, Resch, et al., 2010). Taken together, these results demonstrate that TEA promoter activation plays a central role in targeting primary Vα to Jα rearrangements into the 5′ portion of the Jα array. Moreover, the accessible chromatin state created by TEA and Eα is directly linked to transcriptional elongation across the 5′ Jα gene segments, because the promoter-deletion phenotype could be reproduced by the introduction of transcription terminators downstream of TEA (Abarrategui & Krangel, 2006, 2007; Ji, Little, et al., 2010). Transcriptional elongation from the TEA promoter also appears to facilitate transcription initiating at Jα58–Jα56, perhaps as a consequence of chromatin remodeling (Abarrategui & Krangel, 2007; Hawwari et al., 2005; Villey et al., 1997). Chromatin remodeling and RAG recruitment are presumably highest across the most 5′ Jα gene segments because this is the region with the highest density of elongating RNA pol II (Abarrategui & Krangel, 2006).

Although initial Jα rearrangements are targeted primarily to Jα61, Jα58, and Jα57 immediately downstream of TEA, a subset of rearrangements is targeted to Jα50–Jα45 (Guo et al., 2002; Hawwari et al., 2005). This second focus of rearrangements events corresponds to the area in which Jα usage recovers on TEA-deficient alleles. These rearrangements are targeted by a distinct promoter associated with the Jα49 gene segment (Hawwari et al., 2005). Deletion of both the TEA and Jα49 promoters causes early rearrangements to distribute even further 3′, apparently due to the activation of otherwise cryptic promoters associated with more 3′ Jα gene segments. However, this targeting appears to be nonphysiological, since the 3′ Jα promoters are normally suppressed by transcriptional interference on unrearranged alleles carrying the TEA promoter (Abarrategui & Krangel, 2007; Hawwari et al., 2005), as well as on rearranged alleles carrying the promoter of a rearranged Vα gene segment (Hawwari & Krangel, 2007).

Tcra alleles that have undergone primary Vα to Jα recombination physiologically delete the TEA promoter, and in some instances may delete the Jα49 promoter as well. Therefore, the targeting of secondary Vα to Jα rearrangements must rely on other elements. One insight into the targeting of secondary rearrangements was provided by analysis of the relative use of Jα segments on the two alleles of T cell clones isolated from mice heterozygous for the TEA promoter deletion (Mauvieux et al., 2001). It was shown that although similarly positioned Jα segments were used on the two alleles of wild-type mice (mean 6 kb distance) or homozygous TEA promoter-deleted mice, positional coincidence was lost in mice heterozygous for the TEA promoter deletion (mean 16 kb difference), with the TEA-deleted allele invariably using a further 3′ Jα segment than the wild-type allele. This led the authors to hypothesize that the sites of early rearrangements must dictate the sites that subsequently become accessible and rearrange on the same allele, and that the critical regulatory element may be the promoter of the previously rearranged Vα gene segment (Mauvieux et al., 2001). An experimental test of this was provided by mice in which a rearranged *Trav17*-Jα57 replaced the wild-type Jα57, with deletion of 300 kb between the insertion and the endogenous *Trav20* gene segment (Buch et al., 2002). This allele thus mimics a natural primary rearrangement, and in the absence of positive selection, *Trav17*-Jα57 will be replaced by subsequent rounds of Vα to Jα recombination. In contrast to TEA-deleted alleles, the *Trav17*-Jα57 alleles were found to remodel chromatin and target initial rearrangements to the Jα segments immediately downstream of the introduced cassette. This is consistent with roles for the rearranged V segment promoter in providing accessibility to nearby Jα segments and in suppressing accessibility to more distant Jα segments (Hawwari & Krangel, 2007). Together, the data support a model in which primary rearrangements are directed by the TEA and Jα49 promoters, and subsequent rearrangements are directed in stepwise fashion to successively more 3′ Jα gene segments by chromatin remodeling directed by successively introduced Vα promoters.

4.3.3 Regulation of Vα Accessibility

Studies of Vα accessibility are complicated by some of the same issues that have compromised studies of the Vα repertoire. Nevertheless, a survey of V gene segment histone modifications revealed that V gene segment chromatin regulation varied as a function of V gene segment position in the array (Hawwari & Krangel, 2005). For those V gene segments in the distal 1.1 Mb of the array, histone H3ac and H4ac varied from one V segment to another, but for any particular V gene segment (or V gene family), did not vary between DN and DP thymocytes. In contrast, for those V gene segments in the proximal 0.4 Mb of the V array, these histone marks were dramatically upregulated in DP thymocytes, and this upregulation was abolished on Eα-deleted alleles. Eα regulation was documented as far upstream as *Trav8-2*, 525 kb from Eα (Hawwari & Krangel, 2005). Robust activation of proximal Vα gene segments by Eα offers a potential mechanism to explain how these V gene segments could serve as preferred targets for primary Vα to Jα rearrangement events.

Critical questions remain unanswered. It is presumed that V gene segment promoters vary in their intrinsic abilities to generate Eα-independent accessibility. However, it is not known whether accessibility of the Eα-independent V gene segments is determined solely by their promoters, or in collaboration with other long-distance *cis*-regulatory elements. What defines the discrete region over which Eα functions is not understood. However, the three-dimensional organization of the locus could play an important role (see below). Finally, it should be noted that the above study only analyzed the V array chromatin on an unrearranged allele. It remains unclear how *Tcrd* or primary *Tcra* rearrangement might modify the extent of Eα influence, and in this way influence the targeting of V gene segments during subsequent rounds of recombination.

5. SUBNUCLEAR POSITIONING

5.1 Overview

The spatial organization of genes within the eukaryotic nucleus is understood to be an important determinant of gene expression, and would therefore be expected to influence the dynamics of V(D)J recombination at antigen receptor loci. Within the interphase nucleus, individual chromosomes occupy discrete regions, or chromosome territories, and genomic regions can loop out from those territories in a developmentally regulated fashion (Fraser & Bickmore, 2007; Sutherland & Bickmore, 2009). Such looping may facilitate interactions with foci of RNA Pol II or RNA splicing factors as well as with other actively transcribed genes. In certain cases, transient colocalization of genes has been implicated in the trans-regulation of epigenetic states, as in the processes of X chromosome inactivation and genomic imprinting (Fraser & Bickmore, 2007; Schneider & Grosschedl, 2007). Along these lines, homologous *Igh–Igh* and *Igk–Igk* allelic pairings and heterologous *Igh–Igk* allelic pairings have all been documented in developing B cells; such pairings have been suggested to coordinate regulatory events on the two alleles, including conformational changes, subnuclear positioning, and allelic exclusion (Hewitt et al., 2008, 2009).

The nuclear lamina provides a distinct environment that is thought to be suppressive for gene expression; gene activation is generally associated with relocalization away from the nuclear lamina (Kind & van Steensel, 2010; Peric-Hupkes et al., 2010; Pickersgill et al., 2006). Indeed, forced tethering of genes to the nuclear lamina can inhibit gene expression, at least in part due to interactions with transcriptional repressors and corepressors (Reddy, Zullo, Bertolino, & Singh, 2008; Zullo et al., 2012). Studies have shown that antigen receptor loci undergo developmentally regulated changes in their positioning with respect to the nuclear lamina. Because *Igh* and *Igk* are localized at the nuclear periphery in early lymphoid progenitors, but are centrally located in pro-B cells, it was proposed that dissociation from the nuclear lamina might be a prerequisite for locus transcription and recombination (Kosak et al., 2002; Reddy et al., 2008). Oddly, the *Tcrb* locus stochastically associates with the nuclear lamina in DN thymocytes, which is precisely the developmental stage that supports *Tcrb* gene transcription and rearrangement. However, analysis of the rearrangement status of *Tcrb* alleles showed that rearrangement is substantially more frequent on alleles that are either fully or partially dissociated from the lamina (Chan et al., 2013; Schlimgen, Reddy, Singh, & Krangel, 2008). Suppression of V(D)J recombination at the nuclear lamina may occur, at least in part, by segregation of *Tcrb* alleles from RAG2 (Chan et al., 2013).

Inactive genes are also found associated with foci of pericentromeric heterochromatin (PCH; Kosak & Groudine, 2004; Su, Sridharan, & Smale, 2005). Nonexpressed *Igh* and *Igk* alleles are associated with PCH in pre-B and activated mature B cells (Goldmit et al., 2005; Roldán et al., 2005; Skok et al., 2001). This association has been thought to mediate allelic exclusion, but *Igk* and *Igh* allelic exclusion remains intact in mice carrying *Igk* alleles that lack the silencer in the intervening sequence (Sis), even though associations of both genes with PCH are substantially reduced (Xiang, Zhou, Hewitt, Skok, & Garrard, 2011). Whether gene associations with PCH are the cause or the consequence of their silencing remains uncertain.

3D-FISH provides a valuable tool for single cell analysis, but it is important to bear in mind its limitations. First, the scoring of positioning or colocalization requires objective criteria, but these criteria may vary between laboratories and may not always be explicitly stated. Second, analysis of locus positioning by 3D-FISH, due to limited resolution, cannot be used to infer molecular contacts; indeed, it is challenging to understand mechanisms for *trans*-effects between alleles that may be separated by 1 μm. Third, heterogeneity at the single cell level cannot be assumed to reflect distinct and stable states. 3D-FISH of fixed cells provides a snapshot of the nucleus, but cannot capture the dynamics of spatial relationships. For example, even though many nuclei in a population may display an apparently monoallelic state, the distribution of states in the population of nuclei may actually be determined stochastically (Schlimgen et al., 2008). In this regard, use of live cell imaging has recently shown that interactions with the nuclear lamina may be stochastically reshuffled with each cell division, such that many more sites have a history of interaction with the lamina than might be scored at any one time (Kind et al., 2013). Positional or conformational heterogeneity as determined by 3D-FISH may reflect a dynamic equilibrium between multiple states (Giorgetti et al., 2014).

5.2 Subnuclear Positioning of the *Tcra/Tcrd* Locus

Subnuclear positioning of the *Tcra/Tcrd* locus has been examined in two studies (Schlimgen et al., 2008; Skok et al., 2007). Both documented a predominantly peripheral localization in pro-B cells and a more central localization in DN and DP thymocytes. Localization frequencies were lower in all cell types in the study by Schlimgen et al. because scoring relied on overlap of locus and nuclear lamina signals (Schlimgen et al., 2008). Both studies also showed *Tcra/Tcrd* alleles to have reduced associations with pericentromeric heterochromatin in DN and DP thymocytes as compared to B cells, but as above, frequencies of association with pericentromeric heterochromatin were lower in the study by Schlimgen et al. What regulates dissociation from these repressive compartments is unclear. Knock-in of Eα into the *Tcrb* locus reduces its associations with both compartments. However, the positioning of Eα-deleted *Tcra/Tcrd* locus alleles is unaltered in DN and DP thymocytes, arguing that other elements must play a dominant role at the endogenous locus (Schlimgen et al., 2008).

The relationship between *Tcra* recombination and *Tcra/Tcrd* locus positioning in DP thymocytes has recently been assessed (Chaumeil et al., 2013). One *Tcra/Tcrd* locus allele was found to loop out from the chromosome 14 territory in about 30% of DP thymocytes. Monoallelic looping was associated with locus transcription and RAG activity, as assessed by γ-H2AX DNA repair foci. Moreover, uncleaved alleles tended to be associated with pericentromeric heterochromatin. Interestingly, looping from the chromosome territory was dependent on RAG protein, but not on RAG catalytic activity; how RAG influences this process is unknown. In addition, about 20% of *Tcra/Tcrd* alleles were paired (within 1 μm) in DP thymocytes in a RAG protein-dependent manner. Nevertheless, *Tcra/Tcrd* alleles displayed a continuous distribution of distances from each other and substantial differences in regulation on paired versus unpaired alleles were not apparent. Thus, the biological significance of allelic pairing is uncertain. The authors also showed that the association of *Tcra/Tcrd* alleles (either paired or unpaired) with pericentromeric heterochromatin depended on ATM. In this way, they linked signaling downstream of RAG cleavage to enforcement of a suppressive environment on the uncleaved allele, and suggested that such changes might provide a mechanism to limit rearrangements to one allele at a time (Chaumeil et al., 2013). This is an appealing model. However, the fact that DNA repair foci are monoallelic in many DP thymocytes does not, by itself, signify a regulated process. Indeed, bi-allelic DNA repair foci were detected in this and a previous study (Chaumeil et al., 2013; Chen et al., 2000). In this regard, a recent analysis of DN thymocytes revealed frequencies of contemporaneously cleaved *Tcrb*, *Tcrg*, and *Tcra/Tcrd* loci that were even higher than those that would be predicted based on frequencies of cleavage at the individual loci assuming a stochastic, unregulated model (Bowen, Wangsa, Ried, Livak, & Hodes, 2013). Taken at face value, this result is incompatible with a general role for ATM in signaling to suppress RAG activity following an initial cleavage event. Nevertheless, just as ATM deficiency led increased frequencies of DNA repair foci at two *Tcra/Tcrd* alleles in DP thymocytes (Chaumeil et al., 2013), it also led to increased frequencies of multiple DNA repair foci at combinations of *Tcrb*, *Tcrg,* and *Tcra/Tcrd* alleles in DN thymocytes (Bowen et al., 2013). In the former study, this dysregulation was attributed to precocious RAG cleavage (Chaumeil et al., 2013), whereas in the latter study, it was attributed to defective repair of DSBs (Bowen et al., 2013), a known consequence of ATM deficiency (Bredemeyer et al., 2006; Huang et al., 2007). Additional work will be needed to more fully appreciate the role of ATM as a regulator of allelic localization and RAG cleavage at the *Tcra/Tcrd* locus.

6. LOCUS CONFORMATION

6.1 Overview

Antigen receptor gene rearrangement events may occur between gene segments that are separated by large distances within a particular locus. Studies over the last decade have made clear that antigen receptor loci must adopt conformations that support long-distance V(D)J recombination events, and that these conformations may themselves be regulated in ways that impart developmental control to the recombination process (Bossen et al., 2012; Jhunjhunwala et al., 2009; Seitan, Krangel, & Merkenschlager, 2012; Shih & Krangel, 2013). Similarly, antigen receptor loci carry *cis*-regulatory elements such as enhancers that function as long-distance regulators of accessibility. The three-dimensional organization of antigen receptor loci may be critical enablers of these regulatory interactions as well. The most comprehensively studied antigen receptor locus is *Igh*. Studies using 3D-FISH have demonstrated that the *Igh* locus adopts a relatively extended configuration in non-B lineage cells and undergoes a contraction event in pro-B cells that brings the two ends of the locus into proximity (Fuxa et al., 2004; Jhunjhunwala et al., 2008; Kosak et al., 2002; Roldán et al., 2005; Sayegh, Jhunjhunwala, Riblet, & Murre, 2005). Alleles that had not undergone V_H to D_HJ_H recombination in pro-B cells subsequently extend, or decontract, in pre-B cells, presumably to limit V_H to D_HJ_H recombination at this stage (Roldán et al., 2005). *Igh* contraction allows distal V_H gene segments to compete effectively with proximal V_H gene segment for rearrangement to D_HJ_H gene segments (Jhunjhunwala et al., 2008). In the absence of locus contraction, there is a strong bias towards rearrangement of only the most proximal V_H gene segments (Fuxa et al., 2004; Liu, Schmidt-Supprian, et al., 2007; Roldán et al., 2005).

3D-FISH provides only a low-resolution view of antigen receptor locus conformation. In recent years, a variety of approaches based on chromosome conformation capture (3C) technology have allowed unprecedented high-resolution views of the folding and organizational properties of mammalian genomes (Bickmore & van Steensel, 2013; de Wit & de Laat, 2012; Gibcus & Dekker, 2013). 3C-based technologies all rely on formaldehyde cross-linking to capture DNA fragments interacting over long distances, followed by restriction digestion and ligation to covalently link the interacting fragments. In 3C, interactions between two-specific fragments can be interrogated by PCR using a pair of primers, one in each of the two candidate fragments. 4C relies on circularization of ligation products, which allows the entire interactome of a "viewpoint" fragment to be interrogated by inverse PCR followed by deep sequencing. 5C and Hi-C allow analysis of complete interactomes, either locus wide (5C) or genome wide (Hi-C) without the limitation of a particular "viewpoint." Use of Hi-C, in particular, has shown the mammalian genome to be organized into megabase-sized topological units called topological associating domains (TADs), which are enriched in shorter-distance, intradomain DNA looping interactions.

CTCF is a ubiquitously expressed, zinc-finger protein that binds broadly across the genome to a conserved GC-rich motif and mediates looping between its binding sites (Merkenschlager & Odom, 2013; Ong & Corces, 2014). CTCF binding sites are enriched at TAD borders, but are distributed broadly within TADs as well. CTCF can create DNA loops that segregate enhancers from promoters and in this way insulate promoters from unwanted regulatory influences. Alternatively, CTCF can create DNA loops that bring together enhancers with target promoters and thereby facilitate promoter activation. Cohesin, long studied for its role in the cohesion of sister chromatids during mitosis, is recruited by CTCF to a large proportion of CTCF binding sites across the genome and is necessary for CTCF to carry out its DNA looping and insulation functions (Merkenschlager & Odom, 2013; Ong & Corces, 2014).

Antigen receptor loci are highly enriched for the binding of CTCF and cohesin, which, together with additional transcriptional/architectural proteins, are thought to dictate the chromatin loop organization of these loci (Bossen et al., 2012; Chaumeil & Skok, 2012; Ribeiro de Almeida, Stadhouders, Thongjuea, Soler, & Hendriks, 2012; Seitan et al., 2012;

Shih & Krangel, 2013). At the *Igh* and *Igk* loci, CTCF-mediated looping was shown to insulate proximal V gene segments from the influence of downstream enhancers, and thus prevent their dysregulated rearrangement and overrepresentation in the B cell repertoire (Guo, Yoon, et al., 2011; Ribeiro de Almeida et al., 2011). The *Igh* locus appears to be organized by a combination of local and long-distance loop networks, with CTCF dominating local loop organizations, and transcription factors Pax5 and YY1 more critical for long-range, locus-spanning interactions that contribute to locus contraction (Degner et al., 2011; Ebert et al., 2011; Guo, Gerasimova et al., 2011; Medvedovic et al., 2013; Verma-Gaur et al., 2012).

6.2 *Tcra/Tcrd* Locus Conformation and Loop Organization

Making use of a set of four probes spanning the *Tcra/Tcrd* locus, it has been shown by 3D-FISH that the locus adopts a relatively extended configuration in control B cells and a fully contracted configuration in DN thymocytes (Shih & Krangel, 2010). However, analysis of the DP compartment revealed the locus to be contracted across the 3′ region (defined with a probe hybridizing to proximal V gene segments and a probe hybridizing downstream of Cα), but to adopt an extended configuration across the 5′ region (defined with a probe hybridizing to the most distal V segments and a probe hybridizing to the central duplicated portion of the V array). Moreover, the 5′ extension of the locus in DP thymocytes appeared to fully involve the central V segments, because this region was detected as a multiple foci on a proportion of alleles in DP thymocytes (and control B cells) but not in DN thymocytes. Hence the *Tcra/Tcrd* locus appears to adopt distinct contracted configurations in DN and DP thymocytes, with the 3′ end contracted in both compartments, and the 5′ end contracted in DN thymocytes but extended in DP thymocytes. Within the limits of resolution by 3D-FISH, the 3′ contracted domain in DP thymocytes corresponds roughly to the 500 kb portion of the locus known to be regulated by Eα (Hawwari & Krangel, 2005). This, coupled with the fact that locus conformation is identical on Eα-deleted alleles, suggests that the conformation of the locus might define the Eα-regulatory domain (Shih & Krangel, 2010). Locus conformation was similarly unperturbed on Eδ-deleted alleles. Enhancer independence of the contracted state is consistent with a recent analysis of the *Tcrb* locus (Majumder et al., 2015). Different studies have indicated *Igh* locus contraction to be Eμ dependent (Guo, Gerasimova, et al., 2011) or Eμ independent (Medvedovic et al., 2013).

The unique conformational program of the *Tcra/Tcrd* locus can be rationalized on the basis of the distinct recombination programs that the locus supports in DN and DP thymocytes. In DN thymocytes, the locus must facilitate the rearrangement of Vδ gene segments, some of which are proximal to Dδ and Jδ gene segments, and some of which are quite distant. Moreover, the locus can undergo only a single round of rearrangement on each allele, because Vδ to Dδ to Jδ rearrangement would eliminate all Dδ gene segments. Thus, much like at the *Igh* locus, proximal and distal V gene segments must all be able to compete effectively for contact with the D–J recombination center to insure a diverse repertoire. This would be facilitated by a globally contracted *Tcra/Tcrd* locus conformation in DN thymocytes. As discussed previously, there can be multiple cycles of Vα to Jα rearrangement in DP thymocytes, with initial (primary) rearrangements preferentially involving the most proximal Vα gene segments and the most 5′ Jα gene segments. The 5′ extended but 3′ contracted locus configuration observed in DP thymocytes should bias these early rearrangements towards use of the proximal Vα gene segments, thereby saving more distal Vα gene segments for subsequent rounds of recombination. That said, it should be understood that conformational studies in DN and DP thymocytes apply to the unrearranged locus on a RAG-deficient background. Any potential influence of prior *Tcrd* rearrangement on the starting conformation for primary *Tcra* rearrangement, or of primary *Tcra* rearrangement on starting conformation for secondary *Tcra* rearrangement, cannot be inferred from this work.

It should be noted that the results of the above study were inconsistent with those of a prior study (Skok et al., 2007), which reported an extended *Tcra/Tcrd* locus conformation in DN thymocytes, a contracted conformation in DP thymocytes, and subsequent decontraction in more mature, single-positive thymocytes. The discrepancy in DN thymocytes is particularly vexing because the DN conformation was studied in the same cell population ($Rag2^{-/-}$) using some of the same probes (discussed in Shih and Krangel (2010)). Comparison of data obtained using distal Vα–Cα and central Vα–Cα probe combinations did lead the authors to conclude that locus contraction in DP thymocytes was more efficient at the 3′ end, but results for distal Vα—central Vα measurements were not specifically reported (Skok et al., 2007). However, a report by the same laboratory of locus looping from the chromosome territory in DP thymocytes, and in particular, the depiction of an extended configuration by locus painting, are consistent with a 5′ decontracted locus configuration (Chaumeil et al., 2013).

CTCF and cohesin bind to a large number of sites in the *Tcra/Tcrd* locus and play critical roles in long-distance interactions, transcription, and V(D)J recombination (Seitan et al., 2011; Shih et al., 2012). Indeed, the distribution of binding sites at the *Tcra/Tcrd* locus is rather unique among TCR and Ig loci. At the *Tcra/Tcrd* locus, CTCF and cohesin mark the majority of cis-regulatory elements in the locus, including Eα, the TEA promoter, and the promoters of most V gene segments; a small number of CTCF sites are intergenic (Shih et al., 2012). Although the *Igh* locus displays prominent CTCF binding at the intergenic regulatory region IGCR1 and the 3′ regulatory region 3′RR that flank the D_H–J_H–C_H portion of the locus, those associated with proximal V_H segments are downstream of RSSs, and those in the distal portion of the V_H array are at intergenic sites, including so-called PAIR elements (Degner, Wong, Jankevicius, & Feeney, 2009; Ebert et al., 2011). CTCF sites in the *Igk* locus are primarily intergenic, including the Sis and Cer elements (Degner et al., 2009; Ribeiro de Almeida et al., 2011), as are most CTCF sites in the *Tcrb* locus (Majumder et al., 2015). This suggests that the logic of CTCF-dependent regulation may vary among the different loci.

Analysis by 3C of long-distance looping at the *Tcra/Tcrd* locus documented a network of interactions within the contracted 3′ portion of the locus in DP thymocytes (Shih et al., 2012; Fig. 4). Eα was shown to contact the TEA promoter in a developmental stage-specific way, and was similarly found to contact proximal Vα gene segments over distances of several

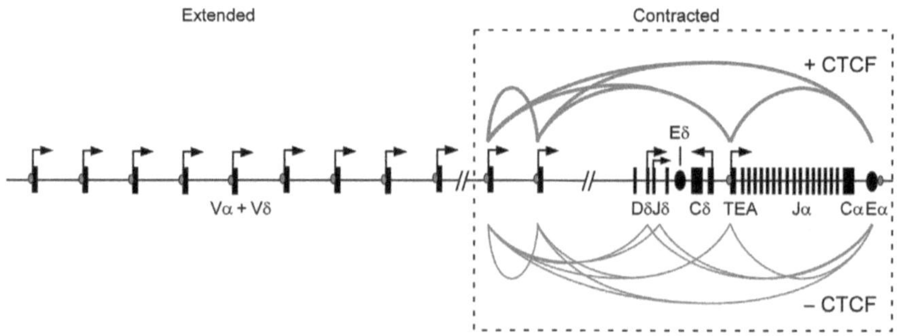

Figure 4 Chromatin loop organization of the *Tcra/Tcrd* locus in DP thymocytes as defined by 3C. Arrows denote promoters, large black ovals denote enhancers, small gray ovals denote CTCF binding sites. High-frequency looping interactions facilitated by CTCF are indicated by thick gray lines above the locus drawing. Low-frequency looping interactions in the absence of CTCF are indicated by thin gray lines below the locus drawing. The absence of CTCF reduces the specificity of the interaction network. All detected interactions take place within the contracted 3′ portion of the locus.

hundred kilobases, consistent with its roles in transcriptional activation and chromatin remodeling at these sites. Importantly, the TEA promoter was also shown to contact the proximal Vα gene segments, and these interactions were shown to be dependent on Eα. This implies that proximal Vα gene segments, the TEA promoter, and Eα are likely all brought together in 3D nuclear space, a scenario that would not only facilitate Eα-dependent accessibility at proximal Vα and 5′ Jα gene segments, but their synapsis for Vα to Jα recombination as well. Interactions between Eα and more distal portions of the Vα/Vδ array were not detected. Thus, the Eα interactome is concordant with the domain over which it is documented to influence transcription and chromatin structure (Hawwari & Krangel, 2005). The data are consistent with the notion that physical interactions of Eα are constrained within a roughly 500 kb, contracted domain encompassing the proximal portion of the *Tcra/Tcrd* locus in DP thymocytes (Shih & Krangel, 2010; Shih et al., 2012) and provide a satisfying physical and regulatory model that may explain the process of primary *Tcra* recombination.

What roles are played by CTCF and cohesin? To address the role of cohesin, the gene encoding the Rad21 subunit of cohesin was conditionally deleted in DP thymocytes using a CD4-Cre transgene (Seitan et al., 2011). This led to reduced TEA promoter-Eα interactions, along with reduced transcription, H3K4me3 and RAG protein recruitment to Jα segments. Analysis of Vα to Jα recombination revealed normal frequencies of primary rearrangement but a selective defect in secondary rearrangements to more 3′ Jα gene segments, which could be attributed to a defect in RAG cleavage. Unperturbed primary rearrangement was proposed to reflect the fact that *Rad21* gene deletion was only partial, and Rad21 protein levels near normal, in $CD71^+$ early DP thymocytes. Thus, in this system, Rad21 protein was depleted only after primary rearrangement had occurred. Earlier deletion of *Rad21* would have been problematic, given that early DP thymocytes are highly proliferative and cohesin plays a critical role in cell division.

A role for CTCF was assayed by conditional deletion using an Lck-Cre transgene, which initiates *Ctcf* deletion in DN thymocytes (Shih et al., 2012). Lck-Cre-driven deletion of *Ctcf* only modestly suppressed the number of DP thymocytes and allowed isolation of early DP thymocytes that were largely depleted of CTCF protein. This resulted in substantial defects in both primary and secondary Vα to Jα recombination. Further, these recombination defects were associated with reduced interactions between Eα and the TEA promoter, between Eα and proximal Vα gene segments, and between the TEA promoter and proximal Vα gene segments (Fig. 4).

Notably, although CTCF-deficiency reduced Eα–TEA promoter interactions and TEA transcription, Eα interacted more frequently with Dδ and Jδ gene segments and stimulated their transcription and recombination in DP thymocytes. Dδ and Jδ segments, unlike TEA and many Vα gene segment promoters, lack CTCF sites. This led to the conclusion that CTCF helps to specify Eα targets in DP thymocytes, biasing it to interact with and regulate those that are CTCF bound and to ignore those that are CTCF free. The ability of CTCF to bias Eα to ignore Dδ and Jδ gene segments in DP thymocytes may be biologically important, because in thymocytes newly entering the DP compartment, as many as 30% of *Tcra/Tcrd* alleles have not undergone Vδ to Dδ rearrangement and could serve as substrates for Vα to Dδ rearrangement (Shih et al., 2012).

Recent studies using 4C have defined a distinct network of interactions that organizes the 3′ region of the *Tcra/Tcrd* locus in DN thymocytes (Chen, Carico, Shih, & Krangel, 2015; Fig. 5). The central players in this interaction network are the TEA promoter CTCF site and two intergenic CTCF sites (denoted INT1 and INT2), which are separated from each other by about 5 kb and are situated about 80 kb upstream of TEA. The TEA CTCF site was found to interact at high frequency with INT2, forming a relatively stable structural loop that segregates a subset of *Tcrd* gene segments (*Trdv4*, Dδ, Jδ, Cδ, and *Trdv5*) from the remainder of the locus. Eδ and its interactome were contained within this loop as well, consistent with the known limits of

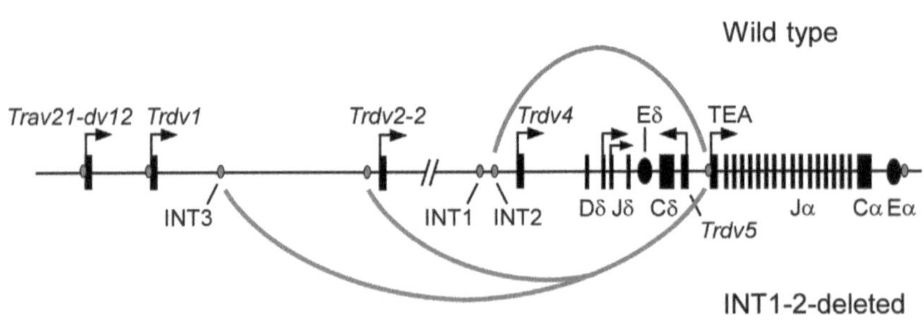

Figure 5 Chromatin loop organization of the *Tcra/Tcrd* locus in DN thymocytes as defined by 3C and 4C. Arrows denote promoters, large black ovals denote enhancers, small gray ovals denote CTCF binding sites. High-frequency looping between the INT2 and TEA CTCF sites on wild-type alleles is indicated by the thick gray line above the locus drawing. With the INT1-2 region deleted, TEA loops to additional CTCF sites located further upstream. This change in looping brings *Trdv2-2* into the same chromatin loop as Dδ and Jδ gene segments and facilitates *Trdv2-2* to Dδ–Jδ recombination. For the sake of clarity, the broad interactome of the INT1 CTCF site is not depicted.

Eδ influence (Hao & Krangel, 2011). Remarkably, although the INT2 CTCF site interacted in restricted fashion with TEA and sites within the 80 kb loop, the INT1 CTCF site, located immediately upstream of INT2, displayed a completely different interactome, looping at low frequencies to multiple upstream and downstream sites spanning the entire proximal 500 kb of the locus (Chen et al., 2015). This suggests a more dynamic tethering function for INT1.

Elimination of the INT1 and INT2 CTCF sites by gene targeting showed that the DN loop organization functions to diversify the *Tcrd* repertoire (Chen et al., 2015). With INT1 and INT2 deleted, the *Tcrd* repertoire was dominated by the rearrangement of *Trdv2-2*, the first Vδ gene segment upstream of the INT1 and INT2 CTCF sites. Newly formed loops from TEA to sites further upstream bring *Trdv2-2* into the same chromatin loop as the Dδ and Jδ segments (Fig. 5); this must allow *Trdv2-2* to outcompete other Vδ gene segments for Dδ substrates. Selective mutation of the INT2 CTCF site suggested that INT1 and INT2 can display partial redundancy; *Trdv2-2* rearrangement was elevated, but only partially, because increased looping between TEA and INT1 limited more long-distance interactions that facilitate highest levels of *Trdv2-2* rearrangement (Chen et al., 2015).

In some respects, the INT1–INT2 deletion phenotype parallels that described for *Igh* alleles lacking the intergenic IGCR1 CTCF sites (Guo, Yoon, et al., 2011). IGCR1-dependent looping separates V_H gene segments from D_H substrates and suppresses rearrangement of the most proximal V_H genes. However, absence of IGCR1 results in hyperactivation of proximal V_H transcription and accessibility by Eμ, implying that IGCR1 normally suppresses proximal V_H gene rearrangement by serving as an insulator. In contrast, deletion of INT1 and INT2 facilitates contacts between *Trdv2-2* and the Dδ–Jδ region, with no apparent changes in *Trdv2-2* transcription and accessibility (Chen et al., 2015). The simplest interpretation is that the INT2-TEA loop influences the Vδ repertoire by limiting the opportunities for synapsis between accessible *Trdv2-2* and Dδ RSSs.

Finally, this work revealed an intriguing link between the *Tcrd* and *Tcra* repertoires (Chen et al., 2015). Earlier, we noted that widely distributed Vα gene segments can rearrange to 5′ Jα gene segments during primary *Tcra* rearrangements. Notably, in mice lacking INT1 and INT2, the unusually homogeneous and proximally biased *Tcrd* repertoire is associated with reduced diversity of primary *Tcra* rearrangements: usage of the most proximal Vα gene segments is enhanced, whereas usage of more distal Vα gene segments is suppressed. This result is consistent with the notion that diversity of *Tcrd* rearrangements in DN thymocytes plays an important physiological role in enhancing *Tcra* combinatorial diversity, because *Tcrd* rearrangements will delete portions of the V array prior to the onset of *Tcra* recombination. In this sense, *Tcrd* rearrangements set the stage for subsequent *Tcra* rearrangements.

7. SUMMARY AND FUTURE DIRECTIONS

Current data suggest the following model for chromatin dynamics at the *Tcra*/*Tcrd* locus. At some point during the process of commitment to the T cell lineage, the *Tcra*/*Tcrd* locus moves away from the nuclear periphery and undergoes large-scale contraction in preparation for *Tcrd* gene rearrangement. What instigates these changes is currently unknown. Activation of Eδ likely results in the assembly of a recombination center at Dδ and Jδ gene segments, which may then capture Vδ gene segments made accessible, in most cases, by Eδ-independent promoter activity. Within the contracted structure, a stable CTCF-dependent loop between INT2 and TEA shapes the *Tcrd* repertoire, insuring broad utilization of Vδ gene segments. The dynamic tethering function of INT1 may contribute to repertoire development by recruiting Vδ gene segments into the recombination center, but this has not been tested directly. More fundamentally, the basis for assembly of the CTCF-dependent interaction network is not understood. The interactome that characterizes DN thymocytes is absent in B lineage cells, even though occupancy of CTCF sites is largely intact (Chen et al., 2015). This suggests that the interaction network requires unknown cofactors.

In thymocytes that undergo β-selection and enter the αβ T cell lineage, pre-TCR signaling induces Eα, the distal portion of the locus decontracts, and the locus loops out from its chromosome territory. Whether looping from the chromosome territory occurs in DN thymocytes has not been analyzed. Within the contracted 3′ portion of the locus, CTCF-directed looping of Eα to TEA results transcriptional activation of the TEA promoter, transcription-dependent 5′ Jα accessibility, and transcription-dependent assembly of a recombination center at 5′ Jα gene segments. A CTCF- and Eα-dependent network of interactions, perhaps built upon the INT2-TEA scaffold, also brings proximal Vα segments into contact with Eα and 5′ Jα gene segments, provokes proximal Vα accessibility, and stimulates primary Vα to Jα rearrangement. RAG-mediated cleavage may temporarily suppress further recombination to allow testing of thymocytes based on their specificity for self-peptide:MHC complexes. However, in the absence of positive selection, Vα to Jα recombination progresses down the Jα array on both alleles in relatively coordinated fashion, with accessibility created in stepwise fashion by the promoters of rearranged Vα gene segments. Vα usage is less coordinated than Jα usage, at least in part based on the history of prior *Tcrd* rearrangement. From the perspective of locus conformation, several questions remain unanswered. What provokes 5′-decontraction of the locus in DP thymocytes? Does the 3′-contracted, 5′-decontracted conformation of the unrearranged locus change following *Tcrd* rearrangement in DN thymocytes, or following primary *Tcra* rearrangement in DP thymocytes? How does the loop organization change once TEA, a key contact point in the pre-rearrangement structure, is deleted by primary *Tcra* rearrangement? We hope that answers to many of these questions will emerge in the near future.

ACKNOWLEDGMENTS

Work in the authors' laboratory has been supported by NIH grant R37 GM41052. The authors would like to thank Danielle Dauphars and Liang Chen for helpful comments on the text of this chapter.

REFERENCES

Abarrategui, I., & Krangel, M. S. (2006). Regulation of T cell receptor-[alpha] gene recombination by transcription. *Nature Immunology*, 7(10), 1109–1115. http://www.nature.com/ni/journal/v7/n10/suppinfo/ni1379_S1.html.

Abarrategui, I., & Krangel, M. S. (2007). Noncoding transcription controls downstream promoters to regulate T-cell receptor α recombination. *The EMBO Journal, 26*(20), 4380–4390. http://dx.doi.org/10.1038/sj.emboj.7601866.

Abarrategui, I., & Krangel, M. (2009). Germline transcription: A key regulator of accessibility and recombination. In P. Ferrier (Ed.), *Vol. 650. V(D)J recombination* (pp. 93–102). New York, NY: Springer.

Arden, B., Clark, S., Kabelitz, D., & Mak, T. (1995). Mouse T-cell receptor variable gene segment families. *Immunogenetics*, *42*(6), 501–530. http://dx.doi.org/10.1007/BF00172177.

Aude-Garcia, C., Gallagher, M., Marche, P. N., & Jouvin-Marche, E. (2001). Preferential ADV-AJ association during recombination in the mouse T-cell receptor alpha/delta locus. *Immunogenetics*, *52*(3–4), 224–230. http://dx.doi.org/10.1007/s002510000266.

Bain, G., Romanow, W. J., Albers, K., Havran, W. L., & Murre, C. (1999). Positive and negative regulation of V(D)J recombination by the E2A proteins. *The Journal of Experimental Medicine*, *189*(2), 289–300.

Balmelle, N., Zamarreño, N., Krangel, M. S., & Hernández-Munain, C. (2004). Developmental activation of the TCR α enhancer requires functional collaboration among proteins bound inside and outside the core enhancer. *The Journal of Immunology*, *173*(8), 5054–5063.

Bassing, C. H., Alt, F. W., Hughes, M. M., D'Auteuil, M., Wehrly, T. D., Woodman, B. B., et al. (2000). Recombination signal sequences restrict chromosomal V(D)J recombination beyond the 12/23 rule. *Nature*, *405*(6786), 583–586. http://www.nature.com/nature/journal/v405/n6786/suppinfo/405583a0_S1.html.

Bassing, C. H., Swat, W., & Alt, F. W. (2002). The mechanism and regulation of chromosomal V(D)J recombination. *Cell*, *109*(2, Supplement 1), S45–S55. http://dx.doi.org/10.1016/S0092-8674(02)00675-X.

Bassing, C. H., Tillman, R. E., Woodman, B. B., Canty, D., Monroe, R. J., Sleckman, B. P., et al. (2003). T cell receptor (TCR) α/δ locus enhancer identity and position are critical for the assembly of TCR δ and α variable region genes. *Proceedings of the National Academy of Sciences of the United States of America*, *100*(5), 2598–2603. http://dx.doi.org/10.1073/pnas.0437943100.

Bevington, S., & Boyes, J. (2013). Transcription-coupled eviction of histones H2A/H2B governs V(D)J recombination. *The EMBO Journal*, *32*(10), 1381–1392. http://dx.doi.org/10.1038/emboj.2013.42.

Bickmore, W. A., & van Steensel, B. (2013). Genome architecture: Domain organization of interphase chromosomes. *Cell*, *152*(6), 1270–1284. http://dx.doi.org/10.1016/j.cell.2013.02.001.

Bolotin, D. A., Shugay, M., Mamedov, I. Z., Putintseva, E. V., Turchaninova, M. A., Zvyagin, I. V., et al. (2013). MiTCR: Software for T-cell receptor sequencing data analysis. *Nature Methods*, *10*(9), 813–814. http://dx.doi.org/10.1038/nmeth.2555. http://www.nature.com/nmeth/journal/v10/n9/abs/nmeth.2555.html#supplementary-information.

Bonneville, M., O'Brien, R. L., & Born, W. K. (2010). γδ T cell effector functions: A blend of innate programming and acquired plasticity. *Nature Reviews. Immunology*, *10*(7), 467–478.

Borgulya, P., Kishi, H., Uematsu, Y., & von Boehmer, H. (1992). Exclusion and inclusion of α and β T cell receptor alleles. *Cell*, *69*(3), 529–537. http://dx.doi.org/10.1016/0092-8674(92)90453-J.

Bosc, N., & Lefranc, M.-P. (2003). The mouse (Mus musculus) T cell receptor alpha (TRA) and delta (TRD) variable genes. *Developmental & Comparative Immunology*, *27*(6–7), 465–497. http://dx.doi.org/10.1016/S0145-305X(03)00027-2.

Bossen, C., Mansson, R., & Murre, C. (2012). Chromatin topology and the regulation of antigen receptor assembly. *Annual Review of Immunology*, *30*(1), 337–356. http://dx.doi.org/10.1146/annurev-immunol-020711-075003.

Boudil, A., Matei, I. R., Shih, H.-Y., Bogdanoski, G., Yuan, J. S., Chang, S. G., et al. (2015). IL-7 coordinates proliferation, differentiation and Tcra recombination during thymocyte [beta]-selection. *Nature Immunology*, *16*, 397–405. http://dx.doi.org/10.1038/ni.3122. http://www.nature.com/ni/journal/v16/n4/abs/ni.3122.html.

Bowen, S., Wangsa, D., Ried, T., Livak, F., & Hodes, R. J. (2013). Concurrent V(D)J recombination and DNA end instability increase interchromosomal trans-rearrangements in ATM-deficient thymocytes. *Nucleic Acids Research*, *41*(8), 4535–4548. http://dx.doi.org/10.1093/nar/gkt154.

Brady, B. L., Steinel, N. C., & Bassing, C. H. (2010). Antigen receptor allelic exclusion: An update and reappraisal. *The Journal of Immunology*, *185*(7), 3801–3808.

Brändle, D., Müller, C., Rülicke, T., Hengartner, H., & Pircher, H. (1992). Engagement of the T-cell receptor during positive selection in the thymus down-regulates RAG-1 expression. *Proceedings of the National Academy of Sciences of the United States of America*, *89*(20), 9529–9533.

Bredemeyer, A. L., Sharma, G. G., Huang, C.-Y., Helmink, B. A., Walker, L. M., Khor, K. C., et al. (2006). ATM stabilizes DNA double-strand-break complexes during V(D)J recombination. *Nature*, *442*(7101), 466–470. http://www.nature.com/nature/journal/v442/n7101/suppinfo/nature04866_S1.html.

Buch, T., Rieux-Laucat, F., Förster, I., & Rajewsky, K. (2002). Failure of HY-specific thymocytes to escape negative selection by receptor editing. *Immunity*, *16*(5), 707–718. http://dx.doi.org/10.1016/S1074-7613(02)00312-6.

Capone, M., Hockett, R. D., & Zlotnik, A. (1998). Kinetics of T cell receptor β, γ, and δ rearrangements during adult thymic development: T cell receptor rearrangements are present in CD44(+)CD25(+) Pro-T thymocytes. *Proceedings of the National Academy of Sciences of the United States of America*, *95*(21), 12522–12527.

Capone, M., Watrin, F., Fernex, C., Horvat, B., Krippl, B., Wu, L., et al. (1993). TCR beta and TCR alpha gene enhancers confer tissue- and stage-specificity on V(D)J recombination events. *The EMBO Journal*, *12*(11), 4335–4346.

Carabana, J., Ortigoza, E., & Krangel, M. S. (2005). Regulation of the murine Dδ2 promoter by upstream stimulatory factor 1, Runx1, and c-Myb. *The Journal of Immunology*, *174*(7), 4144–4152.

Chan, E. A. W., Teng, G., Corbett, E., Choudhury, K. R., Bassing, C. H., Schatz, D. G., et al. (2013). Peripheral subnuclear positioning suppresses Tcrb recombination and segregates Tcrb alleles from RAG2. *Proceedings of the National Academy of Sciences of the United States of America*, *110*(48), E4628–E4637. http://dx.doi.org/10.1073/pnas.1310846110.

Chaumeil, J., Micsinai, M., Ntziachristos, P., Deriano, L., Wang, J. M. H., Ji, Y., et al. (2013). Higher-order looping and nuclear organization of antigen receptor loci facilitate targeted RAG cleavage and regulated rearrangement in recombination centers. *Cell Reports*, *3*(2), 359–370. http://dx.doi.org/10.1016/j.celrep.2013.01.024.

Chaumeil, J., & Skok, J. A. (2012). The role of CTCF in regulating V(D)J recombination. *Current Opinion in Immunology*, *24*(2), 153–159. http://dx.doi.org/10.1016/j.coi.2012.01.003.

Chen, H. T., Bhandoola, A., Difilippantonio, M. J., Zhu, J., Brown, M. J., Tai, X., et al. (2000). Response to RAG-mediated V(D)J cleavage by NBS1 and γ-H2AX. *Science*, *290*(5498), 1962–1964.

Chen, L., Carico, Z., Shih, H.-Y., & Krangel, M. S. (2015). A discrete chromatin loop in the mouse Tcra-Tcrd locus shapes the TCRδ and TCRα repertoires. *Nature Immunology*. (in press). http://dx.doi.org/10.1038/ni.3232.

Chien, Y. H., Iwashima, M., Wettstein, D. A., Kaplan, K. B., Elliott, J. F., Born, W., et al. (1987). T-cell receptor delta gene rearrangements in early thymocytes. *Nature*, *330*(6150), 722–727.

Cieslak, A., Le Noir, S., Trinquand, A., Lhermitte, L., Franchini, D.-M., Villarese, P., et al. (2014). RUNX1-dependent RAG1 deposition instigates human TCR-δ locus rearrangement. *The Journal of Experimental Medicine*, *211*(9), 1821–1832. http://dx.doi.org/10.1084/jem.20132585.

Ciofani, M., & Zúñiga-Pflücker, J. C. (2010). Determining γδ versus αβ T cell development. *Nature Reviews. Immunology*, *10*(9), 657–663.

Cobb, R. M., Oestreich, K. J., Osipovich, O. A., & Oltz, E. M. (2006). Accessibility control of V(D)J recombination. In W. A. Frederick (Ed.), *Vol. 91. Advances in immunology* (pp. 45–109). Waltham, MA: Academic Press.

Couedel, C., Lippert, E., Bernardeau, K., Bonneville, M., & Davodeau, F. (2004). Allelic exclusion at the TCRδ locus and commitment to γδ lineage: Different modalities apply to distinct human γδ subsets. *The Journal of Immunology*, *172*(9), 5544–5552.

Criscitiello, M. F., Saltis, M., & Flajnik, M. F. (2006). An evolutionarily mobile antigen receptor variable region gene: Doubly rearranging NAR-TcR genes in sharks. *Proceedings of the National Academy of Sciences of the United States of America*, *103*(13), 5036–5041. http://dx.doi.org/10.1073/pnas.0507074103.

Davodeau, F., Difilippantonio, M., Roldan, E., Malissen, M., Casanova, J.-L., Couedel, C., et al. (2001). The tight interallelic positional coincidence that distinguishes T-cell receptor Jα usage does not result from homologous chromosomal pairing during VαJα rearrangement. *The EMBO Journal, 20*(17), 4717–4729. http://dx.doi.org/10.1093/emboj/20.17.4717.

De Chasseval, R., & De Villartay, J.-P. (1993). Functional characterization of the promoter for the human germ-line T cell receptor J α (TEA) transcript. *European Journal of Immunology, 23*(6), 1294–1298. http://dx.doi.org/10.1002/eji.1830230616.

Degner, S. C., Verma-Gaur, J., Wong, T. P., Bossen, C., Iverson, G. M., Torkamani, A., et al. (2011). CCCTC-binding factor (CTCF) and cohesin influence the genomic architecture of the Igh locus and antisense transcription in pro-B cells. *Proceedings of the National Academy of Sciences, 108*(23), 9566–9571.

Degner, S. C., Wong, T. P., Jankevicius, G., & Feeney, A. J. (2009). Cutting edge: Developmental stage-specific recruitment of cohesin to CTCF Sites throughout immunoglobulin loci during B lymphocyte development. *The Journal of Immunology, 182*(1), 44–48.

del Blanco, B., Angulo, Ú., Krangel, M. S., & Hernández-Munain, C. (2015). T-cell receptor α enhancer is inactivated in αβ T lymphocytes. *Proceedings of the National Academy of Sciences, 112*(14), E1744–E1753.

del Blanco, B., García-Mariscal, A., Wiest, D. L., & Hernández-Munain, C. (2012). Tcra enhancer activation by inducible transcription factors downstream of Pre-TCR signaling. *The Journal of Immunology, 188*(7), 3278–3293.

del Blanco, B., Roberts, J. L., Zamarreño, N., Balmelle-Devaux, N., & Hernández-Munain, C. (2009). Flexible stereospecific interactions and composition within nucleoprotein complexes assembled on the TCRα gene enhancer. *The Journal of Immunology, 183*(3), 1871–1883.

de Villartay, J. P., Hockett, R. D., Coran, D., Korsmeyer, S. J., & Cohen, D. I. (1988). Deletion of the human T-cell receptor [delta]-gene by a site-specific recombination. *Nature, 335*(6186), 170–174.

de Villartay, J. P., Lewis, D., Hockett, R., Waldmann, T. A., Korsmeyer, S. J., & Cohen, D. I. (1987). Deletional rearrangement in the human T-cell receptor alpha-chain locus. *Proceedings of the National Academy of Sciences, 84*(23), 8608–8612.

de Villartay, J.-P., Mossalayi, D., de Chasseval, R., Dalloul, A., & Debré, P. (1991). The differentiation of human pro-thymocytes along the TCR-α/β pathway in vitro is accompanied by the site-specific deletion of the TCR-δ locus. *International Immunology, 3*(12), 1301–1305.

de Wit, E., & de Laat, W. (2012). A decade of 3C technologies: Insights into nuclear organization. *Genes & Development, 26*(1), 11–24.

Diaz, P., Cado, D., & Winoto, A. (1994). A locus control region in the T cell receptor α/δ locus. *Immunity, 1*(3), 207–217. http://dx.doi.org/10.1016/1074-7613(94)90099-X.

Dik, W. A., Pike-Overzet, K., Weerkamp, F., de Ridder, D., de Haas, E. F. E., Baert, M. R. M., et al. (2005). New insights on human T cell development by quantitative T cell receptor gene rearrangement studies and gene expression profiling. *The Journal of Experimental Medicine, 201*(11), 1715–1723. http://dx.doi.org/10.1084/jem.20042524.

Du, H., Ishii, H., Pazin, M. J., & Sen, R. (2008). Activation of 12/23-RSS-dependent RAG cleavage by hSWI/SNF complex in the absence of transcription. *Molecular Cell, 31*(5), 641–649. http://dx.doi.org/10.1016/j.molcel.2008.08.012.

Ebert, A., McManus, S., Tagoh, H., Medvedovic, J., Salvagiotto, G., Novatchkova, M., et al. (2011). The distal VH gene cluster of the Igh locus contains distinct regulatory elements with Pax5 transcription factor-dependent activity in Pro-B cells. *Immunity, 34*(2), 175–187. http://dx.doi.org/10.1016/j.immuni.2011.02.005.

Elliott, J. F., Rock, E. P., Patten, P. A., Davis, M. M., & Chien, Y. H. (1988). The adult T-cell receptor δ-chain is diverse and distinct from that of fetal thymocytes. *Nature, 331*(6157), 627–631.

Fischer, C., Bouneau, L., Ozouf-Costaz, C., Crnogorac-Jurcevic, T., Weissenbach, J., & Bernot, A. (2002). Conservation of the T-Cell receptor α/δ linkage in the teleost fish

tetraodon nigroviridis. *Genomics, 79*(2), 241–248. http://dx.doi.org/10.1006/geno.2002.6688.
Flajnik, M. F., & Kasahara, M. (2010). Origin and evolution of the adaptive immune system: Genetic events and selective pressures. *Nature Reviews. Genetics, 11*(1), 47–59. http://dx.doi.org/10.1038/nrg2703.
Fraser, P., & Bickmore, W. (2007). Nuclear organization of the genome and the potential for gene regulation. *Nature, 447*(7143), 413–417.
Fuschiotti, P., Pasqual, N., Hierle, V., Borel, E., London, J., Marche, P. N., et al. (2007). Analysis of the TCR α-chain rearrangement profile in human T lymphocytes. *Molecular Immunology, 44*(13), 3380–3388. http://dx.doi.org/10.1016/j.molimm.2007.02.017.
Fuxa, M., Skok, J., Souabni, A., Salvagiotto, G., Roldan, E., & Busslinger, M. (2004). Pax5 induces V-to-DJ rearrangements and locus contraction of the immunoglobulin heavy-chain gene. *Genes & Development, 18*(4), 411–422.
Gallagher, M., Candéias, S., Martinon, C., Borel, E., Malissen, M., Marche, P. N., et al. (1998). Use of TCR ADV gene segments by the delta chain is independent of their position and of CD3 expression. *European Journal of Immunology, 28*(11), 3878–3885.
Gallagher, M., Obeïd, P., Marche, P. N., & Jouvin-Marche, E. (2001). Both TCRα and TCRδ chain diversity Are regulated during thymic ontogeny. *The Journal of Immunology, 167*(3), 1447–1453.
Garman, R. D., Doherty, P. J., & Raulet, D. H. (1986). Diversity, rearrangement, and expression of murine T cell gamma genes. *Cell, 45*(5), 733–742. http://dx.doi.org/10.1016/0092-8674(86)90787-7.
Genolet, R., Stevenson, B. J., Farinelli, L., Østerås, M., & Luescher, I. F. (2012). Highly diverse TCRα chain repertoire of pre-immune CD8+ T cells reveals new insights in gene recombination. *The EMBO Journal, 31*(21), 4247–4248. http://dx.doi.org/10.1038/emboj.2012.277.
Gibcus, J. H., & Dekker, J. (2013). Connecting the genome: Dynamics and stochasticity in a new hierarchy for chromosome conformation. *Molecular Cell, 49*(5), 773–782. http://dx.doi.org/10.1016/j.molcel.2013.02.011.
Giese, K., Kingsley, C., Kirshner, J. R., & Grosschedl, R. (1995). Assembly and function of a TCR alpha enhancer complex is dependent on LEF-1-induced DNA bending and multiple protein-protein interactions. *Genes & Development, 9*(8), 995–1008.
Gill, L. L., Karjalainen, K., & Zaninetta, D. (1991). A transcriptional enhancer of the mouse T cell receptor δ gene locus. *European Journal of Immunology, 21*(3), 807–810. http://dx.doi.org/10.1002/eji.1830210339.
Giorgetti, L., Galupa, R., Nora, E. P., Piolot, T., Lam, F., Dekker, J., et al. (2014). Predictive polymer modeling reveals coupled fluctuations in chromosome conformation and transcription. *Cell, 157*(4), 950–963. http://dx.doi.org/10.1016/j.cell.2014.03.025.
Glusman, G., Rowen, L., Lee, I., Boysen, C., Roach, J. C., Smit, A. F. A., et al. (2001). Comparative genomics of the human and mouse T cell receptor Loci. *Immunity, 15*(3), 337–349. http://dx.doi.org/10.1016/S1074-7613(01)00200-X.
Golding, A., Chandler, S., Ballestar, E., Wolffe, A. P., & Schlissel, M. S. (1999). Nucleosome structure completely inhibits in vitro cleavage by the V(D)J recombinase. *The EMBO Journal, 18*(13), 3712–3723. http://dx.doi.org/10.1093/emboj/18.13.3712.
Goldmit, M., Ji, Y., Skok, J., Roldan, E., Jung, S., Cedar, H., et al. (2005). Epigenetic ontogeny of the Igk locus during B cell development. *Nature Immunology, 6*(2), 198–203. http://www.nature.com/ni/journal/v6/n2/suppinfo/ni1154_S1.html.
Gomos-Klein, J., Harrow, F., Alarcón, J., & Ortiz, B. D. (2007). CTCF-independent, but Not CTCF-dependent, elements significantly contribute to TCR-α locus control region activity. *The Journal of Immunology, 179*(2), 1088–1095.
Guo, C., Gerasimova, T., Hao, H., Ivanova, I., Chakraborty, T., Selimyan, R., et al. (2011). Two forms of loops generate the chromatin conformation of the immunoglobulin

heavy-chain gene locus. *Cell, 147*(2), 332–343. http://dx.doi.org/10.1016/j.cell.2011.08.049.

Guo, J., Hawwari, A., Li, H., Sun, Z., Mahanta, S. K., Littman, D. R., et al. (2002). Regulation of the TCR[alpha] repertoire by the survival window of CD4+CD8+ thymocytes. *Nature Immunology, 3*(5), 469–476.

Guo, C., Yoon, H. S., Franklin, A., Jain, S., Ebert, A., Cheng, H.-L., et al. (2011). CTCF-binding elements mediate control of V(D)J recombination. *Nature, 477*(7365), 424–430. http://www.nature.com/nature/journal/v477/n7365/abs/nature10495.html#supplementary-information.

Haks, M. C., Lefebvre, J. M., Lauritsen, J. P. H., Carleton, M., Rhodes, M., Miyazaki, T., et al. (2005). Attenuation of γδTCR signaling efficiently diverts thymocytes to the αβ lineage. *Immunity, 22*(5), 595–606. http://dx.doi.org/10.1016/j.immuni.2005.04.003.

Hao, B., & Krangel, M. S. (2011). Long-distance regulation of fetal Vδ gene segment TRDV4 by the tcrd enhancer. *The Journal of Immunology, 187*(5), 2484–2491.

Harrow, F., & Ortiz, B. D. (2005). The TCRα locus control region specifies thymic, but Not peripheral, patterns of TCRα gene expression. *The Journal of Immunology, 175*(10), 6659–6667.

Havran, W. L., & Allison, J. P. (1988). Developmentally ordered appearance of thymocytes expressing different T-cell antigen receptors. *Nature, 335*(6189), 443–445.

Hawwari, A., Bock, C., & Krangel, M. S. (2005). Regulation of T cell receptor [alpha] gene assembly by a complex hierarchy of germline J[alpha] promoters. *Nature Immunology, 6*(5), 481–489. http://www.nature.com/ni/journal/v6/n5/suppinfo/ni1189_S1.html.

Hawwari, A., & Krangel, M. S. (2005). Regulation of TCR δ and α repertoires by local and long-distance control of variable gene segment chromatin structure. *The Journal of Experimental Medicine, 202*(4), 467–472.

Hawwari, A., & Krangel, M. S. (2007). Role for rearranged variable gene segments in directing secondary T cell receptor α recombination. *Proceedings of the National Academy of Sciences, 104*(3), 903–907.

Hayes, S. M., Li, L., & Love, P. E. (2005). TCR signal strength influences αβ/γδ lineage fate. *Immunity, 22*(5), 583–593. http://dx.doi.org/10.1016/j.immuni.2005.03.014.

Heilig, J. S., & Tonegawa, S. (1986). Diversity of murine gamma genes and expression in fetal and adult T lymphocytes. *Nature, 322*(6082), 836–840.

Helmink, B. A., & Sleckman, B. P. (2012). The response to and repair of RAG-mediated DNA double-strand breaks. *Annual Review of Immunology, 30*(1), 175–202. http://dx.doi.org/10.1146/annurev-immunol-030409-101320.

Hernandez-Munain, C., & Krangel, M. S. (1994). Regulation of the T-cell receptor delta enhancer by functional cooperation between c-Myb and core-binding factors. *Molecular and Cellular Biology, 14*(1), 473–483.

Hernández-Munain, C., Lauzurica, P., & Krangel, M. S. (1996). Regulation of T cell receptor delta gene rearrangement by c-Myb. *The Journal of Experimental Medicine, 183*(1), 289–293.

Hernandez-Munain, C., Roberts, J. L., & Krangel, M. S. (1998). Cooperation among multiple transcription factors is required for access to minimal T-cell receptor α-enhancer chromatin in vivo. *Molecular and Cellular Biology, 18*(6), 3223–3233.

Hernández-Munain, C., Sleckman, B. P., & Krangel, M. S. (1999). A developmental switch from TCRδ enhancer to TCRα enhancer function during thymocyte maturation. *Immunity, 10*(6), 723–733. http://dx.doi.org/10.1016/S1074-7613(00)80071-0.

Hewitt, S. L., Farmer, D., Marszalek, K., Cadera, E., Liang, H.-E., Xu, Y., et al. (2008). Association between the Igk and Igh loci mediated by the 3' Igk enhancer induces decontraction of the Igh locus in pre-B cells. *Nature Immunology, 9*(4), 396–404. http://dx.doi.org/10.1038/ni1567.

Hewitt, S. L., Yin, B., Ji, Y., Chaumeil, J., Marszalek, K., Tenthorey, J., et al. (2009). RAG1 and ATM coordinate monoallelic recombination and nuclear positioning of immunoglobulin loci. *Nature Immunology, 10*(6), 655–664. http://dx.doi.org/10.1038/ni.1735.

Hirano, M., Das, S., Guo, P., & Cooper, M. D. (2011). Chapter 4—The evolution of adaptive immunity in vertebrates. In W. A. Frederick (Ed.), *Vol. 109. Advances in immunology* (pp. 125–157). Waltham, MA: Academic Press.

Ho, I. C., Yang, L.-H., Morle, G., & Leiden, J. M. (1989). A T-cell-specific transcriptional enhancer element 3′ of Cα in the human T-cell receptor α locus. *Proceedings of the National Academy of Sciences of the United States of America, 86*(17), 6714–6718. http://dx.doi.org/10.2307/34605.

Hockett, R. D., de Villartay, J. P., Pollock, K., Poplack, D. G., Cohen, D. I., & Korsmeyer, S. J. (1988). Human T-cell antigen receptor (TCR) delta-chain locus and elements responsible for its deletion are within the TCR alpha-chain locus. *Proceedings of the National Academy of Sciences, 85*(24), 9694–9698.

Huang, C.-Y., & Kanagawa, O. (2001). Ordered and coordinated rearrangement of the TCR α locus: Role of secondary rearrangement in thymic selection. *The Journal of Immunology, 166*(4), 2597–2601.

Huang, C.-Y., Sharma, G. G., Walker, L. M., Bassing, C. H., Pandita, T. K., & Sleckman, B. P. (2007). Defects in coding joint formation in vivo in developing ATM-deficient B and T lymphocytes. *The Journal of Experimental Medicine, 204*(6), 1371–1381.

Huang, C.-Y., & Sleckman, B. P. (2007). Developmental stage-specific regulation of TCR-α-chain gene assembly by intrinsic features of the TEA promoter. *The Journal of Immunology, 179*(1), 449–454.

Ikuta, K., Kina, T., MacNeil, I., Uchida, N., Peault, B., Chien, Y. H., et al. (1990). A developmental switch in thymic lymphocyte maturation potential occurs at the level of hematopoietic stem cells. *Cell, 62*(5), 863–874. http://dx.doi.org/10.1016/0092-8674(90)90262-D.

Itohara, S., Mombaerts, P., Lafaille, J., Iacomini, J., Nelson, A., Clarke, A. R., et al. (1993). T cell receptor δ gene mutant mice: Independent generation of αβ T cells and programmed rearrangements of γδ TCR genes. *Cell, 72*(3), 337–348. http://dx.doi.org/10.1016/0092-8674(93)90112-4.

Jhunjhunwala, S., van Zelm, M. C., Peak, M. M., Cutchin, S., Riblet, R., van Dongen, J. J. M., et al. (2008). The 3D structure of the immunoglobulin heavy-chain locus: Implications for long-range genomic interactions. *Cell, 133*(2), 265–279. http://dx.doi.org/10.1016/j.cell.2008.03.024.

Jhunjhunwala, S., van Zelm, M. C., Peak, M., & Murre, C. (2009). Chromatin architecture and the generation of antigen receptor diversity. *Cell, 138*(3), 435–448. http://dx.doi.org/10.1016/j.cell.2009.07.016.

Ji, Y., Little, A. J., Banerjee, J. K., Hao, B., Oltz, E. M., Krangel, M. S., et al. (2010). Promoters, enhancers, and transcription target RAG1 binding during V(D)J recombination. *The Journal of Experimental Medicine, 207*(13), 2809–2816.

Ji, Y., Resch, W., Corbett, E., Yamane, A., Casellas, R., & Schatz, D. G. (2010). The in vivo pattern of binding of RAG1 and RAG2 to antigen receptor loci. *Cell, 141*(3), 419–431. http://dx.doi.org/10.1016/j.cell.2010.03.010.

Jones, J. M., & Gellert, M. (2002). Ordered assembly of the V(D)J synaptic complex ensures accurate recombination. *The EMBO Journal, 21*(15), 4162–4171. http://dx.doi.org/10.1093/emboj/cdf394.

Jouvin-Marche, E., Morgado, M., Trede, N., Marche, P., Couez, D., Hue, I., et al. (1989). Complexity, polymorphism, and recombination of mouse T-cell receptor α gene families. *Immunogenetics, 30*(2), 99–104. http://dx.doi.org/10.1007/BF02421537.

Khor, B., Wehrly, T. D., & Sleckman, B. P. (2005). Chromosomal excision of TCRδ chain genes is dispensable for αβ T cell lineage commitment. *International Immunology*, *17*(3), 225–232.

Kind, J., Pagie, L., Ortabozkoyun, H., Boyle, S., de Vries, S. S., Janssen, H., et al. (2013). Single-cell dynamics of genome-nuclear lamina interactions. *Cell*, *153*(1), 178–192. http://dx.doi.org/10.1016/j.cell.2013.02.028.

Kind, J., & van Steensel, B. (2010). Genome–nuclear lamina interactions and gene regulation. *Current Opinion in Cell Biology*, *22*(3), 320–325. http://dx.doi.org/10.1016/j.ceb.2010.04.002.

Ko, L. J., Yamamoto, M., Leonard, M. W., George, K. M., Ting, P., & Engel, J. D. (1991). Murine and human T-lymphocyte GATA-3 factors mediate transcription through a cis-regulatory element within the human T-cell receptor delta gene enhancer. *Molecular and Cellular Biology*, *11*(5), 2778–2784.

Kondilis-Mangum, H. D., Cobb, R. M., Osipovich, O., Srivatsan, S., Oltz, E. M., & Krangel, M. S. (2010). Transcription-dependent mobilization of nucleosomes at accessible T cell receptor gene segments in vivo. *Journal of Immunology (Baltimore, Md: 1950)*, *184*(12), 6970–6977. http://dx.doi.org/10.4049/jimmunol.0903923.

Kosak, S. T., & Groudine, M. (2004). Form follows function: The genomic organization of cellular differentiation. *Genes & Development*, *18*(12), 1371–1384.

Kosak, S. T., Skok, J. A., Medina, K. L., Riblet, R., Le Beau, M. M., Fisher, A. G., et al. (2002). Subnuclear compartmentalization of immunoglobulin loci during lymphocyte development. *Science*, *296*(5565), 158–162.

Kouskoff, V., Vonesch, J.-L., Benoist, C., & Mathis, D. (1995). The influence of positive selection on RAG expression in thymocytes. *European Journal of Immunology*, *25*(1), 54–58. http://dx.doi.org/10.1002/eji.1830250111.

Krangel, M. S. (2003). Gene segment selection in V(D)J recombination: Accessibility and beyond. *Nature Immunology*, *4*(7), 624–630.

Krangel, M. S. (2009). Mechanics of T cell receptor gene rearrangement. *Current Opinion in Immunology*, *21*(2), 133–139. http://dx.doi.org/10.1016/j.coi.2009.03.009.

Krangel, M. S., Yssel, H., Brocklehurst, C., & Spits, H. (1990). A distinct wave of human T cell receptor gamma/delta lymphocytes in the early fetal thymus: Evidence for controlled gene rearrangement and cytokine production. *The Journal of Experimental Medicine*, *172*(3), 847–859.

Kreslavsky, T., Garbe, A. I., Krueger, A., & von Boehmer, H. (2008). T cell receptor–instructed αβ versus γδ lineage commitment revealed by single-cell analysis. *The Journal of Experimental Medicine*, *205*(5), 1173–1186.

Kreslavsky, T., & von Boehmer, H. (2010). γδTCR ligands and lineage commitment. *Seminars in Immunology*, *22*(4), 214–221. http://dx.doi.org/10.1016/j.smim.2010.04.001.

Kubota, T., Wang, J. Y., Göbel, T. W. F., Hockett, R. D., Cooper, M. D., & Chen, C. l. H. (1999). Characterization of an avian (Gallus Gallus domesticus) TCR αδ gene locus. *The Journal of Immunology*, *163*(7), 3858–3866.

Kwon, J., Imbalzano, A. N., Matthews, A., & Oettinger, M. A. (1999). Accessibility of nucleosomal DNA to V(D)J cleavage is modulated by RSS positioning and HMG1. *Molecular Cell*, *2*(6), 829–839.

Lauzurica, P., & Krangel, M. S. (1994a). Enhancer-dependent and -independent steps in the rearrangement of a human T cell receptor delta transgene. *The Journal of Experimental Medicine*, *179*(1), 43–55.

Lauzurica, P., & Krangel, M. S. (1994b). Temporal and lineage-specific control of T cell receptor alpha/delta gene rearrangement by T cell receptor alpha and delta enhancers. *The Journal of Experimental Medicine*, *179*(6), 1913–1921.

Lauzurica, P., Zhong, X.-P., Krangel, M. S., & Roberts, J. L. (1997). Regulation of T cell receptor δ gene rearrangement by CBF/PEBP2. *The Journal of Experimental Medicine*, *185*(7), 1193–1202.

Lee, Y. N., Alt, F. W., Reyes, J., Gleason, M., Zarrin, A. A., & Jung, D. (2009). Differential utilization of T cell receptor TCRα/TCRδ locus variable region gene segments is mediated by accessibility. *Proceedings of the National Academy of Sciences*, *106*(41), 17487–17492.

Lefranc, M.-P. (2001). IMGT, the international ImMunoGeneTics database. *Nucleic Acids Research*, *29*(1), 207–209.

Li, Z., Dordai, D. I., Lee, J., & Desiderio, S. (1996). A conserved degradation signal regulates RAG-2 accumulation during cell division and links V(D)J recombination to the cell cycle. *Immunity*, *5*(6), 575–589. http://dx.doi.org/10.1016/S1074-7613(00)80272-1.

Liu, H., Schmidt-Supprian, M., Shi, Y., Hobeika, E., Barteneva, N., Jumaa, H., et al. (2007). Yin Yang 1 is a critical regulator of B-cell development. *Genes & Development*, *21*(10), 1179–1189.

Liu, Y., Subrahmanyam, R., Chakraborty, T., Sen, R., & Desiderio, S. (2007). A plant homeodomain in Rag-2 that binds hypermethylated lysine 4 of histone H3 is necessary for efficient antigen-receptor-gene rearrangement. *Immunity*, *27*(4), 561–571. http://dx.doi.org/10.1016/j.immuni.2007.09.005.

Livak, F., Petrie, H. T., Crisps, I. N., & Schatz, D. G. (1995). In-frame TCR δ gene rearrangements play a critical role in the α β/$\gamma\delta$ T cell lineage decision. *Immunity*, *2*(6), 617–627. http://dx.doi.org/10.1016/1074-7613(95)90006-3.

Livák, F., Tourigny, M., Schatz, D. G., & Petrie, H. T. (1999). Characterization of TCR gene rearrangements during adult murine T cell development. *The Journal of Immunology*, *162*(5), 2575–2580.

Loh, E. Y., Cwirla, S., Serafini, A. T., Phillips, J. H., & Lanier, L. L. (1988). Human T-cell-receptor delta chain: Genomic organization, diversity, and expression in populations of cells. *Proceedings of the National Academy of Sciences of the United States of America*, *85*(24), 9714–9718.

Magdinier, F., Yusufzai, T. M., & Felsenfeld, G. (2004). Both CTCF-dependent and -independent insulators Are found between the mouse T cell receptor α and Dad1 genes. *Journal of Biological Chemistry*, *279*(24), 25381–25389.

Majumder, K., Koues, O. I., Chan, E. A. W., Kyle, K. E., Horowitz, J. E., Yang-Iott, K., et al. (2015). Lineage-specific compaction of Tcrb requires a chromatin barrier to protect the function of a long-range tethering element. *The Journal of Experimental Medicine*, *212*(1), 107–120.

Malissen, M., Trucy, J., Jouvin-Marche, E., Cazenave, P. A., Scollay, R., & Malissen, B. (1992). Regulation of TCR α and β gene allelic exclusion during T-cell development. *Immunology Today*, *13*(8), 315–322. http://dx.doi.org/10.1016/0167-5699(92)90044-8.

Malissen, M., Trucy, J., Letourneur, F., Rebai, N., Dunn, D. E., Fitch, F. W., et al. (1988). A T cell clone expresses two T cell receptor α genes but uses one $\alpha\beta$ heterodimer for allorecognition and self MHC-restricted antigen recognition. *Cell*, *55*(1), 49–59. http://dx.doi.org/10.1016/0092-8674(88)90008-6.

Matthews, A. G. W., Kuo, A. J., Ramón-Maiques, S., Han, S., Champagne, K. S., Ivanov, D., et al. (2007). RAG2 PHD finger couples histone H3 lysine 4 trimethylation with V(D)J recombination. *Nature*, *450*(7172), 1106–1110. http://dx.doi.org/10.1038/nature06431.

Mauvieux, L., Villey, I., & de Villartay, J.-P. (2001). T early alpha (TEA) regulates initial TCRVAJA rearrangements and leads to TCRJA coincidence. *European Journal of Immunology*, *31*(7), 2080–2086. http://dx.doi.org/10.1002/1521-4141(200107)31:7<2080::AID-IMMU2080>3.0.CO;2-H.

Mauvieux, L., Villey, I., & de Villartay, J.-P. (2003). TEA regulates local TCR-Jα accessibility through histone acetylation. *European Journal of Immunology, 33*(8), 2216–2222. http://dx.doi.org/10.1002/eji.200323867.

Mayall, T. P., Sheridan, P. L., Montminy, M. R., & Jones, K. A. (1997). Distinct roles for P-CREB and LEF-1 in TCR alpha enhancer assembly and activation on chromatin templates in vitro. *Genes & Development, 11*(7), 887–899.

McMurry, M. T., & Krangel, M. S. (2000). A role for histone acetylation in the developmental regulation of V(D)J recombination. *Science, 287*(5452), 495–498.

Medvedovic, J., Ebert, A., Tagoh, H., Tamir, I. M., Schwickert, T. A., Novatchkova, M., et al. (2013). Flexible long-range loops in the VH gene region of the Igh locus facilitate the generation of a diverse antibody repertoire. *Immunity, 39*(2), 229–244. http://dx.doi.org/10.1016/j.immuni.2013.08.011.

Merkenschlager, M., & Odom, D. T. (2013). CTCF and cohesin: Linking gene regulatory elements with their targets. *Cell, 152*(6), 1285–1297. http://dx.doi.org/10.1016/j.cell.2013.02.029.

Michie, A. M., & Zúñiga-Pflücker, J. C. (2002). Regulation of thymocyte differentiation: Pre-TCR signals and β-selection. *Seminars in Immunology, 14*(5), 311–323. http://dx.doi.org/10.1016/S1044-5323(02)00064-7.

Monroe, R. J., Sleckman, B. P., Monroe, B. C., Khor, B., Claypool, S., Ferrini, R., et al. (1999). Developmental regulation of TCRδ locus accessibility and expression by the TCRδ enhancer. *Immunity, 10*(5), 503–513. http://dx.doi.org/10.1016/S1074-7613(00)80050-3.

Naik, A. K., Hawwari, A., & Krangel, M. S. (2015). Specification of Vδ and Vα usage by tcra/tcrd locus V gene segment promoters. *The Journal of Immunology, 194*(2), 790–794.

Nakajima, P. B., Menetski, J. P., Roth, D. B., Gellert, M., & Bosma, M. J. (1995). V-D-J rearrangements at the T cell receptor σ locus in mouse thymocytes of the αβ lineage. *Immunity, 3*(5), 609–621. http://dx.doi.org/10.1016/1074-7613(95)90132-9.

Narayan, K., & Kang, J. (2010). Disorderly conduct in γδ versus αβ T cell lineage commitment. *Seminars in Immunology, 22*(4), 222–227. http://dx.doi.org/10.1016/j.smim.2010.04.003.

Okazaki, K., & Sakano, H. (1988). Thymocyte circular DNA excised from T cell receptor alpha-delta gene complex. *The EMBO Journal*, 7(6), 1669–1674.

Ong, C.-T., & Corces, V. G. (2014). CTCF: An architectural protein bridging genome topology and function. *Nature Reviews. Genetics, 15*(4), 234–246. http://dx.doi.org/10.1038/nrg3663.

Ortiz, B. D., Cado, D., Chen, V., Diaz, P. W., & Winoto, A. (1997). Adjacent DNA elements dominantly restrict the ubiquitous activity of a novel chromatin-opening region to specific tissues. *The EMBO Journal, 16*(16), 5037–5045. http://dx.doi.org/10.1093/emboj/16.16.5037.

Osipovich, O., Milley Cobb, R., Oestreich, K. J., Pierce, S., Ferrier, P., & Oltz, E. M. (2007). Essential function for SWI-SNF chromatin-remodeling complexes in the promoter-directed assembly of Tcrb genes. *Nature Immunology, 8*(8), 809–816. http://www.nature.com/ni/journal/v8/n8/suppinfo/ni1481_S1.html.

Osipovich, O. A., Subrahmanyam, R., Pierce, S., Sen, R., & Oltz, E. M. (2009). Cutting edge: SWI/SNF mediates antisense Igh transcription and locus-wide accessibility in B cell precursors. *The Journal of Immunology, 183*(3), 1509–1513.

Parra, Z. E., Baker, M. L., Schwarz, R. S., Deakin, J. E., Lindblad-Toh, K., & Miller, R. D. (2007). A unique T cell receptor discovered in marsupials. *Proceedings of the National Academy of Sciences of the United States of America, 104*(23), 9776–9781. http://dx.doi.org/10.1073/pnas.0609106104.

Parra, Z. E., Ohta, Y., Criscitiello, M. F., Flajnik, M. F., & Miller, R. D. (2010). The dynamic TCRδ: TCRδ chains in the amphibian Xenopus tropicalis utilize antibody-like

V genes. *European Journal of Immunology, 40*(8), 2319–2329. http://dx.doi.org/10.1002/eji.201040515.

Pasqual, N., Gallagher, M., Aude-Garcia, C., Loiodice, M., Thuderoz, F., Demongeot, J., et al. (2002). Quantitative and qualitative changes in V-J α rearrangements during mouse thymocytes differentiation: Implication for a limited T cell receptor α chain repertoire. *The Journal of Experimental Medicine, 196*(9), 1163–1174.

Patenge, N., Elkin, S. K., & Oettinger, M. A. (2004). ATP-dependent remodeling by SWI/SNF and ISWI proteins stimulates V(D)J cleavage of 5 S arrays. *Journal of Biological Chemistry, 279*(34), 35360–35367.

Peric-Hupkes, D., Meuleman, W., Pagie, L., Bruggeman, S. W. M., Solovei, I., Brugman, W., et al. (2010). Molecular maps of the reorganization of genome-nuclear lamina interactions during differentiation. *Molecular Cell, 38*(4), 603–613. http://dx.doi.org/10.1016/j.molcel.2010.03.016.

Petrie, H. T., Livak, F., Burtrum, D., & Mazel, S. (1995). T cell receptor gene recombination patterns and mechanisms: Cell death, rescue, and T cell production. *The Journal of Experimental Medicine, 182*(1), 121–127.

Pickersgill, H., Kalverda, B., de Wit, E., Talhout, W., Fornerod, M., & van Steensel, B. (2006). Characterization of the Drosophila melanogaster genome at the nuclear lamina. *Nature Genetics, 38*(9), 1005–1014. http://www.nature.com/ng/journal/v38/n9/suppinfo/ng1852_S1.html.

Ramond, C., Berthault, C., Burlen-Defranoux, O., de Sousa, A. P., Guy-Grand, D., Vieira, P., et al. (2014). Two waves of distinct hematopoietic progenitor cells colonize the fetal thymus. *Nature Immunology, 15*(1), 27–35. http://dx.doi.org/10.1038/ni.2782. http://www.nature.com/ni/journal/v15/n1/abs/ni.2782.html#supplementary-information.

Rast, J. P., Anderson, M. K., Strong, S. J., Luer, C., Litman, R. T., & Litman, G. W. (1997). α, β, γ, and δ T cell antigen receptor genes arose early in vertebrate phylogeny. *Immunity, 6*(1), 1–11. http://dx.doi.org/10.1016/S1074-7613(00)80237-X.

Reddy, K. L., Zullo, J. M., Bertolino, E., & Singh, H. (2008). Transcriptional repression mediated by repositioning of genes to the nuclear lamina. *Nature, 452*(7184), 243–247. http://www.nature.com/nature/journal/v452/n7184/suppinfo/nature06727_S1.html.

Redondo, J. M., Hata, S., Brocklehurst, C., & Krangel, M. S. (1990). A T cell-specific transcriptional enhancer within the human T cell receptor delta, locus. *Science, 247*, 1225–1229.

Redondo, J. M., Pfohl, J. L., Hernandez-Munain, C., Wang, S., Speck, N. A., & Krangel, M. S. (1992). Indistinguishable nuclear factor binding to functional core sites of the T-cell receptor delta and murine leukemia virus enhancers. *Molecular and Cellular Biology, 12*(11), 4817–4823.

Redondo, J. M., Pfohl, J. L., & Krangel, M. S. (1991). Identification of an essential site for transcriptional activation within the human T-cell receptor delta enhancer. *Molecular and Cellular Biology, 11*(11), 5671–5680.

Ribeiro de Almeida, C., Stadhouders, R., de Bruijn, M. J. W., Bergen, I. M., Thongjuea, S., Lenhard, B., et al. (2011). The DNA-binding protein CTCF limits proximal Vκ recombination and restricts κ enhancer interactions to the immunoglobulin κ light chain locus. *Immunity, 35*(4), 501–513. http://dx.doi.org/10.1016/j.immuni.2011.07.014.

Ribeiro de Almeida, C., Stadhouders, R., Thongjuea, S., Soler, E., & Hendriks, R. W. (2012). DNA-binding factor CTCF and long-range gene interactions in V(D)J recombination and oncogene activation. *Blood, 119*(26), 6209–6218.

Robins, H. (2013). Immunosequencing: Applications of immune repertoire deep sequencing. *Current Opinion in Immunology, 25*(5), 646–652. http://dx.doi.org/10.1016/j.coi.2013.09.017.

Roldán, E., Fuxa, M., Chong, W., Martinez, D., Novatchkova, M., Busslinger, M., et al. (2005). Locus 'decontraction' and centromeric recruitment contribute to allelic exclusion of the immunoglobulin heavy-chain gene. *Nature Immunology*, *6*(1), 31–41. http://dx.doi.org/10.1038/ni1150.

Roth, M. E., Holman, P. O., & Kranz, D. M. (1991). Nonrandom use of J alpha gene segments. Influence of V alpha and J alpha gene location. *The Journal of Immunology*, *147*(3), 1075–1081.

Rytkönen, M. A., Hurwitz, J. L., Thompson, S. D., & Pelkonen, J. (1996). Restricted onset of T cell receptor α gene rearrangement in fetal and neonatal thymocytes. *European Journal of Immunology*, *26*(8), 1892–1896. http://dx.doi.org/10.1002/eji.1830260833.

Rytkönen, M., Hurwitz, J. L., Tolonen, K., & Pelkonen, J. (1994). Evidence for recombinatorial hot spots at the T cell receptor Jα locus. *European Journal of Immunology*, *24*(1), 107–115. http://dx.doi.org/10.1002/eji.1830240117.

Sayegh, C., Jhunjhunwala, S., Riblet, R., & Murre, C. (2005). Visualization of looping involving the immunoglobulin heavy-chain locus in developing B cells. *Genes & Development*, *19*(3), 322–327.

Schatz, D. G., & Ji, Y. (2011). Recombination centres and the orchestration of V(D)J recombination. *Nature Reviews Immunology*, *11*(4), 251–263.

Schatz, D. G., & Swanson, P. C. (2011). V(D)J recombination: Mechanisms of initiation. *Annual Review of Genetics*, *45*(1), 167–202. http://dx.doi.org/10.1146/annurev-genet-110410-132552.

Schlimgen, R. J., Reddy, K. L., Singh, H., & Krangel, M. S. (2008). Initiation of allelic exclusion by stochastic interaction of Tcrb alleles with repressive nuclear compartments. *Nature Immunology*, *9*(7), 802–809. http://dx.doi.org/10.1038/ni.1624.

Schneider, R., & Grosschedl, R. (2007). Dynamics and interplay of nuclear architecture, genome organization, and gene expression. *Genes & Development*, *21*(23), 3027–3043.

Seitan, V. C., Hao, B., Tachibana-Konwalski, K., Lavagnolli, T., Mira-Bontenbal, H., Brown, K. E., et al. (2011). A role for cohesin in T-cell-receptor rearrangement and thymocyte differentiation. *Nature*, *476*(7361), 467–471. http://www.nature.com/nature/journal/v476/n7361/abs/nature10312.html#supplementary-information.

Seitan, V. C., Krangel, M. S., & Merkenschlager, M. (2012). Cohesin, CTCF and lymphocyte antigen receptor locus rearrangement. *Trends in Immunology*, *33*(4), 153–159. http://dx.doi.org/10.1016/j.it.2012.02.004.

Shih, H.-Y., & Krangel, M. S. (2010). Distinct contracted conformations of the Tcra/Tcrd locus during Tcra and Tcrd recombination. *The Journal of Experimental Medicine*, *207*(9), 1835–1841.

Shih, H.-Y., & Krangel, M. S. (2013). Chromatin architecture, CCCTC-binding factor, and V(D)J recombination: Managing long-distance relationships at antigen receptor loci. *The Journal of Immunology*, *190*(10), 4915–4921.

Shih, H.-Y., Verma-Gaur, J., Torkamani, A., Feeney, A. J., Galjart, N., & Krangel, M. S. (2012). Tcra gene recombination is supported by a Tcra enhancer- and CTCF-dependent chromatin hub. *Proceedings of the National Academy of Sciences*, *109*(50), E3493–E3502.

Shimazaki, N., Tsai, A. G., & Lieber, M. R. (2009). H3K4me3 stimulates the V(D)J RAG complex for both nicking and hairpinning in trans in addition to tethering in cis: Implications for translocations. *Molecular Cell*, *34*(5), 535–544. http://dx.doi.org/10.1016/j.molcel.2009.05.011.

Shimizu, T., Takeshita, S., Muto, M., Kubo, E., Sado, T., & Yamagishi, H. (1993). Mouse germline transcript of TCR α joining region and temporal expression in ontogeny. *International Immunology*, *5*(2), 155–160.

Singer, P. A., McEvilly, R. J., Balderas, R. S., Dixon, F. J., & Theofilopoulos, A. N. (1988). T-cell receptor alpha-chain variable-region haplotypes of normal and autoimmune laboratory mouse strains. *Proceedings of the National Academy of Sciences, 85*(20), 7729–7733.

Skok, J. A., Brown, K. E., Azuara, V., Caparros, M.-L., Baxter, J., Takacs, K., et al. (2001). Nonequivalent nuclear location of immunoglobulin alleles in B lymphocytes. *Nature Immunology, 2*(9), 848–854. http://www.nature.com/ni/journal/v2/n9/suppinfo/ni0901-848_S1.html.

Skok, J. A., Gisler, R., Novatchkova, M., Farmer, D., de Laat, W., & Busslinger, M. (2007). Reversible contraction by looping of the Tcra and Tcrb loci in rearranging thymocytes. *Nature Immunology, 8*(4), 378–387. http://www.nature.com/ni/journal/v8/n4/suppinfo/ni1448_S1.html.

Sleckman, B. P., Bardon, C. G., Ferrini, R., Davidson, L., & Alt, F. W. (1997). Function of the TCRα enhancer in αβ and γδ T cells. *Immunity*, 7(4), 505–515. http://dx.doi.org/10.1016/S1074-7613(00)80372-6.

Sleckman, B. P., Bassing, C. H., Hughes, M. M., Okada, A., D'Auteuil, M., Wehrly, T. D., et al. (2000). Mechanisms that direct ordered assembly of T cell receptor β locus V, D, and J gene segments. *Proceedings of the National Academy of Sciences, 97*(14), 7975–7980.

Sleckman, B. P., Khor, B., Monroe, R., & Alt, F. W. (1998). Assembly of productive T cell receptor δ variable region genes exhibits allelic inclusion. *The Journal of Experimental Medicine, 188*(8), 1465–1471.

Spicuglia, S., Payet, D., Tripathi, R. K., Rameil, P., Verthuy, C., Imbert, J., et al. (2000). TCRα enhancer activation occurs via a conformational change of a pre-assembled nucleo-protein complex. *The EMBO Journal, 19*(9), 2034–2045.

Stanhope-Baker, P., Hudson, K. M., Shaffer, A. L., Constantinescu, A., & Schlissel, M. S. (1996). Cell type–specific chromatin structure determines the targeting of V(D)J recombinase activity in vitro. *Cell, 85*(6), 887–897. http://dx.doi.org/10.1016/S0092-8674(00)81272-6.

Steinel, N. C., Lee, B.-S., Tubbs, A. T., Bednarski, J. J., Schulte, E., Yang-Iott, K. S., et al. (2013). The ataxia telangiectasia mutated kinase controls Igκ allelic exclusion by inhibiting secondary Vκ-to-Jκ rearrangements. *The Journal of Experimental Medicine, 210*(2), 233–239. http://dx.doi.org/10.1084/jem.20121605.

Su, R.-C., Sridharan, R., & Smale, S. T. (2005). Assembly of silent chromatin during thymocyte development. *Seminars in Immunology, 17*(2), 129–140. http://dx.doi.org/10.1016/j.smim.2005.01.003.

Sutherland, H., & Bickmore, W. A. (2009). Transcription factories: Gene expression in unions? *Nature Reviews. Genetics, 10*(7), 457–466.

Taghon, T., Yui, M. A., Pant, R., Diamond, R. A., & Rothenberg, E. V. (2006). Developmental and molecular characterization of emerging β- and γδ-selected Pre-T cells in the adult mouse thymus. *Immunity, 24*(1), 53–64. http://dx.doi.org/10.1016/j.immuni.2005.11.012.

Takeshita, S., Toda, M., & Yamagishi, H. (1989). Excision products of the T cell receptor gene support a progressive rearrangement model of the alpha/delta locus. *The EMBO Journal, 8*(11), 3261–3270.

Tani-ichi, S., Satake, M., & Ikuta, K. (2011). The pre-TCR signal induces transcriptional silencing of the TCRγ locus by reducing the recruitment of STAT5 and Runx to transcriptional enhancers. *International Immunology, 23*(9), 553–563.

Taniuchi, I., Osato, M., Egawa, T., Sunshine, M. J., Bae, S.-C., Komori, T., et al. (2002). Differential requirements for runx proteins in CD4 repression and epigenetic silencing during T lymphocyte development. *Cell, 111*(5), 621–633. http://dx.doi.org/10.1016/S0092-8674(02)01111-X.

Thompson, S. D., Pelkonen, J., & Hurwitz, J. L. (1990). First T cell receptor alpha gene rearrangements during T cell ontogeny skew to the 5' region of the J alpha locus. *The Journal of Immunology*, *145*(7), 2347–2352.

Thuderoz, F., Simonet, M.-A., Hansen, O., Pasqual, N., Dariz, A., Baum, T. P., et al. (2010). Numerical modelling of the V-J combinations of the T cell receptor TRA/TRD locus. *PLoS Computational Biology*, *6*(2), e1000682. http://dx.doi.org/10.1371/journal.pcbi.1000682.

Toda, M., Fujimoto, S., Iwasato, T., Takeshita, S., Tezuka, K., Ohbayashi, T., et al. (1988). Structure of extrachromosomal circular DNAs excised from T-cell antigen receptor alpha and delta-chain loci. *Journal of Molecular Biology*, *202*(2), 219–231. http://dx.doi.org/10.1016/0022-2836(88)90453-6.

Travis, A., Amsterdam, A., Belanger, C., & Grosschedl, R. (1991). LEF-1, a gene encoding a lymphoid-specific protein with an HMG domain, regulates T-cell receptor alpha enhancer function [corrected]. *Genes & Development*, *5*(5), 880–894.

Turka, L. A., Schatz, D. G., Oettinger, M. A., Chun, J. J., Gorka, C., Lee, K., et al. (1991). Thymocyte expression of RAG-1 and RAG-2: Termination by T cell receptor cross-linking. *Science*, *253*, 778–781.

Verma-Gaur, J., Torkamani, A., Schaffer, L., Head, S. R., Schork, N. J., & Feeney, A. J. (2012). Noncoding transcription within the Igh distal V(H) region at PAIR elements affects the 3D structure of the Igh locus in pro-B cells. *Proceedings of the National Academy of Sciences of the United States of America*, *109*(42), 17004–17009. http://dx.doi.org/10.1073/pnas.1208398109.

Vettermann, C., & Schlissel, M. S. (2010). Allelic exclusion of immunoglobulin genes: Models and mechanisms. *Immunological Reviews*, *237*(1), 22–42. http://dx.doi.org/10.1111/j.1600-065X.2010.00935.x.

Villey, I., Caillol, D., Selz, F., Ferrier, P., & de Villartay, J.-P. (1996). Defect in rearrangement of the most 5′ TCR-Jα following targeted deletion of T early α (TEA): Implications for TCR α locus accessibility. *Immunity*, *5*(4), 331–342. http://dx.doi.org/10.1016/S1074-7613(00)80259-9.

Villey, I., Quartier, P., Selz, F., & de Villartay, J.-P. (1997). Germ-line transcription and methylation status of the TCR-Jα locus in its accessible configuration. *European Journal of Immunology*, *27*(7), 1619–1625. http://dx.doi.org/10.1002/eji.1830270705.

Wang, K., Gan, L., Kunisada, T., Lee, I., Yamagishi, H., & Hood, L. (2001). Characterization of the Japanese pufferfish (Takifugu rubripes) T-cell receptor alpha locus reveals a unique genomic organization. *Immunogenetics*, *53*, 31–42.

Wang, F., Huang, C.-Y., & Kanagawa, O. (1998). Rapid deletion of rearranged T cell antigen receptor (TCR) Vα-Jα segment by secondary rearrangement in the thymus: Role of continuous rearrangement of TCR α chain gene and positive selection in the T cell repertoire formation. *Proceedings of the National Academy of Sciences*, *95*(20), 11834–11839.

Wang, X., Xiao, G., Zhang, Y., Wen, X., Gao, X., Okada, S., et al. (2008). Regulation of Tcrb recombination ordering by c-Fos-dependent RAG deposition. *Nature Immunology*, *9*(7), 794–801. http://www.nature.com/ni/journal/v9/n7/suppinfo/ni.1614_S1.html.

Warmflash, A., & Dinner, A. R. (2006). A model for TCR gene segment Use. *The Journal of Immunology*, *177*(6), 3857–3864.

Weber-Arden, J., Wilbert, O. M., Kabelitz, D., & Arden, B. (2000). Vδ repertoire during thymic ontogeny suggests three novel waves of γδ TCR expression. *The Journal of Immunology*, *164*(2), 1002–1012.

Winoto, A., & Baltimore, D. (1989a). A novel, inducible and T cell-specific enhancer located at the 3' end of the T cell receptor alpha locus. *The EMBO Journal*, *8*(3), 729–733.

Winoto, A., & Baltimore, D. (1989b). αβ lineage-specific expression of the α T cell receptor gene by nearby silencers. *Cell*, *59*(4), 649–655. http://dx.doi.org/10.1016/0092-8674(89)90010-X.

Xiang, Y., Zhou, X., Hewitt, S. L., Skok, J. A., & Garrard, W. T. (2011). A multifunctional element in the mouse Igκ locus that specifies repertoire and Ig loci subnuclear location. *Journal of Immunology (Baltimore, Md: 1950), 186*(9), 5356–5366. http://dx.doi.org/10.4049/jimmunol.1003794.

Xiong, N., & Raulet, D. H. (2007). Development and selection of γδ T cells. *Immunological Reviews, 215*(1), 15–31. http://dx.doi.org/10.1111/j.1600-065X.2006.00478.x.

Yancopoulos, G. D., & Alt, F. W. (1985). Developmentally controlled and tissue-specific expression of unrearranged VH gene segments. *Cell, 40*(2), 271–281. http://dx.doi.org/10.1016/0092-8674(85)90141-2.

Yang, X., Liu, D., Lv, N., Zhao, F., Liu, F., Zou, J., et al. (2015). TCRklass: A new K-string-based algorithm for human and mouse TCR repertoire characterization. *Journal of Immunology (Baltimore, Md: 1950), 194*(1), 446–454.

Yannoutsos, N., Wilson, P., Yu, W., Chen, H. T., Nussenzweig, A., Petrie, H., et al. (2001). The role of recombination activating gene (RAG) reinduction in thymocyte development in vivo. *The Journal of Experimental Medicine, 194*(4), 471–480.

Zhong, X.-P., & Krangel, M. S. (1999). Enhancer-blocking activity within the DNase I hypersensitive site 2 to 6 region between the TCR α and Dad1 genes. *The Journal of Immunology, 163*(1), 295–300.

Zullo, J. M., Demarco, I. A., Piqué-Regi, R., Gaffney, D. J., Epstein, C. B., Spooner, C. J., et al. (2012). DNA sequence-dependent compartmentalization and silencing of chromatin at the nuclear lamina. *Cell, 149*(7), 1474–1487. http://dx.doi.org/10.1016/j.cell.2012.04.035.

CHAPTER EIGHT

Long-Range Control of V(D)J Recombination & Allelic Exclusion: Modeling Views

Pernelle Outters, Sébastien Jaeger, Nancy Zaarour, Pierre Ferrier[1]

Centre d'Immunologie de Marseille-Luminy, Aix-Marseille Université UM2, Inserm, U1104, CNRS UMR7280, 13288 Marseille, France

[1]Corresponding author: e-mail address: ferrier@ciml.univ-mrs.fr

Contents

Abstract

Allelic exclusion of immunoglobulin (Ig) and T-cell receptor (TCR) genes ensures the development of B and T lymphocytes operating under the mode of clonal selection. This phenomenon associates asynchronous V(D)J recombination events at Ig or TCR alleles and inhibitory feedback control. Despite years of intense research, however, the mechanisms that sustain asymmetric choice in random Ig/TCR dual allele usage and the production of Ig/TCR monoallelic expressing B and T lymphocytes remain unclear and open for debate. In this chapter, we first recapitulate the biological evidence

Advances in Immunology, Volume 128
ISSN 0065-2776
http://dx.doi.org/10.1016/bs.ai.2015.08.002

© 2015 Elsevier Inc.
All rights reserved.

that almost from the start appeared to link V(D)J recombination and allelic exclusion. We review the theoretical models previously proposed to explain this connection. Finally, we introduce our own mathematical modeling views based on how the developmental dynamics of individual lymphoid cells combine to sustain allelic exclusion.

1. INTRODUCTION

The vast majority of peripheral B and T lymphocytes express a unique form of antigen (Ag) receptor: immunoglobulin (Ig) or T-cell receptor (TCR), respectively. This phenomenon, first identified in the mid 1960s by surface membrane analysis of B-lymphocytes using Ig-allotype-specific antisera (Cebra, Colberg, & Dray, 1966; Pernis, Chiappino, Kelus, & Gell, 1965), is termed allelic exclusion. Its discovery provided a molecular foundation to the clonal selection concept of acquired immunity—"one lymphocyte, one Ag receptor"—formulated earlier on (Burnet, 1959; Jerne, 1955). An intimate link between allelic exclusion and the process of Ig/TCR gene somatic rearrangement—or V(D)J recombination—during B- and T-cell early development became increasingly clear and led to a congruous evolution of knowledge in both fields (Alt, Enea, Bothwell, & Baltimore, 1980; Alt, Rosenberg, Lewis, Thomas, & Baltimore, 1981; Beck-Engeser, Jack, & Wabl, 1987; Coleclough, 1983; Coleclough, Perry, Karjalainen, & Weigert, 1981; Wabl & Steinberg, 1982; reviewed by Alt et al., 1992; Bergman, 1999; Gorman & Alt, 1998; Tonegawa et al., 1981). In particular, much attention surrounded the importance of the underlying chromosomal template and epigenetics in the regulation of both V(D)J recombination and allelic exclusion (Bergman & Cedar, 2004; Krangel, 2003; Mostoslavsky, Alt, & Rajewsky, 2004; Schlissel, 2002). More recently, decisive breakthroughs have been made regarding large-scale topological changes occurring at Ig/TCR genomic loci during the assembly of the variable gene region, with ostensible implications for allelic exclusion (Bossen, Mansson, & Murre, 2012; Ebert, Medvedovic, Tagoh, Schwickert, & Busslinger, 2013; Johnson, Chaumeil, & Skok, 2010). However, despite undeniable progress and considerable efforts made over half a century, how lymphoid cells truly discriminate individual Ig/TCR alleles in order to achieve allelic exclusion still lacks a convincing explanation.

The sophistication of V(D)J recombination and its numerous regulatory facets, with an often overwhelming *prima facie* number of pieces to the puzzle, have since its discovery driven this subject's appeal for tentative modeling, some of which were attempted long before the secret of discrete biochemical nodes and molecular players would (and could) be decrypted (Alt & Baltimore, 1982). Many models and original visions have been put forward—each with ingenuity and passion—to explain the regulation of allelic exclusion (Brady, Steinel, & Bassing, 2010; Brandt, Hewitt, & Skok, 2010; Cedar & Bergman, 2008; Chaumeil & Skok, 2013; Cohn & Langman, 1990; Coleclough, 1990; Gorman & Alt, 1998; Levin-Klein & Bergman, 2014; Vettermann & Schlissel, 2010). With time, models with a hefty dose of stochasticity in rearrangement events initially popular among computational biologists gradually gave way to purely deterministic models envisaged by many molecular immunologists. Here, we present our current vision of a dynamic regulation of allelic exclusion based on the association of both deterministic and probabilistic features whereby epigenetic-driven controls in individual developing lymphocytes likely play a central role. Accordingly, epigenetics would contribute, along with V(D)J recombination, to the uniqueness of individual B and T cells. However, the dynamical integration of individual probabilistic behaviors at the level of a whole cell population would result in a robust system able to enforce allelic exclusion at minimal regulatory expense. Our models, which are sustained by mathematical simulations, have little experimental certification at this point. As such, they should be taken as working hypotheses that we hope will stimulate new research directions. This review article is divided into two main parts. The first part recapitulates the recognized features and reviews more recent findings, regarding the genes involved in V(D)J recombination and allelic exclusion, as well as the associated epigenetics. The second part more specifically examines the mathematical modeling studies performed to investigate the underlying regulatory controls. Frequently, we will refer to the IgH and TCRβ gene loci to illustrate these various points, as many recent studies have focused on these two particular loci, whose regulation of allelic exclusion remains specially enigmatic given the differing behaviors of their D–J and V gene segment-containing domains (i.e., allelically included vs. allelically excluded, respectively).

2. Ig AND TCR GENES AND ASSOCIATED REGULATORY ELEMENTS

Three Ig and four TCR genes organized into a total of six distinct chromosomal loci are subjected to V(D)J recombination assembly during

B- or T-cell development [namely, the IgH, IgLκ, and IgLλ genes/loci; and the TCRβ, TCRγ, TCRα, and TCRδ genes, the latter two intermingled within a unique TCRα/δ locus (Aschauer, Outters, & Ferrier, 2015; Stevens, 2015); note that this situation primarily applies to the well-studied mouse and human species]. All loci, with the notable exception of the TCRα locus, are also subjected to allelic exclusion, though with varying stringency [in the case of TCRδ, the subject remains debatable, with possible differences between the human and the mouse, and among distinct γδ T-cell subsets (Couedel, Lippert, Bernardeau, Bonneville, & Davodeau, 2004; Davodeau et al., 1993; Sleckman, Khor, Monroe, & Alt, 1998)]. Following V(D)J assembly, these genes altogether encode polypeptide chains that pair up to form the IgH/Lκ or IgH/Lλ and the TCRαβ or TCRγδ immune receptors of developed B and T cells, respectively. Figure 1 shows a schematic representation of the various Ig and TCR gene loci in the mouse germline (GL), where—despite huge differences in size and complexity—a shared foundation is revealed. Thus, V gene segments, the number of which varies from a few (1–4) units to a hundred depending on the locus, lie within the upstream (5′) part of the given locus, with successive downstream (3′) sets (or, at times, a unique member) of D (when present) and/or J gene segments, followed by the constant region (C) encoding domain(s). Depending on the given gene, Ig or TCR variable region assembly thus involves V, D, and J gene segments (IgH, TCRβ, and TCRδ genes), or V and J gene segments only (IgLκ/λ, TCRα, and TCRγ genes).

Rearrangeable Ig/TCR gene segments are flanked by either one of two types of recombination signal sequences (so-called 12-RSS and 23-RSS, respectively; e.g., Sakano, Kurosawa, Weigert, & Tonegawa, 1981). Depending on the orientation of their RSSs, *cis*-rearrangement between gene segment dual partners may occur by either chromosomal deletion (opposite orientation) or inversion (same orientation), with the consequence being that the intervening sequences are either eliminated from or maintained in the given locus, respectively (the latter situation applies on the largest scale in the IgLκ locus). At some loci, isolated RSS-like sequences can also be involved in the removal or reshuffling of a large part of the given locus (e.g., the RS element at the mouse IgLκ locus, which could possibly be involved in the phenomenon of IgLκ/λ isotype switching—reviewed by Gorman & Alt, 1998); and the *δRec* and ψJ_α elements at the TCRα/δ locus, which could contribute toward the deletion of the δ locus in αβ-committed early T cells (De Villartay, Hockett, Coran, Korsmeyer, & Cohen, 1988; Hockett et al., 1988). Although phylogenetically conserved, the 12- and

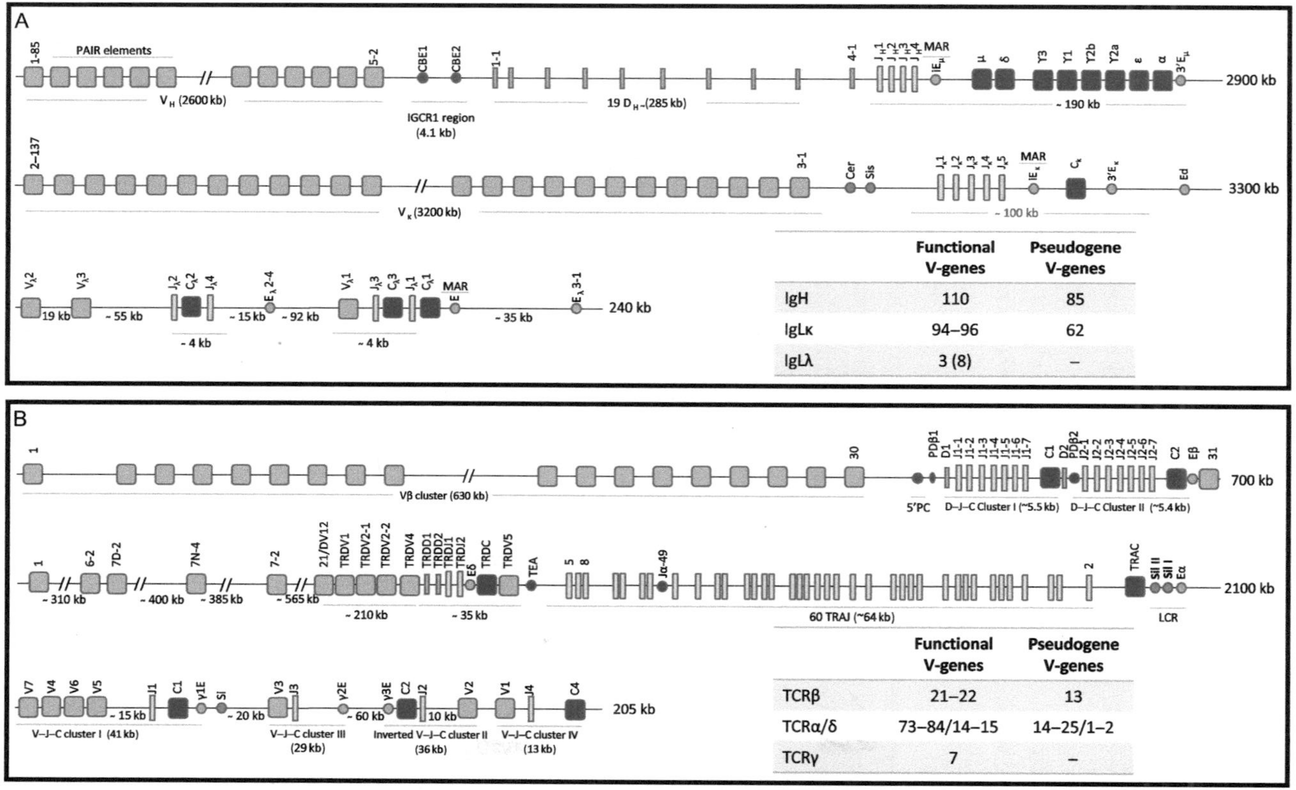

	Functional V-genes	Pseudogene V-genes
IgH	110	85
IgLκ	94–96	62
IgLλ	3 (8)	–

	Functional V-genes	Pseudogene V-genes
TCRβ	21–22	13
TCRα/δ	73–84/14–15	14–25/1–2
TCRγ	7	–

Figure 1 Schematic representation of the mouse TCR and Ig gene loci. (A and B) Schematic representation of the IgH, IgLκ, and IgLλ loci, and of the TCRβ, TRCα/δ, and TCRγ loci, respectively. The various gene segments and associated elements are indicated as follows: orange square boxes indicate V gene segments; green rectangular boxes, D gene segments; yellow rectangular boxes, J gene segments; red square boxes, C genes; blue circles, E enhancers; green circles, silencing elements; red circle, a promoter; and purple circle, CCCTC-binding factor (CTCF)/cohesin TFBS-containing elements. The various boxed and circled elements are not drawn to scale. (See the color plate.)

23-RSSs of V(D)J recombination display a relatively high degree of sequence variability. Statistical computations of the correlation structure that underlies sequence variability led to the definition of an information content score, which rates the potential function in terms of the recombination efficiency of any individual RSS (Cowell, Davila, Ramsden, & Kelsoe, 2004; Cowell, Davila, Yang, Kepler, & Kelsoe, 2003). Using this score, a number of V_H and Vβ gene flanking RSSs were predicted and experimentally confirmed to recombine on average less frequently when compared to those flanking the Vκ/λ or Vα gene segments (Cowell et al., 2003; Liang et al., 2002). Even the farthest recombinational impacts of RSSs have occasionally been demonstrated, as exemplified first at the TCRβ locus for which 5′ Dβ 12-RSSs, but not the Jβ 12-RSSs, were shown to target rearrangement of 23-RSS flanked Vβ gene segments—a restriction referred to as Beyond (B)12/23 that in practice precludes Vβ-to-Jβ direct joining (Bassing et al., 2000; Franchini, Benoukraf, Jaeger, Ferrier, & Payet-Bornet, 2009; Hughes, Tillman, Wehrly, White, & Sleckman, 2003; Sleckman et al., 2000; Tillman, Wooley, Hughes, Khor, & Sleckman, 2004; Tillman et al., 2003). At the same locus, the recruitment of the V(D)J recombinase to the 3′ Dβ1 23RSS was later shown to occur via interaction with the transcription factor (TF) c-fos bound at a specific site within this RSS, with concomitant inhibition in the usage of the adjacent 5′ Dβ1 12RSS (Wang et al., 2008).

Regulatory elements at Ig/TCR loci first include promoters and enhancers, which altogether implement a prime level of transcriptional control, in a mostly lineage- and/or locus-specific manner (Andersson, 2015; Ernst & Smale, 1995; Kim & Shiekhattar, 2015; Roy, Sen, & Roeder, 2011; Smale, 2001). Briefly, promoters flank the individual V gene segments of the various Ig/TCR loci, where they act locally, in an orientation-dependent manner to organize the preinitiation complex of transcription, including general TFs of the RNA polymerase II (Pol II) holoenzyme. Thus, promoters and bound factors carry out the transcriptional activation both of Ig/TCR genes assembled from the corresponding V segments and of unrearranged V-region genes prior to recombination (GL transcription). Regularly, additional promoters for GL transcripts are also found adjacent to D and/or J gene segments. Enhancers are generally positioned on the 3′ part of a particular locus; in Ig loci, there may be two such elements (e.g., the $_i$Eκ and 3′Eκ enhancers at the IgLκ locus), whereas in TCR loci, there is usually only one (e.g., the Eβ and Eα enhancers at the TCRβ and TCRα loci, respectively). Classically described as short areas [a few hundred base pairs (bp)] of densely packed TF-binding sites (TFBS), enhancers function in an orientation- and distance-independent manner to stimulate transcriptional initiation at target promoters and mostly act in *cis*. Besides promoters and enhancers, other regulatory elements of gene expression involved in higher-level interactions among chromatin structures and/or between chromatin and matrix structures from the nucleus have occasionally been described at the distinct Ig/TCR loci. These include scaffold/matrix attachment regions [SAR/MARs; e.g., adjacent to the intronic Eμ and $_i$Eκ enhancers, respectively (Cockerill & Garrard, 1986; Cockerill, Yuen, & Garrard, 1987)], silencers/antisilencers [IgLκ, TCRγ, and TCRα loci (Ferradini et al., 1996; Ishida et al., 1990; Liu et al., 2006; Ortiz, Cado, & Winoto, 1999)], and locus control regions [e.g., on the 3′ end of the TCRα locus (Diaz, Cado, & Winoto, 1994)]. Recently, V-to-(D)J intergenic regions and associated elements that overlap with a TFBS(s) for the CCCTC-binding factor (CTCF)/cohesion proteins have attracted attention for their hypothesized role in chromosomal domain folding and/or insulation, and their attachment to potentially suppressive heterochromatin compartments [e.g., the IRG1 region including the CB1 and CBE2 elements within the IgH locus (Degner et al., 2011; Guo, Yoon, et al., 2011; Lin, Guo, Su, Zhang, & Alt, 2015); the DNase I-hypersensitive 1–6 containing region including the cer and Sis elements within the IgLκ locus (Liu et al., 2006; Xiang, Park, & Garrard, 2014; Xiang, Zhou, Hewitt, Skok, & Garrard, 2011); and the sequence named *5′Prss2*-CTCF site (5′PC) within the TCRβ locus (Carabana, Watanabe, Hao, & Krangel, 2011; Majumder, Koues, et al., 2015)]. Paradoxically regarding the large size of (at least some) V gene regions, few new putative regulatory elements have been discovered within these areas—besides the V specific promoters. One exception within the distal Ig V_H gene regions consists in conserved repeat sequences with Pax5 TF-dependent activity in pro-B cells called Pax5-activated intergenic repeat elements (Ebert et al., 2011).

3. V(D)J RECOMBINATION AND ALLELIC EXCLUSION: GENERAL FEATURES

3.1 V(D)J Recombination

From a biochemical standpoint, V(D)J recombination is initiated and primarily carried out by the lymphoid-specific recombination-activating gene (RAG)1/RAG2 nuclease (Oettinger, Schatz, Gorka, & Baltimore, 1990; Schatz, Oettinger, & Baltimore, 1989), which cleaves DNA at the asymmetric 12-RSS and 23-RSS of gene segment pairs [thus defining the so-called

12/23 rule of V(D)J recombination (Sakano et al., 1981; reviewed by Brandt & Roth, 2004; Gellert, 2002; Jones & Gellert, 2004; Oettinger, 2004; Schatz, 2004; Schatz & Spanopoulou, 2005)]. The resultant DNA double-stranded breaks are resolved primarily by components of the non-homologous end-joining (NHEJ) DNA repair pathway to yield separate coding- and signal-joints (CJs and SJs) (De Villartay, 2002; Lieber, Ma, Pannicke, & Schwarz, 2003; Roth & Roth, 2000). CJs often display limited deletion and/or nontemplated addition of nucleotides. Randomization of CJ sequences, while contributing to immune cell diversity, ultimately only yields a fraction (1/3 on average) of V(D)J exons assembled in the proper translational reading frame [i.e., productive or in-frame rearrangements; *NB*: at individual loci this fraction may be slightly lower or higher depending on structural or functional features—such as deleterious gene segments or the occurrence of successive *cis*-rearrangements, respectively (for details see Melchers, Ten Boekel, Yamagami, Andersson, & Rolink, 1999; Rajewsky, 1996)].

The above-presented brief overview simply skims the basics of V(D)J recombination. An abundant literature, masterly reviewed recently by the abovementioned and other authors, is available for a comprehensive, up-to-date description of all sequential biochemical events composing this unique DNA transaction process and of the molecular players involved [Deriano & Roth, 2013; Lieber, 2010; Matthews & Oettinger, 2009; Schatz & Ji, 2011; Schatz & Swanson, 2011; also see recent articles on the long-sought crystal structure of the core RAG1/RAG2 complex (Kim, Lapkouski, Yang, & Gellert, 2015) and the molecular foundation of the 12/23 rule (Lapkouski, Chuenchor, Kim, Gellert, & Yang, 2015)].

3.2 Allelic Exclusion

Allelic exclusion of Ig and TCR genes is established during rearrangement of V, D, and J gene segments in early developing B or T cells, respectively. Consequently, at the corresponding loci of most differentiated B/T lymphocytes, one allele carries a productively assembled V–D–J (or V–J) variable gene region (hereafter denoted VDJ+ or VJ+, respectively), while the homologous allele maintains a nonrearranged GL configuration, displays an incomplete D–J rearranged structure (DJ), or carries a nonproductively rearranged V–D–J (or V–J) variable gene region (VDJ− or VJ−) and as such is kept or rendered nonfunctional, respectively (Bergman, 1999; Gorman & Alt, 1998; Khor & Sleckman, 2002; Schlissel, 2002).

These genetic profiles have formed the basis of our understanding of allelic exclusion in B and T lymphoid cells for decades, yet the precise mechanism(s) through which they are achieved and their significance to the functioning of the immune system remain unclear. Monoallelic gene expression represents an essential phenomenon in developmental cell biology, both physiologically and clinically. Classically studied examples of monoallelic gene expression include genomic imprinting (Bartolomei & Ferguson-Smith, 2011), in which genes are expressed exclusively from either the paternal or maternal allele (nonrandom monoallelic expression); and X-chromosome inactivation in female mammalian cells (Gendrel & Heard, 2011), in which genes on a single, randomly selected X chromosome are silenced [random monoallelic expression; to which allelic exclusion was often likened (e.g., Cedar & Bergman, 2008)]. Lately, genome-wide studies have provided evidence that random monoallelic expression is more frequent than originally anticipated, affecting up to 25% of the autosomes in a cell lineage-specific manner, and usually including Ig and TCR genes (Chess, 2012). However, besides the characteristic trait of somatic DNA V(D)J rearrangement in particular as it relates to these latter genes, the two phenomena can be further distinguished by several additional divergent properties. Thus, distinct from random monoallelic expressed genes for which transcription occurs from just one of the two alleles, both alleles (prior to and following VDJ assembly) may be transcribed in lymphoid cells (Daly, Licence, Nanou, Morgan, & Martensson, 2007; Jia, Kondo, & Zhuang, 2007; Singh, Bergman, Cedar, & Chess, 2003). Furthermore, whereas defects in random monoallelic expression systems such as X-inactivation often result in cell developmental and pathological disorders, much in line with the strictness of their regulation (Chaligné & Heard, 2014; Gendrel et al., 2014), allelically included [VDJ+/VDJ+] B cells, can develop without apparent defect or clinical impact (Sonoda et al., 1997). Indeed, when sought-after in normal wild-type lymphoid B and T cells, small fractions of allelically included Ig and TCR genes can consistently be found not only at the phenotypic level of Ig/TCR antigen receptor chain expression (1–10% depending on the locus) but also, and generally at a higher incidence, at the genotypic level involving Ig/TCR variable gene region rearrangement (Brady, Steinel, et al., 2010). Clearly, although allelic exclusion and random monoallelic gene expression share regulatory features, their genomic and cellular outcomes differ.

4. V(D)J RECOMBINATION AND ALLELIC EXCLUSION: REGULATION

4.1 Cell Lineage, Developmental Stage, and Allelic Specificities of V(D)J Recombination

Assembly of Ig/TCR genes by V(D)J recombination is regulated at several distinct levels, including lymphoid cell lineage, stage of cell development, and, as assessed here, allele usage. Taking the TCRβ gene locus as a symbolic example, Vβ–Dβ–Jβ DNA rearrangements are normally restricted to the T-cell lineage and proceed in a stepwise manner during $CD4^-CD8^-$ [double negative (DN)] thymic cell development. Namely, Dβ-to-Jβ rearrangements occur first, prior to Vβ-to-DJβ assembly. Based on average profiling of recombination products, Dβ-to-Jβ joining is initiated during the $CD4^-CD8^-CD44^+CD25^+$ (DN2) thymic cell stage and concurrently involves both TCRβ alleles. The Vβ-to-DJβ assembly is completed in the following $CD4^-CD8^-CD44^-CD25^+$ (DN3) thymic cells. From this stage onward, in-frame VβDJβ rearranged CJ products are generally detected at only one of the two TCRβ alleles, in line with the broadly held assumption that the V-to-(D)J joining process is the regulated step in monoallelic productive assembly (reviewed by Bonnet, Ferrier, & Spicuglia, 2009; Sikes & Oltz, 2012). Notably, interallelic ordering of TCRβ chain gene assembly is regulated in a manner that is not biased toward completion of all possible VDJβ rearrangements on one allele before activating complete VDJβ rearrangements on the alternate allele (Khor & Sleckman, 2005), though evidence also exists that a VDJ joint using DJβ1 has a positive (nonnegligible) probability of being removed by a secondary Vβ–DJβ *cis*-rearrangement involving a DJβ2 unit (Brady, Oropallo, et al., 2010; Farcot et al., 2010). Importantly, rearrangements of the TCRβ gene are tightly linked with T-cell differentiation. Indeed, following transcription of the VβDJβ rearranged gene and translation of the RNA products, pairing of the resulting TCRβ chains with invariant pre-Tα chains forms, in association with CD3 components, the pre-TCR. Upon assembly, the pre-TCR signals the cells to (i) downregulate RAG activity and mediate TCRβ allelic exclusion, (ii) activate cell-survival, and (iii) proliferate and develop toward the $CD4^-CD8^-CD44^-CD25^-$ (DN4) cell stage and subsequently, the $CD4^+CD8^+$ (double positive (DP)) cell compartment; a series of molecular events collectively referred to as β-selection (Kreslavsky et al., 2012; von Boehmer et al., 1998). RAG activity resumes at the latter DP cell stage, where it appears to be redirected toward the TCRα gene to mediate Vα-to-Jα rearrangement, thus contributing to the strict ordering of TCRβ and TCRα recombination during αβT-cell development. Later on during this stage, β and α chain pairing into mature αβ-TCR as well as TCR-dependent and threshold-sensitive signaling trigger DP cells to pass through further checkpoints during which they are negatively or positively selected. In the latter case, the cells eventually develop into either $CD4^{+}CD8^{-}$ or $CD4^{-}CD8^{+}$ (single positive) terminally differentiated T cells (Kisielow & von Boehmer, 1995; Singer, Adoro, & Park, 2008).

Alike the TCRβ gene, other Ig and TCR loci are also regulated with respect to V(D)J recombination assembly, though not always with equivalent levels of spatiotemporal and/or allelic specificities. Taking Ig genes as an example, assembly of the IgH variable gene region in pro-B cells follows a very similar pattern of events, including biallelic D_H-to-J_H joining prior to allelically excluded V_H-to-DJ_H assembly. These events are then followed in pre-B cells by V-to-J IgLκ assembly and, occasionally, IgLλ gene assembly, overall directing an analogous sequence of regulated pro-B to pre-B and then pre-B to B developmental cell transitions (Alt et al., 1984; Barreto & Cumano, 2000; reviewed by Melchers, 2005). Nonetheless, it is noteworthy that D_H-to-J_H rearrangement does occur in 30–50% of developing T cells and thus escapes the strict lymphoid lineage constraint (Born, White, Kappler, & Marrack, 1988). In addition, V_H replacement, a RAG-mediated secondary rearrangement in which an upstream V_H element recombines using a cryptic RSS within the original V_HDJ_H joint to generate a novel CJ product, can account for 12–20% of the IgH repertoire of mature B cells (Kumar et al., 2015; Sun et al., 2015). Sequential IgLκ V–J rearrangements also occur on each chromosome at significant rates, entailing in-frame Vκ–Jκ joints in ~10% of B cells (Alonso, Hozumi, & Murialdo, 1985; Alt, Enea, et al., 1980; Arakawa, Shimizu, & Takeda, 1996; Kwan, Max, Seidman, Leder, & Scharff, 1981; Levy, Campbell, & Levy, 1989; Zou, Takeda, & Rajewsky, 1993; reviewed by Brady, Steinel, et al., 2010; Gorman & Alt, 1998; Vettermann & Schlissel, 2010). Regarding TCR gene loci, Vα-to-Jα recombination, which exhibits lymphoid T-cell lineage and developmental stage constraints, but not allelic exclusion as already mentioned, likewise yields biallelic in-frame rearrangement in up to 30% of differentiated T cells (Alam & Gascoigne, 1998; Casanova, Romero, Widmann, Kourilsky, & Maryanski, 1991; Krangel, Carabana, Abbarategui, Schlimgen, & Hawwari, 2004; von Boehmer & Melchers, 2010). In contrast to the well-separated timeframes of TCRβ/IgH and

TCRα/IgLκ/λ rearrangements, V(D)J assembly at the TCRδ and TCRγ gene loci mostly overlaps within both the DN and early DP (EDP) thymic compartments (reviewed by Raulet, 1989; Xiong & Raulet, 2007). Furthermore, VδDJδ rearrangements are not strictly ordered, proceeding via assembled DJδ (or DDJδ) as well as VδDδ (or VδDDδ) intermediates; and assembly of productive VδD(D)Jδ region genes frequently exhibits allelic inclusion (Sleckman et al., 1998). Some of the above observations may be relevant to the process of receptor editing via ongoing recombination, the purpose of which is to eliminate/replace nonfunctional or autoreactive antigen receptors (Nemazee, 2006). Yet, how this whole palette of related yet disparate spatiotemporal controls is implemented remains unclear—though it likely involves particular structural and regulatory features each of which impinge upon the recombination dynamics and/or epigenetics associated with to the corresponding Ig/TCR loci.

4.2 The Accessibility Hypothesis

Once a unique recombination apparatus targeting analogous RSSs of Ig and TCR gene loci had been revealed (Yancopoulos, Blackwell, Suh, Hood, & Alt, 1986), the question arose as to how to account for the lineage, developmental stage, and preferential monoallelic profiles of V(D)J recombination. In the mid 1980s, Alt and colleagues formulated their soon-famed "accessibility hypothesis" whereby differential chromosomal access would be the key to interpret such high-level specificities, with modulated changes of the chromatin envelope anticipated to represent the true molecular hint (Alt, Blackwell, & Yancopoulos, 1987). Two groups of evidence, collectively documented in a wealth of studies spread over the next two decades, would strongly and consistently support this proposition. First, ongoing recombination at given gene segments and/or loci correlated, in a mutually nonexclusive manner, with focused marks of accessible DNA and "open" chromatin, including (i) GL transcription, both sense and antisense transcription; (ii) hypomethylation of CpG dinucleotides; (iii) hypersensitivity to DNA nucleases and to restriction enzymes; and (iv) histone tags generally associated with the activation of gene expression [including histone H3/H4 acetylation (H3ac; H4ac) and di-/trimethylation of Lys4 of histone H3 (H3K4me2/me3)]. Conversely, the opposite proved true when studying recombination-quiescent or -inactivated Ig/TCR loci, which were generally found transcriptionally silent, hypermethylated, protected from nucleases activity, and decorated with suppressive histone marks [including di-/trimethylation of Lys9 or Lys 27 of histone H3 (H3K9me2, H3K27me3), respectively] (Bolland, Wood, & Corcoran, 2009; Cobb, Oestreich, Osipovich, & Oltz, 2006; Jung, Giallourakis, Mostoslavsky, & Alt, 2006; Krangel, 2007). Second, regulatory elements known to promote local changes in accessibility during the activation of gene expression (i.e., enhancers and promoters) were shown to critically impact on the patterns of V(D)J *cis*-recombination, as predicted (Cobb et al., 2006; Jung et al., 2006; Krangel, 2007; Spicuglia, Pekowska, Zacarias-Cabeza, & Ferrier, 2010). Importantly, enhancers were found to exert long range, generally activating effects throughout a whole domain area and/or a whole locus, while V- and (D)J-associated promoters exhibited a more localized range of action, limited to adjacent sequences and thus mimicking their respective transcriptional activation properties. Mirroring the above findings, genetic manipulations of the mouse genome to intentionally modulate GL transcription and/or the targeting of chromatin-modifying enzymes at specific Ig/TCR loci, effectively impinged on V(D)J recombination of linked chromosomal gene segments, and in parallel overrode the recombination-activating function of associated *cis*-regulatory elements (e.g., Abarrategui & Krangel, 2006, 2007; Osipovich et al., 2007, 2004).

The abovementioned studies, while highly supportive of the accessibility hypothesis, remained correlative, as they did not directly address the access of the RAG1/2 apparatus to discrete RSS targets, in connection with chromosomal accessibility. Several observations and novel findings then filled this gap. First, using an *in vitro* system in which intact nuclei were incubated with a nuclear extract supplemented with purified RAG1 protein, Schlissel and colleagues demonstrated that chromatinized RSSs flanking Ig and TCR gene segments indeed represent differential targets for RAG cleavage depending upon the lineage and developmental stage origins of the reacted nuclei (Stanhope-Baker, Hudson, Shaffer, Constantinescu, & Schlissel, 1996). Later, *in vitro* biochemical studies demonstrated that the nucleosome, when containing RSS substrates, represents a significant barrier to RAG-mediated initiation of V(D)J recombination (Baumann, Mamais, Mcblane, Xiao, & Boyes, 2003; Du, Ishii, Pazin, & Sen, 2008; Golding, Chandler, Ballestar, Wolffe, & Schlissel, 1999; Kwon, Imbalzano, Matthews, & Oettinger, 1998; Kwon, Morshead, Guyon, Kingston, & Oettinger, 2000). However, the reality at an endogenous locus may be more subtle. Indeed, a detailed analysis of the nucleosome organization of accessible and inaccessible TCRβ and TCRα alleles in primary murine thymocytes found no consistent positioning of nucleosomes with respect to individual RSSs (Kondilis-Mangum et al., 2010). Instead, this study suggested a model system whereby enhancer-/promoter-dependent transcription generates optimal recombinase substrates in which some nucleosomes are missing and others are covalently modified.

4.3 Regulation of Allelic Exclusion: Views Integrating Recombinational Accessibility

Allelic exclusion has primarily been observed at IgH, IgL, and TCRβ loci (Bergman, 1999; Gorman & Alt, 1998; Mostoslavsky et al., 2004). Among the various models that have been proposed to explain how allelic exclusion occurs at these loci, one—termed the "regulated model"—has now gained wide agreement. As originally stated, the model—first formulated to explain allelic exclusion at the IgLκ locus—proposes that V–J rearrangement is ordered, proceeding on one chromosome at a time, and that protein products from a functional rearrangement ([VJ +], one that encodes an IgL chain which pairs with the preexisting IgH chain to make surface Ig) play a critical role in mediating IgL allelic exclusion through feedback inhibition of further IgL locus rearrangement and freezing of the second allele in the GL status. The model also predicts that cells having made a nonfunctional [VJ −] joint on the first allele go on to rearrange the second allele, possibly leading to rescued cells harboring VJ rearrangement on both alleles, one productive and one nonproductive ([VJ +/VJ −] cells; *NB*: [VJ −/VJ −] cells, inevitably also generated through this process though not observed by current experimental techniques, would be rapidly purged from the developmental system). Extended to the IgH and TCRβ loci, the model assumes that the V-to-DJ junction is the regulated allele-excluding step at these loci, while initial D–J joining occurs simultaneously on both alleles (Alt, Bothwell, et al., 1980; Alt et al., 1981; Kisielow & von Boehmer, 1995). Notably, given junctional diversification mechanisms of V(D)J recombination and the ~1/3 fraction of V-DJ rearrangements predicted to be in frame (i.e., enabling the generation of Ig or TCR chains), the regulated model is in agreement with the observed ratio of approximately [60%] and [40%] lymphocytes displaying at the corresponding gene loci the [V(D)J +/GL or DJ] and [V(D)J +/V(D)J −] genotypes, respectively (Malissen et al., 1992; Yancopoulos & Alt, 1986). Moreover, this model was supported by a wealth of experimental results from transgenic mice as well as genetically engineered knockout and knockin (KO, KI) mice, which demonstrated that premature expression or, conversely, ablation of Ig/TCR surface antigen receptors or downstream signaling components, respectively, induces and disrupts allelic exclusion (reviewed by Brady, Steinel, et al., 2010; Mostoslavsky et al., 2004; Vettermann & Schlissel, 2010). Parenthetically, a separate study demonstrating that KI mice carrying two distinct [VDJ+] IgH alleles give rise to allelically included B cells (Sonoda et al., 1997) refuted another proposed model—termed the "cellular selection model"—postulating that such [VDJ+/VDJ+] cells are generated frequently but via counter-selection are eliminated from the B cell repertoire (Beck-Engeser et al., 1987; Wabl & Steinberg, 1982).

In the context of the original regulated model, allelic exclusion was proposed to encompass two separate steps (Alt, Enea, et al., 1980): (i) rearrangement should be initiated at only one of the two homologous Ig or TCR loci (initiation of allelic exclusion); and (ii) rearrangement on the second allele should be prevented by expression of functional Ig or TCR chains encoded by the first rearranged allele (maintenance of allelic exclusion). Largely, these preconditions have been regarded through the prism of chromosomal accessibility. In particular, during the initiation step of allelic exclusion, it was assumed that chromosomal accessibility must be regulated such that only one allele at a given locus becomes accessible to the VDJ recombinase during the time period in which feedback inhibition is achieved. Two types of model concepts, stochasticity and determinism, were evoked in a mutually exclusive manner to explain this assumption. Tenants of the stochastic concept have relied on the notion of "V-to-(D) J recombination inefficiency," commonly considering into the filigree a low probability p for Ig/TCR alleles to acquire an accessible chromosomal structure, possibly in connection with additional intrinsic Ig/TCR gene segment properties, i.e., inherent inefficiency of V_H and $V\beta$ RSSs [(Liang, Hsu, Cado, & Schlissel, 2004; Schlimgen, Reddy, Singh, & Krangel, 2008; reviewed by Vettermann & Schlissel, 2010); also of note, the earliest stochastic models of allelic exclusion, in which p instead related to the infrequent generation of a productive joint by—at this time—an unknown putative recombinase (e.g., Coleclough, 1983; Coleclough et al., 1981), had to be abandoned due to obvious incompatibility with subsequently accumulated experimental results; also see below]. On the other hand, tenants of the deterministic/instructive concept argued that opposite alleles are non-equivalent substrates for the VDJ recombinase and that this regulated choice directs the ordering, on a strictly sequential mode, of subsequent allelic rearrangements (Hewitt et al., 2009; Mostoslavsky et al., 2001; reviewed by Goldmit & Bergman, 2004; Hewitt, Chaumeil, & Skok, 2010; Levin-Klein & Bergman, 2014).

Feedback inhibition mediated by receptor-induced signaling marks the onset of the maintenance phase of allelic exclusion. Various underlying mechanisms have been evoked to explain this inhibitory phase, including downregulation of RAG expression (e.g., Grawunder et al., 1995), developmental stage-specific inhibition of accessibility to GL V segments (e.g., Chowdhury & Sen, 2003; Tripathi, Jackson, & Krangel, 2002), and/or the suppression of juxtaposition of V and (D)–J gene segments [i.e., locus decontraction (Jackson, Kondilis, Khor, Sleckman, & Krangel, 2005; Jackson & Krangel, 2005; Roldan et al., 2005); and see below]. However, when questioned individually, the implication of one or other of the processes in feedback inhibition often appeared less critical than expected (Li, Dordai, Lee, & Desiderio, 1996; Majumder, Rupp, et al., 2015) or appeared to impact V gene segments differently depending on locus location (Roldan et al., 2005; Yang-Iott et al., 2010). Cooperative functionality involving distinct processes is likely (e.g., Kondilis-Mangum, Shih, Mahowald, Sleckman, & Krangel, 2011); and additional, complementary mechanism(s) or pathway(s) have been suggested involving, e.g., developmental stage-specific factors that control RAG binding to RSSs of DJ complex intermediates (Wang et al., 2008); or the Ataxia Telangiectasia mutated (ATM)-mediated damage response (Steinel, Fisher, Yang-Iott, & Bassing, 2014; Steinel et al., 2013). Both the initiation phase and the maintenance phase of allelic exclusion therefore retain part of their mystery. Further to that, temporal interconnection between these two phases has not been explored thoroughly.

4.4 The Notion of "Recombination Centers"

Combinatorial covalent modifications on nucleosomes establish a "histone code" that is specifically recognized by conserved domains shared by many chromatin-associated regulatory proteins (Jenuwein & Allis, 2001; Ruthenburg, Li, Patel, & Allis, 2007). Importantly, RAG-2 was known to contain a noncanonical plant homeodomain (PHD) finger in a part of the protein that is dispensable for RAG-mediated DNA cleavage *in vitro* (RAG-2 noncore region) (Callebaut & Mornon, 1998; Elkin et al., 2005). In a number of proteins associated with epigenetic regulation, a similar domain has been found to specifically bind H3K4me2 or H3K4me3. Two studies have shown that this noncanonical PHD finger mediates direct binding of RAG-2 to H3K4me2 and (preferentially) H3K4me3. The functional significance of the latter interaction was demonstrated when (i) mutations in the PHD finger that abolished H3K4me3 recognition severely impaired V(D)J recombination; and (ii) DNA binding and recombination was found to depend upon the amount of H3K4me3 deposition. Therefore, recognition of the posttranslational H3K4me3 epigenetic mark by RAG-2—possibly enhanced by dimethylation of arginine 2 (H3R2me2)—appears critical to V(D)J recombination (Liu, Subrahmanyam, Chakraborty, Sen, & Desiderio, 2007; Matthews et al., 2007; Ramon-Maiques et al., 2007). Strikingly, the function of this interaction may not be confined to recombinase recruitment. Indeed, building on these previous findings, biochemical investigations have indicated that recognition of H3K4me3 by RAG2 also stimulates the catalytic activity of the RAG nuclease (Grundy, Yang, & Gellert, 2010; Shimazaki, Tsai, & Lieber, 2009), likely via the allosteric control of an auto-inhibitory domain the action of which is relieved by engagement of active chromatin (Lu, Ward, Bettridge, Liu, & Desiderio, 2015); and possibly also stabilizes the newly excised recombination ends within the post-cleavage complex and facilitates their transfer to the NHEJ repair machinery (Wang, Dhar, Swanson, Levitus, & Chang, 2012). Separately, RAG1 may itself, through a RING domain within its N-terminal noncore region, preferentially interact with and promote monoubiquitylation of histone H3, resulting in increased DNA cleavage by the complete RAG1/2 apparatus (Deng, Liu, & Liu, 2015; Grazini et al., 2010; Jones et al., 2011). The emerging picture thus depicts the whole V(D)J recombination process being assisted—even accelerated—with pinpoint accuracy through such self-sustaining chain reaction of punctual epigenetic modulations.

To which RSS-containing domain(s) of endogenous antigen receptor loci does the RAG nuclease actually bind? This issue was recently addressed in studies using transgenic mice expressing an active site-mutated RAG-1 protein that could bind DNA normally (and interact with RAG-2) but which lacked catalytic activity, thus avoiding the formation of recombination products that might have complicated the interpretation of chromatin immuno-precipitation (ChIP) data at Ig and TCR loci. Interestingly, RAG protein binding occurred in a focal manner across small genomic regions that were highly enriched for activating histone modifications (H3Ac, H3K4me3), referred to as "recombination centers" (Ji, Resch, et al., 2010). Notably, recombination centers comprise regions encompassing Igκ and TCRα J gene segments and IgH and TCRβ J and J-proximal D gene segments. While RAG-1 appeared to predominantly interact with RSS-containing regions, RAG-2 more broadly bound to H3K4me-rich sequences throughout the mouse genome (Ji, Resch, et al., 2010; Teng et al., 2015). A complementary study using mutant TCRα and TCRβ alleles found that enhancers control global occupancy of RAG1 at Jα or Dβ–Jβ gene segments, while promoters direct its local binding, consistent with the V(D)J recombinational function of these "accessibility control elements" (ACEs) defined in prior knockout studies (Ji, Little, et al., 2010). Overall, "recombination centers" have been interpreted as specialized sites of high local RAG concentration that allow facilitated RSS binding and synapsis as well as an efficient regulation of recombination order and fidelity (Schatz & Ji, 2011). In agreement with this view, analysis of DJ_H-recombined alleles has provided evidence that DJ_H junctions are selectively epigenetically marked to allow the binding of RAG proteins, which in turn permits the DJ_H-5′ intact RSS to initiate the second (V-to-DJ) step of IgH gene assembly (Subrahmanyam et al., 2012).

4.5 V(D)J Recombination, Chromosomal Location, and 3D Architecture

Control of RAG access to chromatin-embedded RSSs, while undoubtedly a prerequisite for V(D)J recombination, may not be sufficient to effectively initiate this process at the 3D level of chromosomal topology. Indeed, conditions were reported in which seemingly fully accessible V gene segments did not proceed with recombination (Hesslein et al., 2003; Fuxa et al., 2004; Jackson et al., 2005; Jackson & Krangel, 2005; discussed in Jackson & Krangel, 2006). Most progress made addressing this question has involved the use of fluorescent *in situ* hybridization (FISH) techniques and imaging approaches alongside complementary chromosomal conformation capture techniques. These have contributed toward unveiling additional mechanisms operating to impact the developmental regulation of locus recombination events, namely (i) repositioning relative to chromatin repressive environments [including the nuclear periphery and/or pericentromeric heterochromatin (Bank & Gruenbaum, 2011; Steiner & Henikoff, 2015)]; and (ii) locus folding/unfolding via reversible large-scale compaction.

Transcriptionally inactive but not transcriptionally active genes had been shown to associate with heterochromatin foci in interphase nuclei of lymphoid cells, suggesting such association as a general feature of heritable gene silencing (Brown, Baxter, Graf, Merkenschlager, & Fisher, 1999; Brown et al., 1997). In the case of Ig genes in splenic B cells, this apparently involves one single IgH allele, one single IgLκ allele, and, in most cells analyzed, the two endogenous IgLλ alleles, raising the possibility that heterochromatin association may help maintain Ig monoallelic expression in developed B cells (Skok et al., 2001). Then the question arose of whether an analogous process might be associated with earlier developmental events, including ongoing V(D)J recombination. Singh and colleagues were the first to provide a FISH view of the subnuclear compartmentalization of Ig loci in developing lymphocytes, emphasizing preferential positioning of both IgH and IgLκ loci at the nuclear periphery in hematopoietic progenitors and pro-T cells but of their central configuration in pro-B cells, indicating that subnuclear positioning may represent a means of regulating Ig gene expression and recombination during early lymphoid cell development (Kosak et al., 2002). Interestingly, complementary analyses of FISH signals also notably implied that (i) Ig gene recombination is not required for the central localization of the IgH and Igκ loci in pro-B nuclei (as it was also observed using RAG-deficient pro-B cells), which, at least in the case of IgLκ, may occur prior to GL transcription (not yet activated in pro-B cells); and (ii) central configuration of IgH loci in pro-B nuclei is in parallel marked by large-scale locus compaction as demonstrated by the relatively lower resolution of V_H distal and proximal FISH probes compared with those observed at peripheral alleles in T-cell nuclei (Kosak et al., 2002). Later studies, using different mouse genetic models as a source of lymphoid cell nuclei, confirmed and extended these findings at the IgH and/or IgLκ loci (Fuxa et al., 2004; Goldmit et al., 2005; Roldan et al., 2005; Sayegh, Jhunjhunwala, Riblet, & Murre, 2005) as well as at the TCRβ and TCRα/δ loci (Schlimgen et al., 2008; Skok et al., 2007). Overall, these data contributed toward establishing two new consensual paradigms. First, V-to-(D)J rearranging Ig/TCR loci are found preferentially dissociated from the nuclear lamina and pericentromeric heterochromatin compartments. Second, these loci undergo reversible large-scale contraction likely allowing long-range internal interactions to be mediated by looping of individual Ig/TCR subdomains, in a lymphoid cell lineage- and developmental stage-specific manner. Additional analyses also led to the proposition that central repositioning and decontraction of the IgH locus in developing B cells depend on interchromosomal association with the IgLκ locus at the transition between these two loci's recombination, a regulated process involving the 3′Eκ enhancer (Hewitt et al., 2008). Independently, homologous IgH and IgLκ alleles were proposed to also pair stage-specifically, in a manner mirroring the recombination patterns of these loci. In these cases, pairing would depend first on RAG1 binding. Introduction of DNA breaks on a single Ig allele would then induce ATM-dependent repositioning of the opposite allele to pericentromeric heterochromatin (Hewitt et al., 2009).

Obviously, these new paradigms were mostly regarded in reference to Ig/TCR monoallelic gene expression and recombination and intended two phases—initiation and maintenance—of allelic exclusion. Indeed, both dissociation from and juxtaposition with heterochromatin suppressive areas, coupled to locus compaction and subsequent unfolding, could explain the sequential stages of activation versus inhibition of gene expression and recombination at homologous alleles predicted by allelic exclusion models. However, the issue may not be as straightforward as it would seem and caution should be taken to avoid being drawn into making oversimplistic deductions. Indeed, considering data from separate studies concerning both IgLκ and TCRβ loci and implying biallelic locus expression in the corresponding pre-B and DN cells (Jia et al., 2007; Singh et al., 2003), a scenario directly linking heterochromatin juxtaposition to global inactivation at the locus is unlikely. Likewise, images portraying fractions of >50% of the counted Ig and TCR alleles being monoallelically associated with heterochromatin were generally interpreted as representing the deterministic/instructive model of the initiation of allelic exclusion, whereby one allele would be targeted to heterochromatin prior to recombination (Goldmit et al., 2005; Hewitt et al., 2009; Roldan et al., 2005; Skok et al., 2007). However, reexamination of monoallelic versus biallelic association of the TCRβ locus with heterochromatin compartments in RAG2-deficient DN thymocytes instead statistically implied stochastic allelic behaviors at this locus (Schlimgen et al., 2008). Using the accumulation of the DNA repair protein 53BP1 to identify recently rearranged TCRβ alleles, this same group later confirmed that V(D)J recombination is suppressed on peripheral as compared with central TCRβ alleles, and also provided evidence that TCRβ alleles recombine asynchronously in DN thymocytes, both during the Dβ-to-Jβ and Vβ-to-DJβ joining stages (Chan et al., 2013).

The discovery of reversible Ig/TCR chromosomal folding during B/T-cell development raised the question of how large antigen receptor loci are structured in 3D space. The determination of spatial distances between genomic markers spanning the entire IgH locus and comparison to computer simulations of chromatin arrangements predicted an organization into compartments containing clusters of loops separated by linkers with conformational changes involving V_H and D_H–J_H containing domains during B cell development. Notably, in pro-B cells, the entire repertoire of V_H regions appeared to have merged and juxtaposed to the D_H elements, likely providing equal opportunity for V_H segments to be engaged in recombination with RAG1/2 loaded DJ_H intermediates (Jhunjhunwala et al., 2008;

reviewed by Bossen et al., 2012; Lieberman Aiden & Casellas, 2015). Chromosomal conformation capture analyses provided further evidence of the IgH locus being folded into multilooped domains, each brought into close spatial proximity allowing distal V_H clusters to undergo interactions with the more rigidly structured proximal domain (Gopalakrishnan et al., 2013; Guo, Gerasimova, et al., 2011; Medvedovic et al., 2013). Building on these assets through a tour-de-force study in which the distal V_H and D_H–J_H regions were marked in order to trace their 3D trajectories, it was revealed that, in pro-B cells, these elements display nonlinear diffusion trajectories best described by "fractional Langevin subdiffusive motion" (a form of movement whereby the viscoelastic hindrance from the surrounding network predisposes a moving particle to bounce back toward its previous position) (Lucas, Zhang, Dudko, & Murre, 2014). Using fractional Langevin dynamics modeling, the authors found that, with high probability, D_H–J_H elements reached a V_H element within minutes; they thus proposed a model system in which coding elements and associated regulatory elements bounce back and forth until specific genomic interactions are established, with spatial confinement of topological domains determining first-passage times for genomic interactions (Lucas et al., 2014). In terms of its principle, this model system is likely extendable to other large antigen receptor loci. Indeed, previous examination of the TCRα/δ locus indicated distinct contracted conformations during TCRδ and TCRα gene recombination in DN cells and in DP cells, respectively (Shih & Krangel, 2010; Skok et al., 2007).

4.6 Epigenetic Changes Favoring the Activation of V(D)J Recombination

Epigenetics plays a central role in gene activation events and the maintenance of gene expression, including during hematopoietic cell development (Natoli, 2010; Winter & Amit, 2014). Although the exact dynamics of epigenetics are yet to be revealed, epigenetic phenomena appear to be graded and time-related changes that likely occur both within and across cell generations (Burggren, 2015). Little is known about the detailed epigenetic changes that accompany the activation of antigen receptor loci in differentiating lymphocytes, either in terms of their form or their kinetics. However, past analyses have helped gain an overall picture of general tendencies. Focusing on the IgH and TCRβ loci, for example, it seems clear that the earliest stages of their activation are marked by a focal decrease in histone marks classically associated with gene silencing (such as H3K9me2 or H3K27me3), along with a parallel acquisition of activating marks

(H3K9ac, H3K4me2, or H3K36me3). This balance of epigenetic changes primarily targets the J- and D-containing regions of the locus (at the large—2.5 Mbp—IgH locus, only the most 5′ and 3′ D gene segments), while the V region maintains a suppressive landscape until its later activation, when distinct behaviors are possible depending on whether proximal or distal V genes are considered (Chakraborty et al., 2007; Chowdhury & Sen, 2001; Morshead, Ciccone, Taverna, Allis, & Oettinger, 2003; Spicuglia et al., 2002; Xu & Feeney, 2009; Xu, Schaffer, Head, & Feeney, 2008). Also, whereas the very initial changes occur independently from associated enhancer and/or promoter activities, those that follow specifically require these functioning elements [e.g., H3K9me2 is replaced by H3K9ac in all parts of the D_H–Cμ region in pro-B cells; loss of H3K9me2 is independent of the intronic enhancer Eμ; and gain of H3K9ac is Eμ dependent, correlating with the role of this regulatory element in targeting D_H–J_H *cis*-recombination (Afshar, Pierce, Bolland, Corcoran, & Oltz, 2006; Chakraborty et al., 2009)]. Best studied at the IgH locus, GL transcription including antisense transcription, initiates before D-to-J recombination throughout the D_H–J_H region and starts near to and with dependency on Eμ, before extending into the V_H domain, where it possibly facilitates locus compaction (Bolland et al., 2007, 2004; Giallourakis et al., 2010; Inoue et al., 2015; Verma-Gaur et al., 2012). "Hot spots" of H3K4me2 and H3K4me3 and the nucleosome-remodeling factor Brg1 are concurrently localized at the active D–J domains of both the IgH and TCRβ beta loci (Bonnet, Huang, et al., 2009; Bossen et al., 2015; Morshead et al., 2003). At the latter locus, they expand to form a specialized epigenetic marking characterized by high H3K4me3 levels throughout the Dβ–Jβ–Cβ gene segments, with parallel enrichment of phosphorylated serine-5 and serine-2 C-terminal domain forms of Pol II (Ser^{5P}/Ser^{2P} Pol II), and short-RNA transcripts (Zacarias-Cabeza et al., 2015). Ultimate molecular events toward a fully RAG-accessible status may include enhancer/promoter-dependent reorganization of nucleosomal areas, characterized by diminished abundance of certain nucleosomes, repositioning of others, and, possibly, transcription-mediated eviction of H2A/H2B dimers to form hexasomes (Bevington & Boyes, 2013; Kondilis-Mangum et al., 2010); and eventually localized demethylation of specific CpGs (Selimyan et al., 2013; Whitehurst, Schlissel, & Chen, 2000). Evidence also exists for the hierarchical progression of both chromosomal accessibility and structural organization along the given Ig/TCR loci, via one or more multi-step pathways mediated by distinct nuclear factors (e.g., Gerasimova et al., 2015; Bossen et al., 2015; Reynaud, et al., 2008; Agata et al., 2007; Liu, Schmidt-Supprian, et al., 2007; Fuxa et al., 2004).

Figure 2 integrates the above series of epigenetic changes leading to V(D)J recombination-favoring accessibility into a common framework. The resulting schematic view explains how allele asynchrony is ensured during early differentiation of individual lymphocytes, from the initiating steps of allelic activation onward. Overall, the developing process evolves from an inert status involving antigen receptor gene alleles displaying extended topology and their association with suppressive heterochromatin (inner nuclear membrane), to a stage where one allele folds into a compact rosette-like structure thereby permitting high-dynamic interactions between a RAG-loaded recombination center and any—with equal opportunity—upstream V segments, as proposed (Bossen et al., 2012; Jhunjhunwala et al., 2008; Lucas et al., 2014). In our opinion, homologous alleles—in principle from any Ig/TCR locus including the IgH and/or TCRβ loci to which many of the abovementioned data are relevant—are considered as independent and are stochastically recruited into this process

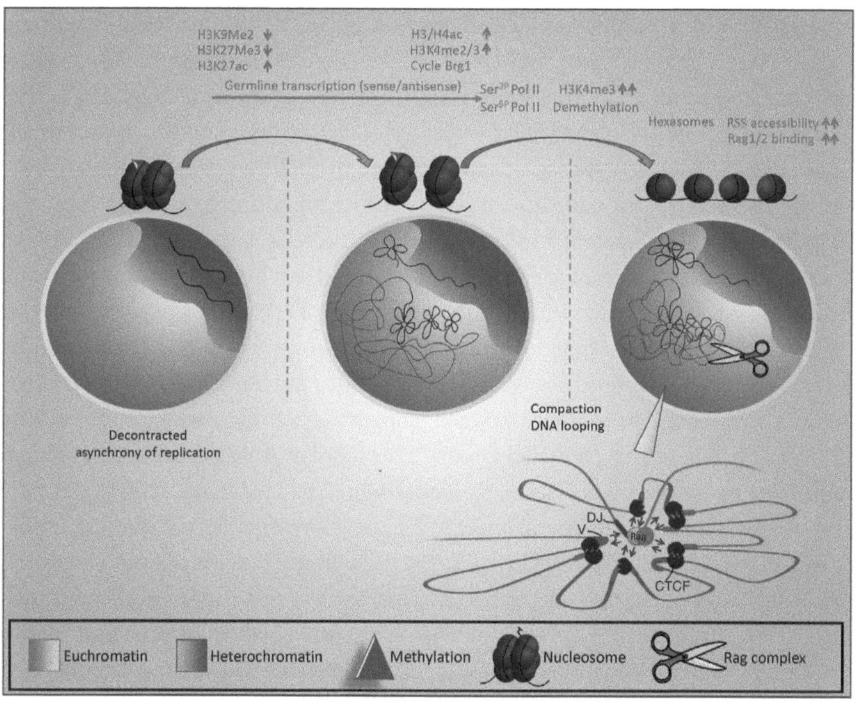

Figure 2 Schematic representation of epigenetic changes and variations in chromosomal topology at an antigen receptor gene locus during early lymphoid cell development (see text for details). (See the color plate.)

via noisy epigenetics (Eldar & Elowitz, 2010; Pujadas & Feinberg, 2012), entailing their asynchronous dissociation from the heterochromatin compartment and concurrent topological folding (i.e., allele asynchrony that is stochastically introduced during the very first epigenetic initiating steps is thus propagated along the whole developmental process). Deterministic features are conceivable as integral parts of the system, including possibly a prime epigenetic change(s) favored via replication asynchrony that would promote the initial, random monoallelic choice (Mostoslavsky et al., 2001). Although still hypothetical at this point, we will describe in the following section a mathematical model—parameterized to actual biological results—which strongly supports this original view.

5. MATHEMATICAL MODELING OF THE REGULATION OF ALLELIC EXCLUSION

5.1 General Considerations

5.1.1 Mathematical Modeling in Biology: Why, How, and When

Regularly used in fundamental and applied physics, mathematical modeling has long been proved a powerful tool and has formed an integral role in most breakthroughs made in these fields. During the past three decades, it has become increasingly clear that the complementary application of methodologies from especially statistics and dynamic modeling can likewise contribute to a better description and understanding of intricate interactions and resulting behaviors in cell biology (De Jong, 2002; Servedio et al., 2014). Recent dual advances made on the one hand in biological research and on the other in computer technology have created an invigorated level of enthusiasm. Thus, the development of increasingly inclusive experimental approaches has prompted a switch from traditional research focusing on isolated biochemical processes, toward a more systemic exploration at the whole cell level, aiming toward genetic, molecular, functional, and temporal completeness (Karr, Takahashi, & Funahashi, 2015; Kitano, 2002). In parallel, relentless progress in computing power now enables dynamical investigation of key questions and paradigms in biological systems, the structural and behavioral complexities of which may far exceed those of physical systems (De Jong, 2006; May, 1976). A number of mathematical tools, inherited from physics and, notably, the deterministic chaos theory and investigations of nonlinear dynamical systems (Ventsel, 1977), are available (e.g., including differential equations, discrete/continuous-time models, Markov models, and Boolean and Bayesian networks). While some, if not all, may be more specifically suited to address one particular issue (e.g., Andrew, Baker, & Bocharov, 2007), several options are usually possible, which can be selected on a case-by-case basis (depending on the anticipated evolutions of the model, modeler's preferences, etc.). As it relates to modeling objectives, one can roughly distinguish three broad categories as follows. Predictive modeling generally is adapted to systems for which the most critical mechanisms have been determined with high levels of confidence, such that the model is chosen to provide the probability of an outcome given a set amount of input data (e.g., industrial prototype modeling and weather forecasting). Integrative modeling aims to collect all available knowledge on a given system, including not only the system's components but also their recognized functional partners. Usually, the resulting models are mostly descriptive, as they comprise high numbers of generally poorly calibrated parameters. Finally, explanatory modeling aims toward a better comprehension of cellular and/or molecular processes for which knowledge is only partial. Of special interest here, their specificities and conditions for relevance and accuracy are assessed below.

5.1.2 Specificity and Suitability of Exploratory Models

Explanatory models are those preferentially used in fundamental research, including biological researches. Following proper calibration and through analytical results or numerical simulations, they first permit an evaluation to be made on the applicability of specified assumptions. Ideally, and provided that their results and predictions are overall compatible with available experimental data for a given set(s) of parameter values, they also help pinpoint informative experiments to validate ostensibly relevant postulates. Otherwise, if analytical results and simulations do not fit the biological data regardless of parameters' tunings, they allow a deduction that the particular hypothesis is wrong or incomplete. Importantly, while also displaying predictive value for the tested assumptions, explanatory models always integrate prior knowledge and established features of the explored system though not necessarily aspiring to reproduce *in silico* its finest details. In practice, explanatory models should preferably remain as simple as possible, gathering hypotheses and mechanisms of the given system that appear essential and ignoring those that seem ancillary. Thus, the lower the degree of freedom of—in other words the number of free parameters in—the modeled system, the higher the eventual significance of the modeling views. Indeed, the model appears most suitable to reveal crucial governing features when this entity is at its lowest while allowing a satisfactory description of the modeled system. Conversely, models endowed with a high degree of freedom—particularly if the number of free parameters is largely superior to that of reference data used for their calibration (degenerated models)—generally permit fitting of the system's behaviors but only at a low confidence level with regard to inferred conclusions and their meaning to the explored subject (Ellner & Guckenheimer, 2011).

Additional complementary features to be considered relate to the fact that the modeled processes inevitably share a dynamic—if not noisy/stochastic—dimension (not always taken into account when considering biological data sets, primarily due to technical limitations). Therefore, modeling approaches need to pay attention to the timing properties particular to the investigated system, which may be impacted by phenomena acting on very different timescales. In this context and for the sake of straightforwardness in keeping the model as informative as possible regarding the main investigated issue, it may be useful—at least initially on a first modeling attempt—to leave out any phase that would potentially occur during a much shorter or longer period of time relative to the characteristic duration (named "characteristic time" by modelers) of the whole system. In this sense, theoretical explanatory modeling generally applies a research strategy much alike that often used by experimentalists, considering initially the system of interest from a simplified angle in order to infer more general assumption(s) which could then be refined and explored in further detail thereafter.

5.1.3 Why Mathematical Modeling of Allelic Exclusion?

As reviewed above, V(D)J recombination and especially allelic exclusion offer beautiful examples of intricate and puzzling phenomena, for which controversial issues continue to rise concerning the potential/putative multilayer controls that are imposed throughout the timeframe of lymphoid cell differentiation. Obvious dissimilarities in geno-/phenotypic allelic exclusion outcomes between the individual Ig/TCR genes raise the question of whether their regulations significantly differ depending on the locus considered. Yet, a common framework—hence potentially basic underlying mechanism(s)—is regularly recognized whereby the individual alleles can be activated separately for expression and recombination, a feature that appears to be conserved from the earliest vertebrates with an immune system based on V(D)J rearrangement (e.g., sharks; Malecek et al., 2008; Zhu et al., 2011). Commonly held views to explain this unique property—whether based on highly stochastic or deterministic/instructive settings (Chaumeil & Skok, 2013; Krangel, 2009; Levin-Klein & Bergman, 2014; Vettermann & Schlissel, 2010)—assume a "one allele at a time" option of rearrangement (relying in the former view on a notion of "low efficiency" of the recombination process such that simultaneous rearrangements are improbable), with the nonrearranging allele meanwhile maintaining a suppressive structure. Taken at face value, one main caveat with these views is that neither of them account for the frequent situation of biallelic complete V–DJ rearrangement (≥40% of differentiated B/T cells), which would imply all of a sudden (re)activation of the second allele if the prime V–(D)J recombination attempt failed to yield a functional CJ product. Besides, the notion of one entire allele (the late replicating allele) being instructed to keep away from recombination (Farago et al., 2012) is also difficult to understand and reconcile with the course of recombination at tripart IgH and TCRβ loci mostly displaying biallelic DJ rearrangements (also see Alves-Pereira et al., 2014). Additional mechanisms, proposed by both highly stochastic and deterministic/instructive models to circumvent these limitations, would involve cells that make a nonproductive rearrangement persisting in the corresponding developmental stages until they acquire, presumably through the epigenetic-activating events, the ability to rearrange the second allele (Fig. 2). However, such [V(D)J −/GL or DJ] cells have never been observed so far, thus raising doubt on their survival for long periods of time. Experimentally, only in situations where mutant mice displaying a defect in V(D)J recombination are bred with mice expressing an antiapoptotic Bcl-2 transgene has cell death been prevented (e.g., Mathieu et al., 2003). More recently, other elaborate mechanisms and supplementary controls ensuing recombination at a first rearranging allele were proposed to separately contribute to a suppressive effect targeting the opposite homologue that would thus enforce, possibly in a reversible manner, the interallelic asynchrony and/or the maintenance of allelic exclusion. These may involve RAG1-mediated allele pairing, followed by RAG1/2-mediated initial dual RSS cleavage at one selected allele and the targeting of the remaining uncleaved allele to heterochromatin; and/or the inhibition of *rag1* and *rag2* gene expression through the activation of the ATM signaling pathway (Chaumeil et al., 2013; Hewitt et al., 2009; Steinel et al., 2014, 2013; reviewed in Brady, Steinel, et al., 2010; Chaumeil & Skok, 2013); and/or the recognition of productive versus nonproductive gene rearrangements through differences in the stability of ensuing-mRNA sensed via the nonsense-mediated mRNA decay pathway, resulting in the induction of a trans-dominant suppressive effect

(Frischmeyer-Guerrerio et al., 2011; Lutz et al., 2011). Without disputing the reality of such mechanisms, we note that the inactivation of the corresponding pathways or processes at best results in limited effects on the rate of allelic inclusion; and also that their character of being functionally activated following the onset of a first rearrangement would predispose them to playing a more operative role in the maintenance phase rather than in the initiation phase of allele asynchrony. It is also worth stressing that none of these recently proposed mechanisms truly account, any more than the former highly stochastic or deterministic/instructive options, for the regular occurrence, aside the two prominent [V(D)J+/GL or DJ] and [VDJ+/VDJ−] genotypic profiles, of small but significant fractions of allelically included [VDJ+/VDJ+] cells—unless the occasional defect or lessening of one of these various mechanisms is assumed. In such a clouded landscape, mathematical approaches and especially—we assumed—those based on dynamical systems modeling may help obtain a clearer and more objective view on this whole conundrum. In the remaining part of this review article, we will first review the few probabilistic models that have already addressed the issue of allelic exclusion in the past. We will then provide our current views in light of recent mathematical modeling analysis of the TCRβ gene recombination process based on probabilistic transitions. Finally, we will describe our most recent model developed to test the assumption that noisy kinetics in the initiation of primary Dβ-to-Jβ rearrangements promote and significantly contribute toward enforcing TCRβ allelic exclusion, and explain how we developed this rather unorthodox hypothesis. Importantly, in our latter studies, the [VDJ+/VDJ+] allelically included outliers were considered rare, yet authentic outcomes of the recombination developmental processes, i.e., distribution tails of a dynamic system. In particular, we assumed that such systems directly rely on the importance of epigenetics and associated "noisy" behaviors in mammalian cells (Eldar & Elowitz, 2010; Pujadas & Feinberg, 2012).

5.2 Probabilistic Models of Allelic Exclusion

5.2.1 Earliest and Pure Probabilistic Models

The two earliest types of pure probabilistic models of allelic exclusion were proposed at the beginning of the 1980s: both used Markovian formalisms involving no (Coleclough, 1983; Coleclough et al., 1981) or limited dynamical features (Claverie & Langman, 1984). Also, both preferentially addressed allelic and isotypic exclusions in the context of IgLκ and IgLλ gene rearrangements; and—at a lesser extent—allelic exclusion in the context of IgH gene rearrangement, based on better knowledge at this time of the former two loci and corresponding pre-B cell developmental stage. They featured either a single probabilistic parameter for each allele/locus (i.e., probability p of yielding a functional rearrangement; thus three parameters in total (Coleclough et al., 1981; Coleclough, 1983)) or two types of parameter [i.e., probabilities of (i) initiating rearrangement and (ii) yielding functional rearrangements, thus four to six parameters in total depending on whether the model considered IgLκ/λ loci alone or in conjunction with the IgH locus (Claverie & Langman, 1984)]. Furthermore, the latter model tested two possible scenarios of ending the developmental process: (i) either gene rearrangements at a given locus stop when a productive one has occurred (i.e., there is a feedback stop signal) or (ii) gene rearrangements only stop when no more GL alleles remain (i.e., there is no feedback stop signal). Comprehensibly, both types of model suffered from the lack of existing knowledge on V(D)J recombination [for instance, the Coleclough model assumed "a rearrangement process so inefficient that the probability of achieving productive rearrangements within both alleles in one cell might be very small"; i.e., proposed $p \approx 0.1$ (Coleclough, 1983; Coleclough et al., 1981)]. Nonetheless, despite rudimentary knowledge and tools at the time, each study arrived at some appealing predictions regarding some key features of the system. These included predictions of a feedback loop at the end of the recombination series, of a competing course between the IgLκ and IgLλ loci that most frequently turns to the advantage of the former locus, and of the conceivable existence of a unique recombination machinery in both B and T cells [then deduced from existing data on partial D–J_H rearrangements in a few T cells (Coleclough, 1983)]. These earlier models also predicted the existence of developing systems that associate probabilistic rearrangements in a locus autonomous manner with a functionality on the order of 1/3, as well as feedback stop signaling (Claverie & Langman, 1984; we note that due to their lack of knowledge at this time of sequential, ordered rearrangements at the IgH locus and frequent outcome of the [V_HDJ_H+/DJ_H] genotype, this feedback signaling was assumed by these authors to involve the final IgH/L complete BCR only). In summary, while data from these studies were preliminary, it is important to recognize that they provided valuable research foundations when both V(D)J recombination and allelic exclusion were mostly in their infancy. Notably, both earlier models envisaged a dynamic system comprised of an initiation phase based on probabilistic features and a termination phase based on feedback stop signaling.

5.2.2 Probabilistic, High-Parameterized Models

For this category of models, we will assess a series of papers published in the late 1990s to the early 2000s, by the groups of Mehr (Kalmanovich & Mehr, 2003; Mehr, Shannon, & Litwin, 1999; Piper, Litwin, & Mehr, 1999) and Weigert (Louzoun, Friedman, Luning Prak, Litwin, & Weigert, 2002) to better understand the properties that may lead to differences regarding dual allele behaviors at the allelically excluded IgLκ gene locus (Kalmanovich & Mehr, 2003; Louzoun et al., 2002; Mehr et al., 1999; also considering the IgH locus in the latter) versus the allelically included TCRα gene locus (Piper et al., 1999). In all cases, Markov formalism was again used (we note that the model by Louzoun et al., 2002 was derived from the one by Mehr et al., 1999). The novelty here was that concurrent rearrangement processes in T and B cell precursors were modeled using Monte-Carlo simulations of the fate of a finite number of cells. Most models of rearrangement of Ig or TCR genes by Mehr and colleagues assumed that lymphocyte precursors would try all rearrangement possibilities before they matured or died (Kalmanovich & Mehr, 2003; Mehr et al., 1999; Piper et al., 1999). Contrarily, Louzoun et al. (2002) was the first to introduce a "rigid time window," assuming that B cell precursors would only make a limited number of rearrangement attempts. In all cases, these authors also aimed to integrate in several of their models concurrent features regarding B- or T-cell differentiation in addition to rearrangement features in the true sense (cell death and/or selection and/or cell cycle progression). The resulting attractive real-like and comprehensive models, nevertheless, each had an Achilles' heel due to them comprising high numbers of parameters (nearly 20 parameters in Mehr et al., 1999; Piper et al., 1999; 10 in Louzoun et al., 2002; 14 in Kalmanovich & Mehr, 2003). As stressed above, the strength of a model is inversely related to parameter number, with slight fluctuations in parameter values more likely to yield discordant results, thus reducing robustness. Overall, such models would *in fine* support (or reject) a proposed hypothesis on the condition that all or a majority of steady-state parameter values would fit the reality, but the level of confidence to approaching such a precondition may be low. Though using related models, opposite conclusions were drawn with respect to the preference of either targeting secondary IgLκ rearrangements on the allele that productively rearranged first (Louzoun et al., 2002; Mehr et al., 1999) or choosing at random the location of the next rearrangement (Kalmanovich & Mehr, 2003), whereas consistent conclusions were reached regarding preferential early usage of most 5′ Jκ1/Jκ2 gene segments in comparison to downstream Jκ3/Jκ4 gene segments

(Kalmanovich & Mehr, 2003; Louzoun et al., 2002; Mehr et al., 1999). These models also predicted that productive rearrangement and subsequent cell surface expression of one allele of the IgH gene probably lead to the prevention of rearrangement completion on the other allele, and that additional mechanisms, such as peripheral selection disfavoring cells with two productively IgH genes, may also play a role (Kalmanovich & Mehr, 2003); they also separately supported the proposition that differences in cell proliferation can explain at least in part the balance between α/β versus γ/δ T cells (Piper et al., 1999).

5.2.3 Probabilistic and Dynamic Continuous-Time Markov Models

Sepulveda, Boucontet, Pereira, and Carneiro (2005) proposed a stochastic modeling of TCRγ gene rearrangement using data sets from a significant number of γδT-cell hybridomas expressing either Vγ4Jγ1 or Vγ1Jγ4 rearranged loci. Parallel analysis of these hybridomas showed that assembly of TCRγ V-region genes displays properties of allelic exclusion and, mostly, of isotypic inclusion at the genotypic level (Boucontet, Sepulveda, Carneiro, & Pereira, 2005). In these contexts, the specific goal was to better comprehend the mechanisms that may control termination of V–J recombination at TCRγ loci. Briefly, using continuous-time Markov modeling as the basis for a construction, three possible scenarios were tested, namely (i) probabilistic rearrangements within a recombination window (termed the TWM model); (ii) probabilistic rearrangements with feedback (PFM model); and (iii) probabilistic rearrangements with feedback and allele asynchrony from the initiation stage onward (FACAM model). Analytical calculations strongly favored the later FACAM model over the PFM model and, based on the strikingly different from expected distributions yielded by the TWM model, enabled the confident retaining of the former model and rejection of the TWM model, respectively. In the FACAM model, a total of four parameters were considered, in addition to a recombination window (which without loss of generality was set to 1). One interesting feature in the FACAM model is the finding that, within the pool of precursors that have remodeled the chromatin to make γ loci accessible, almost 60% cells have both alleles accessible and actively rearranging. Remarkably, this pattern remains compatible with Vγ–Jγ allelic/iostypic exclusion, thus providing evidence that V(D)J recombination outcomes can be reached without strictly calling for the "one allele at the time" traditional option, in line with modern views on probabilistic features (Ventsel, 1977). The data also suggest that a combination of stochastic and deterministic features alone might be sufficient to provide a solid basis for the allelic exclusion phenomena, a hypothesis which we have further strengthened and extended to the TCRβ locus, as described below.

Farcot et al. (2010): Building on the work by Sepulveda et al. (2005) and using TCRβ allelic exclusion as a well-documented reference system, we assembled an exploratory model to test whether time-related combinations of three separate concepts—allele independency, regulated feedback, and probabilistic behaviors—account for the distribution of TCRβ genotypes emerging from early T-cell development. We mainly wished to link these basic concepts in a sort of rally (oriented toward a productive VDJ joint) with stochastic transitions impinging on molecular developments throughout the whole process, including the feedback closing step; we also aimed to eventually integrate individual allele behaviors at the level of a whole population of differentiating DN T cells, as modeled using the continuous-time Markov chain formalism (Fig. 3A). In this system, deterministic features would simply include the basic attribute of randomness in V(D)J recombination (1/3 productive–2/3 nonproductive rearrangement outcomes) and, with regard to TCRβ gene recombination, the sequential Dβ–Jβ and Vβ–DJβ rearrangements with no direct Vβ–Jβ joining (B12/23 constraint), followed by and ending with feedback inhibition. All conditions were comprised within an invariant recombination window common to every single cell (defined by the deterministic parameter t_s in Fig. 3A; also set to 1 as in the above model). In this window, molecular changeovers are characterized by transition rates that are expressed in terms of the mean duration of a given rearrangement step and the average time lapse to achieve inhibition, respectively; hence, a model system depending on three probabilistic parameters (t_x, τ, and ρ in Fig. 3A). Stochasticity relies on temporal variations (based on Poisson distributions) in the execution of these changes, both between homologous alleles and individual cells (three bottom right curves in Fig. 3A).

Following calibration of parameter values using analytical calculations based on available data sets, this model system—when integrated at the level of a whole cell population—makes it possible to readily predict the distinct TCRβ cell genotypes and allelic exclusion/inclusion profiles from wild-type and mutant mice, both in developing DN thymocytes and in differentiated T cells (Farcot et al., 2010). While keeping allele cross talk to a minimum, the model thus comprehensively satisfies the allelic exclusion landmark, suggesting that its basic precepts capture the nature of the underlying controls at the TCRβ locus. Estimates of parameter rates (as calculated from the

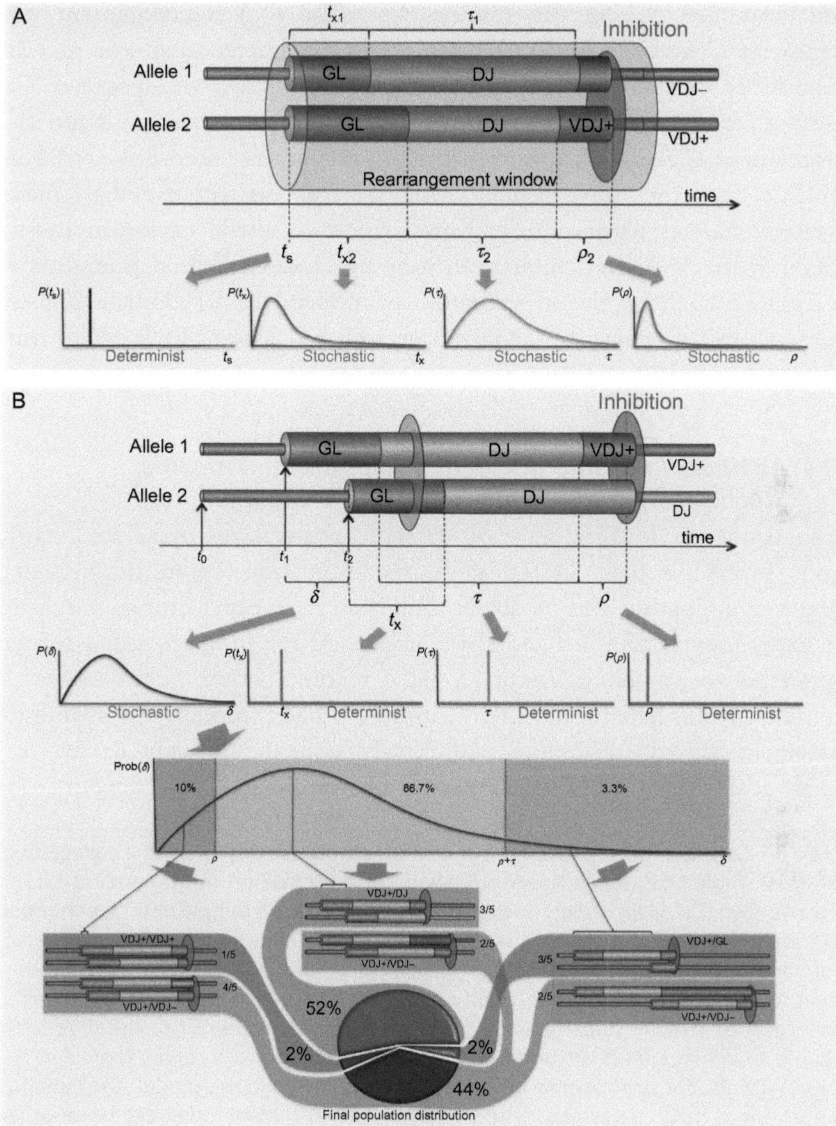

Figure 3 Dynamic combination of deterministic and stochastic features enforces allelic exclusion at V–D–J containing gene loci. (A) Schematic depiction of stochastic modeling of DNA rearrangement at V–D–J containing antigen receptor alleles based on the Markov formalism, as described by Farcot et al. (2010). The recombination window (transparent surrounding cylinder) starting at time t_s defines the time period along developmental maturation of a single lymphoid cell during which all deterministic requirements for sequential rearrangements could be met (hence during which recombination of the given locus could occur). The blue, green, and red/gray sections

(Continued)

model) implied that in this setting, the initial D–J rearrangement step (parameter t_x) and the feedback closing step (parameter ρ) as well may be substantially shorter than the intervening V–DJ joining step (parameter τ) (also see Fig. 3A), consistent with the current notion of the initiation and enforcement of allelic exclusion mainly involving this latter step. One bottom line from the above results would be that—as anticipated by many investigators—regulatory mechanisms critical to allelic exclusion mostly pertain to the V–DJ recombination step, i.e., hidden behind parameter τ, for instance enabling the structuration of monoallelic Vβ domain higher-order looping in contact to the inner part of RAG-loaded TCRβ loci comprising the DJβ intermediates (but also see below).

5.2.4 Additional Considerations and Hypothesis: Assessing Stochasticity in the Initiation of Allelic Exclusion

Though this previous model was met with approval, we ourselves initially found surprising that TCRβ gene recombination could be modeled using recombination and feedback inhibition rates only, fitting straight to the experimental data at hand. By comparison for instance, using related approaches to modeling Vγ4Jγ1/Vγ1Jγ4 recombination, Sepulveda et al. eventually implemented their initial PFM model (probabilistic rearrangements + feedback) to additionally simulate differential chromosomal accessibility (FACAM model), in order to best account for

Figure 3—Cont'd represent the time course ofmonoallelic behaviors in a collection of individual single cells, and respectively show (i) the transition from germline (GL) to DJ-rearranged (DJ) allelic status, (ii) the transition from DJ-rearranged to VDJ-rearranged (VDJ) allelic status, and (iii) the residence of productively rearranged [VDJ+] alleles (red) or of nonproductively rearranged [VDJ −] alleles (gray) until the recombination window comes to an end. Transition times t_x and τ correspond to the time lapses required (in the given case) to achieve steps (i) and (ii), respectively, ρ corresponds to the time lapse required to achieve feedback inhibition (red disc). The bottom curves represent the types of distributions/probability laws of t_s, t_x, τ, and ρ used in this model. The initiation time (t_s) is a fully deterministic parameter constant from one cell to another. Rearrangement times (t_x and τ) as well as the inhibitory time lapse ρ are stochastic parameters (taking values according to their probability law in each individual cell). (B) Alternative model in which stochasticity exclusively concentrates over the very first initiating stage of allele activation. Times t_1 and t_2 define stochastic initiation timing of recombinational accessibility at each allele, leading to the definition of a probabilistic time gap δ specific to each individual cell; all remaining parameters are deterministic. The implications of this alternate model regarding genotype distribution (which depends on the relative values of δ, ρ, and $\rho+\tau$) within a developing lymphoid cell population are depicted on the bottom part. (See the color plate.)

experimental findings at these loci (Boucontet et al., 2005; Sepulveda et al., 2005). Incidentally, it was noticed that higher estimates of the Vγ–Jγ rearrangement rates were obtained with the FACAM model in comparison to those with the PFM model, highlighting possible parameter interconnection in such dynamical systems. Differences in modeling parameterization may reflect supplementary biological constraints specific to the particular loci, including the allelic/isotypic exclusion constraint at the TCRγ1/γ4 loci and the sequential D–J/V–DJ dual rearrangement constraint at the TCRβ locus. Regarding the latter restriction, however, it is noteworthy that as long as the key issue rests on a mechanism capable of generating a time lag in the assembly of homologous alleles, a two-step recombination process should mechanically allow this outcome not only in reality at the given locus (possibly) but also in mathematical simulation (definitely).

As reviewed above, there is compelling evidence that activating modulations of the chromatin template at the TCRβ locus first initiate within the Dβ–Jβ clusters, and set the stage for TCRβ gene expression and likely prime Dβ-to-Jβ recombination. These well-established biological properties, coupled to the above modeling considerations, prompted us to reconsider the premise that TCRβ allele asynchrony might be primarily introduced from the D-to-J recombination initiating phase onward (and hence also highlight that our continuous-time Markov modeling might have overlooked this particular phase). To evaluate this alternative hypothesis, we developed an original model based on a distinct formalism (discrete time modeling) and which (i) can dispense with the artificial constraint of a recombination window that interrupts the developmental process before eventual equilibrium (a factor introducing potential instability to the model system); and (ii) enable the integration of the system's cellular inflows/outflows, i.e., arrival of TCRβ-unrearranged precursor cells in and departure of β-selectable derivatives from the DN thymic cell compartment (not feasible using Markov modeling; Jaeger et al., submitted for publication). The mathematical formulations in this new model comply with the same basic principles as before—allele independency, regulated feedback, and stochasticity—except that the latter exclusively concentrates over the very first initiating stage of allele activation (i.e., prior to the D–J rearrangement step) (Fig. 3B). Thus, stochastic asynchrony entailed between the two homologous alleles (interval δ in Fig. 3B) propagates along the whole developmental process until an asymptotic status is achieved. By tracing *in silico* all possible single-cell histories with respect to TCRβ allelic rearrangement outcomes and, from there, corresponding distributions on a developing

T-cell population level, we found that this model consistently accounted for the genotypic profiles typically associated with TCRβ allelic exclusion. Considering all rearranging cells in this system, 75% on average are predicted to undergo concurrent rearrangements of the two TCRβ alleles, far from the notion of a "low efficiency" of the recombination process. Furthermore, prime simulations implied that, still with the new settings, biallelic DJ/DJ prevalence peaks prior to the first V-to-DJ joining attempt. Quite remarkably for a single-parameter model, it also enables a satisfactory simulation of a variety of published data involving the TCRβ locus from wild-type or engineered mutant mice (i.e., as reported in Brady, Oropallo, et al., 2010; Carpenter et al., 2009; Chan et al., 2013; Jaeger et al., submitted for publication).

Many theoretical models have been proposed over the years to explain the connection between V(D)J recombination and allelic exclusion. Our own mathematical models consider how the developmental dynamics of individual lymphoid cells combine to sustain allelic exclusion. Increasing recognition is given to cell fate decisions and underlying genetic circuits being subjected to stochastic fluctuations to generate nongenetic cellular diversity. At the single-cell level, stochasticity relies on temporal variations in the execution of these changes (in other words, noise), both between cells (extrinsic noise) and alleles (intrinsic noise), with epigenetics likely occupying the front stage. Based on these concepts, our latest modeling approach—when integrated at the level of a whole cell population—makes it possible to readily predict the distinct TCRβ cell genotypes and allelic exclusion/inclusion profiles from wild-type and mutant mice (including notably those at KI TCRβ alleles carrying either a rearranged DJ or a rearranged VDJ gene segment; S. Jaeger & P. Ferrier, unpublished data). In keeping with the same concepts, we anticipate that corresponding features at other Ig/TCR gene loci could likewise be interpreted. Indeed, based on this modeling view, we speculate that varying epigenetics in allelic activation may account for dissimilarities in allelic exclusion/inclusion outcomes between the individual Ig/TCR genes. Assuming that recombination centers represent the first genomic areas within the Ig/TCR loci to experience higher-order allelic looping (e.g., Oestreich et al., 2006), we further speculate that any modification of the epigenetic properties in these primary topological domains can impinge on these outcomes. With this consideration in mind, it may prove interesting to further explore a series of studies reporting disruption of allelic exclusion following engineered mutations which may have prematurely connected one or a few V gene segments (and associated *cis*-regulatory ACEs, including V promoters) with such primary domains

(Bates, Cado, Nolla, & Schlissel, 2007; Guo, Yoon, et al., 2011; Jackson et al., 2005; Lin et al., 2015; Sieh & Chen, 2001). This may have been case in a recent study reporting that deletion of the 3′ regulatory region (3′RR) from the IgH locus resulted in an accumulation not only of V_HDJ_H recombination final products but also, intriguingly, of DJ_H intermediates (Braika et al., 2015).

6. CONCLUSION AND PERSPECTIVES

In conclusion, modeling studies have highlighted the possibility that within a dynamical system combining allele independence, inhibitory feedback, and stochasticity, the impact of this latter element on the very initiating phase of TCRβ locus activation significantly contributes to the enforcement of allelic exclusion at this locus. This apparent iconoclastic inference is nevertheless strongly supported by mathematical analysis, acting here as a proof-of-concept test. It complies with observed features and reported genotypic distributions in wild-type and mutant mice and, presumably, does not require additional sophisticated controls aside those already recognized, with potential economical value at the cellular level in terms of molecular costs. We note that recent, previously unanticipated data implying (i) that both steps of TCRβ recombination, D–J and V–D–J, are likely to occur asynchronously on TCRβ alleles (Chan et al., 2013); and (ii) that interactions between DJ elements and remote V gene segments occur rapidly within merged chromatin domains (Lucas et al., 2014), appear to be pointing in the same direction (alike this latest modeling view, they suggest longer and shorter timeframes of the corresponding rearrangement events). This unorthodox option also sheds new light on (D)–J rooted recombination centers, associated regulatory elements (e.g., Eβ and pDβ elements), and flanking boundary/insulator sequences (e.g., TCRβ 5′PC), in respect to the organization/orchestration of initial allele asynchrony. In brief, such variety of potential implications offers new perspectives and research directions to continue exploring the mystery of allelic exclusion.

ACKNOWLEDGMENTS

Work in the PF laboratory is supported by institutional grants from "Institut National de la Santé et de la Recherche Médicale" (Inserm) and "Centre National de la Recherche Scientifique" (CNRS), and by dedicated grants from the Commission of the European Communities, the "Agence Nationale de la Recherche" (ANR), the "Institut National du Cancer" (INCa), the "ITMO Cancer—Alliance Nationale pour les Sciences de la Vie et de la Santé" (AVIESAN), and the "Fondation Princesse Grace de la Principauté de Monaco." We apologize to those colleagues whose work we could not acknowledge owing to space restrictions.

Disclosures: The authors have no financial conflicts of interests.

REFERENCES

Abarrategui, I., & Krangel, M. S. (2006). Regulation of T cell receptor-a gene recombination by transcription. *Nature Immunology*, *7*, 1109–1115.

Abarrategui, I., & Krangel, M. S. (2007). Noncoding transcription controls downstream promoters to regulate T-cell receptor a recombination. *The EMBO Journal*, *26*, 4380–4390.

Afshar, R., Pierce, S., Bolland, D. J., Corcoran, A., & Oltz, E. M. (2006). Regulation of IgH gene assembly: Role of the intronic enhancer and 5′DQ52 region in targeting DHJH recombination. *Journal of Immunology*, *176*, 2439–2447.

Agata, Y., Tamaki, N., Sakamoto, S., Ikawa, T., Masuda, K., Kawamoto, H., et al. (2007). Regulation of T cell receptor beta gene rearrangements and allelic exclusion by the helix-loop-helix protein, E47. *Immunity*, *27*, 871–884.

Alam, S. M., & Gascoigne, N. R. (1998). Posttranslational regulation of TCR Valpha allelic exclusion during T cell differentiation. *Journal of Immunology*, *160*, 3883–3890.

Alonso, A., Hozumi, N., & Murialdo, H. (1985). Mouse immunoglobulin gene rearrangements. The sequence of a nonexpressed lambda 3-chain gene. *Journal of Immunology*, *135*, 614–619.

Alt, F. W., & Baltimore, D. (1982). Joining of immunoglobulin heavy chain gene segments: Implications from a chromosome with evidence of three D-JH fusions. *Proceedings of the National Academy of Sciences of the United States of America*, *79*, 4118–4122.

Alt, F. W., Blackwell, T. K., & Yancopoulos, G. D. (1987). Development of the primary antibody repertoire. *Science*, *238*, 1079–1087.

Alt, F. W., Bothwell, A. L., Knapp, M., Siden, E., Mather, E., Koshland, M., et al. (1980). Synthesis of secreted and membrane-bound immunoglobulin mu heavy chains is directed by mRNAs that differ at their 3′ ends. *Cell*, *20*, 293–301.

Alt, F. W., Enea, V., Bothwell, A. L., & Baltimore, D. (1980). Activity of multiple light chain genes in murine myeloma cells producing a single, functional light chain. *Cell*, *21*, 1–12.

Alt, F. W., Oltz, E. M., Young, F., Gorman, J., Taccioli, G., & Chen, J. (1992). VDJ recombination. *Immunology Today*, *13*, 306–314.

Alt, F., Rosenberg, N., Lewis, S., Thomas, E., & Baltimore, D. (1981). Organization and reorganization of immunoglobulin genes in A-MULV-transformed cells: Rearrangement of heavy but not light chain genes. *Cell*, *27*, 381–390.

Alt, F. W., Yancopoulos, G. D., Blackwell, T. K., Wood, C., Thomas, E., Boss, M., et al. (1984). Ordered rearrangement of immunoglobulin heavy chain variable region segments. *The EMBO Journal*, *3*, 1209–1219.

Alves-Pereira, C. F., De Freitas, R., Lopes, T., Gardner, R., Marta, F., Vieira, P., et al. (2014). Independent recruitment of Igh alleles in V(D)J recombination. *Nature Communications*, *5*, 5623.

Andersson, R. (2015). Promoter or enhancer, what's the difference? Deconstruction of established distinctions and presentation of a unifying model. *Bioessays*, *37*, 314–323.

Andrew, S. M., Baker, C. T. H., & Bocharov, G. A. (2007). Rival approaches to mathematical modelling in immunology. *Journal of Computational and Applied Mathematics*, *205*, 669–686.

Arakawa, H., Shimizu, T., & Takeda, S. (1996). Re-evaluation of the probabilities for productive arrangements on the kappa and lambda loci. *International Immunology*, *8*, 91–99.

Aschauer, C., Outters, P., & Ferrier, P. (2015). Organization and rearrangement of TCR loci. In M. Ratcliffe (Ed.), *Encyclopedia immunobiology* (1st ed.). Langford Lane, Kidlington, Oxford, United Kingdom: Elsevier. in press.

Bank, E. M., & Gruenbaum, Y. (2011). The nuclear lamina and heterochromatin: A complex relationship. *Biochemical Society Transactions*, *39*, 1705–1709.

Barreto, V., & Cumano, A. (2000). Frequency and characterization of phenotypic Ig heavy chain allelically included IgM-expressing B cells in mice. *Journal of Immunology*, *164*, 893–899.

Bartolomei, M. S., & Ferguson-Smith, A. C. (2011). Mammalian genomic imprinting. *Cold Spring Harbor Perspectives in Biology*. *3*. http://dx.doi.org/10.1101/cshperspect.a002592 pii: a002592. PMID: 21576252.

Bassing, C. H., Alt, F. W., Hughes, M. M., D'auteuil, M., Wehrly, T. D., Woodman, B. B., et al. (2000). Recombination signal sequences restrict chromosomal V(D)J recombination beyond the 12/23 rule. *Nature*, *405*, 583–586.

Bates, J. G., Cado, D., Nolla, H., & Schlissel, M. S. (2007). Chromosomal position of a VH gene segment determines its activation and inactivation as a substrate for V(D)J recombination. *The Journal of Experimental Medicine*, *204*, 3247–3256.

Baumann, M., Mamais, A., Mcblane, F., Xiao, H., & Boyes, J. (2003). Regulation of V(D)J recombination by nucleosome positioning at recombination signal sequences. *The EMBO Journal*, *22*, 5197–5207.

Beck-Engeser, G., Jack, H. M., & Wabl, M. (1987). Allelic inclusion in a pre-B-cell line that generates immunoglobulin heavy chain genes *in vitro*. *Proceedings of the National Academy of Sciences of the United States of America*, *84*, 1060–1064.

Bergman, Y. (1999). Allelic exclusion in B and T lymphopoiesis. *Seminars in Immunology*, *11*, 319–328.

Bergman, Y., & Cedar, H. (2004). A stepwise epigenetic process controls immunoglobulin allelic exclusion. *Nature Reviews. Immunology*, *4*, 753–761.

Bevington, S., & Boyes, J. (2013). Transcription-coupled eviction of histones H2A/H2B governs V(D)J recombination. *The EMBO Journal*, *32*, 1381–1392.

Bolland, D. J., Wood, A. L., Afshar, R., Featherstone, K., Oltz, E. M., & Corcoran, A. E. (2007). Antisense intergenic transcription precedes Igh D-to-J recombination and is controlled by the intronic enhancer Emu. *Molecular and Cellular Biology*, *27*, 5523–5533.

Bolland, D. J., Wood, A. L., & Corcoran, A. E. (2009). Large-scale chromatin remodeling at the immunoglobulin heavy chain locus: A paradigm for multigene regulation. *Advances in Experimental Medicine and Biology*, *650*, 59–72.

Bolland, D. J., Wood, A. L., Johnston, C. M., Bunting, S. F., Morgan, G., Chakalova, L., et al. (2004). Antisense intergenic transcription in V(D)J recombination. *Nature Immunology*, *5*, 630–637.

Bonnet, M., Ferrier, P., & Spicuglia, S. (2009). Molecular genetics at the T-cell receptor b locus: Insights into the regulation of V(D)J recombination. *Advances in Experimental Medicine and Biology*, *650*, 116–132.

Bonnet, M., Huang, F., Benoukraf, T., Cabaud, O., Verthuy, C., Boucher, A., et al. (2009). Duality of enhancer functioning mode revealed in a reduced TCR beta gene enhancer knockin mouse model. *Journal of Immunology*, *183*, 7939–7948.

Born, W., White, J., Kappler, J., & Marrack, P. (1988). Rearrangement of IgH genes in normal thymocyte development. *Journal of Immunology*, *140*, 3228–3232.

Bossen, C., Mansson, R., & Murre, C. (2012). Chromatin topology and the regulation of antigen receptor assembly. *Annual Review of Immunology*, *30*, 337–356.

Bossen, C., Murre, C. S., Chang, A. N., Mansson, R., Rodewald, H.-R., & Murre, C. (2015). The chromatin remodeler Brg1 activates enhancer repertoires to establish B cell identity and modulate cell growth. *Nature Immunology*, *16*, 775–784.

Boucontet, L., Sepulveda, N., Carneiro, J., & Pereira, P. (2005). Mechanisms controlling termination of V-J recombination at the TCRgamma locus: Implications for allelic and isotypic exclusion of TCRgamma chains. *Journal of Immunology, 174*, 3912–3919.

Brady, B. L., Oropallo, M. A., Yang-Iott, K. S., Serwold, T., Hochedlinger, K., Jaenisch, R., et al. (2010). Position-dependent silencing of germline Vß segments on TCRß alleles containing preassembled VßDJßCß1 genes. *Journal of Immunology, 185*, 3564–3573.

Brady, B. L., Steinel, N. C., & Bassing, C. H. (2010). Antigen receptor allelic exclusion: An update and reappraisal. *Journal of Immunology, 185*, 3801–3808.

Braika, F.-Z., Conte, C., Moutahir, M., Denizot, Y., Cogné, M., & Khamlichi, A. A. (2015). Developmental switch in the transcriptional activity of a long-range regulatory element. *Molecular and Cellular Biology, 35*, 3370–3380.

Brandt, V. L., Hewitt, S. L., & Skok, J. A. (2010). It takes two: Communication between homologous alleles preserves genomic stability during V(D)J recombination. *Nucleus, 1*, 23–29.

Brandt, V. L., & Roth, D. B. (2004). V(D)J recombination: How to tame a transposase. *Immunological Reviews, 200*, 249–260.

Brown, K. E., Baxter, J., Graf, D., Merkenschlager, M., & Fisher, A. G. (1999). Dynamic repositioning of genes in the nucleus of lymphocytes preparing for cell division. *Molecular Cell, 3*, 207–217.

Brown, K. E., Guest, S. S., Smale, S. T., Hahm, K., Merkenschlager, M., & Fisher, A. G. (1997). Association of transcriptionally silent genes with Ikaros complexes at centromeric heterochromatin. *Cell, 91*, 845–854.

Burggren, W. W. (2015). Dynamics of epigenetic phenomena: Intergenerational and intragenerational phenotype 'washout'. *The Journal of Experimental Biology, 218*, 80–87.

Burnet, F. M. (1959). *The clonal selection theory of acquired immunity*. Cambridge: Cambridge University Press.

Callebaut, I., & Mornon, J. P. (1998). The V(D)J recombination activating protein RAG2 consists of a six-bladed propeller and a PHD fingerlike domain, as revealed by sequence analysis. *Cellular and Molecular Life Sciences, 54*, 880–891.

Carabana, J., Watanabe, A., Hao, B., & Krangel, M. S. (2011). A barrier-type insulator forms a boundary between active and inactive chromatin at the murine TCRbeta locus. *Journal of Immunology, 186*, 3556–3562.

Carpenter, A. C., Yang-Iott, K. S., Chao, L. H., Nuskey, B., Whitlow, S., Alt, F. W., et al. (2009). Assembled DJ beta complexes influence TCR beta chain selection and peripheral V beta repertoire. *Journal of Immunology, 182*, 5586–5595.

Casanova, J. L., Romero, P., Widmann, C., Kourilsky, P., & Maryanski, J. L. (1991). T cell receptor genes in a series of class I major histocompatibility complex-restricted cytotoxic T lymphocyte clones specific for a Plasmodium berghei nonapeptide: Implications for T cell allelic exclusion and antigen-specific repertoire. *The Journal of Experimental Medicine, 174*, 1371–1383.

Cebra, J. J., Colberg, J. E., & Dray, S. (1966). Rabbit lymphoid cells differentiated with respect to alpha-, gamma-, and mu- heavy polypeptide chains and to allotypic markers Aa1 and Aa2. *The Journal of Experimental Medicine, 123*, 547–558.

Cedar, H., & Bergman, Y. (2008). Choreography of Ig allelic exclusion. *Current Opinion in Immunology, 20*, 308–317.

Chakraborty, T., Chowdhury, D., Keyes, A., Jani, A., Subrahmanyam, R., Ivanova, I., et al. (2007). Repeat organization and epigenetic regulation of the DH-Cmu domain of the immunoglobulin heavy-chain gene locus. *Molecular Cell, 27*, 842–850.

Chakraborty, T., Perlot, T., Subrahmanyam, R., Jani, A., Goff, P. H., Zhang, Y., et al. (2009). A 220-nucleotide deletion of the intronic enhancer reveals an epigenetic hierarchy in immunoglobulin heavy chain locus activation. *The Journal of Experimental Medicine, 206*, 1019–1027.

Chaligné, R., & Heard, E. (2014). X-chromosome inactivation in development and cancer. *FEBS Letters*, *588*, 2514–2522.

Chan, E. A., Teng, G., Corbett, E., Choudhury, K. R., Bassing, C. H., Schatz, D. G., et al. (2013). Peripheral subnuclear positioning suppresses Tcrb recombination and segregates Tcrb alleles from RAG2. *Proceedings of the National Academy of Sciences of the United States of America*, *110*, E4628–E4637.

Chaumeil, J., Micsinai, M., Ntziachristos, P., Roth, D. B., Aifantis, I., Kluger, Y., et al. (2013). The RAG2 C-terminus and ATM protect genome integrity by controlling antigen receptor gene cleavage. *Nature Communications*, *4*, 2231.

Chaumeil, J., & Skok, J. A. (2013). A new take on V(D)J recombination: Transcription driven nuclear and chromatin reorganization in Rag-mediated cleavage. *Frontiers in Immunology*, *4*(423), 1–8.

Chess, A. (2012). Mechanisms and consequences of widespread random monoallelic expression. *Nature Reviews Genetics*, *13*, 421–428.

Chowdhury, D., & Sen, R. (2001). Stepwise activation of the immunoglobulin mu heavy chain gene locus. *The EMBO Journal*, *20*, 6394–6403.

Chowdhury, D., & Sen, R. (2003). Transient IL-7/IL-7R signaling provides a mechanism for feedback inhibition of immunoglobulin heavy chain gene rearrangements. *Immunity*, *18*, 229–241.

Claverie, J. M., & Langman, R. (1984). Models for the rearrangements of immunoglobulin genes: A computer view. *Trends in Biochemical Sciences*, *9*, 293–296.

Cobb, R. M., Oestreich, K. J., Osipovich, O. A., & Oltz, E. M. (2006). Accessibility control of V(D)J recombination. *Advances in Immunology*, *91*, 45–109.

Cockerill, P. N., & Garrard, W. T. (1986). Chromosomal loop anchorage of the kappa immunoglobulin gene occurs next to the enhancer in a region containing topoisomerase II sites. *Cell*, *44*, 273–282.

Cockerill, P. N., Yuen, M. H., & Garrard, W. T. (1987). The enhancer of the immunoglobulin heavy chain locus is flanked by presumptive chromosomal loop anchorage elements. *The Journal of Biological Chemistry*, *262*, 5394–5397.

Cohn, M., & Langman, R. E. (1990). The protection: The unit of humoral immunity selected by evolution. *Immunological Reviews*, *115*, 11–147.

Coleclough, C. (1983). Chance, necessity and antibody gene dynamics. *Nature*, *303*, 23–26.

Coleclough, C. (1990). A critique of the Cohn-Langman protection theory. *Immunological Reviews*, *115*, 173–181. discussion 182-179-173.

Coleclough, C., Perry, R. P., Karjalainen, K., & Weigert, M. (1981). Aberrant rearrangements contribute significantly to the allelic exclusion of immunoglobulin gene expression. *Nature*, *290*, 372–378.

Couedel, C., Lippert, E., Bernardeau, K., Bonneville, M., & Davodeau, F. (2004). Allelic exclusion at the TCRd locus and commitment to gd lineage: Different modalities apply to distinct human gd subsets. *Journal of Immunology*, *172*, 5544–5552.

Cowell, L. G., Davila, M., Ramsden, D., & Kelsoe, G. (2004). Computational tools for understanding sequence variability in recombination signals. *Immunological Reviews*, *200*, 57–69.

Cowell, L. G., Davila, M., Yang, K., Kepler, T. B., & Kelsoe, G. (2003). Prospective estimation of recombination signal efficiency and identification of functional cryptic signals in the genome by statistical modeling. *The Journal of Experimental Medicine*, *197*, 207–220.

Daly, J., Licence, S., Nanou, A., Morgan, G., & Martensson, I. L. (2007). Transcription of productive and nonproductive VDJ-recombined alleles after IgH allelic exclusion. *The EMBO Journal*, *26*, 4273–4282.

Davodeau, F., Peyrat, M. A., Houde, I., Hallet, M. M., De Libero, G., Vie, H., et al. (1993). Surface expression of two distinct functional antigen receptors on human gammadelta T cells. *Science*, *260*, 1800–1802.

De Jong, H. (2002). Modeling and simulation of genetic regulatory systems: A literature review. *Journal of Computational Biology*, *9*, 67–103.

De Jong, T. (2006). Computer simulations. Technological advances in inquiry learning. *Science*, *312*, 532–533.

De Villartay, J.-P. (2002). V(D)J recombination and DNA repair: Lessons from human immune deficiencies and other animal models. *Current Opinion in Allergy and Clinical Immunology*, *2*, 473–479.

De Villartay, J. P., Hockett, R. D., Coran, D., Korsmeyer, S. J., & Cohen, D. I. (1988). Deletion of the human T-cell receptor delta-gene by a site-specific recombination. *Nature*, *335*, 170–174.

Degner, S. C., Verma-Gaur, J., Wong, T. P., Bossen, C., Iverson, G. M., Torkamani, A., et al. (2011). CCCTC-binding factor (CTCF) and cohesin influence the genomic architecture of the Igh locus and antisense transcription in pro-B cells. *Proceedings of the National Academy of Sciences of the United States of America*, *108*, 9566–9571.

Deng, Z., Liu, H., & Liu, X. (2015). RAG1-mediated ubiquitylation of histone H3 is required for chromosomal V(D)J recombination. *Cell Research*, *25*, 181–192.

Deriano, L., & Roth, D. B. (2013). Modernizing the nonhomologous end-joining repertoire: Alternative and classical NHEJ share the stage. *Annual Review of Genetics*, *47*, 433–455.

Diaz, P., Cado, D., & Winoto, A. (1994). A locus control region in the T cell receptor alpha/delta locus. *Immunity*, *1*, 207–217.

Du, H., Ishii, H., Pazin, M. J., & Sen, R. (2008). Activation of 12/23-RSS-dependent RAG cleavage by hSWI/SNF complex in the absence of transcription. *Molecular Cell*, *31*, 641–649.

Ebert, A., Mcmanus, S., Tagoh, H., Medvedovic, J., Salvagiotto, G., Novatchkova, M., et al. (2011). The distal V(H) gene cluster of the Igh locus contains distinct regulatory elements with Pax5 transcription factor-dependent activity in pro-B cells. *Immunity*, *34*, 175–187.

Ebert, A., Medvedovic, J., Tagoh, H., Schwickert, T. A., & Busslinger, M. (2013). Control of antigen receptor diversity through spatial regulation of V(D)J recombination. *Cold Spring Harbor Symposia on Quantitative Biology*, *78*, 11–21.

Eldar, A., & Elowitz, M. B. (2010). Functional roles for noise in genetic circuits. *Nature*, *467*, 167–173.

Elkin, S. K., Ivanov, D., Ewalt, M., Ferguson, C. G., Hyberts, S. G., Sun, Z. Y., et al. (2005). A PHD finger motif in the C terminus of RAG2 modulates recombination activity. *The Journal of Biological Chemistry*, *280*, 28701–28710.

Ellner, S. P., & Guckenheimer, J. (2011). *Dynamic models in biology*. Princeton, NJ: Princeton University Press.

Ernst, P., & Smale, S. T. (1995). Combinatorial regulation of transcription I: General aspects of transcriptional control. *Immunity*, *2*, 311–319.

Farago, M., Rosenbluh, C., Tevlin, M., Fraenkel, S., Schlesinger, S., Masika, H., et al. (2012). Clonal allelic predetermination of immunoglobulin-kappa rearrangement. *Nature*, *490*, 561–565.

Farcot, E., Bonnet, M., Jaeger, S., Spicuglia, S., Fernandez, B., & Ferrier, P. (2010). TCR beta allelic exclusion in dynamical models of V(D)J recombination based on allele independence. *Journal of Immunology*, *185*, 1622–1632.

Ferradini, L., Gu, H., De Smet, A., Rajewsky, K., Reynaud, C. A., & Weill, J. C. (1996). Rearrangement-enhancing element upstream of the mouse immunoglobulin kappa chain J cluster. *Science*, *271*, 1416–1420.

Franchini, D. M., Benoukraf, T., Jaeger, S., Ferrier, P., & Payet-Bornet, D. (2009). Initiation of V(D)J recombination by Db-associated recombination signal sequences: A critical control point in TCRb gene assembly. *PLoS One*, *4*, e4575.

Frischmeyer-Guerrerio, P. A., Montgomery, R. A., Warren, D. S., Cooke, S. K., Lutz, J., Sonnenday, C. J., et al. (2011). Perturbation of thymocyte development in nonsense-mediated decay (NMD)-deficient mice. *Proceedings of the National Academy of Sciences of the United States of America, 108*, 10638–10643.

Fuxa, M., Skok, J., Souabni, A., Salvagiotto, G., Roldan, E., & Busslinger, M. (2004). Pax5 induces V-to-DJ rearrangements and locus contraction of the immunoglobulin heavy-chain gene. *Genes & Development, 18*, 411–422.

Gellert, M. (2002). V(D)J recombination: RAG proteins, repair factors, and regulation. *Annual Review of Biochemistry, 71*, 101–132.

Gendrel, A.-V., Attia, M., Chen, C.-J., Diabangouaya, P., Servant, N., Barillot, E., et al. (2014). Developmental dynamics and disease potential of random monoallelic gene expression. *Developmental Cell, 28*, 366–380.

Gendrel, A.-V., & Heard, E. (2011). Fifty years of X-inactivation research. *Development, 138*, 5049–5055.

Gerasimova, T., Guo, C., Ghosh, A., Qiu, X., Montefiori, L., Verma-Gaur, J., et al. (2015). A structural hierarchy mediated by multiple nuclear factors establishes IgH locus conformation. *Genes & Development, 29*, 1683–1695.

Giallourakis, C. C., Franklin, A., Guo, C., Cheng, H. L., Yoon, H. S., Gallagher, M., et al. (2010). Elements between the IgH variable (V) and diversity (D) clusters influence antisense transcription and lineage-specific V(D)J recombination. *Proceedings of the National Academy of Sciences of the United States of America, 107*, 22207–22212.

Golding, A., Chandler, S., Ballestar, E., Wolffe, A. P., & Schlissel, M. S. (1999). Nucleosome structure completely inhibits *in vitro* cleavage by the V(D)J recombinase. *The EMBO Journal, 18*, 3712–3723.

Goldmit, M., & Bergman, Y. (2004). Monoallelic gene expression: A repertoire of recurrent themes. *Immunological Reviews, 200*, 197–214.

Goldmit, M., Ji, Y., Skok, J., Roldan, E., Jung, S., Cedar, H., et al. (2005). Epigenetic ontogeny of the Igk locus during B cell development. *Nature Immunology, 6*, 198–203.

Gopalakrishnan, S., Majumder, K., Predeus, A., Huang, Y., Koues, O. I., Verma-Gaur, J., et al. (2013). Unifying model for molecular determinants of the preselection Vbeta repertoire. *Proceedings of the National Academy of Sciences of the United States of America, 110*, E3206–E3215.

Gorman, J. R., & Alt, F. W. (1998). Regulation of immunoglobulin light chain isotype expression. *Advances in Immunology, 69*, 113–181.

Grawunder, U., Leu, T. J., Schatz, D. G., Werner, A., Rolink, A. G., Melchers, F., et al. (1995). Down-regulation of RAG1 and RAG2 gene expression in pre B cells after functional immunoglobulin heavy chain rearrangement. *Immunity, 3*, 601–608.

Grazini, U., Zanardi, F., Citterio, E., Casola, S., Goding, C. R., & Mcblane, F. (2010). The RING domain of RAG1 ubiquitylates histone H3: A novel activity in chromatin-mediated regulation of V(D)J joining. *Molecular Cell, 37*, 282–293.

Grundy, G. J., Yang, W., & Gellert, M. (2010). Autoinhibition of DNA cleavage mediated by RAG1 and RAG2 is overcome by an epigenetic signal in V(D)J recombination. *Proceedings of the National Academy of Sciences of the United States of America, 107*, 22487–22492.

Guo, C., Gerasimova, T., Hao, H., Ivanova, I., Chakraborty, T., Selimyan, R., et al. (2011). Two forms of loops generate the chromatin conformation of the immunoglobulin heavy-chain gene locus. *Cell, 147*, 332–343.

Guo, C., Yoon, H. S., Franklin, A., Jain, S., Ebert, A., Cheng, H. L., et al. (2011). CTCF-binding elements mediate control of V(D)J recombination. *Nature, 477*, 424–430.

Hesslein, D. G. T., Pflugh, D. L., Chowdhury, D., Bothwell, A. L. M., Sen, R., & Schatz, D. G. (2003). Pax5 is required for recombination of transcribed, acetylated, 5′ IgH V gene segments. *Genes & Development, 17*, 37–42.

Hewitt, S. L., Chaumeil, J., & Skok, J. A. (2010). Chromosome dynamics and the regulation of V(D)J recombination. *Immunological Reviews*, *237*, 43–54.

Hewitt, S. L., Farmer, D., Marszalek, K., Cadera, E., Liang, H.-E., Xu, Y., et al. (2008). Association between the Igk and Igh immunoglobulin loci mediated by the 3′ Igk enhancer induces 'decontraction' of the Igh locus in pre-B cells. *Nature Immunology*, *9*, 396–404.

Hewitt, S. L., Yin, B., Ji, Y., Chaumeil, J., Marszalek, K., Tenthorey, J., et al. (2009). RAG-1 and ATM coordinate monoallelic recombination and nuclear positioning of immunoglobulin loci. *Nature Immunology*, *10*, 655–664.

Hockett, R. D., De Villartay, J. P., Pollock, K., Poplack, D. G., Cohen, D. I., & Korsmeyer, S. J. (1988). Human T-cell antigen receptor (TCR) delta-chain locus and elements responsible for its deletion are within the TCR alpha-chain locus. *Proceedings of the National Academy of Sciences of the United States of America*, *85*, 9694–9698.

Hughes, M. M., Tillman, R. E., Wehrly, T. D., White, J. M., & Sleckman, B. P. (2003). The B12/23 restriction is critically dependent on recombination signal nonamer and spacer sequences. *Journal of Immunology*, *171*, 6604–6610.

Inoue, T., Morita, M., Hijikata, A., Fukuda-Yuzawa, Y., Adachi, S., Isono, K., et al. (2015). CNOT3 contributes to early B cell development by controlling Igh rearrangement and p53 mRNA stability. *The Journal of Experimental Medicine*, *212*, 1465–1479. jem.20150384 [Epub ahead of print].

Ishida, I., Verbeek, S., Bonneville, M., Itohara, S., Berns, A., & Tonegawa, S. (1990). T-cell receptor gamma delta and gamma transgenic mice suggest a role of a gamma gene silencer in the generation of alpha beta T cells. *Proceedings of the National Academy of Sciences of the United States of America*, *87*, 3067–3071.

Jackson, A., Kondilis, H. D., Khor, B., Sleckman, B. P., & Krangel, M. S. (2005). Regulation of T cell receptor beta allelic exclusion at a level beyond accessibility. *Nature Immunology*, *6*, 189–197.

Jackson, A. M., & Krangel, M. S. (2005). Allele-specific regulation of TCRb variable gene segment chromatin structure. *Journal of Immunology*, *175*, 5186–5191.

Jackson, A. M., & Krangel, M. S. (2006). Turning T-cell receptor b recombination on and off: More questions than answers. *Immunological Reviews*, *209*, 129–141.

Jaeger, S., Lima, R., Meyroneinc, A., Bonnet, M., Ulgade, E. & Ferrier, P. *A dynamical model of TCRbeta gene recombination: Assessing stochasticity in the initiation of allelic exclusion.* (Submitted for publication).

Jenuwein, T., & Allis, C. D. (2001). Translating the histone code. *Science*, *293*, 1074–1080.

Jerne, N. K. (1955). The natural-selection theory of antibody formation. *Proceedings of the National Academy of Sciences of the United States of America*, *41*, 849–857.

Jhunjhunwala, S., Van Zelm, M. C., Peak, M. M., Cutchin, S., Riblet, R., Van Dongen, J. J., et al. (2008). The 3D structure of the immunoglobulin heavy-chain locus: Implications for long-range genomic interactions. *Cell*, *133*, 265–279.

Ji, Y., Little, A. J., Banerjee, J. K., Hao, B., Oltz, E. M., Krangel, M. S., et al. (2010). Promoters, enhancers, and transcription target RAG1 binding during V(D)J recombination. *The Journal of Experimental Medicine*, *207*, 2809–2816.

Ji, Y., Resch, W., Corbett, E., Yamane, A., Casellas, R., & Schatz, D. G. (2010). The in vivo pattern of binding of RAG1 and RAG2 to antigen receptor loci. *Cell*, *141*, 419–431.

Jia, J., Kondo, M., & Zhuang, Y. (2007). Germline transcription from T-cell receptor Vbeta gene is uncoupled from allelic exclusion. *The EMBO Journal*, *26*, 2387–2399.

Johnson, K., Chaumeil, J., & Skok, J. A. (2010). Epigenetic regulation of V(D)J recombination. *Essays in Biochemistry*, *48*, 221–243.

Jones, J. M., Bhattacharyya, A., Simkus, C., Vallieres, B., Veenstra, T. D., & Zhou, M. (2011). The RAG1 V(D)J recombinase/ubiquitin ligase promotes ubiquitylation of acetylated, phosphorylated histone 3.3. *Immunology Letters*, *136*, 156–162.

Jones, J. M., & Gellert, M. (2004). The taming of a transposon: V(D)J recombination and the immune system. *Immunological Reviews*, *200*, 233–248.

Jung, D., Giallourakis, C., Mostoslavsky, R., & Alt, F. W. (2006). Mechanism and control of V(D)J recombination at the immunoglobulin heavy chain locus. *Annual Review of Immunology, 24*, 541–570.

Kalmanovich, G., & Mehr, R. (2003). Models for antigen receptor gene rearrangement. III. Heavy and light chain allelic exclusion. *Journal of Immunology, 170*, 182–193.

Karr, J. R., Takahashi, K., & Funahashi, A. (2015). The principles of whole-cell modeling. *Current Opinion in Microbiology, 27*, 18–24.

Khor, B., & Sleckman, B. P. (2002). Allelic exclusion at the TCRbeta locus. *Current Opinion in Immunology, 14*, 230–234.

Khor, B., & Sleckman, B. P. (2005). Intra- and inter-allelic ordering of T cell receptor beta chain gene assembly. *European Journal of Immunology, 35*, 964–970.

Kim, T.-K., & Shiekhattar, R. (2015). Architectural and functional commonalities between enhancers and promoters. *Cell, 162*, 948–959.

Kim, M.-S., Lapkouski, M., Yang, W., & Gellert, M. (2015). Crystal structure of the V(D)J recombinase RAG1-RAG2. *Nature, 518*, 507–511.

Kisielow, P., & von Boehmer, H. (1995). Development and selection of T cells: Facts and puzzles. *Advances in Immunology, 58*, 87–208.

Kitano, H. (2002). Systems biology: A brief overview. *Science, 295*, 1662–1664.

Kondilis-Mangum, H. D., Cobb, R. M., Osipovich, O., Srivatsan, S., Oltz, E. M., & Krangel, M. S. (2010). Transcription-dependent mobilization of nucleosomes at accessible TCR gene segments in vivo. *Journal of Immunology, 184*, 6970–6977.

Kondilis-Mangum, H. D., Shih, H.-Y., Mahowald, G., Sleckman, B. P., & Krangel, M. S. (2011). Regulation of TCRbeta allelic exclusion by gene segment proximity and accessibility. *Journal of Immunology, 187*, 6374–6381.

Kosak, S. T., Skok, J. A., Medina, K. L., Riblet, R., Le Beau, M. M., Fisher, A. G., et al. (2002). Subnuclear compartmentalization of immunoglobulin loci during lymphocyte development. *Science, 296*, 158–162.

Krangel, M. S. (2003). Gene segment selection in V(D)J recombination: Accessibility and beyond. *Nature Immunology, 4*, 624–630.

Krangel, M. S. (2007). T cell development: Better living through chromatin. *Nature Immunology, 8*, 687–694.

Krangel, M. S. (2009). Mechanics of T cell receptor gene rearrangement. *Current Opinion in Immunology, 21*, 133–139.

Krangel, M. S., Carabana, J., Abbarategui, I., Schlimgen, R., & Hawwari, A. (2004). Enforcing order within a complex locus: Current perspectives on the control of V(D)J recombination at the murine T-cell receptor alpha/delta locus. *Immunological Reviews, 200*, 224–232.

Kreslavsky, T., Gleimer, M., Miyazaki, M., Choi, Y., Gagnon, E., Murre, C., et al. (2012). Beta-selection-induced proliferation is required for alpha/beta T cell differentiation. *Immunity, 37*, 840–853.

Kumar, R., Bach, M. P., Mainoldi, F., Maruya, M., Kishigami, S., Jumaa, H., et al. (2015). Antibody repertoire diversification through VH gene replacement in mice cloned from an IgA plasma cell. *Proceedings of the National Academy of Sciences of the United States of America, 112*, E450–E457.

Kwan, S. P., Max, E. E., Seidman, J. G., Leder, P., & Scharff, M. D. (1981). Two kappa immunoglobulin genes are expressed in the myeloma S107. *Cell, 26*, 57–66.

Kwon, J., Imbalzano, A. N., Matthews, A., & Oettinger, M. A. (1998). Accessibility of nucleosomal DNA to V(D)J cleavage is modulated by RSS positioning and HMG1. *Molecular Cell, 2*, 829–839.

Kwon, J., Morshead, K. B., Guyon, J. R., Kingston, R. E., & Oettinger, M. A. (2000). Histone acetylation and hSWI/SNF remodeling act in concert to stimulate V(D)J cleavage of nucleosomal DNA. *Molecular Cell, 6*, 1037–1048.

Lapkouski, M., Chuenchor, W., Kim, M.-S., Gellert, M., & Yang, W. (2015). Assembly pathway and characterization of the RAG1/2-DNA paired and signal-end complexes. *The Journal of Biological Chemistry*, *290*, 14618–14625.

Levin-Klein, R., & Bergman, Y. (2014). Epigenetic regulation of monoallelic rearrangement (allelic exclusion) of antigen receptor genes. *Frontiers in Immunology*, *5*, 625.

Levy, S., Campbell, M. J., & Levy, R. (1989). Functional immunoglobulin light chain genes are replaced by ongoing rearrangements of germline V kappa genes to downstream J kappa segment in a murine B cell line. *The Journal of Experimental Medicine*, *170*, 1–13.

Li, Z., Dordai, D. I., Lee, J., & Desiderio, S. (1996). A conserved degradation signal regulates RAG-2 accumulation during cell division and links V(D)J recombination to the cell cycle. *Immunity*, *5*, 575–589.

Liang, H. E., Hsu, L. Y., Cado, D., Cowell, L. G., Kelsoe, G., & Schlissel, M. S. (2002). The "dispensable" portion of RAG2 is necessary for efficient V-to-DJ rearrangement during B and T cell development. *Immunity*, *17*, 639–651.

Liang, H. E., Hsu, L. Y., Cado, D., & Schlissel, M. S. (2004). Variegated transcriptional activation of the immunoglobulin kappa locus in pre-b cells contributes to the allelic exclusion of light-chain expression. *Cell*, *118*, 19–29.

Lieber, M. R. (2010). The mechanism of double-strand DNA break repair by the nonhomologous DNA end-joining pathway. *Annual Review of Biochemistry*, *79*, 181–211.

Lieber, M. R., Ma, Y., Pannicke, U., & Schwarz, K. (2003). Mechanism and regulation of human non-homologous DNA end-joining. *Nature Reviews Molecular Cell Biology*, *4*, 712–720.

Lieberman Aiden, E., & Casellas, R. (2015). Somatic rearrangement in B cells: it's (mostly) nuclear physics. *Cell*, *162*, 708–711.

Liu, Y., Subrahmanyam, R., Chakraborty, T., Sen, R., & Desiderio, S. (2007). A plant homeodomain in RAG-2 that binds hypermethylated lysine 4 of histone H3 is necessary for efficient antigen-receptor-gene rearrangement. *Immunity*, *27*, 561–571.

Lin, S. G., Guo, C., Su, A., Zhang, Y., & Alt, F. W. (2015). CTCF-binding elements 1 and 2 in the Igh intergenic control region cooperatively regulate V(D)J recombination. *Proceedings of the National Academy of Sciences of the United States of America*, *112*, 1815–1820.

Liu, H., Schmidt-Supprian, M., Shi, Y., Hobeika, E., Barteneva, N., Jumaa, H., et al. (2007). Ying Yang 1 is a critical regulator of B-cell development. *Genes & Development*, *21*, 1179–1189.

Liu, Z., Widlak, P., Zou, Y., Xiao, F., Oh, M., Li, S., et al. (2006). A recombination silencer that specifies heterochromatin positioning and Ikaros association in the immunoglobulin kappa locus. *Immunity*, *24*, 405–415.

Louzoun, Y., Friedman, T., Luning Prak, E., Litwin, S., & Weigert, M. (2002). Analysis of B cell receptor production and rearrangement. Part I. Light chain rearrangement. *Seminars in Immunology*, *14*, 169–190. discussion 221-122-169.

Lu, C., Ward, A., Bettridge, J., Liu, Y., & Desiderio, S. (2015). An autoregulatory mechanism imposes allosteric control on the V(D)J recombinase by histone H3 methylation. *Cell Reports*, *10*, 29–38.

Lucas, J. S., Zhang, Y., Dudko, O. K., & Murre, C. (2014). 3D trajectories adopted by coding and regulatory DNA elements: First-passage times for genomic interactions. *Cell*, *158*, 339–352.

Lutz, J., Heideman, M. R., Roth, E., Van Den Berk, P., Müller, W., Raman, C., et al. (2011). Pro-B cells sense productive immunoglobulin heavy chain rearrangement irrespective of polypeptide production. *Proceedings of the National Academy of Sciences of the United States of America*, *108*, 10644–10649.

Majumder, K., Koues, O. I., Chan, E. A. W., Kyle, K. E., Horowitz, J. E., Yang-Iott, K., et al. (2015). Lineage-specific compaction of Tcrb requires a chromatin barrier to protect the function of a long-range tethering element. *The Journal of Experimental Medicine*, *212*, 107–120.

Majumder, K., Rupp, L. J., Yang-Iott, K. S., Koues, O. I., Kyle, K. E., Bassing, C. H., et al. (2015). Domain-specific and stage-intrinsic changes in Tcrb conformation during thymocyte development. *Journal of Immunology*, *195*, 1262–1272.

Malecek, K., Lee, V., Feng, W., Huang, J. L., Flajnik, M. F., Ohta, Y., et al. (2008). Immunoglobulin heavy chain exclusion in the shark. *PLoS Biology*, *6*, e157.

Malissen, M., Trucy, J., Jouvin-Marche, E., Cazenave, P. A., Scollay, R., & Malissen, B. (1992). Regulation of TCR alpha and beta gene allelic exclusion during T-cell development. *Immunology Today*, *13*, 315–322.

Mathieu, N., Spicuglia, S., Gorbatch, S., Cabaud, O., Fernex, C., Verthuy, C., et al. (2003). Assessing the role of the T cell receptor b gene enhancer in regulating coding joint formation during V(D)J recombination. *The Journal of Biological Chemistry*, *278*, 18101–18109.

Matthews, A. G., Kuo, A. J., Ramon-Maiques, S., Han, S., Champagne, K. S., Ivanov, D., et al. (2007). RAG2 PHD finger couples histone H3 lysine 4 trimethylation with V(D)J recombination. *Nature*, *450*, 1106–1110.

Matthews, A. G., & Oettinger, M. A. (2009). RAG: A recombinase diversified. *Nature Immunology*, *10*, 817–821.

May, R. M. (1976). Simple mathematical models with very complicated dynamics. *Nature*, *261*, 459–467.

Medvedovic, J., Ebert, A., Tagoh, H., Tamir, I. M., Schwickert, T. A., Novatchkova, M., et al. (2013). Flexible long-range loops in the VH gene region of the Igh locus facilitate the generation of a diverse antibody repertoire. *Immunity*, *39*, 229–244.

Mehr, R., Shannon, M., & Litwin, S. (1999). Models for antigen receptor gene rearrangement. I. Biased receptor editing in B cells: Implications for allelic exclusion. *Journal of Immunology*, *163*, 1793–1798.

Melchers, F. (2005). The pre-B-cell receptor: Selector of fitting immunoglobulin heavy chains for the B-cell repertoire. *Nature Reviews. Immunology*, *5*, 578–584.

Melchers, F., Ten Boekel, E., Yamagami, T., Andersson, J., & Rolink, A. (1999). The roles of preB and B cell receptors in the stepwise allelic exclusion of mouse IgH and L chain gene loci. *Seminars in Immunology*, *11*, 307–317.

Morshead, K. B., Ciccone, D. N., Taverna, S. D., Allis, C. D., & Oettinger, M. A. (2003). Antigen receptor loci poised for V(D)J rearrangement are broadly associated with BRG1 and flanked by peaks of histone H3 dimethylated at lysine 4. *Proceedings of the National Academy of Sciences of the United States of America*, *100*, 11577–11582.

Mostoslavsky, R., Alt, F. W., & Rajewsky, K. (2004). The lingering enigma of the allelic exclusion mechanism. *Cell*, *118*, 539–544.

Mostoslavsky, R., Singh, N., Tenzen, T., Goldmit, M., Gabay, C., Elizur, S., et al. (2001). Asynchronous replication and allelic exclusion in the immune system. *Nature*, *414*, 221–225.

Natoli, G. (2010). Maintaining cell identity through global control of genomic organization. *Immunity*, *33*, 12–24.

Nemazee, D. (2006). Receptor editing in lymphocyte development and central tolerance. *Nature Reviews. Immunology*, *6*, 728–740.

Oestreich, K. J., Cobb, R. M., Pierce, S., Chen, J., Ferrier, P., & Oltz, E. M. (2006). Regulation of TCRb gene assembly by a promoter/enhancer holocomplex. *Immunity*, *24*, 381–391.

Oettinger, M. A. (2004). How to keep V(D)J recombination under control. *Immunological Reviews*, *200*, 165–181.

Oettinger, M. A., Schatz, D. G., Gorka, C., & Baltimore, D. (1990). RAG-1 and RAG-2, adjacent genes that synergistically activate V(D)J recombination. *Science*, *248*, 1517–1523.

Ortiz, B. D., Cado, D., & Winoto, A. (1999). A new element within the T-cell receptor alpha locus required for tissue-specific locus control region activity. *Molecular and Cellular Biology*, *19*, 1901–1909.

Osipovich, O., Cobb, R. M., Oestreich, K. J., Pierce, S., Ferrier, P., & Oltz, E. M. (2007). Essential function for SWI-SNF chromatin-remodeling complexes in the promoter-directed assembly of Tcrb genes. *Nature Immunology*, *8*, 809–816.

Osipovich, O., Milley, R., Meade, A., Tachibana, M., Shinkai, Y., Krangel, M. S., et al. (2004). Targeted inhibition of V(D)J recombination by a histone methyltransferase. *Nature Immunology*, *5*, 309–316.

Pernis, B., Chiappino, G., Kelus, A. S., & Gell, P. G. (1965). Cellular localization of immunoglobulins with different allotypic specificities in rabbit lymphoid tissues. *The Journal of Experimental Medicine*, *122*, 853–876.

Piper, H., Litwin, S., & Mehr, R. (1999). Models for antigen receptor gene rearrangement. II. Multiple rearrangement in the TCR: Allelic exclusion or inclusion? *Journal of Immunology*, *163*, 1799–1808.

Pujadas, E., & Feinberg, A. P. (2012). Regulated noise in the epigenetic landscape of development and disease. *Cell*, *148*, 1123–1131.

Rajewsky, K. (1996). Clonal selection and learning in the antibody system. *Nature*, *381*, 751–758.

Ramon-Maiques, S., Kuo, A. J., Carney, D., Matthews, A. G., Oettinger, M. A., Gozani, O., et al. (2007). The plant homeodomain finger of RAG2 recognizes histone H3 methylated at both lysine-4 and arginine-2. *Proceedings of the National Academy of Sciences of the United States of America*, *104*, 18993–18998.

Raulet, D. H. (1989). The structure, function, and molecular genetics of the gamma T cell receptor. *Annual Review of Immunology*, *9*, 175–207.

Reynaud, D., Demarco, I. A., Reddy, K. L., Schjerven, H., Bertolino, E., Chen, Z., et al. (2008). Regulation of B cell fate commitment and immunoglobulin heavy-chain gene rearrangements by Ikaros. *Nature Immunology*, *9*, 927–936.

Roldan, E., Fuxa, M., Chong, W., Martinez, D., Novatchkova, M., Busslinger, M., et al. (2005). Locus 'decontraction' and centromeric recruitment contribute to allelic exclusion of the immunoglobulin heavy-chain gene. *Nature Immunology*, *6*, 31–41.

Roth, D. B., & Roth, S. Y. (2000). Unequal access. Regulating V(D)J recombination through chromatin remodeling. *Cell*, *103*, 699–702.

Roy, A. L., Sen, R., & Roeder, R. G. (2011). Enhancer-promoter communication and transcriptional regulation of Igh. *Trends in Immunology*, *32*, 532–539.

Ruthenburg, A. J., Li, H., Patel, D. J., & Allis, C. D. (2007). Multivalent engagement of chromatin modifications by linked binding modules. *Nature Reviews. Molecular Cell Biology*, *8*, 983–994.

Sakano, H., Kurosawa, Y., Weigert, M., & Tonegawa, S. (1981). Identification and nucleotide sequence of a diversity DNA segment (D) of immunoglobulin heavy-chain genes. *Nature*, *290*, 562–565.

Sayegh, C., Jhunjhunwala, S., Riblet, R., & Murre, C. (2005). Visualization of looping involving the immunoglobulin heavy-chain locus in developing B cells. *Genes & Development*, *19*, 322–327.

Schatz, D. G. (2004). Antigen receptor genes and the evolution of a recombinase. *Seminars in Immunology*, *16*, 245–256.

Schatz, D. G., & Ji, Y. (2011). Recombination centres and the orchestration of V(D)J recombination. *Nature Reviews Immunology*, *11*, 251–263.

Schatz, D. G., Oettinger, M. A., & Baltimore, D. (1989). The V(D)J recombination activating gene, RAG-1. *Cell*, *59*, 1035–1048.

Schatz, D. G., & Spanopoulou, E. (2005). Biochemistry of V(D)J recombination. *Current Topics in Microbiology and Immunology, 290*, 49–85.

Schatz, D. G., & Swanson, P. C. (2011). V(D)J recombination: Mechanisms of initiation. *Annual Review of Genetics, 45*, 167–202.

Schlimgen, R. J., Reddy, K. L., Singh, H., & Krangel, M. S. (2008). Initiation of allelic exclusion by stochastic interaction of Tcrb alleles with repressive nuclear compartments. *Nature Immunology, 9*, 802–809.

Schlissel, M. (2002). Allelic exclusion of immunoglobulin gene rearrangement and expression: Why and how? *Seminars in Immunology, 14*, 207–212. discussion 225-206-207.

Selimyan, R., Gerstein, R. M., Ivanova, I., Precht, P., Subrahmanyam, R., Perlot, T., et al. (2013). Localized DNA demethylation at recombination intermediates during immunoglobulin heavy chain gene assembly. *PLoS Biology, 11*, e1001475.

Sepulveda, N., Boucontet, L., Pereira, P., & Carneiro, J. (2005). Stochastic modeling of T cell receptor gamma gene rearrangement. *Journal of Theoretical Biology, 234*, 153–165.

Servedio, M. R., Brandvain, Y., Dhole, S., Fitzpatrick, C. L., Goldberg, E. E., Stern, C. A., et al. (2014). Not just a theory—The utility of mathematical models in evolutionary biology. *PLoS Biology, 12*, e1002017.

Shih, H. Y., & Krangel, M. S. (2010). Distinct contracted conformations of the Tcra/Tcrd locus during Tcra and Tcrd recombination. *The Journal of Experimental Medicine, 207*, 1835–1841.

Shimazaki, N., Tsai, A. G., & Lieber, M. R. (2009). H3K4me3 stimulates the V(D)J RAG complex for both nicking and hairpinning in trans in addition to tethering in cis: Implications for translocations. *Molecular Cell, 34*, 535–544.

Sieh, P., & Chen, J. (2001). Distinct control of the frequency and allelic exclusion of the V beta gene rearrangement at the TCR beta locus. *Journal of Immunology, 167*, 2121–2129.

Sikes, M. L., & Oltz, E. M. (2012). Genetic and epigenetic regulation of Tcrb gene assembly. *Current Topics in Microbiology and Immunology, 356*, 91–116.

Singer, A., Adoro, S., & Park, J. H. (2008). Lineage fate and intense debate: Myths, models and mechanisms of CD4- versus CD8-lineage choice. *Nature Reviews Immunology, 8*, 788–801.

Singh, N., Bergman, Y., Cedar, H., & Chess, A. (2003). Biallelic germline transcription at the kappa immunoglobulin locus. *The Journal of Experimental Medicine, 197*, 743–750.

Skok, J. A., Brown, K. E., Azuara, V., Caparros, M. L., Baxter, J., Takacs, K., et al. (2001). Nonequivalent nuclear location of immunoglobulin alleles in B lymphocytes. *Nature Immunology, 2*, 848–854.

Skok, J. A., Gisler, R., Novatchkova, M., Farmer, D., De, L. W., & Busslinger, M. (2007). Reversible contraction by looping of the Tcra and Tcrb loci in rearranging thymocytes. *Nature Immunology, 8*, 378–387.

Sleckman, B. P., Bassing, C. H., Hughes, M. M., Okada, A., D'auteuil, M., Wehrly, T. D., et al. (2000). Mechanisms that direct ordered assembly of T cell receptor beta locus V, D, and J gene segments. *Proceedings of the National Academy of Sciences of the United States of America, 97*, 7975–7980.

Sleckman, B. P., Khor, B., Monroe, R., & Alt, F. W. (1998). Assembly of productive T cell receptor delta variable region genes exhibits allelic inclusion. *The Journal of Experimental Medicine, 188*, 1465–1471.

Smale, S. T. (2001). Core promoters: Active contributors to combinatorial gene regulation. *Genes & Development, 15*, 2503–2508.

Sonoda, E., Pewzner-Jung, Y., Schwers, S., Taki, S., Jung, S., Eilat, D., et al. (1997). B cell development under the condition of allelic inclusion. *Immunity, 6*, 225–233.

Spicuglia, S., Kumar, S., Yeh, J.-H., Vachez, E., Chasson, L., Gorbatch, S., et al. (2002). Promoter activation by enhancer-dependent and -independent loading of activator and coactivator complexes. *Molecular Cell, 10*, 1479–1487.

Spicuglia, S., Pekowska, A., Zacarias-Cabeza, J., & Ferrier, P. (2010). Epigenetic control of Tcrb gene rearrangement. *Seminars in Immunology*, *22*, 330–336.
Stanhope-Baker, P., Hudson, K. M., Shaffer, A. L., Constantinescu, A., & Schlissel, M. S. (1996). Cell type-specific chromatin structure determines the targeting of V(D)J recombinase activity in vitro. *Cell*, *85*, 887–897.
Steinel, N. C., Fisher, M. R., Yang-Iott, K. S., & Bassing, C. H. (2014). The ataxia telangiectasia mutated and cyclin D3 proteins cooperate to help enforce TCR beta and IgH allelic exclusion. *Journal of Immunology*, *193*, 2881–2890.
Steinel, N. C., Lee, B.-S., Tubbs, A. T., Bednarski, J. J., Schulte, E., Yang-Iott, K. S., et al. (2013). The ataxia telangiectasia mutated kinase controls Igkappa allelic exclusion by inhibiting secondary Vkappa-to-Jkappa rearrangements. *The Journal of Experimental Medicine*, *210*, 233–239.
Steiner, F. A., & Henikoff, S. (2015). Diversity in the organization of centromeric chromatin. *Current Opinion in Genetics & Development*, *31*, 28–35.
Stevens, S. (2015). Genomic organization of the murine and human Ig light chain loci. In M. Ratcliffe (Ed.), *Encyclopedia of immunobiology*. (1st ed.). Langford Lane, Kidlington, Oxford, United Kingdom: Elsevier Limited (Corporate Office). In press.
Subrahmanyam, R., Du, H., Ivanova, I., Chakraborty, T., Ji, Y., Zhang, Y., et al. (2012). Localized epigenetic changes induced by DH recombination restricts recombinase to DJH junctions. *Nature Immunology*, *13*, 1205–1212.
Sun, A., Novobrantseva, T. I., Coffre, M., Hewitt, S. L., Jensen, K., Skok, J. A., et al. (2015). VH replacement in primary immunoglobulin repertoire diversification. *Proceedings of the National Academy of Sciences of the United States of America*, *112*, E458–E466.
Teng, G., Maman, Y., Resch, W., Kim, M., Yamane, A., Qian, J., et al. (2015). RAG represents a widespread threat to the lymphocyte genome. *Cell*, *162*(4), 751–765. http://dx.doi.org/10.1016/j.cell.2015.07.009. [Epub ahead of print]. pii: S0092-8674(15)00846-6.
Tillman, R. E., Wooley, A. L., Hughes, M. M., Khor, B., & Sleckman, B. P. (2004). Regulation of T-cell receptor b-chain gene assembly by recombination signals: The beyond 12/23 restriction. *Immunological Reviews*, *200*, 36–43.
Tillman, R. E., Wooley, A. L., Khor, B., Wehrly, T. D., Little, C. A., & Sleckman, B. P. (2003). Targeting of Vbeta to Dbeta rearrangement by RSSs can be mediated by the V(D)J recombinase in the absence of additional lymphoid-specific factors. *Journal of Immunology*, *170*, 5–9.
Tonegawa, S., Sakano, H., Make, R., Traunecker, A., Heinrich, G., Roeder, W., et al. (1981). Somatic reorganization of immunoglobulin genes during lymphocyte differentiation. *Cold Spring Harbor Symposia on Quantitative Biology*, *45*(Pt 2), 839–858.
Tripathi, R., Jackson, A., & Krangel, M. S. (2002). A change in the structure of Vb chromatin associated with TCRb allelic exclusion. *Journal of Immunology*, *168*, 2316–2324.
Ventsel, H. (1977). *Probability theory*. Moscow: Edition MIR.
Verma-Gaur, J., Torkamani, A., Schaffer, L., Head, S. R., Schork, N. J., & Feeney, A. J. (2012). Noncoding transcription within the Igh distal VH region at PAIR elements affects the 3D structure of the Igh locus in pro-B cells. *Proceedings of the National Academy of Sciences of the United States of America*, *109*, 17004–17009.
Vettermann, C., & Schlissel, M. S. (2010). Allelic exclusion of immunoglobulin genes: Models and mechanisms. *Immunological Reviews*, *237*, 22–42.
von Boehmer, H., Aifantis, I., Azogui, O., Feinberg, J., Saint-Ruf, C., Zober, C., et al. (1998). Crucial function of the pre-T-cell receptor (TCR) in TCRb selection, TCRb allelic exclusion and alpha beta versus gd lineage commitment. *Immunological Reviews*, *165*, 111–119.
von Boehmer, H., & Melchers, F. (2010). Checkpoints in lymphocyte development and autoimmune disease. *Nature Immunology*, *11*, 14–20.

Wabl, M., & Steinberg, C. (1982). A theory of allelic and isotypic exclusion for immunoglobulin genes. *Proceedings of the National Academy of Sciences of the United States of America*, *79*, 6976–6978.

Wang, G., Dhar, K., Swanson, P. C., Levitus, M., & Chang, Y. (2012). Real-time monitoring of RAG-catalyzed DNA cleavage unveils dynamic changes in coding end association with the coding end complex. *Nucleic Acids Research*, *40*, 6082–6096.

Wang, X., Xiao, G., Zhang, Y., Wen, X., Gao, X., Okada, S., et al. (2008). Regulation of Tcrb recombination ordering by c-Fos-dependent RAG deposition. *Nature Immunology*, *9*, 794–801.

Whitehurst, C. E., Schlissel, M. S., & Chen, J. (2000). Deletion of germline promoter PD beta 1 from the TCR beta locus causes hypermethylation that impairs D beta 1 recombination by multiple mechanisms. *Immunity*, *13*, 703–714.

Winter, D. R., & Amit, I. (2014). The role of chromatin dynamics in immune cell development. *Immunological Reviews*, *261*, 9–22.

Xiang, Y., Park, S.-K., & Garrard, W. T. (2014). A major deletion in the Vkappa-Jkappa intervening region results in hyperelevated transcription of proximal Vkappa genes and a severely restricted repertoire. *Journal of Immunology*, *193*, 3746–3754.

Xiang, Y., Zhou, X., Hewitt, S. L., Skok, J. A., & Garrard, W. T. (2011). A multifunctional element in the mouse Igkappa locus that specifies repertoire and Ig loci subnuclear location. *Journal of Immunology*, *186*, 5356–5366.

Xiong, N., & Raulet, D. H. (2007). Development and selection of gammadelta T cells. *Immunological Reviews*, *215*, 15–31.

Xu, C. R., & Feeney, A. J. (2009). The epigenetic profile of Ig genes is dynamically regulated during B cell differentiation and is modulated by pre-B cell receptor signaling. *Journal of Immunology*, *182*, 1362–1369.

Xu, C. R., Schaffer, L., Head, S. R., & Feeney, A. J. (2008). Reciprocal patterns of methylation of H3K36 and H3K27 on proximal vs. distal IgVH genes are modulated by IL-7 and Pax5. *Proceedings of the National Academy of Sciences of the United States of America*, *105*, 8685–8690.

Yancopoulos, G. D., & Alt, F. W. (1986). Regulation of the assembly and expression of variable-region genes. *Annual Review of Immunology*, *4*, 339–368.

Yancopoulos, G. D., Blackwell, T. K., Suh, H., Hood, L., & Alt, F. W. (1986). Introduced T cell receptor variable region gene segments recombine in pre-B cells: Evidence that B and T cells use a common recombinase. *Cell*, *44*, 251–259.

Yang-Iott, K. S., Carpenter, A. C., Rowh, M. A. W., Steinel, N., Brady, B. L., Hochedlinger, K., et al. (2010). TCR beta feedback signals inhibit the coupling of recombinationally accessible V beta 14 segments with DJ beta complexes. *Journal of Immunology*, *184*, 1369–1378.

Zacarias-Cabeza, J., Belhocine, M., Vanhille, L., Cauchy, P., Koch, F., Pekowska, A., et al. (2015). Transcription-dependent generation of a specialized chromatin structure at the TCRbeta locus. *Journal of Immunology*, *194*, 3432–3443.

Zhu, C., Feng, W., Weedon, J., Hua, P., Stefanov, D., Ohta, Y., et al. (2011). The multiple shark Ig H chain genes rearrange and hypermutate autonomously. *Journal of Immunology*, *187*, 2492–2501.

Zou, Y. R., Takeda, S., & Rajewsky, K. (1993). Gene targeting in the Ig kappa locus: Efficient generation of lambda chain-expressing B cells, independent of gene rearrangements in Ig kappa. *The EMBO Journal*, *12*, 811–820.

INDEX

Note: Page numbers followed by "*f*" indicate figures and "*t*" indicate tables.

D

G

H

K

L

M

V

CONTENTS OF RECENT VOLUMES

Volume 88

Volume 89

Volume 90

Volume 91

Volume 92

Volume 93

Volume 94

Volume 95

Volume 96

Volume 97

Volume 98

Volume 99

Volume 100

Volume 101

Volume 102

Volume 103

Volume 104

Volume 105

Volume 106

Volume 107

Volume 108

Volume 109

Volume 110

Volume 111

Volume 112

Volume 113

Volume 114

Volume 115

Volume 116

Volume 117

Volume 118

Volume 119

Volume 120

Volume 121

Volume 122

Volume 123

Volume 124

Volume 125

Volume 126

Volume 127

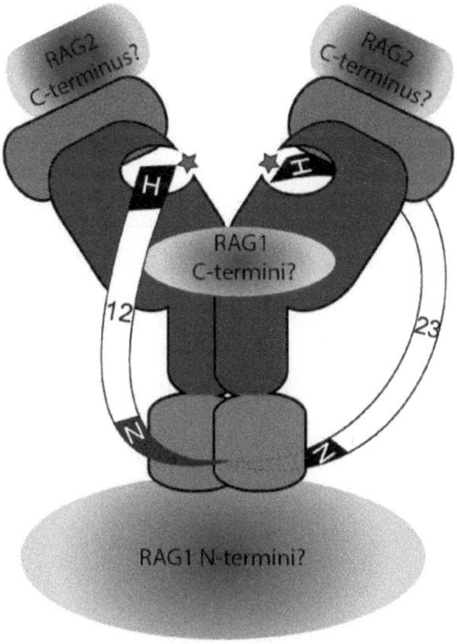

Chapter 1, Figure 2 (See page 4 of this volume.)

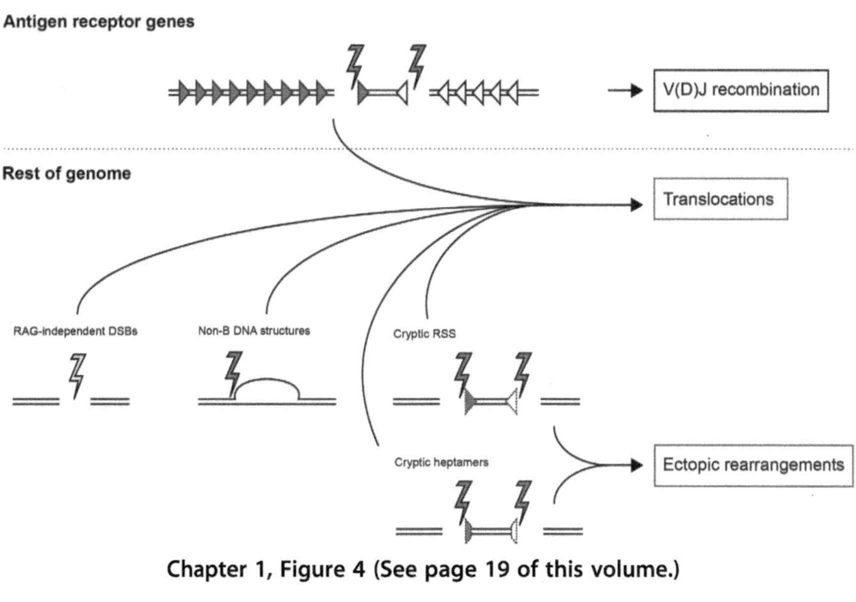

Chapter 1, Figure 4 (See page 19 of this volume.)

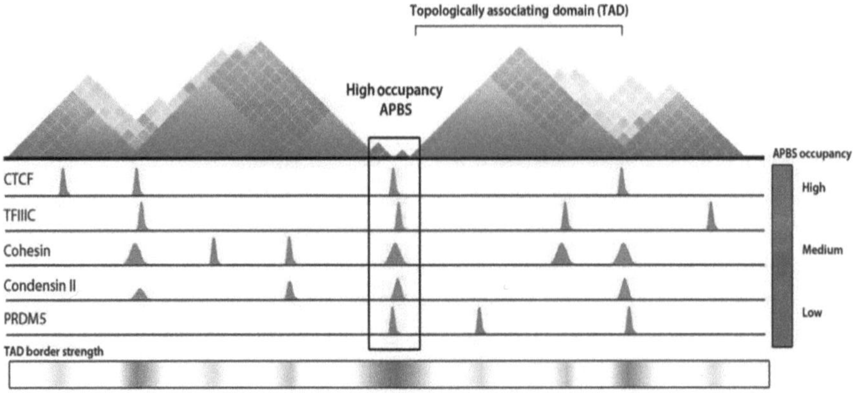

Chapter 2, Figure 1 (See page 48 of this volume.)

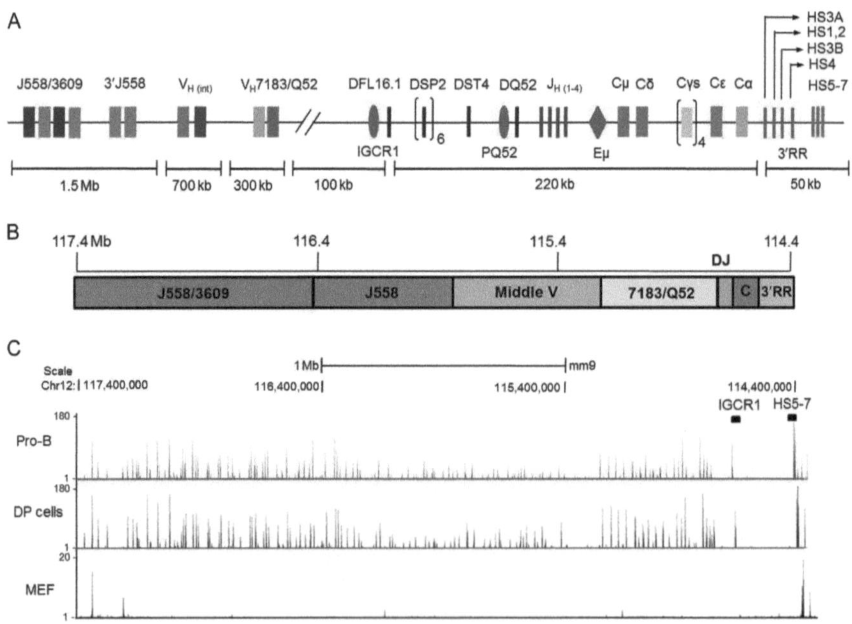

Chapter 2, Figure 2 (See page 52 of this volume.)

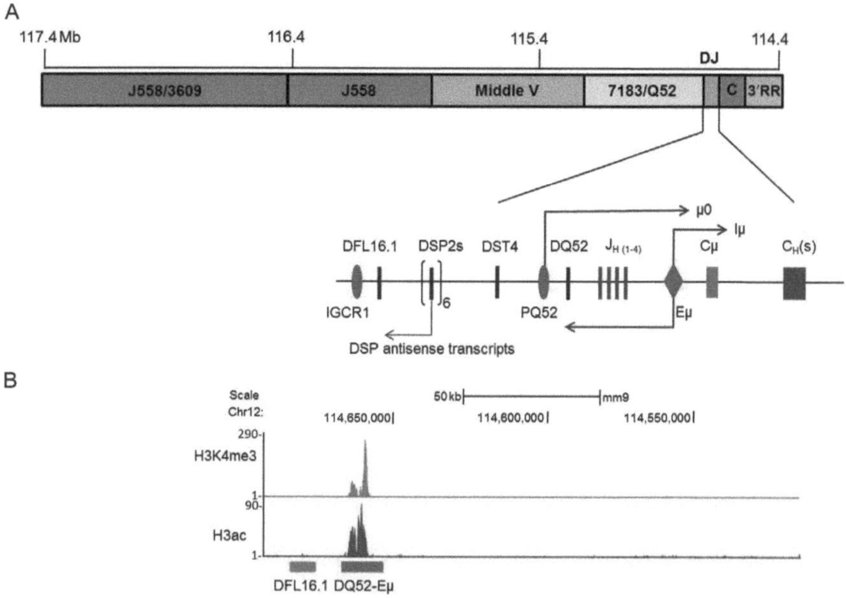

Chapter 2, Figure 3 (See page 58 of this volume.)

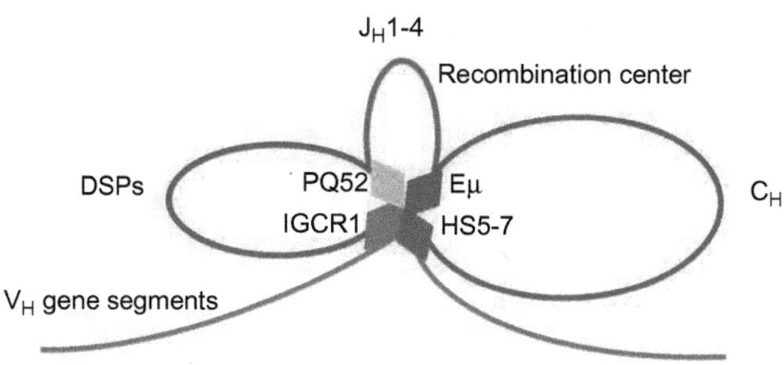

Chapter 2, Figure 4 (See page 64 of this volume.)

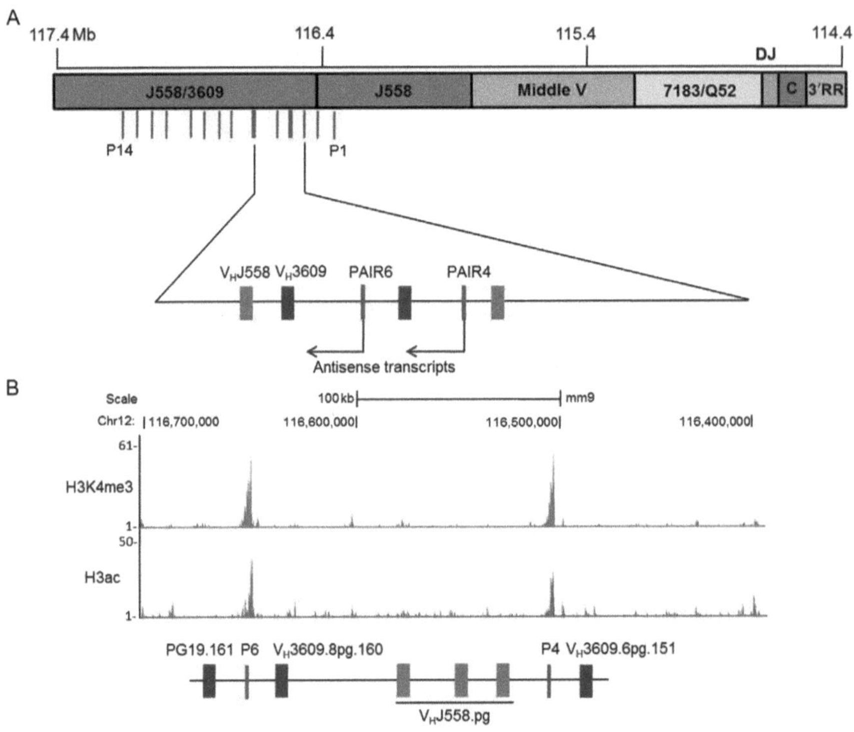

Chapter 2, Figure 5 (See page 72 of this volume.)

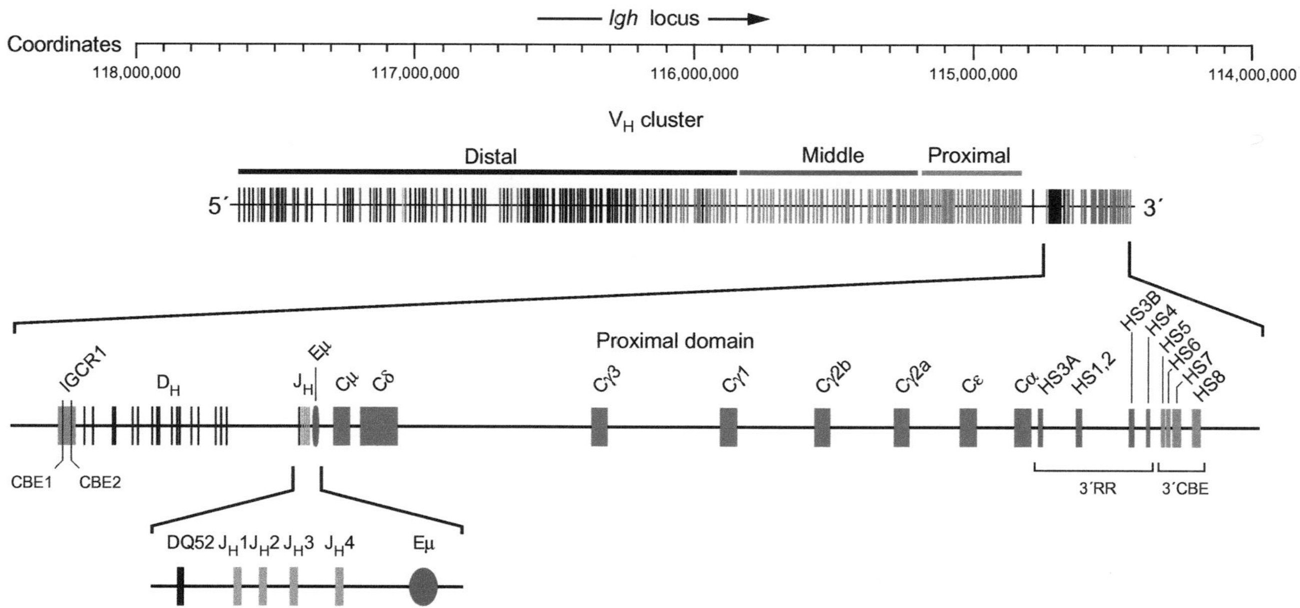

Chapter 3, Figure 1 (See page 96 of this volume.)

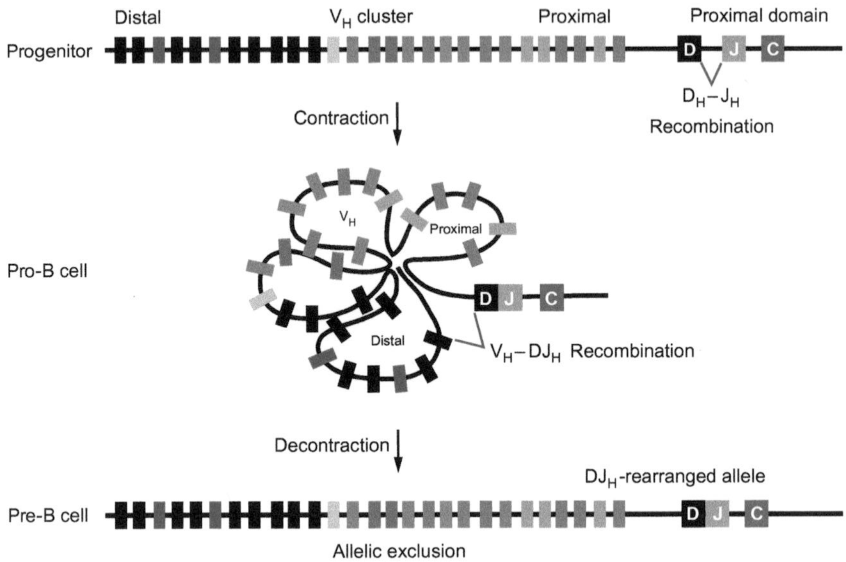

Chapter 3, Figure 2 (See page 99 of this volume.)

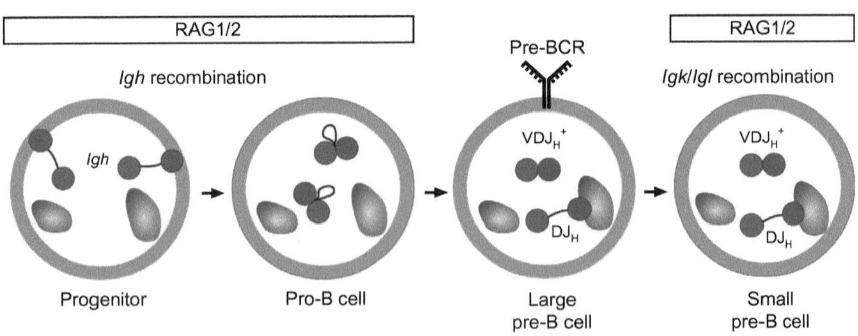

Chapter 3, Figure 3 (See page 101 of this volume.)

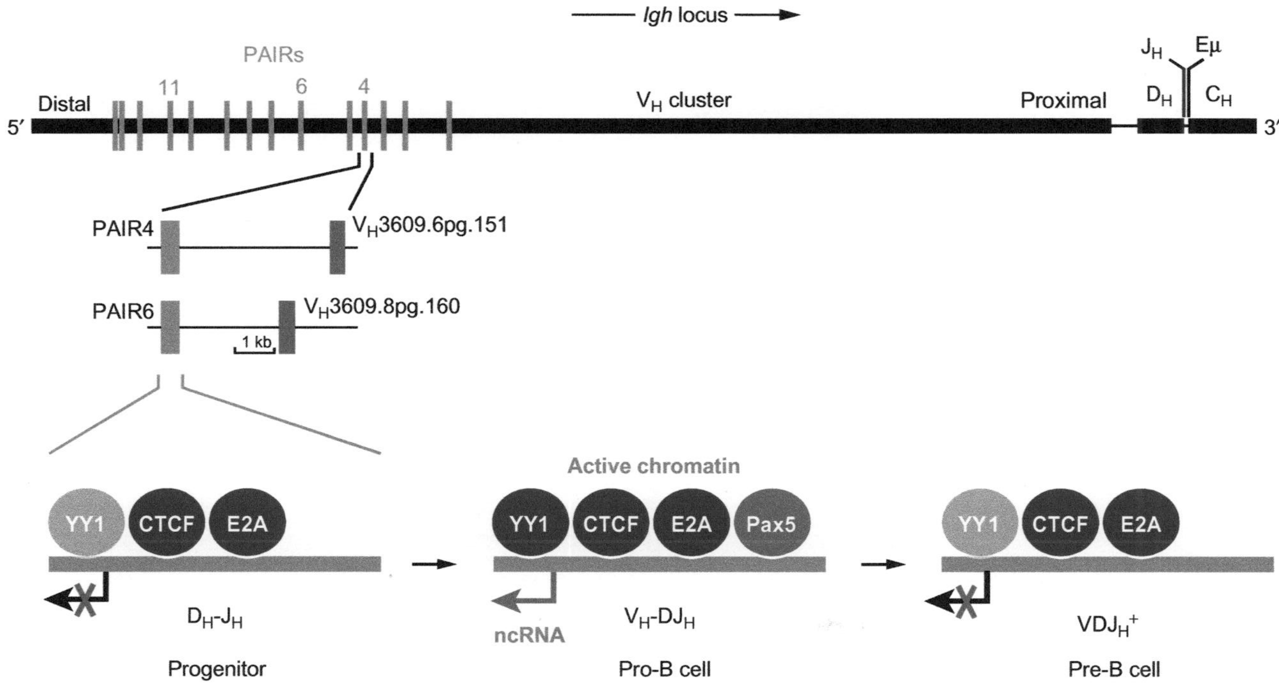

Chapter 3, Figure 4 (See page 104 of this volume.)

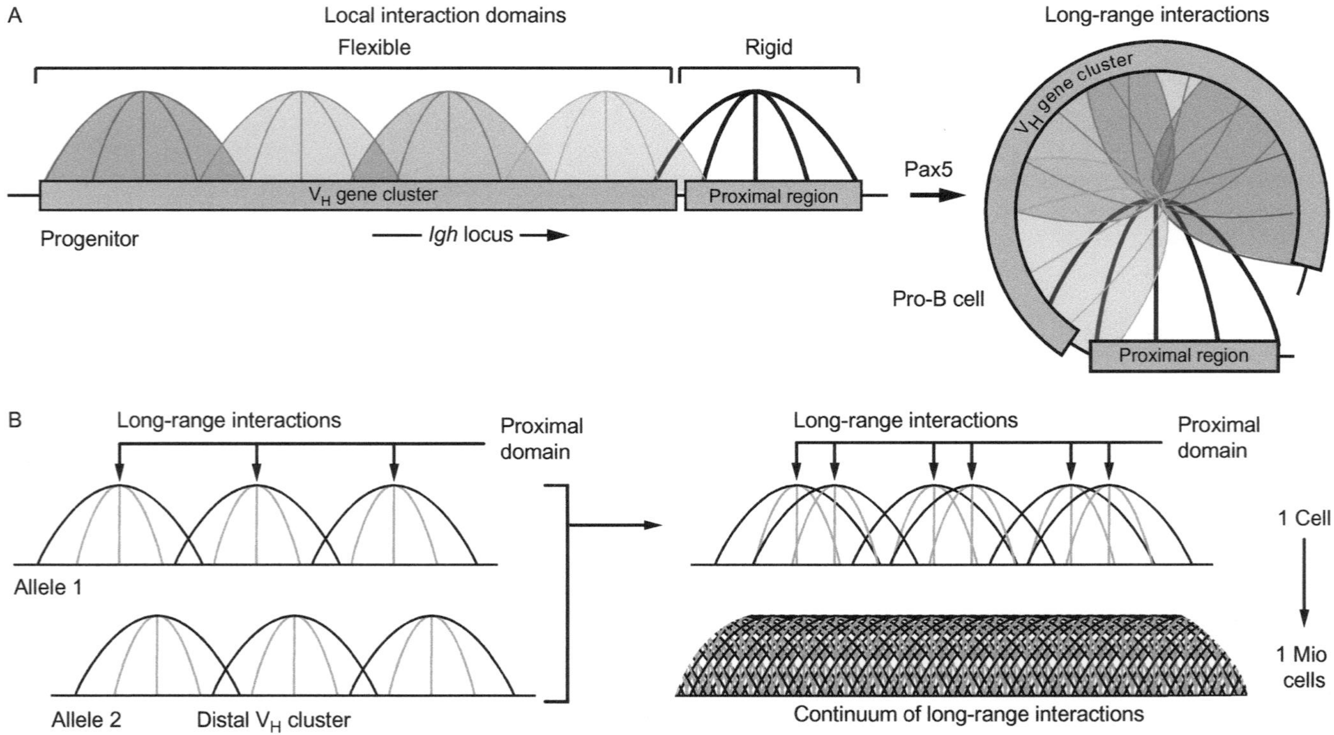

Chapter 3, Figure 5 (See page 106 of this volume.)

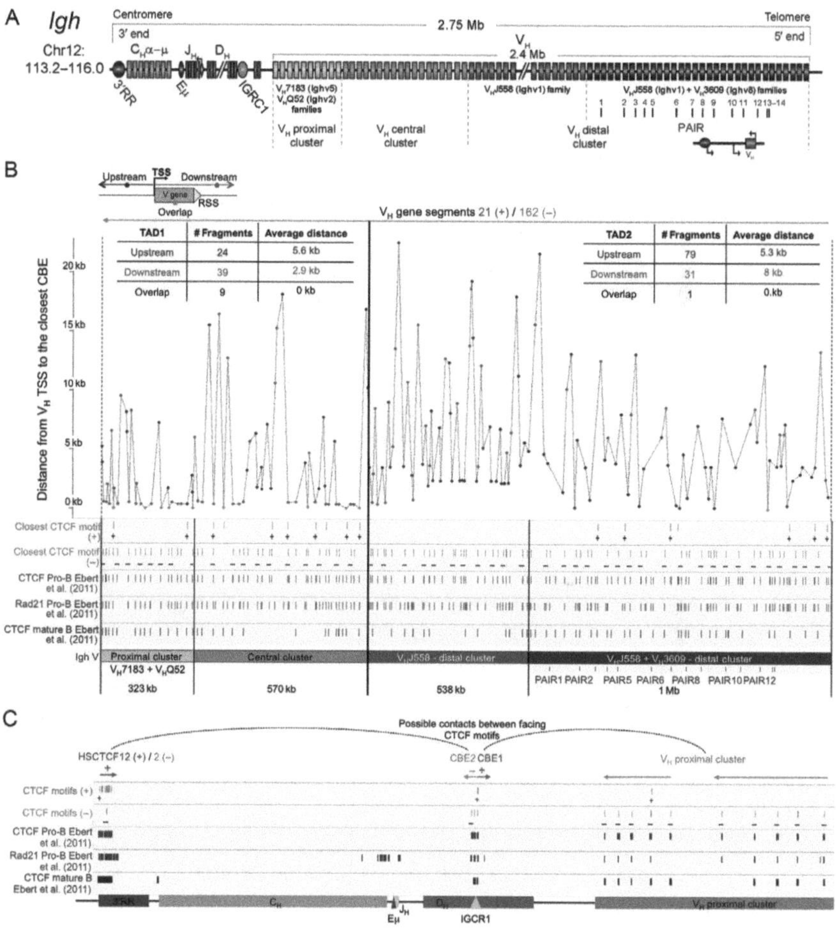

Chapter 4, Figure 2 (See page 128 of this volume.)

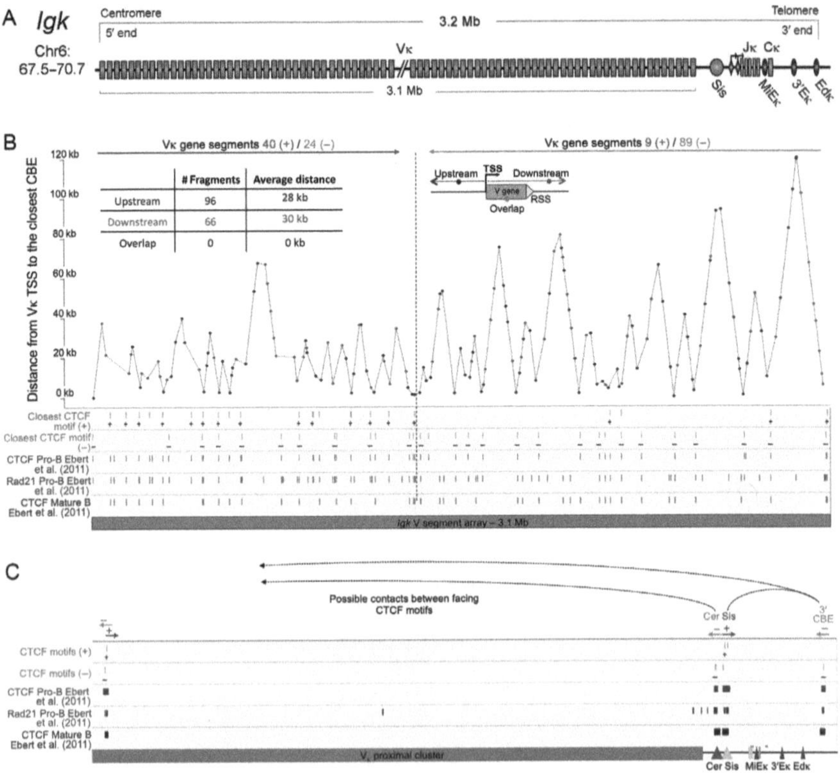

Chapter 4, Figure 3 (See page 130 of this volume.)

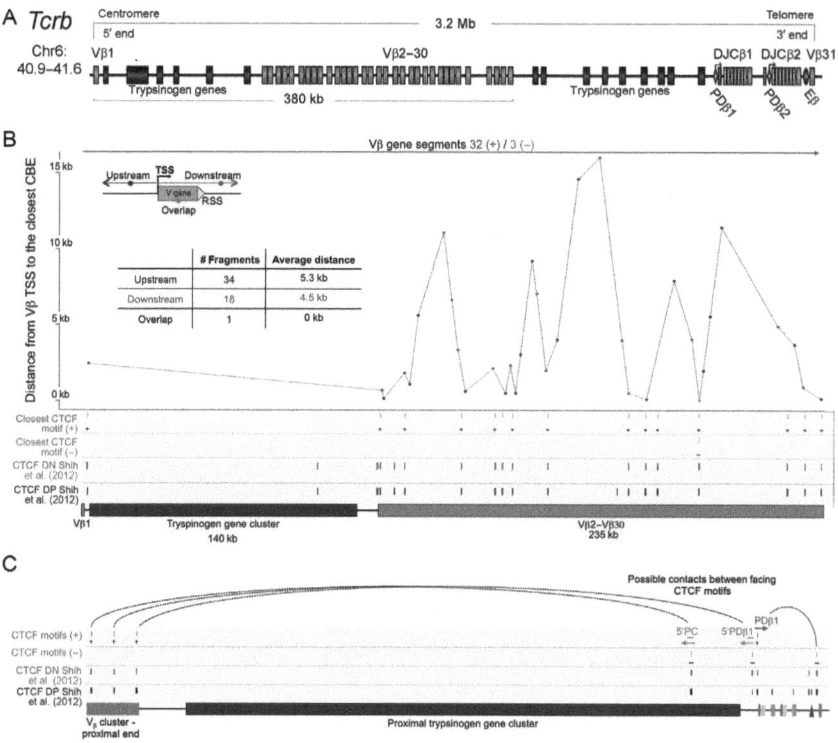

Chapter 4, Figure 4 (See page 132 of this volume.)

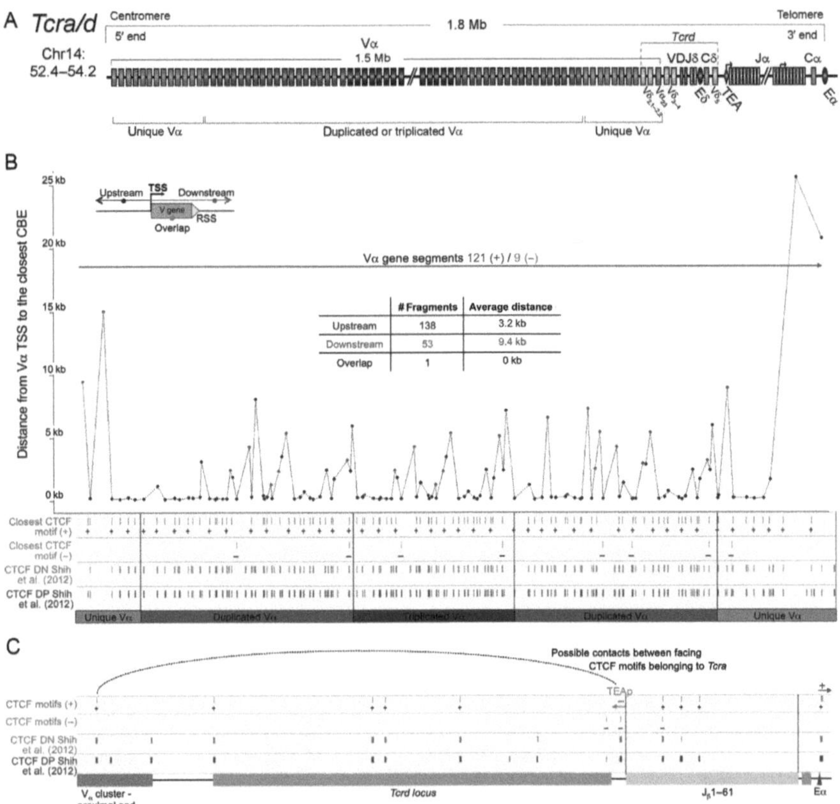

	# Fragments	Average distance
Upstream	138	3.2 kb
Downstream	53	9.4 kb
Overlap	1	0 kb

Chapter 4, Figure 5 (See page 133 of this volume.)

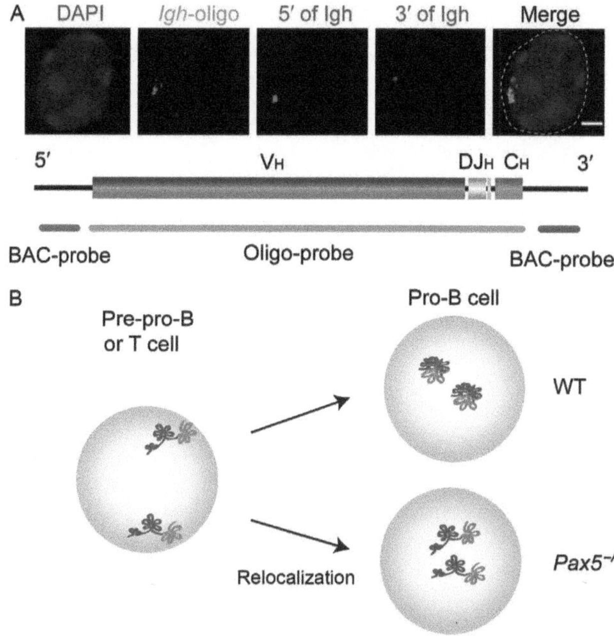

Chapter 4, Figure 6 (See page 135 of this volume.)

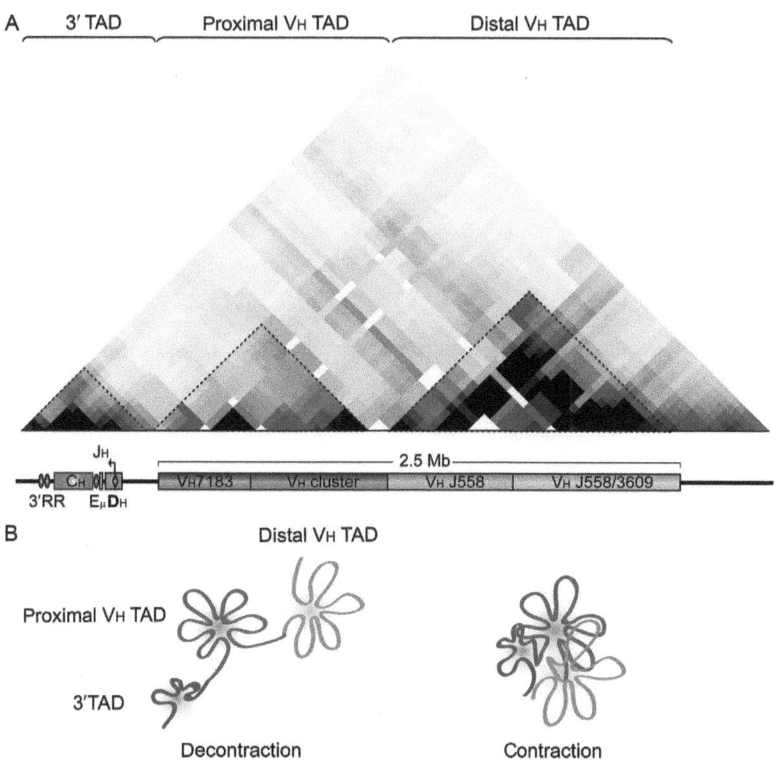

Chapter 4, Figure 7 (See page 137 of this volume.)

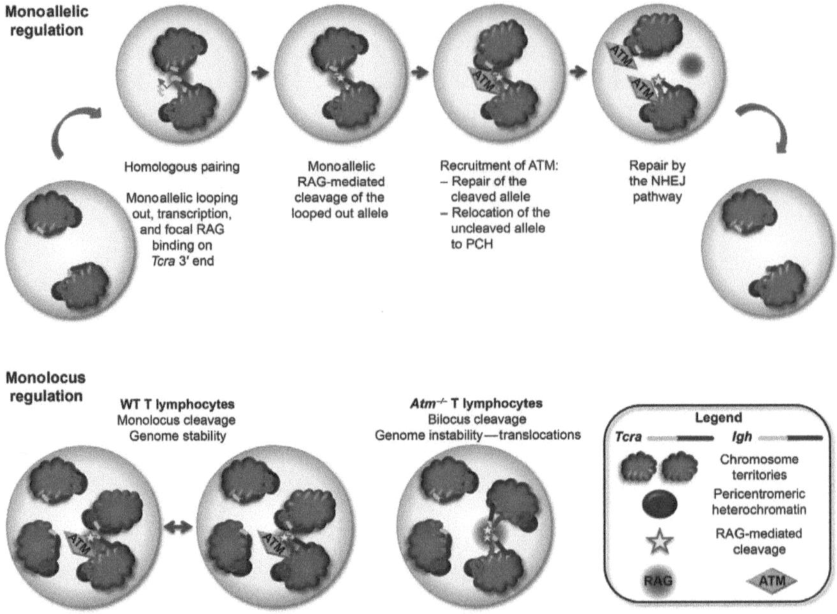

Chapter 4, Figure 8 (See page 139 of this volume.)

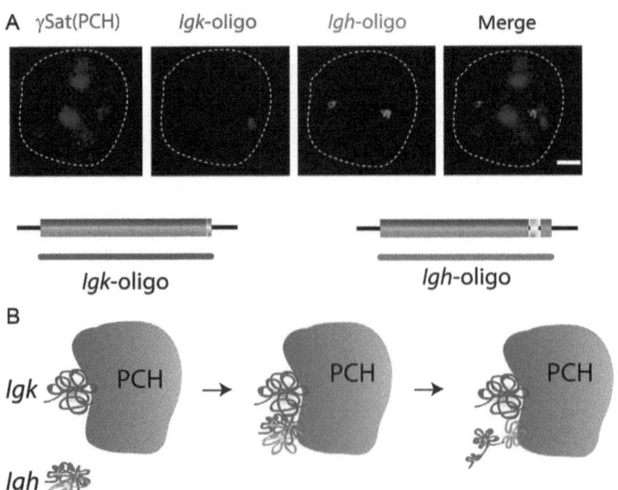

Chapter 4, Figure 9 (See page 141 of this volume.)

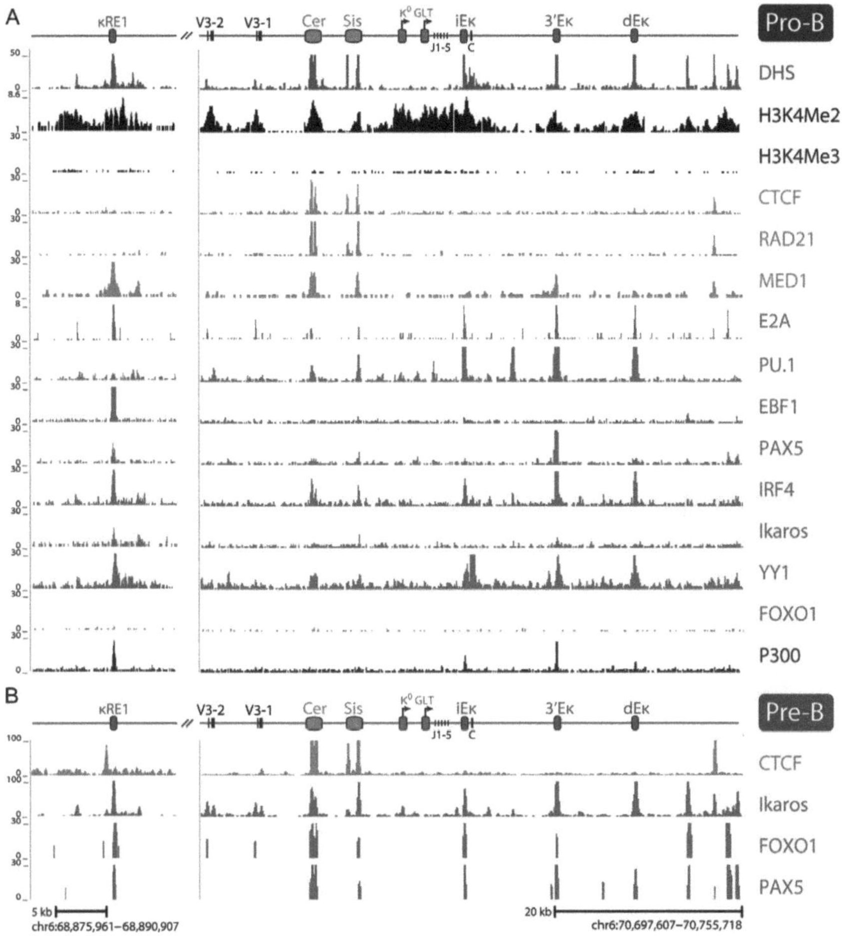

Chapter 5, Figure 3 (See page 216 of this volume.)

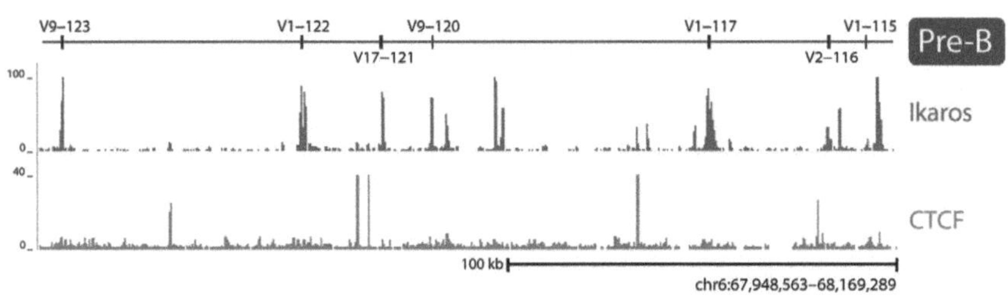

Chapter 5, Figure 4 (See page 217 of this volume.)

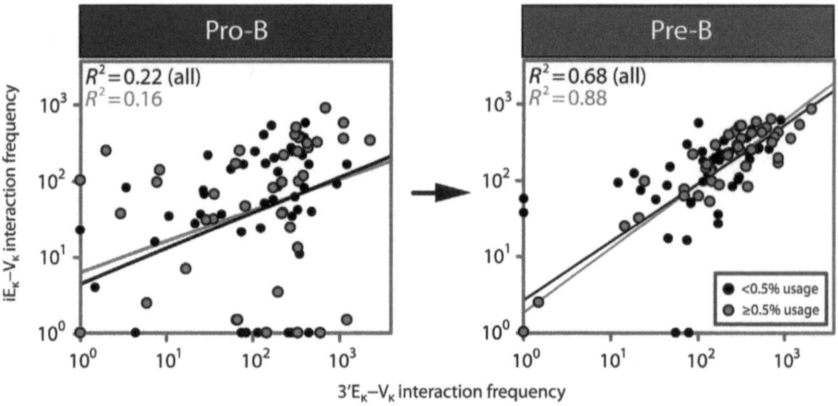

Chapter 5, Figure 5 (See page 245 of this volume.)

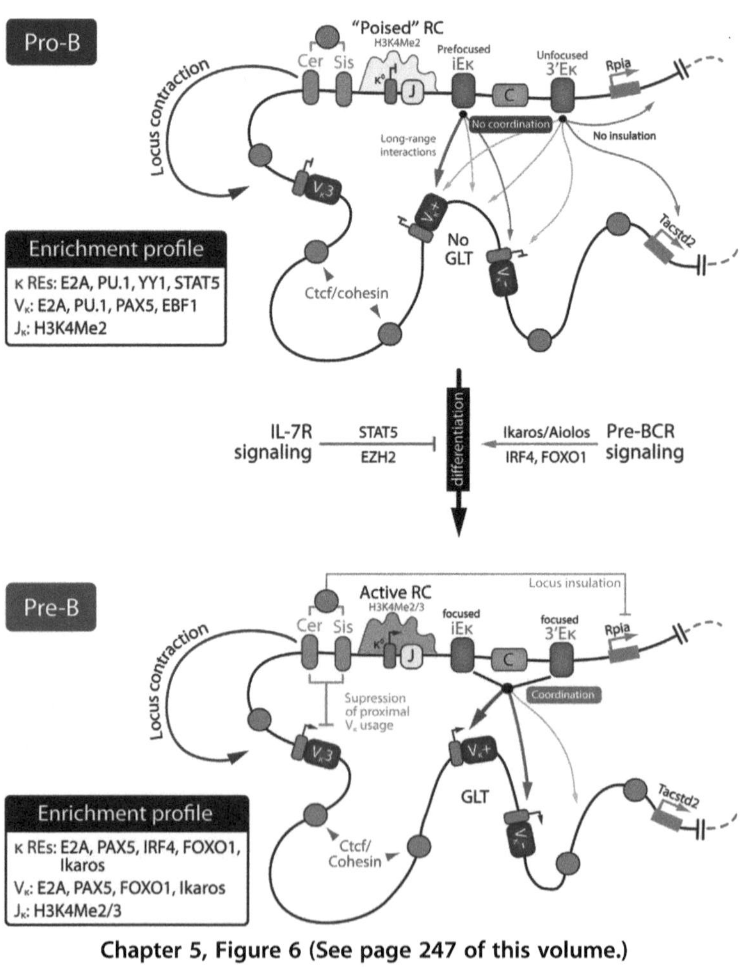

Chapter 5, Figure 6 (See page 247 of this volume.)

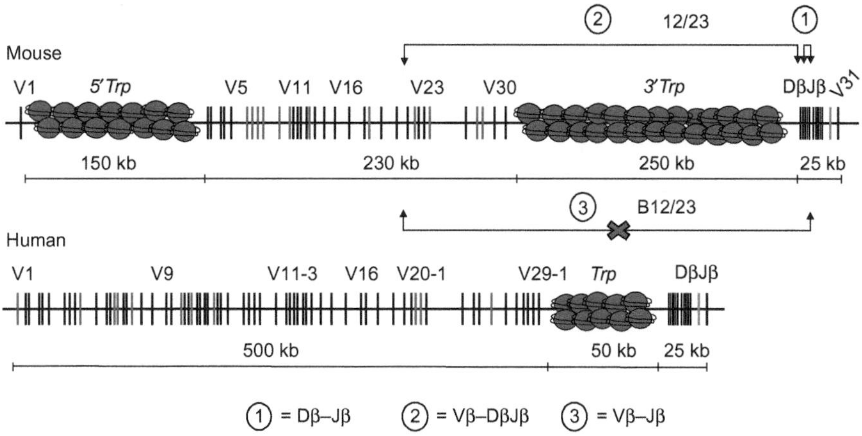

Chapter 6, Figure 1 (See page 276 of this volume.)

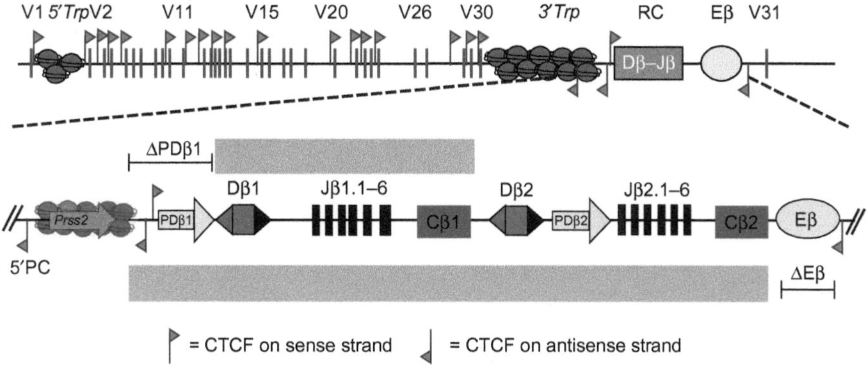

Chapter 6, Figure 2 (See page 277 of this volume.)

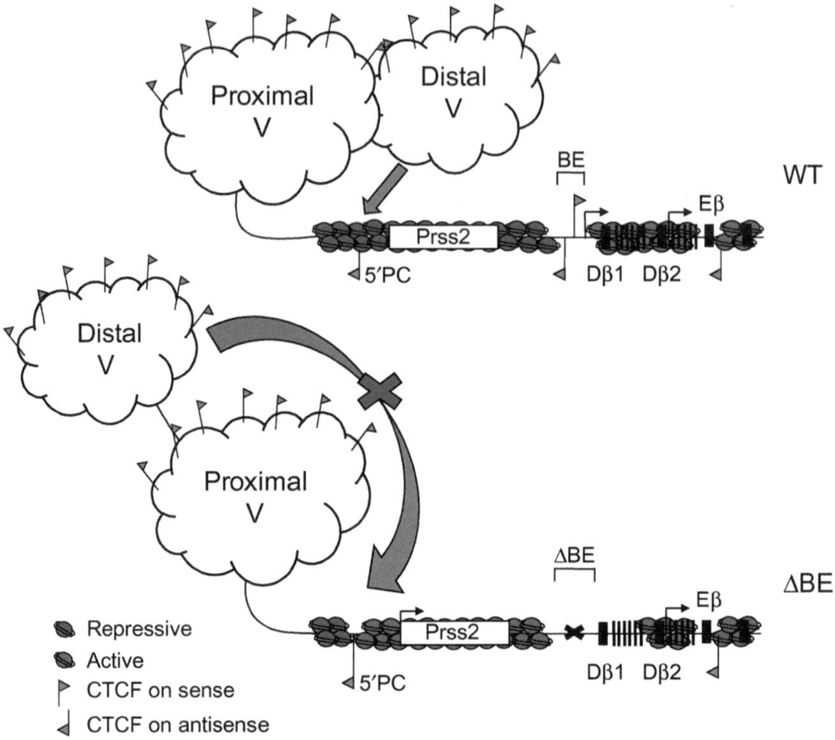

Chapter 6, Figure 3 (See page 286 of this volume.)

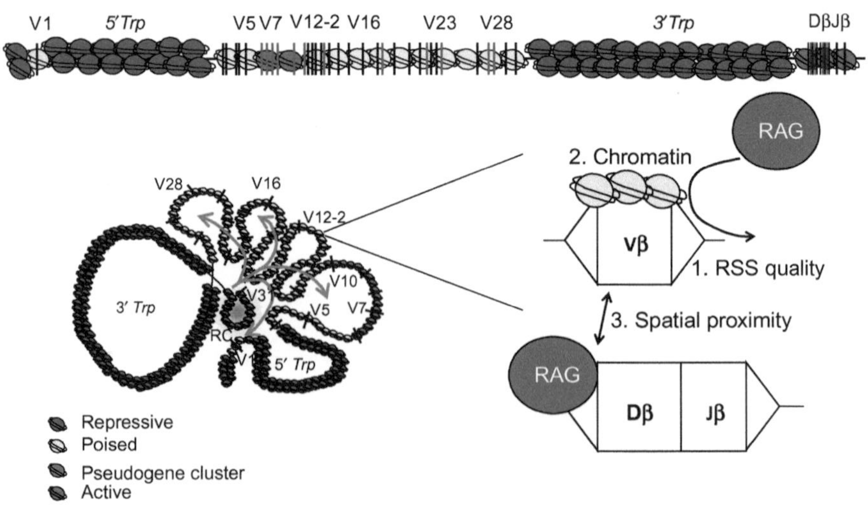

Chapter 6, Figure 4 (See page 293 of this volume.)

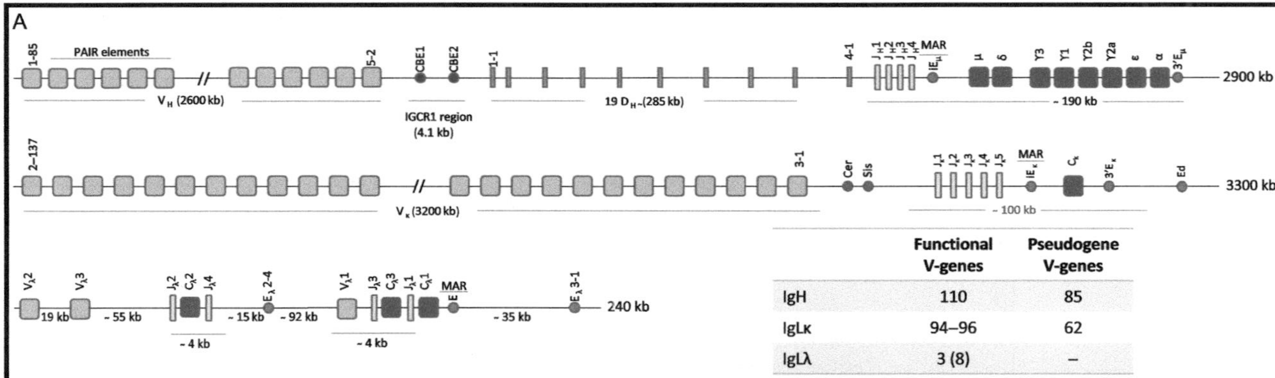

	Functional V-genes	Pseudogene V-genes
IgH	110	85
IgLκ	94–96	62
IgLλ	3 (8)	–

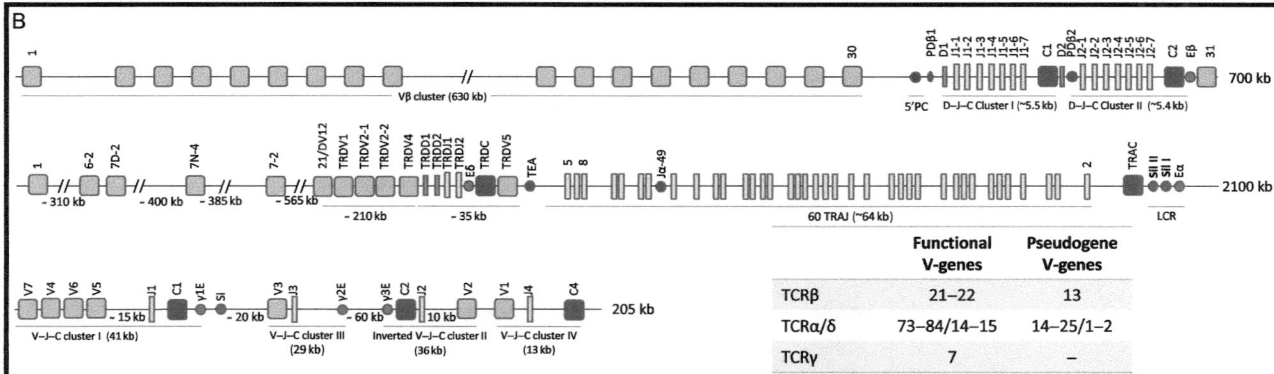

	Functional V-genes	Pseudogene V-genes
TCRβ	21–22	13
TCRα/δ	73–84/14–15	14–25/1–2
TCRγ	7	–

Chapter 8, Figure 1 (See page 367 of this volume.)

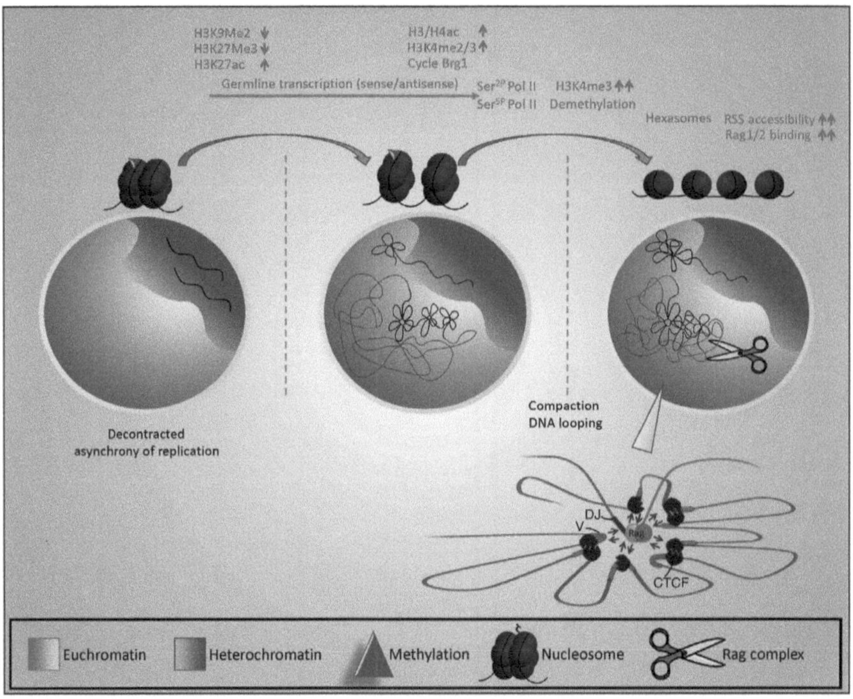

Chapter 8, Figure 2 (See page 385 of this volume.)

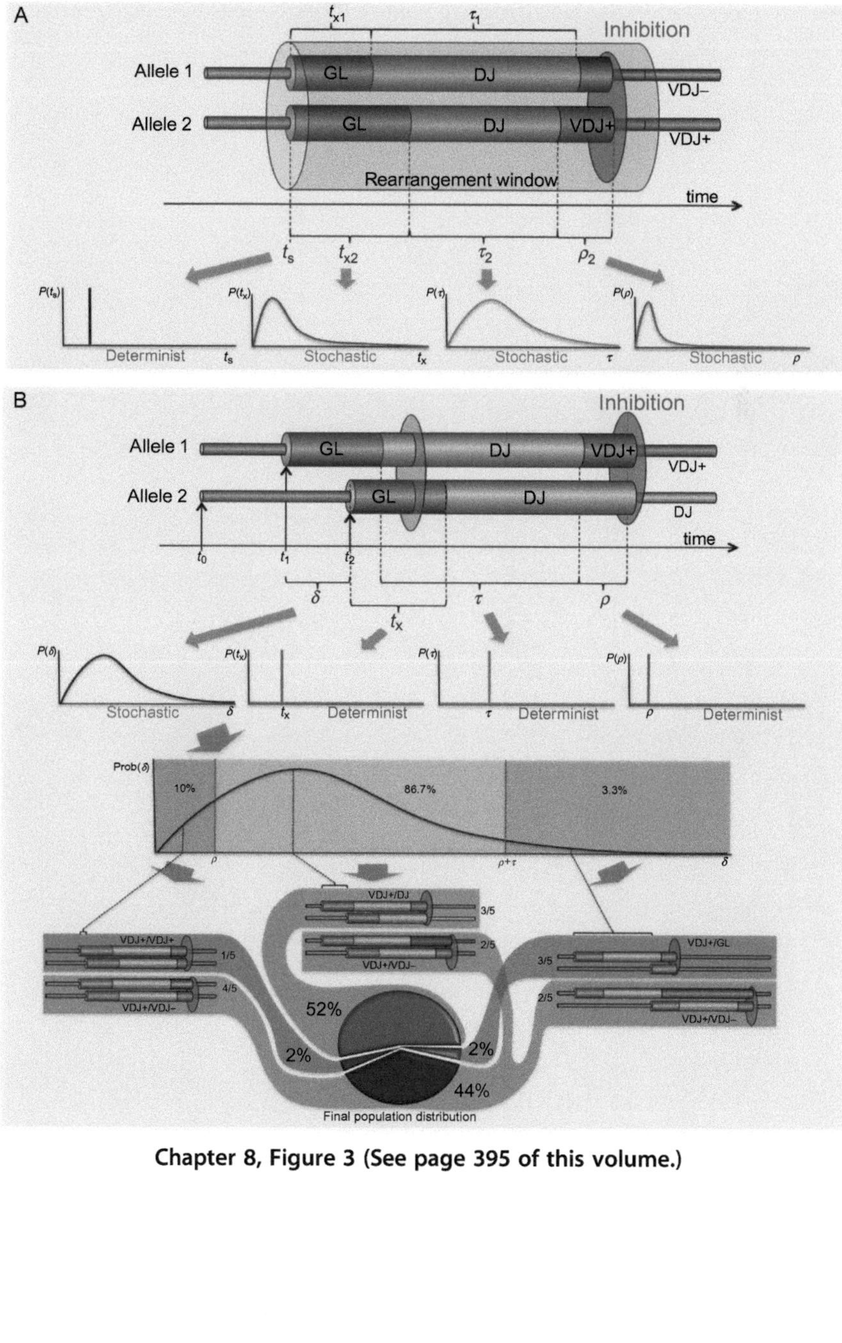

Chapter 8, Figure 3 (See page 395 of this volume.)

FSC
www.fsc.org
MIX
Papier | Fördert
gute Waldnutzung
FSC® C083411

Kolibri 360 GmbH
Ferdinand-Jühlke-Straße 7
99095 Erfurt, Deutschland
produktsicherheit@kolibri360.de